REA's

CHEMISTRY BUILDER

for Admission & Standardized Tests

by the staff of
Research & Education Association

RESEARCH & EDUCATION ASSOCIATION
61 Ethel Road West • Piscataway, New Jersey 08854

REA's CHEMISTRY BUILDER
for Admission & Standardized Tests

Printed in the United States of America

Library of Congress Catalog Card Number 98-66312

International Standard Book Number 0-87891-939-2

Research & Education Association
61 Ethel Road West
Piscataway, New Jersey 08854

REA supports the effort to conserve and protect environmental resources by printing on recycled papers.

ACKNOWLEDGMENTS

We would like to thank the following people for their contributions to the Chemistry Builder:

Rachel Kramer, for compiling and editing the manuscript

Ronald Fikar, for his technical editing of the manuscript

In addition, special recognition is also extended to the following persons:

Dr. Max Fogiel, President, for his overall guidance which has brought this publication to completion

Larry B. Kling, Quality Control Manager of Books in Print, for his editorial direction

Bernadette Brick, Editorial Assistant, for coordinating the revision of the book

Martin Perzan, for typesetting the book

CONTENTS

Chapter 1   ABOUT THE CHEMISTRY BUILDER 1
About This Book 3
How to Use This Book 3

Chapter 2   GASES 9
Diagnostic Test 11
    Answer Key 19
    Detailed Explanations of Answers 20
Gases Review 27
    Answer Key to Drills 54
Glossary 55

Chapter 3   PHYSICAL PROPERTIES OF GASES 57
Diagnostic Test 59
    Answer Key 63
    Detailed Explanations of Answers 64
Physical Properties of Gases Review 72
    Answer Key to Drills 80
Glossary 81

Chapter 4   AVOGADRO'S HYPOTHESIS 83
Diagnostic Test 85
    Answer Key 93
    Detailed Explanations of Answers 94
Avogadro's Hypothesis Review 102
    Answer Key to Drills 117
Glossary 118

Chapter 5	**STOICHIOMETRY**	121
	Diagnostic Test	123
	Answer Key	132
	Detailed Explanations of Answers	133
	Stoichiometry Review	142
	Answer Key to Drills	155
	Glossary	156
Chapter 6	**SOLIDS**	159
	Diagnostic Test	161
	Answer Key	168
	Detailed Explanations of Answers	169
	Solids Review	174
	Answer Key to Drills	185
	Glossary	186
Chapter 7	**PROPERTIES OF LIQUIDS**	189
	Diagnostic Test	191
	Answer Key	197
	Detailed Explanations of Answers	198
	Properties of Liquids Review	206
	Answer Key to Drills	219
	Glossary	220
Chapter 8	**SOLUTION CHEMISTRY**	223
	Diagnostic Test	225
	Answer Key	233
	Detailed Explanations of Answers	234
	Solution Chemsitry Review	243
	Answer Key to Drills	249
	Glossary	250
Chapter 9	**EQUILIBRIUM**	251
	Diagnostic Test	253
	Answer Key	263
	Detailed Explanations of Answers	264
	Equilibrium Review	276
	Answer Key to Drills	298
	Glossary	299

Chapter 10 **ACID AND BASE EQUILIBRIA** 301
    Diagnostic Test... 303
        Answer Key 311
        Detailed Explanations of Answers 312
    Acid and Base Equilibria Review 323
        Answer Key to Drills 356
    Glossary 357

Chapter 11 **THERMODYNAMICS I** 359
    Diagnostic Test... 361
        Answer Key 373
        Detailed Explanations of Answers 374
    Thermodynamics I Review 384
        Answer Key to Drills 406
    Glossary 407

Chapter 12 **THERMODYNAMICS II** 409
    Diagnostic Test... 411
        Answer Key 422
        Detailed Explanations of Answers 423
    Thermodynamics II Review 430
        Answer Key to Drills 437
    Glossary 438

Chapter 13 **ELECTROCHEMISTRY** 439
    Diagnostic Test... 441
        Answer Key 451
        Detailed Explanations of Answers 452
    Electrochemistry Review 460
        Answer Key to Drills 484
    Glossary 485

Chapter 14 **ATOMIC THEORY** 487
    Diagnostic Test... 489
        Answer Key 498
        Detailed Explanations of Answers 499
    Atomic Theory Review 505

Answer Key to Drills 526

Glossary 527

Chapter 15   **QUANTUM CHEMISTRY** 529

Diagnostic Test... 531

     Answer Key 539

     Detailed Explanations of Answers 540

Quantum Chemistry Review 547

     Answer Key to Drills 558

Glossary 559

Chapter 16   **ORGANIC CHEMISTRY** 561

Diagnostic Test... 563

     Answer Key 574

     Detailed Explanations of Answers 575

Organic Chemistry Review 585

     Answer Key to Drills 612

Glossary 613

Chapter 17   **MINI TESTS** 615

Mini Test 1... 617

     Answer Key for Mini Test 1 623

     Detailed Explanations of Answers 624

Mini Test 2... 631

     Answer Key for Mini Test 2 636

     Detailed Explanations of Answers 637

THE PERIODIC TABLE 642

CHAPTER 1

About the Chemistry Builder

About Research & Education Association

Research & Education Association (REA) is an organization of educators, scientists, and engineers specializing in various academic fields. Founded in 1959 with the purpose of disseminating the most recently developed scientific information to groups in industry, government, high schools, and universities, REA has since become a successful and highly respected publisher of study aids, test preps, handbooks, and reference works.

REA's Test Preparation series includes study guides for all academic levels in almost all disciplines. Research & Education Association publishes test preps for students who have not yet completed high school, as well as high school students preparing to enter college. Students from countries around the world seeking to attend college in the United States will find the assistance they need in REA's publications. For college students seeking advanced degrees, REA publishes test preps for many major graduate school admission examinations in a wide variety of disciplines, including engineering, law, and medicine. Students at every level, in every field, with every ambition can find what they are looking for among REA's publications.

Unlike most test preparation books—which present only a few practice tests that bear little resemblance to the actual exams—REA's series presents tests that accurately depict the official exams in both degree of difficulty and types of questions. REA's practice tests are always based upon the most recently administered exams, and include every type of question that can be expected on the actual exams.

REA's publications and educational materials are highly regarded and continually receive an unprecedented amount of praise from professionals, instructors, librarians, parents, and students. Our authors are as diverse as the subjects and fields represented in the books we publish. They are well-known in their respective fields and serve on the faculties of prestigious universities throughout the United States.

About this Book

REA's staff of authors and educators has prepared material, exercises, and tests based on each of the major standardized exams, including the Advanced Placement (AP) Chemistry, Armed Services Vocational Aptitude Battery (ASVAB), CLEP General Chemistry, PRAXIS II Subject Assessment: Chemistry, and SAT II: Chemistry tests. The types of questions represented on these standardized exams have been analyzed in order to produce the most comprehensive preparatory material possible. You will find review material, helpful strategies, and exercises geared to your level of studying. This book will teach as well as review and refresh chemistry skills needed to score high on standardized tests.

How to Use this Book

If your are preparing to take the AP Chemistry, ASVAB, CLEP General Chemistry, GRE Chemistry, MCAT, PRAXIS II Subject Assessment: Chemistry, or the SAT II: Chemistry exam, you will be taking a test that requires excellent knowledge of chemistry. This book comprises a comprehensive chemistry review that can be tailored to your specific test preparation needs.

Locate your test on the chart shown on page 7, and then find the corresponding sections recommended for study. REA suggests that you study the indicated material thoroughly as a review for your exam.

This book will help you prepare for your exam because it includes different types of questions and drills that are representative of what might appear on each exam. The book also includes diagnostic tests so that you can determine your strengths and weaknesses within a specific subject. The explanations are clear and comprehensive, telling you why the answer is correct. The Chemistry Builder gives you practice within a wide range of categories and question types.

The **Gases and Physical Properties of Gases** chapters should be studied for the AP Chemistry, ASVAB, CLEP General Chemistry, GRE Chemistry, MCAT, PRAXIS II Subject Assessment: Chemistry, and SAT II: Chemistry exams. These chapters provide comprehensive reviews in pressure, Boyle's law, Charles' law, mole fractions, Dalton's law of Partial Pressures, and Graham's law of Gaseous Diffusion.

The **Avogadro's Hypothesis** chapter reviews Avogadro's law, atomic and molecular weights, equivalent weights, chemical compounds and the periodic table. This chapter should be used when studying for the AP Chemistry, ASVAB, CLEP General Chemistry, GRE Chemistry, MCAT, PRAXIS II Subject Assessment: Chemistry, and SAT II: Chemistry exams.

The **Stoichiometry** chapter carefully reviews balancing equations, calculations using chemical arithmetic, reactions with limiting reagents, and net ionic equations. These topics should be studied carefully for the AP Chemistry, ASVAB, CLEP General Chemistry, GRE Chemistry, MCAT, PRAXIS II Subject Assessment: Chemistry, and SAT II: Chemistry exams.

The **Solids** chapter carefully reviews phase diagrams, phase equilibrium, ionic crystals and crystal structure. These topics should be studied carefully for the AP Chemistry, CLEP General Chemistry, GRE Chemistry, MCAT, PRAXIS II Subject Assessment: Chemistry, and SAT II: Chemistry tests.

The **Properties of Liquids** chapter should be fully studied for the AP Chemistry, CLEP General Chemistry, GRE Chemistry, MCAT, PRAXIS II Subject Assessment: Chemistry, and SAT II: Chemistry exams.

The **Solution Chemistry** chapter prepares students for questions involving gases on the AP Chemistry, ASVAB, CLEP General Chemistry, GRE Chemistry, MCAT, PRAXIS II Subject Assessment: Chemistry, and SAT II: Chemistry exams. Even if you are not planning to take an exam in which gases are tested, this chapter can be extremely helpful in building your knowledge for more difficult questions.

The **Equilibrium** and **Acid and Base Equilibria** chapters prepare students for questions concerning the physical properties of gases on the AP Chemistry, CLEP General Chemistry, GRE Chemistry, MCAT, PRAXIS II Subject Assessment: Chemistry, and SAT II: Chemistry exams. They include comprehensive reviews of the properties of gases, from the most basic terms to complex questions that require more in-depth knowledge of the physical properties of gases.

The **Thermodynamics I and II** chapters highlight all the terms and topics needed to succeed on questions about Avogadro's Hypothesis on any standardized test. Students taking the AP Chemistry, ASVAB, CLEP General Chemistry, GRE Chemistry, MCAT, PRAXIS II Subject Assessment: Chemistry, and SAT II: Chemistry exams will encounter Avogadro's Hypothesis questions on their test, and should study this chapter thoroughly.

The **Electrochemistry** chapter should be used when studying for the AP Chemistry, ASVAB, CLEP General Chemistry, GRE Chemistry, MCAT, PRAXIS II Subject Assessment: Chemistry, and SAT II: Chemistry exams. This chapter thoroughly reviews the important aspects of electrochemistry which are commonly found on all of the previously stated tests.

The **Atomic Theory** chapter should be studied for the AP Chemistry, ASVAB, CLEP General Chemistry, GRE Chemistry, MCAT, PRAXIS II Subject Assessment: Chemistry, and SAT II: Chemistry exams. The atomic weight and components of atomic structure, valence and electron dot diagrams, ionic and covalent bonding, electronegativity and ionization energy are fully covered.

The **Quantum Chemistry** chapter prepares students for questions on the AP Chemistry, CLEP General Chemistry, GRE Chemistry, MCAT, PRAXIS II Subject Assessment: Chemistry, and SAT II: Chemistry which involve various aspects of quantum chemistry, from the Pauli Exclusion Principle to electron configuration.

The **Organic Chemistry** chapter prepares students for organic chemistry questions on the AP Chemistry, ASVAB, CLEP General Chemistry, GRE Chemistry, MCAT, PRAXIS II Subject Assessment: Chemistry, and SAT II: Chemistry exams. Even if you are not planning to take an exam in which organic chemistry is tested, this chapter can be extremely helpful in building your knowledge for more difficult questions.

Finally, before getting started, here are a few guidelines:

➤ Study full chapters. If you think after a few minutes that the chapter appears easy, continue studying. Many chapters (like the tests themselves) become more difficult as they continue.

➤ Use this guide as a supplement to the review materials provided by the test administrators.

➤ Take the diagnostic test before each review chapter, even if you feel confident that you already know the material well enough to skip a particular chapter. Taking the diagnostic test will put your mind at ease: you will discover either that you absolutely know the material or that you need to review. This will eliminate the panic you might otherwise experience during the test upon discovering that you have forgotten how to approach a certain type of question.

As you prepare for a standardized chemistry test, you will want to review some of the basic concepts. The more familiar you are with the fundamental principles, the better you will do on your test. Our chemistry reviews represent the various topics that appear on chemistry standardized tests or those tests with chemistry sections.

Along with knowledge of these topics, how quickly and accurately you answer chemistry questions will have an effect on your success. All tests have time limits, so the more questions you can answer correctly in the given period of time, the better off you will be. Our suggestion is that you first take each diagnostic test, make sure to complete the drills as you review for extra practice, and then take the Mini Tests when you feel confident with the material. Pay special attention to both the time it takes to complete the diagnostic tests and Mini Tests, and to the number of correct answers you achieve.

The glossary at the end of each chapter will also refresh your memory after you have completed the reviews and drills. Important terms, laws, and principles are clearly defined to enhance your study regimen.

Cross-Referencing Chart

	AP Chemistry	ASVAB	CLEP General Chemistry	GRE Chemistry	MCAT	PRAXIS II Subject Test: Chemistry	SAT II: Chemistry
Chapter 2: Gases	×	×	×	×	×	×	×
Chapter 3: Gas Mix & Other	×		×	×	×	×	×
Chapter 4: Avogadro's Hypothesis	×	×	×	×	×	×	×
Chapter 5: Stoichiometry	×	×	×	×	×	×	×
Chapter 6: Solids	×		×	×	×	×	×
Chapter 7: Properties of Liquids	×		×	×	×	×	×
Chapter 8: Solution Chemistry	×	×	×	×	×	×	×
Chapter 9: Equilibrium	×		×	×	×	×	×
Chapter 10: Acid & Base Equilibria	×	×	×	×	×	×	×
Chapter 11: Thermodynamics I	×	×	×	×	×	×	×
Chapter 12: Thermodynamics II	×		×	×	×	×	×
Chapter 13: Electrochemistry	×		×	×	×	×	×
Chapter 14: Atomic Theory	×	×	×	×	×	×	×
Chapter 15: Quantum Chemistry	×		×	×	×	×	×
Chapter 16: Organic Chemistry	×		×	×	×	×	×

CHAPTER 2

Gases

➤ Diagnostic Test
➤ Gases Review & Drills
➤ Glossary

GASES
DIAGNOSTIC TEST

1. Ⓐ Ⓑ Ⓒ Ⓓ Ⓔ
2. Ⓐ Ⓑ Ⓒ Ⓓ Ⓔ
3. Ⓐ Ⓑ Ⓒ Ⓓ Ⓔ
4. Ⓐ Ⓑ Ⓒ Ⓓ Ⓔ
5. Ⓐ Ⓑ Ⓒ Ⓓ Ⓔ
6. Ⓐ Ⓑ Ⓒ Ⓓ Ⓔ
7. Ⓐ Ⓑ Ⓒ Ⓓ Ⓔ
8. Ⓐ Ⓑ Ⓒ Ⓓ Ⓔ
9. Ⓐ Ⓑ Ⓒ Ⓓ Ⓔ
10. Ⓐ Ⓑ Ⓒ Ⓓ Ⓔ
11. Ⓐ Ⓑ Ⓒ Ⓓ Ⓔ
12. Ⓐ Ⓑ Ⓒ Ⓓ Ⓔ
13. Ⓐ Ⓑ Ⓒ Ⓓ Ⓔ
14. Ⓐ Ⓑ Ⓒ Ⓓ Ⓔ
15. Ⓐ Ⓑ Ⓒ Ⓓ Ⓔ
16. Ⓐ Ⓑ Ⓒ Ⓓ Ⓔ
17. Ⓐ Ⓑ Ⓒ Ⓓ Ⓔ
18. Ⓐ Ⓑ Ⓒ Ⓓ Ⓔ

19. Ⓐ Ⓑ Ⓒ Ⓓ Ⓔ
20. Ⓐ Ⓑ Ⓒ Ⓓ Ⓔ
21. Ⓐ Ⓑ Ⓒ Ⓓ Ⓔ
22. Ⓐ Ⓑ Ⓒ Ⓓ Ⓔ
23. Ⓐ Ⓑ Ⓒ Ⓓ Ⓔ
24. Ⓐ Ⓑ Ⓒ Ⓓ Ⓔ
25. Ⓐ Ⓑ Ⓒ Ⓓ Ⓔ
26. Ⓐ Ⓑ Ⓒ Ⓓ Ⓔ
27. Ⓐ Ⓑ Ⓒ Ⓓ Ⓔ
28. Ⓐ Ⓑ Ⓒ Ⓓ Ⓔ
29. Ⓐ Ⓑ Ⓒ Ⓓ Ⓔ
30. Ⓐ Ⓑ Ⓒ Ⓓ Ⓔ
31. Ⓐ Ⓑ Ⓒ Ⓓ Ⓔ
32. Ⓐ Ⓑ Ⓒ Ⓓ Ⓔ
33. Ⓐ Ⓑ Ⓒ Ⓓ Ⓔ
34. Ⓐ Ⓑ Ⓒ Ⓓ Ⓔ
35. Ⓐ Ⓑ Ⓒ Ⓓ Ⓔ

GASES DIAGNOSTIC TEST

This diagnostic test is designed to help you determine your strengths and your weaknesses in gases. Follow the directions and check your answers.

Study this chapter for the following tests:
ACT, AP Chemistry, ASVAB, CLEP General
Chemistry, GRE Chemistry, MCAT, PRAXIS II
Subject Assessment: Chemistry, SAT II: Chemistry

---

35 Questions

DIRECTIONS: Choose the correct answer for each of the following problems. Fill in each answer on the answer sheet.

1. A gas has a volume of 10 liters at 50°C and 200 mm Hg pressure. What correction factor is needed to give a volume at STP?

   (A) $\dfrac{0}{50} \times \dfrac{200}{760}$

   (B) $\dfrac{0}{50} \times \dfrac{760}{200}$

   (C) $\dfrac{273}{323} \times \dfrac{200}{760}$

   (D) $\dfrac{273}{323} \times \dfrac{760}{200}$

   (E) $\dfrac{323}{273} \times \dfrac{760}{200}$

2. Which of the following will cause the pressure of a gas to decrease?

   (A) Decreasing the volume while maintaining the temperature

   (B) Decreasing the volume while increasing the temperature

   (C) Maintaining the volume while increasing the temperature

   (D) Increasing the volume by 1 liter while increasing the temperature by 1°K at STP

   (E) Decreasing the volume by 1 liter while decreasing the temperature by 1°K at STP

3. The relationship between the absolute temperature and volume of a gas at constant pressure is given by

   (A) Boyle's law.

   (B) Charles' law.

(C) the Combined gas law.   (D) Avogadro's law.

(E) None of the above.

4. How many molecules are present in 0.20 g of hydrogen gas? (1 mole of H = 1.0008 g)

(A) $(0.20/1.008) (6.02 \times 10^{23})$

(B) 0.20 (1.008)

(C) 0.20 (2.016)

(D) $(0.20/2.016) (6.02 \times 10^{23})$

(E) Cannot tell

5. A sample of hydrogen gas is in a closed container, at 1.0 atmosphere pressure and 27°C. If the sample is heated to 127°C, the pressure will be approximately which of the following?

(A) 4.0 atm   (D) 0.67 atm

(B) 1.3 atm   (E) 0.25 atm

(C) 0.75 atm

6. Five hundred ml of a gas experiences a pressure change from 760 mm Hg pressure to a barometric pressure of 800 mm Hg. If all other laboratory factors are held constant, its new volume in ml is

(A) 400   (D) 525

(B) 425   (E) 595

(C) 475

7. Under standard conditions, 44.8 liters of a gas are collected in a lab. The number of molecules in this volume is:

(A) 112,000   (D) 18.06×10^{23}

(B) 6.02×10^{23} (E) 1,112,000

(C) 12.04×10^{23}

8. Atmospheric pressure is measured by a

(A) barometer.   (D) thermocouple.

(B) spectrophotometer.   (E) thermometer.

(C) sphygmomanometer.

9. A 300 ml volume of gas experiences a pressure change from 1,000 mm Hg to 760 mm Hg, with other laboratory factors held constant. Its new volume in ml is

(A) 131.6   (D) 500.8

(B) 228.0   (E) 1,000.0

(C) 394.7

10. The density of gaseous SO_3 at 15°C, 50 mm Hg and $R = 0.0821$ liter-atm/mole-K is

(A) 0.223 g/l   (D) 0.357 l/g

(B) 0.223 $l\,g^{-1}$ (E) 0.357 g/l

(C) 169.17 g/l

11. A gas at 25°C occupies a 10 liter volume at P atm pressure. The gas is allowed to expand to a volume of 15 liters at 377°C. What is the new pressure?

(A) 1.45   (D) 1.07

(B) 1.45P (E) 1.07/P

(C) 1.45/P

12. A sample of a pure gas occupied a volume of 500 ml at a temperature of 27°C and a pressure of 0.4 atm. The number of moles present in this sample is

(A) 0.045   (D) 8.13

(B) 0.008   (E) 0.182

(C) 0.091

13. The molecular weight of 0.25 g of a gas that occupies a volume of 100 ml at a pressure of 2.5 atm and a temperature of 25°C is

(A) 28.0   (D) 20.05

(B) 12.2   (E) 24.4

(C) 2.05

14. If a certain number of moles of a gas occupied a volume of 200 ml at 25°C, what volume (in ml) would it occupy if it was heated to 40°C at a constant pressure?

(A) 210   (B) 320

(C) 230

(D) 500

(E) 120

15. In the Van der Waal's equation

$$\left(P + \frac{n^2 a}{V^2}\right)(V - nb) = nRT,$$

the constant a is best described as a correction factor due to

(A) temperature.

(B) intermolecular attractions of real gases.

(C) the molecular weights of gases.

(D) the volume of the actual gas molecules.

(E) the specific heat of the gas molecules.

16. The molecular weight of a gas is 16. At STP, 4.48 liters of this gas weighs

(A) 2.3 g.

(D) 4.1 g.

(B) 2.7 g.

(E) 4.9 g.

(C) 3.2 g.

17. The fractions by which the original volume of a gas at STP is multiplied to correct its volume to new conditions are

$$\frac{298}{273} \times \frac{760}{800}$$

What are the new conditions?

(A) 25°C, 1 atm

(D) 0°C, 800 torr

(B) 25°C, 800 torr

(E) 298°K, 1 atm

(C) 0°C, 1 atm

18. The relation $P_1 V_1 = P_2 V_2$ is known as

(A) Boyle's law.

(D) the Combined gas law.

(B) Charles's law.

(E) the Ideal gas law.

(C) Van der Waal's law.

19. Standard conditions using a Kelvin thermometer are

(A) 760 torr, 273°K.

(B) 760 torr, 273°K, 1 liter.

(C) 760 torr, 0°K.   (D) 0 torr, 0°K.

(E) 0 torr, 273°K, 1 liter.

20. What is the resulting volume if 10 liters of a gas at 546 K and 2 atm is brought to standard conditions?

(A) 5 l (D) 20 l

(B) 10 l (E) 25 l

(C) 15 l

21. What is the density of bromine vapor at STP?

(A) 5.2 g/l (D) 4.9 g/l

(B) 2.9 g/l (E) 7.1 g/l

(C) 3.6 g/l

22. What volume would 16g of oxygen gas occupy at STP?

(A) 5.6 l (D) 33.6 l

(B) 11.2 l (E) 44.8 l

(C) 22.4 l

23. An ideal gas is most likely to be found under conditions of

(A) low temperatures.   (D) both (B) and (C).

(B) high temperatures.   (E) high pressure.

(C) low pressure.

Use the following diagrams to answer questions 24–26.

(A)

(D)

(B)

(E)

(C)

Which of the above plots would be obtained if one were to experimentally investigate?

24. Boyle's law?

25. Charles' law?

26. The relationship of P and T?

27. How much does 1 liter of a gas weigh if its molecular weight is 254 g/mol?

   (A) 11.3 g   (D) 76.5 g

   (B) 25.4 g   (E) 254 g

   (C) 30.6 g

28. What is the density of a diatomic gas whose gram-molecular weight is 80g?

   (A) 1.9 g/l (D) 4.3 g/l

   (B) 2.8 g/l (E) 5.0 g/l

   (C) 3.6 g/l

29. How many grams of oxygen gas are required to fill a box of volume 44.8 liters to 1 atmosphere pressure at STP?

   (A) 8 g   (D) 48 g

   (B) 16 g   (E) 64 g

   (C) 32 g

30. How many atoms are present in 22.4 liters of oxygen gas at STP?

   (A) 3×10^{23} (D) 12×10^{23}

   (B) 6×10^{23} (E) 15×10^{23}

   (C) 9×10^{23}

31. Select the characteristic that is *not* a standard condition for comparing gas volumes.

   (A) Pressure: 31 inches

   (B) Pressure: 760 torr

   (C) Temperature: 0° Centigrade

   (D) Temperature: 32° Fahrenheit

   (E) Temperature: 273° Kelvin

32. Which of the following represents the Real gas law?

    (A) $nRT = (P + a/V^2)(V - b)$

    (B) $nRT = (P - a/V^2)(V - b)$

    (C) $nRT = (P + a/V^2)(V - nb)$

    (D) $nRT = (P - a/V^2)(V + nb)$

    (E) $nRT = (P + n^2a/V^2)(V - nb)$

33. Which of the following most closely represents an ideal gas?

    (A) H_2 (D) CO_2

    (B) He   (E) Ne

    (C) O_2

34. The graphs below show the behavior of various gases. Which of these gases exhibit behavior that deviates most significantly from that expected of an ideal gas?

    (A)

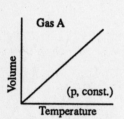

    (D)

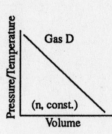

    (B)

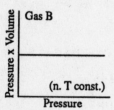

    (E)

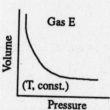

    (C)

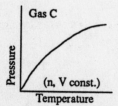

35. The relation between the pressure, volume, and absolute temperature is given by

(A) Boyle's law.

(D) the Ideal gas law.

(B) Charles' law.

(E) None of the above.

(C) the Combined gas law.

GASES DIAGNOSTIC TEST

ANSWER KEY

1. (C)	11. (B)	21. (E)	31. (A)
2. (D)	12. (B)	22. (B)	32. (E)
3. (B)	13. (E)	23. (D)	33. (B)
4. (D)	14. (A)	24. (B)	34. (D)
5. (B)	15. (B)	25. (C)	35. (C)
6. (C)	16. (C)	26. (C)	
7. (C)	17. (B)	27. (A)	
8. (A)	18. (A)	28. (C)	
9. (C)	19. (A)	29. (E)	
10. (A)	20. (B)	30. (D)	

DETAILED EXPLANATIONS
OF ANSWERS

1. **(C)**   This problem can be solved by applying the combined gas law.

$$\frac{P_1 V_1}{T_1} = \frac{P_2 V_2}{T_2}$$

Rearranging the equation gives:

$$V_2 = V_1 \times \frac{T_2}{T_1} \times \frac{P_1}{P_2}$$

By substituting in the given values we get:

$$V_2 = 10 \times \frac{273}{373} \times \frac{200}{760}$$

This then gives the correction factor for changing V_1 to V_2.

$$= \frac{273}{323} \times \frac{200}{760}$$

2. **(D)**   By rearranging the Combined gas law

$$\frac{P_1 V_1}{T_1} = \frac{P_2 V_2}{T_2},$$

we obtain

$$P_2 = P_1 \times \frac{V_1}{V_2} \times \frac{T_2}{T_1}.$$

   Thus, we see that decreasing the volume (i.e., $V_1/V_2 < 1$) while maintaining the temperature (i.e., $T_2/T_1 = 1$) increases the pressure. Similarly, decreasing the volume while increasing the temperature (i.e., $T_2/T_1 > 1$) will also increase the pressure. Maintaining the volume while increasing the temperature also leads to an increase in pressure. On the other hand, increasing the volume by 1 liter and increasing the temperature by 1°K at STP gives

$$P_2 = P_1 \times \frac{22.4}{23.4} \times \frac{274}{273} = 0.96 \, P_1$$

Since $P_2/P_1 = 0.96$ (a number less than 1) the pressure has decreased.
   Using the same method, a volume decrease of 1 liter and a temperature

increase of 1°K at STP results in a P_2/P_1 value of 1.04, a number larger than 1. This signifies an increase in pressure.

3.   **(B)**   Charles' law states that the volume of a gas varies directly with the absolute temperature at constant pressure. On the other hand, Boyle's law states that the volume of a gas varies inversely with the pressure at constant temperature. The combined gas law is a combination of Boyle's law and Charles's law that does not involve constant pressure. Avogadro's law states that equal volumes of all gases contain equal numbers of molecules under the same conditions of temperature and pressure.

4.   **(D)**   Since hydrogen is a diatomic gas, the molecular weight of hydrogen gas (H_2) is 2.016 g mol⁻¹. Hence, 0.20 g of H_2 contains

$$\frac{0.20}{2.016}\left(6.02 \times 10^{23}\right) \text{molecules}.$$

5.   **(B)**   We can use the ideal gas law, $PV = nRT$. Since V and n are constant, an increase in T will cause a proportional increase in P. Remember that we must use the absolute (Kelvin) temperature scale. The temperature increases from 300°K to 400°K, thus the pressure will also increase by a third from 1.0 to 1.3 atm. We might express this in the equation below:

$$\frac{T_1}{T_2} = \frac{P_1}{P_2}$$

6.   **(C)**   Boyle's law predicts an inverse pressure-volume relationship. A pressure increase means a volume decrease. The multiplied fraction is thus less than one:

$$500 \times \frac{760}{800} = 475$$

7.   **(C)**   One mole at STP is occupied by 22.4 liters of a gas. This mole contains 6.02×10^{23} molecules, Avogadro's number. The given volume in this problem is twice that.

8.   **(A)**   Air pressure is measured by the height of a column of mercury in a tube; this is known as a barometer.

9.   **(C)**   Boyle's law states that gas volume varies by inverse proportion with a pressure change. Pressure is reduced here so the gas will expand.

$$300 \times \frac{1,000}{760} = 394.7 \text{ ml}$$

10. **(A)** For n moles of gas:

$$PV = nRT$$

$$PV = (m/M)RT$$

$$M = \text{molecular weight}$$

$$m = \text{mass of gas}$$

$$\text{density} = \text{mass/volume}$$

$$d = (m/V) = (PM/RT)$$

Molecular weight of SO_3 is

$$M_{so} = 32 + 3\,(16) = 80 \text{ g/mol}$$

$$P = 50 \text{ mm Hg} = 50/760 \text{ atm} = 0.066 \text{ atm}$$

$$T = 15 + 273 = 288°K$$

$$R = 0.0821 \text{ liter-atm/(mole-°K)}$$

$$d = (0.066 \times 80) / (0.0821 \times 288) = 0.223 \text{ g/liter}$$

11. **(B)**

$$P_1V_1/T_1 = P_2V_2/T_2$$

where $P_1 = P$, $V_1 = 10\ l$, $T_1 = 298$ K, $T_2 = 650$ K, $V_2 = 15\ l$
Rearrange and solve for P_2.

$$P_2 = 1.45\ P$$

12. **(B)** We can use the Ideal gas law, $PV = nRT$ to solve for the number of moles. By rearranging we have

$$n = \frac{PV}{RT}$$

To solve this equation we must use the correct units. Using $R = 0.082$ l-atm/mol-°K, we must express the pressure in atmospheres, the volume in liters, and the temperature in K.

$$V = 500 \text{ ml} = 0.5 \text{ L}$$

$$P = 0.4 \text{ atm}$$

$$T = 27°C + 273 = 300°K$$

Substituting:

$$n = 0.4 \text{ atm} \times \frac{0.5l}{0.82\,\frac{1\,\text{atm}}{\text{mol}°K} \times 300°\,K} = 0.008 \text{ moles}$$

13. **(E)** Using the Ideal gas law $PV = nRT$ and rearranging:

$$n = \frac{PV}{RT}$$

and since $n = m/M$, we now have

$$\frac{m}{M} = \frac{PV}{RT}$$

and solving for M gives us

$$M = \frac{RT}{PV}$$

Using the Ideal gas law constant $R = 0.082\ 1$ atm/mol°K we make the substitutions:

$$m = 0.25$$

$$\text{pressure} = 2.5 \text{ atm}$$

$$\text{volume} = 1\ l$$

$$\text{temperature} = 25°C = 298°K$$

$$M = \frac{0.25\ g \times 0.082 \frac{1\ atm}{mol°K} \times 298°\,K}{2.5\ atm \times 0.1\ l}$$

$$M = 24.4 \text{ g/mol}$$

14. **(A)** Charles' law states that the volume is directly proportional to temperature at constant pressure. Therefore

$$\frac{V}{T} = k$$

In this problem

$$V_1 = 200 \text{ ml}$$

$$T_1 = 25°C + 273 = 298°K$$

$$V_2 = \text{unknown}$$

$$T_2 = 40°C + 273 = 313°K$$

Thus,

$$\frac{200 \text{ ml}}{298°\,K} = \frac{V_2}{313°\,K}$$

and $\quad V_2 = \dfrac{200 \text{ ml} \times 313° \text{K}}{298° \text{K}} = 210 \text{ ml}$

15. **(B)** The Ideal gas law equation, $PV = nRT$, derived from kinetic and molecular theory neglects two important factors in regard to real gases. First, it neglects the actual volume of gas molecules, and second, it does not take into account the intermolecular forces that real gases exhibit.

The factor b in the Van der Waal's equation is a correction for the actual volume of the gas molecules and the factor a is a correction for the pressure due to the intermolecular attractions that occur in real gases.

16. **(C)** A molecular weight of 16 g tells us that a volume of 22.4 liters (molar volume) of that gas weighs 16 g. To determine the weight of a 4.48 l sample, we multiply:

$$4.48 \, l = \dfrac{16 \text{ g}}{22.4 \, l} = 3.2 \text{ g}$$

17. **(B)** The sample of gas is originally at STP (0°C, 760 torr). By examining the factors used to calculate the volume change, we see that the new conditions are 298°K and 800 torr. Remembering that $T_c = T_k + 273$, we have the new conditions 25°C, 800 torr.

18. **(A)** Boyle's law shows that the volume of a gas varies inversely with the pressure at constant temperature.

19. **(A)** Standard conditions are 1 atm or 760 torr and 273°K or 0°C.

20. **(B)** Using the Combined gas law

$$\dfrac{P_1 V_1}{T_1} = \dfrac{P_2 V_2}{T_2}$$

and rearranging

$$V_2 = V_1 \times \dfrac{P_1}{P_2} \times \dfrac{T_2}{T_1}$$

we obtain

$$V_2 = 10 \times \dfrac{2}{1} \times \dfrac{273}{546} = 10 \text{ liters}.$$

21. **(E)** Since the volume of one mole of an ideal gas is 22.4 liters, we have

$$\frac{160 \text{ g of Br}_2}{1 \text{ mole of Br}_2} \times \frac{1 \text{ mole of Br}_2}{22.4 \text{ liters}} = \frac{7.1 \text{ g of Br}_2}{\text{liter}}$$

22. **(B)** Oxygen, O_2, has a molecular weight of 32 g/mol. Taking the volume of one mole of an ideal gas to occupy 22.4 liters, we have

$$16 \text{ g} \times \frac{1 \text{ mole}}{32 \text{ g}} \times \frac{22.4 \text{ liters}}{1 \text{ mole}} = 11.2 \text{ liters}$$

23. **(D)** A gas behaves ideally under conditions where it encounters no other gas molecules with which to interact. Gas molecules have the largest distances between themselves under conditions of high temperature and low pressure.

24. **(B)** Boyle's law states that the product of P and V is a constant (at a constant temperature). Since $P \times V =$ constant, then a plot of P vs. V (and of V vs. P) will result in a graph as displayed in (B).

25. **(C)** Charles' law states that the ratio of V/T is a constant (at constant pressure). Since $V/T =$ constant, then a plot of V vs. T, and of T vs. V, will result in a graph similar to (C).

26. **(C)** The pressure of a gas at constant volume varies directly with the absolute temperature. Therefore, $P/T =$ constant and a plot of P vs. T or T vs. P gives a graph similar to (C).

27. **(A)** Recalling that one mole of an ideal gas occupies a volume of 22.4 liters, we calculate

$$\frac{254 \text{ g}}{1 \text{ mole}} \times \frac{1 \text{ mole}}{22.4 \text{ liters}} = 11.3 \text{ g}/l$$

28. **(C)** Recalling that density $= m/v$ and substituting

$$p = \frac{80 \text{ g/mol}}{22.4 \text{ liters/mol}} = 3.6 \text{ g}/l$$

29. **(E)** The molecular weight of O_2 gas is 32 grams/mol and the volume occupied by one mole of a gas at STP is 22.4 liters/mol. Thus,

$$44.8 \text{ liters} \times \frac{1 \text{ mole}}{22.4 \text{ liters}} \times \frac{32 \text{ g } O_2}{1 \text{ mole}} = 64 \text{ grams of } O_2.$$

30. **(D)**

$$22.4 \text{ liters} \times \frac{1 \text{ mole}}{22.4 \text{ liters}} \times \frac{6.023 \times 10^{23} \text{ molecules}}{1 \text{ mole}} \times \frac{2 \text{ atoms}}{\text{molecule}} = 12 \times 10^{23}$$

31. **(A)** All three standard temperature choices express identical intensities from three different measurement scales. At the standard atmospheric pressure of sea level, a mercury barometer reveals an elevated mercury column of 760 mm (torr) or 29.92 inches.

32. **(E)**

$$nRT = (P + n^2a/V^2)\,(V - nb) \text{ represents the real gas law.}$$

33. **(B)** An ideal gas is composed of molecules which have no attraction for each other and do not occupy any space in the containing vessel. These criteria are best satisfied by a small monatomic molecule with a complete electronic valence shell.

34. **(D)** Graphs (A) and (E) illustrate Charles' law (V proportional to T) and Boyle's law (V proportional to $1/P$), respectively. Graphs (B) and (C) are both in accordance with the ideal gas law ($PV = nRT$). Graph (D) is the only one that diverges significantly from this law. If n is constant, pressure/temperature will vary inversely with the volume: P/T proportional to $1/V$; therefore, the curve should approach both axes asymptotically.

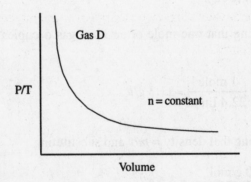

35. **(C)** The Combined gas law is a combination of Boyle's law and Charles' law, taking the form

$$\frac{P_1 V_1}{T_1} = \frac{P_2 V_2}{T_2}$$

and $P_2 V_2 = nRT_2$,

GASES REVIEW

John Dalton, in the early nineteenth century, proposed that all matter is composed of small subunits that he called "atoms." This was to explain such phenomena as the behavior of gases and the changes in the physical state of matter. For example, the atoms of solids are packed quite closely together, where the only possible motion is vibration. As the solid is heated, the vibrations become more intense, causing the atoms to move further apart (remember, heat makes things expand). If the solid is heated enough, the atoms will finally break loose from one another, and we say the solid has melted, or has become a liquid. In liquids the atoms are free to move around and travel fairly rapidly. This is why liquids flow, and unlike solids, they have no definite shape; they assume the shape of their container. As the liquid is heated, the atoms move even faster, and further apart until the liquid becomes a gas. In gases, the atoms are quite far apart and travel very fast. As a result, unlike solids and liquids, gases have no definite volume. Their atoms are so far apart they can be pushed closer together with relatively little effort. Hence, a gas will assume the volume of its container, as well as the container's shape.

1. Pressure

Pressure is defined as force per unit area. Atmospheric pressure is measured using a barometer and describes the pressure exerted by the air above the earth's surface.

Atmospheric pressure is directly related to the length (h) of the column of mercury in a barometer and is expressed in mm or cm of mercury (Hg).

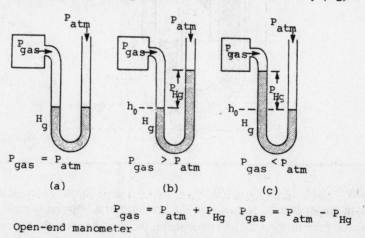

Open-end manometer

Figure 1.1

A sample of gas may be measured using either an open-end manometer (above) or a closed-end manometer as in Figure 1.2.

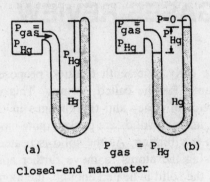

$$P_{gas} = P_{Hg} \quad \text{(b)}$$

Closed-end manometer

Figure 1.2

Standard atmospheric pressure is expressed in several ways: 14.7 pounds per square inch (psi), 760 mm of mercury, 760 torr, or simply 1 "atmosphere" (1 atm).

PROBLEM

Given the setup in the following figure, what would be the pressure of the gas (in atm) if P_{atm} is 745 torr and P_{liq} is the equivalent of a mercury column 30 cm high?

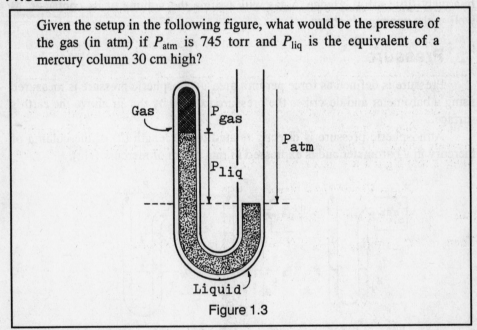

Figure 1.3

SOLUTION

A manometer is used to measure the pressure of a trapped sample of gas. If the right-hand tube is open to the atmosphere, the pressure which is exerted in the right-hand surface is atmospheric pressure, P_{atm}. If the liquid level is the same in both arms of the tube, the pressures must be equal; otherwise, there would be a flow of liquid from one arm to the other. At the level indicated by the dashed line in the figure, the pressure in the left arm is equal to the pressure of the trapped gas, P_{gas}, plus the pressure of the column of liquid above the dashed line, P_{liq}. One can write

$$P_{gas} = P_{atm} - P_{liq}.$$

Here, one is given that P_{atm} is 745 torr and that P_{liq} is equivalent to a mercury column 3.0 cm high. One wishes to find P_{gas} in atm, thus, one must convert 745 torr to atm and find the P_{liq}. There are 760 torr in 1 atm, which means

$$\text{no. of atm in } P_{atm} = \frac{745 \text{ torr}}{760 \text{ torr} / 1 \text{ atm}} = .98 \text{ atm}.$$

One atmosphere pressure supports 76 cm of mercury, thus 3 cm of mercury supports 3/76 atm. Therefore, P_{liq} = 3 cm/76 cm/atm = .039 atm.
Solving for P_{gas}:

$$P_{gas} = .980 \text{ atm} - .039 \text{ atm} = .941 \text{ atm}.$$

PROBLEM

> Consider the manometer, illustrated below, first constructed by Robert Boyle. When $h = 40$ mm, what is the pressure of the gas trapped in the volume, V_{gas}. The temperature is constant, and atmospheric pressure is $P_{atm} = 1$ atm.

SOLUTION

We do not need to know any gas law to solve this problem. All we must realize is that the pressure exerted on the gas, P_{total}, is equal to the sum of the pressure exerted by the mercury, P_{Hg}, and the pressure exerted by the air, P_{atm}. Since 1 mm Hg = 1 torr and 1 atm = 760 torr,

$$P_{Hg} = 40 \text{ mm Hg} = 40 \text{ torr}$$

and $\qquad P_{atm} = 1 \text{ atm} = 760 \text{ torr}.$

Then $\qquad P_{total} = P_{Hg} + P_{atm} = 40 \text{ torr} + 760 \text{ torr} = 800 \text{ torr}.$

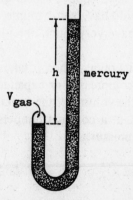

Figure 1.4

Drill 1: Pressure

1. If atmospheric pressure is 0.95 atm, how much is atmospheric pressure in mm Hg?

    (A) 950   (D) 760

    (B) 800   (E) 722

    (C) 133

2. If a gas exerts a pressure of 6.00 psi, how many atmospheres does it exert?

    (A) 0.816   (D) 760

    (B) 0.408   (E) 816

    (C) 14.7

3. In the following setup, what would atmospheric pressure be (in mm Hg) if $P_{gas} = 0.890$ atm and P_{liq} is equivalent to a mercury column 6.0 cm high?

    (A) 890

    (B) 676

    (C) 760

    (D) 736

    (E) 312

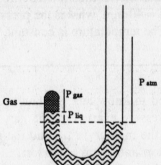

2. Boyle's Law

This and the following laws apply to ideal gases, which do not exist. The molecules of an ideal gas would have zero volume, and their collisions would be totally elastic. Real gases behave very much like ideal gases under low pressures and high temperatures.

A gas has no shape of its own; rather, it takes the shape of its container. It has no fixed volume, but is compressed or expanded as its container changes in size. The volume of a gas is the volume of the container in which it is held.

Boyle's law states that, at a constant temperature, the volume of a gas is inversely proportional to the pressure.

$$V \alpha \frac{1}{P} \text{ or } V \text{ constant} \times \frac{1}{P} \text{ or } PV = \text{constant}$$

$$P_iV_i = P_vV_f$$

$$V_f = V_i\left(\frac{P_i}{P_f}\right)$$

A hypothetical gas that would follow Boyle's law under all conditions is called an ideal gas. Deviations from Boyle's law that occur with real gases represent non-ideal behavior. The following diagram shows a graphical representation of Boyle's law.

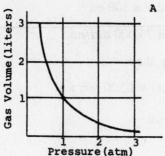

Figure 2.1

PROBLEM

One hundred ml. of gas are enclosed in a cylinder under a pressure of 760 torr. What would the volume of the same gas be at a pressure of 1,520 torr?

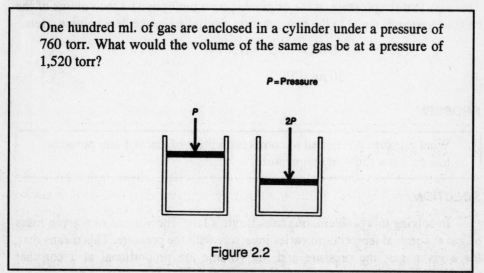

Figure 2.2

SOLUTION

Since this problem deals with the pressure and volume of a gas at a constant temperature, Boyle's law can be used. Boyle's law states that the volume, V, of a given mass of gas, at constant temperature, varies inversely with the pressure, P. It can be stated as

$$V = k \times \frac{1}{P},$$

where k is a constant. Hence,

$$k = PV.$$

For a particular system, at constant temperature, k is constant. Therefore, if either the pressure or the volume is changed, the other must adjust accordingly.
Here,

$$P = 760 \text{ torr and } V = 100 \text{ ml}$$

so $\qquad k = 760 \times 100 = 76{,}000 \text{ torr-ml.}$

If P is doubled to 1,520 torr, then

$$k = 76{,}000 \text{ torr-ml} = 1{,}520 \text{ torr} \times V$$

$$V = \frac{76{,}000 \text{ torr-ml}}{1{,}250 \text{ torr}} = 50 \text{ ml}$$

Since k is a constant for a given system, another form of Boyle's law can be expressed as

$$P_1 V_1 = P_2 V_2.$$

This says that the pressure of the original system multiplied by the volume of the original system is equal to the new pressure multiplied by the new volume. Here,

$$760 \text{ torr} \times 100 \text{ ml} = 1{,}520 \text{ torr} \times V_2$$

$$50 \text{ ml} = V_2$$

PROBLEM

What pressure is required to compress 5 liters of gas at 1 atm pressure to 1 liter at a constant temperature?

SOLUTION

In solving this problem, one uses Boyle's law: The volume of a given mass of gas at constant temperature varies inversely with the pressure. This means that, for a given gas, the pressure and the volume are proportional at a constant temperature, and their product equals a constant.

$$P \times V = k$$

where P is the pressure, V is the volume, and k is a constant. From this one can propose the following equation

$$P_1 V_1 = P_2 V_2,$$

where P_1 is the original pressure, V_1 is the original volume, P_2 is the new pressure and V_2 is the new volume.

In this problem, one is asked to find the new pressure and is given the original pressure and volume and the new volume.

$$P_1V_1 = P_2V_2$$

$$P_1 = 1 \text{ atm}$$

$$V_1 = 5 \text{ liters}$$

$$P_2 = ?$$

$$V_2 = 1 \text{ liter}$$

$$1 \text{ atm} \times 5 \text{ liters} = P_2 \times 1 \text{ liter}$$

$$\frac{1 \text{ atm} \times 5 \text{ liters}}{1 \text{ liter}} = P_2$$

$$5 \text{ atm} = P_2$$

Drill 2: Boyle's Law

1. A mass of gas is under a pressure of 760 mm Hg and occupies a volume of 525 ml. If the pressure were doubled, what volume would the gas occupy? Assume the temperature is constant.

    (A) 525 ml   (D) 345.4 ml

    (B) 760.1 ml   (E) 104.3 ml

    (C) 262.5 ml

2. A gaseous sample of neon, maintained at constant temperature, occupies 500 ml at 2.0 atm. What will the volume be if the pressure is changed to 1.8×10^{-3} ton?

    (A) 2.3×10^{-6} ml   (D) $6.25 \times 10^5 \, l$

    (B) $4.22 \times 10^5 \, l$ (E) 760 ml

    (C) $40 \times 10^9 \, l$

3. In the situation described in question 2, what would the volume be if the pressure were changed to 4.00 atm?

    (A) 6.00 l (D) 0.50 l

    (B) 1.50 l (E) 0.75 l

    (C) 0.25 l

3. Charles' Law

Charles' law states that at constant pressure, the volume of a given quantity of a gas varies directly with the temperature

$$V \alpha T$$

or $$\frac{V}{T} = \text{constant}$$

$$\frac{V_1}{T_1} = \frac{V_2}{T_2}$$

or $$\frac{V_1}{V_2} = \frac{T_1}{T_2}$$

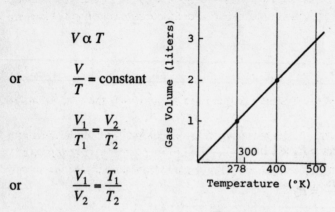

Figure 3.1

If Charles' law were strictly obeyed, gases would not condense when they are cooled. This means that gases behave in an ideal fashion only at relatively high temperatures and low pressures.

PROBLEM

A sample of gaseous krypton, maintained at constant pressure, is found to have a volume of 10.5 l at 25°C. If the system is heated to 250°C, what is the resulting volume?

SOLUTION

We need a relationship between volume and temperature. Such a relationship is provided by Charles' law, which states that volume V and absolute temperature T are proportional, or, as an equality, $V = kT$, where k is a constant.

We can determine k for our system from the initial volume and temperature. Thus,

$$k = \frac{V}{T} = \frac{10.5\,l}{25°C} = \frac{10.5\,l}{298.15°K} = 0.0352\,l - K^{-1}$$

The volume corresponding to a temperature of 250°C, which is 523.15°K, is $V = kT = 0.0352\,l - K^{-1} \times 523.15°K = 18.41\,l.$

PROBLEM

A certain gas occupies a volume of 100 ml at a temperature of 20°C. What will its volume be at 10°C, if the pressure remains constant?

SOLUTION

In a gaseous system, when the volume is changed by increasing the temperature, keeping the pressure constant, Charles' law can be used to determine the new volume. Charles' law states that, at a constant pressure, the volume of a given mass of gas is directly proportional to the absolute temperature. Charles' law may also be written

$$\frac{V_1}{T_1} = \frac{V_2}{T_2}$$

where V_1 is the volume at the original temperature T_1, and V_2 is the volume at the new temperature T_2.

To use Charles' law, the temperature must be expressed on the absolute scale. The absolute temperature is calculated by adding 273 to the temperature in degrees Centigrade. In this problem, the centigrade temperatures are given and must be converted to the absolute scale.

$$T_1 = 20°C + 273 = 293°K$$

$$T_2 = 10°C + 273 = 283°K$$

Using Charles' law,

$$V_1 = 100 \text{ ml}$$

$$T_1 = 293°K$$

$$T_2 = 283°K$$

$$V_2 = ?$$

$$\frac{V_1}{T_1} = \frac{V_2}{T_2}$$

$$V_2 = \frac{V_1 T_2}{T_1}$$

$$V_2 = \frac{(100 \text{ ml})(283°K)}{293°K}$$

$$V_2 = 96.6 \text{ ml}$$

Drill 3: Charles' Law

1. Assume that one cubic foot of air near a thermonuclear explosion is heated from 0°C to 546,000°C. To what volume does the air expand?

    (A) 273 ft³

    (D) 4,004 ft³

    (B) 1,000 ft³

    (E) 2,001 ft³

    (C) 320 ft³

2. Four hundred ml of a gas at 25°C expands to 600 ml. What would the resulting temperature be if all other factors remain constant.

   (A) 447°C   (D) 121°C

   (B) 200°C   (E) 32°C

   (C) 174°C

3. According to Charles' law when the temperature of an ideal gas is increased, volume

   (A) increases   (D) is always 1 liter.

   (B) decreases   (E) cannot be determined.

   (C) remains the same

4. The Law of Gay-Lussac

The law of Gay-Lussac states that at constant volume, the pressure exerted by a given mass of gas varies directly with the absolute temperature.

$$P \alpha T \text{ (where volume and mass of gas are constant)}$$

$$\frac{P_1}{T_1} = \frac{P_2}{T_2}$$

Gay-Lussac's law of combining volumes states that when reactions take place in the gaseous state, under conditions of constant temperature and pressure, the volumes of reactants and products can be expressed as ratios of small whole numbers.

PROBLEM

The air in a tank has a pressure of 640 mm of Hg at 23°C. When placed in sunlight, the temperature rose to 48°C. What was the pressure in the tank?

SOLUTION

The law of Gay-Lussac deals with the relationship existing between pressure and the absolute temperature (°C + 273°), for a given mass of gas at constant volume. The relationship is expressed in the law of Gay-Lussac: volume constant, the pressure exerted by a given mass of gas varies directly with the absolute temperature. That is:

$$P \alpha T \text{ (volume and mass of gas constant)}$$

The variation that exists between pressure and temperature at different states can be expressed as

$$\frac{P_1}{T_1} = \frac{P_2}{T_2}$$

where P_1 = pressure of original state, T_1 = absolute temperature of original state, P_2 = pressure of final state, and T_2 = absolute temperature of final state.

Thus this problem is solved by substituting the given values into Gay-Lussac's law.

$$P_1 = 640 \text{ mm Hg}$$

$$T_1 = 23°C + 273° = 296°K$$

$$P_2 = ?$$

$$T_2 = 48°C + 273° = 321°K$$

Substituting and solving,

$$\frac{640 \text{ mm}}{296°\text{K}} = \frac{P_2}{321°\text{K}}$$

$$P_2 = 640 \text{ mm} \times \frac{321°\text{K}}{296°\text{K}}$$

$$= 694 \text{ mm of Hg}$$

PROBLEM

A sealed glass bulb contained helium at a pressure of 750 mm Hg and 27°C. The bulb was packed in dry ice at − 73°C. What was the resultant pressure of the helium?

SOLUTION

The only parameters mentioned in this problem are pressure (P) and temperature (T). One is given the initial pressure and temperature and the final temperature. One is asked to determine the pressure at − 73°C. The law of Gay-Lussac relates temperature and pressure. It can be stated mathematically

$$\frac{P_1}{T_1} = \frac{P_2}{T_2}$$

where P_1 is the initial pressure, T_1 is the initial absolute temperature, P_2 is the final pressure, and T_2 is the final absolute temperature.

The temperature in °C is converted to K by adding 273.

$$T_1 = 27°C + 273 = 300°K$$

$$T_2 = -73°C + 273 = 200°K$$

Solving for P_2,

$$\frac{755 \text{ mm Hg}}{300° \text{K}} = \frac{P_2}{200° \text{K}}$$

$$P_2 = \frac{(750 \text{ mm Hg})(200° \text{K})}{300° \text{K}} = 500 \text{ mm Hg}$$

Drill 4: The Law of Gay-Lussac

1. A gas at 43°C has a pressure of 680 torr. If the temperature is raised to 55°C, what will the new pressure of the gas be? (Assume all other variables remain constant.)

    (A) 735 torr   (D) 870 torr

    (B) 706 torr   (E) 374 torr

    (C) 158 torr

2. If the pressure of a gas at 298°K is tripled, what will the new temperature be? (Assume all other variables are kept constant.)

    (A) 447°K   (D) 99°K

    (B) 894°K   (E) 495°K

    (C) 298°K

3. If the temperature of a constant quantity of gas at constant volume and 2.0 atm changes from 30°C to 24°C, what will the pressure change to?

    (A) 1.6 atm   (D) 1.0 atm

    (B) 2.0 atm   (E) 6.2 atm

    (C) 8.0 atm

5. The Combined Gas Law

Combining the previously discussed laws, (Boyle's, Charles', and Gay-Lussac's laws) gives us the Combined gas law.

The Combined gas law states that for a given mass of gas, the volume is inversely proportional to the pressure and directly proportional to the absolute temperature. This law can be written

$$\frac{P_1 V_1}{T_1} = \frac{P_2 V_2}{T_2}$$

where P_1 is the original pressure, V_1 is the original volume, T_1 is the original absolute temperature, P_2 is the new pressure, V_2 is the new volume, and T_2 is the new absolute temperature.

PROBLEM

A gas occupies a volume of 1.0 liter at a temperature of 27°C and 500 torr pressure. Calculate the volume of the gas if the temperature is changed to 60°C and the pressure to 700 torr.

SOLUTION

For a given mass of gas, the volume is inversely proportional to the pressure and directly proportional to the absolute temperature. This combined law can also be written:

$$\frac{P_1 V_1}{T_1} = \frac{P_2 V_2}{T_2}$$

where P_1 is the original pressure, V_1 is the original volume, T_1 is the original absolute temperature, P_2 is the new temperature, V_2 is the new volume, and T_2 is the new absolute temperature.

In this problem, one is given the original pressure, temperature, and volume and the new temperature and pressure. One is asked to calculate the new volume. The temperatures are given in °C; they must be converted to the absolute scale before using the combined law. This can be done by adding 273 to the temperature in °C.

Converting the temperature:

$$T_1 = 27° + 273 = 330°K$$

$$T_2 = 60° + 273 = 333°K$$

Using the combined law:

$$\frac{P_1 V_1}{T_1} = \frac{P_2 V_2}{T_2}$$

$$P_1 = 500 \text{ torr}$$

$$P_2 = 700 \text{ torr}$$

$$V_1 = 1.0 \text{ liter}$$

$$V_2 = ?$$

$$T_1 = 300°K$$

$$T_2 = 333°K$$

$$\frac{(500 \text{ torr}) (1.0 \text{ liter})}{(300° \text{ K})} = \frac{(700 \text{ torr}) (V_2)}{(333° \text{ K})}$$

$$V_2 = \frac{(500 \text{ torr}) (1.0 \text{ liter}) (333° \text{ K})}{(300° \text{ K}) (700 \text{ torr})}$$

$$V_2 = 0.79 \text{ liter}.$$

PROBLEM

Calculate the pressure required to compress 2 liters of a gas at 700 mm pressure and 20°C into a container of 0.1 liter capacity at a temperature of –150°C.

SOLUTION

One is dealing with changing volumes, pressures, and temperatures of a gas. Therefore, this problem can be solved using the combined gas law. It states that as the pressure increases, the volume decreases and that as the temperature increases, the volume increases. These factors are related by the equation

$$\frac{P_1 V_1}{T_1} = \frac{P_2 V_2}{T_2}$$

where P_1, V_1, and T_1 are the initial pressure, volume, and temperature and P_2, V_2, and T_2 are the final values.

For any problem dealing with gases, the first step always involves converting all of the temperatures to Kelvin by the equation

$$K = °C + 273.$$

For this question

$$T_1 = 20°C = 20 + 273 = 293°K$$

$$T_2 = - 150°C = - 150 + 273 = 123°K$$

This seems to indicate that the pressure would decrease. But one is also told that the volume decreases, which would have the effect of increasing the pressure. Therefore, one cannot predict the final change in volume.

For the sake of clarity, set up a table as given below.

$$P_1 = 700 \text{ mm}$$

$$P_2 = ?$$

$$V_1 = 2 l$$

$$V_2 = 0.1 l$$

$$T_1 = 293°K$$

$$T_2 = 123°K$$

Since one is given five of the six values, it is possible to use the combined gas law equation to determine P_2.

$$\frac{P_1V_1}{T_1} = \frac{P_2V_2}{T_2}$$

$$P_2 = \frac{T_2V_1P_1}{T_1V_2}$$

$$= \frac{123°\,K\,(2l)\,(700\text{ mm})}{293°\,K\,(0.1l)}$$

$$= 5,877\text{ mm}$$

Drill 5: The Combined Gas Law

1. Seven hundred fifty ml of gas at 300 torr pressure and 50°C is heated until the volume of the gas is 2,000 ml at a pressure of 700 torr. What is the final temperature of the gas?

    (A) 2,451°C

    (B) 282°C

    (C) 300°C

    (D) 1,737°C

    (E) 369°C

2. Five hundred liters of a gas at 27°C and 700 torr would occupy what volume at STP?

    (A) 17.1 *l*

    (B) 529 *l*

    (C) 419 *l*

    (D) 494 *l*

    (E) 1.54 *l*

3. A chemist has a certain amount of gas under a pressure of 33.3 atm; it occupies 30 *l* at 273°C. For his research, however, the gas must be at standard conditions. Under standard conditions what will the volume of the gas be?

    (A) 499.5 *l*

    (B) 999 *l*

    (C) 1,998 *l*

    (D) 259 *l*

    (E) 854 *l*

6. Avogadro's Law and the Mole Concept

The gram-atomic weight of any element is defined as the mass, in grams, which contains one mole of atoms of that element.

For example, approximately 12.0 g of carbon, 16.0 g of oxygen, and 32.1 g of sulfur, each contain 1 mole of atoms.

The term "one mole" in the above definition is a certain number of atoms. Just as a dozen always means 12, and a gross is always 144, a mole is always 6.02×10^{23}. The value 6.02×10^{23} is also known as Avogadro's number, which by definition is the number of atoms in one gram of pure hydrogen.

Avogadro's law states that under conditions of constant temperature and pressure, equal volumes of different gases contain equal numbers of molecules.

If the initial and final pressure and temperature are the same, then the relationship between the number of molecules, N, and the volume, V, is

$$\frac{V_f}{V_i} = \frac{N_f}{N_i}.$$

PROBLEM

What is the atomic weight of lead if 207.2 grams of the metal are found to have as many atoms as 1 gram of hydrogen (Avogadro's number)?

SOLUTION

If we have Avogadro's number of atoms present, that is 6.02×10^{23}, we then have 1 mole of atoms present. This gives us 207.2 grams/per 1 mole, which satisfies the above definition of atomic weight. The answer is 207.2 Daltons, or 207.2 amu (atomic mass units). Note the types of units used.

PROBLEM

For a single year, the motor vehicles in a large city produced a total of 9.1×10^6 kg of the poisonous gas carbon monoxide (CO). How many moles of CO does this correspond to?

SOLUTION

The number of moles of a substance is equal to the quotient of the mass (in grams) of that substance and its molecular weight (in g/mol), or

$$\frac{\text{mass (g)}}{\text{molecular weight (g/mol)}}.$$

The mass of CO is 9.1×10^6 kg $= 9.1 \times 10^6$ kg $\times$ 1,000 g/kg $= 9.1 \times 10^9$ g. The molecular weight of CO is the sum of the atomic weight of C and the atomic weight of O, or

molecular weight of (CO) = atomic weight (C) + atomic weight (O)

$$= 12 \text{ g/mol} + 16 \text{ g/mol}$$

$$= 28 \text{ g/mol}$$

Hence,

$$\text{moles} = \frac{\text{mass (g)}}{\text{molecular weight (g/mol)}}$$

$$= \frac{9.1 \times 10^9 \text{ g}}{28 \text{ g/mol}}$$

$$= 3.3 \times 10^8 \text{ moles}$$

Drill 6: Avogadro's Law and the Mole Concept

1. A flask containing H_2 at 0°C was sealed off at a pressure of 1 atm and the gas was found to weigh .451 2g (AW H = 1.0079). How many moles of H_2 were present?

    (A) 0.2238   (D) 4.151

    (B) 0.4413   (E) 0.1692

    (C) 2.016

2. How many moles of CO are there in 46 g of CO? (aw C = 12, O = 16)

    (A) 0.61   (D) 2.9

    (B) 1,344   (E) 1.6

    (C) 74

3. If 9 grams of the fictional element X contains 0.12 moles, what is the atomic weight of X?

    (A) 1.08 g/mol   (D) 81 g/mol

    (B) 75 g/mol   (E) 9.12 g/mol

    (C) 0.013 g/mol

7. The Ideal Gas Law

According to Boyle's law,

$$V \propto \frac{1}{P};$$

according to Charles' law,

$$V \propto T;$$

and from Gay-Lussac's law,

$$V \propto n.$$

This leads to the law

$$V \propto \frac{nT}{P}.$$

Using the gas constant R as the proportionality constant, we get

$$PV = nRT.$$

This is known as the Ideal gas law.

The hypothetical ideal gas obeys exactly the mathematical statement of the ideal gas law. This statement is also called the equation of state of an ideal gas because it relates the variables (P, V, n, T) that specify properties of the gas. Molecules of ideal gases have no attraction for one another and have no intrinsic volume; they are "point particles." Real gases act in a less than ideal way, especially under conditions of increased pressure and/or decreased temperature. Real gas behavior approaches that of ideal gases as the gas pressure becomes very low. The ideal gas law is thus considered a "limiting law."

When using the ideal gas law, the term R is always a constant. If the volume is in liters, the temperature in Kelvin, and the pressure is in atmospheres, R has a value of 0.082 and units of liter × atm/mol Kelvin. R may also be given in other units. For example, R is also equal to 8.314 J/mol × K. Remember, this value of R can only be used under the stated conditions.

PROBLEM

How many moles of hydrogen gas are present in a 50 liter steel cylinder if the pressure is 10 atmospheres and the temperature is 27°C? $R =$ .082 liter-atm/mole°K.

SOLUTION

In this problem, one is asked to find the number of moles of hydrogen gas present where the volume, pressure, and temperature are given. This would indicate that the Ideal gas law should be used because this law relates these quantities to each other. The ideal gas law can be stated:

$PV = nRT,$

where P is the pressure, V is the volume, n is the number of moles, R is the gas constant (.082 liter-atm/mole °K), and T is the absolute temperature. Here, one is given the temperature in °C, which means it must be converted to the absolute scale. P, V, and R are also known. To convert a temperature in °C to the absolute scale, add 273 to the temperature in °C.

$T = 27 + 273 = 300°K$

Using the Ideal gas Law:

$PV = nRT$

or $n = \dfrac{PV}{RT}$

$P = 10$ atm

$V = 50\,l$

$R = .082\,l\text{-atm/mole °K}$

$T = 300$ K

n = number of moles of H_2 present

$n = \dfrac{(10\,\text{atm})\,(50l)}{(.082l\text{--atm/mol °K})\,(300°\,K)}$

$= 20$ moles

PROBLEM

A sample of a gas exhibiting ideal behavior occupies a volume of 350 ml at 25.0°C and 600 torr. Determine its volume at 600°C and 25.0 torr.

SOLUTION

We can solve this problem by determining the number of moles, n, of gas from the ideal gas equation, $PV = nRT$ (where P = pressure, V = volume, n = number of moles, R = gas constant, and T = absolute temperature) by using the first set of conditions and then substituting this value of n along with P and T from the second set of conditions into the ideal gas equation and solving for V. However, we can save useless calculation by denoting the first and second sets of conditions by subscripts "1" and "2," respectively, to obtain

$P_1V_1 = nRT_1,$

and $P_2V_2 = nRT_2,$

where n is the same in both cases. Dividing the second of these equations by the first, we obtain

$$\frac{P_2V_2}{P_1V_1} = \frac{nRT_2}{nRT_1}$$

or $$\frac{P_2V_2}{P_1V_1} = \frac{T_2}{T_1},$$

where we have cancelled n and R (both constant). Solving for V_2, we obtain

$$V_2 = \frac{P_1V_1}{P_2} \times \frac{T_2}{T_1}$$

$$= \frac{600 \text{ torr} \times 350 \text{ ml}}{25.0 \text{ torr}} \times \frac{873.15° \text{ K}}{298.15° \text{ K}}$$

$$= 24.6 \times 103 \text{ ml}$$

$$= 24.6 \; l$$

Drill 7: The Ideal Gas Law

1. The barometric pressure on the lunar surface is about 10.0^{-10} torr. At a temperature of 100 K, what volume of lunar atmosphere contains 10.0^6 molecules of gas?

    (A) $1.316 \times 10^{-13} \; l$ (D) $6.24 \times 10^{10} \; l$

    (B) $3.13 \times 10^5 \; l$ (E) $1.04 \times 10^{-4} \; l$

    (C) $2.21 \times 10^{11} \; l$

2. A sample of gas occupies $14.3 \; l$ at 19°C and 790 mm Hg. How many moles of gas are present?

    (A) .2516 moles   (D) 6.213 moles

    (B) 1.341 moles   (E) .0135 moles

    (C) 0.6194 moles

3. Using the information from the above question, what volume will this same amount of gas occupy at 190°C and 79.0 mm Hg?

    (A) 226.68 l (D) 221.41 l

    (B) 300.15 l (E) 62.311 l

    (C) 273.15 l

4. What is the molecular weight of a gas if 4.50 g of it occupies 4.00 liters at 950 torr and 182°C? ($R = 0.082$ l atm-mol°K)

    (A) 0.0289   (D) 7.69

    (B) 33.6   (E) 55.3

    (C) 0.130

5. A cylinder contains oxygen at a pressure of 10 atm and a temperature of 300°K. The volume of the cylinder is 10 l. What is the mass of the oxygen?

    (A) 87.467 grams   (D) 130.08 grams

    (B) 76.876 grams   (E) 18.361 grams

    (C) 145.32 grams

8. Density of Gases

At STP, standard temperature and pressure, 0°C and 760 mm of mercury pressure, 1 mole of an ideal gas occupies 22.4 liters. (The "molar volume" of the gas at STP is 22.4 l/mole.)

The density of a gas can be converted to molecular weight using the 22.4 liters/mole relationship:

$$M = \text{(density) (molar volume)}$$

$$\left(\frac{g}{mol}\right) = \left(\frac{g}{l}\right)\left(\frac{l}{mol}\right)$$

PROBLEM

> The density of a gas is measured as 0.222 g/l at 20.0°C and 800 torr. What volume will 10.0 g of this gas occupy under standard conditions?

SOLUTION

Since we know the density at a given temperature and pressure, we can apply the equation $P/(\rho T) = $ constant, which holds for ideal gases of density r at pressure P and absolute temperature T. We must first determine the constant for our system and then use this value along with standard pressure (760 torr) and temperature (273.15°K) to determine the density of the gas under standard conditions. From this density, we obtain the volume that 10.0 g of the gas occupies under standard conditions.

The value of the constant is, as previously indicated,

$$\frac{P}{\rho T} = \frac{800 \text{ torr}}{0.222 \text{ g/}l \times 20.0^\circ C}$$

$$= \frac{800 \text{ torr}}{0.222 \text{ g/}l \times 293.15 \text{ K}}$$

$$= 12.28 \frac{\text{torr-}l}{\text{g-K}}.$$

Hence, for our gas,

$$\frac{P}{\rho T} = 12.28 \frac{\text{torr-}l}{\text{g-K}}$$

or

$$\rho = \frac{P}{T} \times \frac{1}{12.28 \text{ torr-}l\text{/g-K}}.$$

Under standard conditions, the density of our gas is then

$$\rho = \frac{P}{T} \times \frac{1}{12.28 \text{ torr-}l\text{/g-K}}$$

$$= \frac{760 \text{ torr}}{273.15 \text{ K}} \times \frac{1}{12.28 \text{ torr-}l\text{/g-K}}$$

$$= 0.2265 \text{ g/}l$$

Density is defined as ρ = mass/volume. Hence, volume = mass/ρ, and the volume occupied by 10.0 g of our gas under standard conditions is

$$\text{volume} = \frac{\text{mass}}{\rho} = \frac{10.0 \text{ g}}{.2265 \text{ g/}l} = 44.15 \, l.$$

PROBLEM

At standard conditions, it is found that 28.0 g of carbon monoxide occupies 22.4 l. What is the density of carbon monoxide at 20°C and 600 torr?

SOLUTION

The density of carbon monoxide under STP or standard conditions (pressure = 1 atm or 760 torr and temperature = 0°C or 273.15°K) is ρ = mass/volume. 28 g of CO, carbon monoxide, represents one mole, and a mole of any gas occupies 22.4 liters of volume under standard conditions. Thus, ρ = 28 g/22.4 l = 1.25 g/l.

We can apply the equation $P/(\rho T)$ = constant, which holds for ideal gases of density ρ at pressure P and absolute temperature T. We will first determine the constant for our system and then use this value to calculate the density at $20°C = 293.15°K$ and 600 torr.

The value of the constant for our system is

$$\text{constant} = \frac{P}{\rho T} = \frac{760 \text{ torr}}{1.25 \text{ g/l} \times 273.15°K}$$

$$= 2.225 \frac{\text{torr-}l}{\text{g-K}}$$

Hence, for our system,

$$\frac{P}{\rho T} = 2.225 \frac{\text{torr-}l}{\text{g-K}}$$

or $$\rho = \frac{P}{T} \times \frac{1}{2.225 \text{ torr-}l/\text{g-K}}$$

At $20°C = 293.15°K$ and 600 torr, the density of CO is

$$\rho = \frac{P}{T} \times \frac{1}{2.225 \text{ torr-}l/\text{g- K}}$$

$$= \frac{600 \text{ torr}}{293.15 \text{ K}} \times \frac{1}{2.225 \text{ torr-}l/\text{g-K}}$$

$$= .9198 \text{ g/}l$$

Drill 8: Density of Gases

1. What is the approximate density of fluorine gas at STP?

   (A) 0.8 g/l   (D) 2.0 g/l

   (B) 1.0 g/l   (E) 2.3 g/l

   (C) 1.7 g/l

2. Determine the density of methane gas (MW = 16.0) at 25°C and 6.00 atm.

   (A) 3.77 g/l   (D) 3.92 g/l

   (B) 10.4 g/l   (E) 1.21 g/l

   (C) 46.8 g/l

3. At 100°C and 720 mm Hg, what is the density of carbon dioxide, CO_2?

   (A) 32.3 g/l   (D) 23.6 g/l

   (B) 4.40 g/l   (E) 51.6 g/l

   (C) 1.32 g/l

9. Real Gases

Real gases fail to obey the Ideal gas law under most conditions of temperature and pressure.

Real gases have a finite (non-zero) molecular volume; i.e., they are not true "point particles." The volume within which the molecules may not move is called the excluded volume. The real volume (volume of the container) is therefore slightly larger than the ideal volume (the volume the gas would occupy if the molecules themselves occupied no space):

$$V_{real} = V_{ideal} + nb,$$

where b is the excluded volume per mole and n is the number of moles of gas. The ideal pressure, that is, the pressure the gas could exert in the absence of intermolecular attractive forces, is higher than the actual pressure by an amount that is directly proportional to n^2/V^2:

$$P_{ideal} = P_{real} + \frac{n^2 a}{V^2},$$

where a is a proportionality constant that depends on the strength of the intermolecular attractions. Therefore,

$$\left(P + \frac{n^2 a}{V^2}\right)(V - nb) = nRT$$

is the Van der Waals equation of state for a real gas.

The values of the constants a and b depend on the particular gas and are tabulated for many real gases.

PROBLEM

Using the data from the accompanying figure, calculate the pressure exerted by .250 moles of CO_2 in 0.275 liters at 100°C. Compare this value with that expected for an ideal gas.

Van der Waals Constants

Gas	a, liter^2 atm/mol^2	b, liter/mol
Helium	0.0341	0.0237
Argon	1.35	0.0322
Nitrogen	1.39	0.0391
Carbon dioxide	3.59	0.0427
Acetylene	4.39	0.0514
Carbon tetrachloride	20.39	0.1383

SOLUTION

To solve this problem you must understand the concept of an "ideal gas" and the formulas associated with it. Boyle's law states that PV = constant, if the temperature is fixed. In other words, if T is constant, then a fixed mass of gas occupies a volume inversely proportional to the pressure exerted on it. If a gas obeys this law, it is termed an "ideal gas." When a gas is ideal, $PV = nRT$, where P = pressure, V = volume, n = moles, R = universal gas constant, and T = temperature in Kelvin. When a gas is not ideal, it doesn't obey Boyle's law; and instead of the Ideal gas law you use Van der Waal's equation,

$$\left(P + \frac{n^2 a}{V^2}\right)(V - nb) = nRT,$$

where, for CO_2, a = 3.59 liter^2-atm/mol^2 and b = .0427 liter/mol. These values are called Van der Waals constants. Substitute the values into these equations and solve for P. As such,

$$\left(P + \frac{(.250)^2 (3.59)}{(.275)^2}\right)[.275 - (.250)(.0427)] = (.250)(.08206)(373).$$

Solving for P, you obtain P = 26.0 atm.

If you had considered the gas as ideal, then $PV = nRT$, and P = 27.8 atm.

PROBLEM

Of the following two pairs, which member will more likely deviate from ideal gas behavior? (1) N_2 versus CO, (2) CH_4 versus C_2H_6.

SOLUTION

Characteristics of ideal gases include: (1) Gases are composed of molecules such that the actual volume of the molecules is negligible compared with the empty space between them. (2) There are no attractive forces between molecules. (3) Molecules do not lose net kinetic energy in collisions. (4) The average kinetic energy is directly proportional to the absolute temperature.

From these assumptions came the ideal equation of state: $PV = nRT$, where P = pressure, V = volume, n = moles, R = universal gas constant, and T = temperature. No gas is ideal; they don't obey these assumptions absolutely. To reflect these limitations the ideal gas law can be written as:

$$\left(P + \frac{n^2 a}{V^2}\right)(V - nb) = nRT.$$

The term $V = nb$ can be thought of as representing the free volume minus the volume occupied by the gas molecules themselves. The magnitude of b is proportional to a gas molecule's size. Thus, the greater the size of a gas molecule, the greater the nb is, and, the greater the deviation, since the volume is assumed to equal exactly V in the ideal gas law. $V - nb$ is smaller than V. The greater nb is, the smaller it becomes.

The term n/V, a concentration term, when squared, gives probability of collisions. a gives a measure of the cohesive force between molecules. Real molecules *do* have an attraction for each other. The greater the value of a, the larger n^2a/V^2 becomes, and the larger $(P + n^2a/V^2)$ gets. The ideal gas law assumes a value of exactly P. Thus, as a increases, the more deviation.

To determine which gas deviates the most, look at the size of the gas molecules and their dipole moment. The dipole moment is an indication of unlike charges separated by a given amount of distance. Unlike charges on separate molecules will be attracted to each other. Thus, a higher dipole moment indicates greater cohesive attraction among molecules and greater deviation. Proceed as follows:

(1) N_2 versus CO. The CO gas has a net dipole moment, while N_2 does not. This is because O is more electromagnetic than C. In N_2 both of the atoms are the same and thus their electronegativities are equal and no net dipole moment exists. Thus, CO deviates to a greater extent.

(2) CH_4 versus C_2H_6. C_2H_6 is a much larger molecule and occupies more volume. Therefore, it is more likely to deviate from ideal gas behavior.

Drill 9: Real Gases

1. The Van der Waals equation of state has two constant terms (a and b) that are not present in the Combined gas law. The following is true of the terms a and b:

    (A) b corrects for the intermolecular forces of interaction and a corrects for the included volume of the molecules.

    (B) a corrects for the external pressure of the gas and b for the internal pressure of the gas.

    (C) a corrects for the force of intermolecular interactions and b for the excluded volume of the molecules.

    (D) b corrects for the free volume and a for the intermolecular repulsion between molecules.

    (E) None of the above.

2. Using the Van der Waals equation, what is the pressure exerted by one mole of carbon dioxide at 0°C in a volume of 1.00 liter? (For CO_2 $a = 3.592$, $b = 0.04267$.)

    (A) 23.38 atm   (D) 14.361 atm

    (B) 6.732 atm   (E) 19.79 atm

    (C) 3.592 atm

3. Using the information in the previous question, what would the pressure be if the gas occupies a volume of 0.05 liter?

(A) 1,617 atm

(D) 0.08211 atm

(B) 23.38 atm

(E) 4,125 atm

(C) 3,054 atm

GASES DRILLS

ANSWER KEY

Drill 1 — Pressure
1. (E)
2. (B)
3. (D)

Drill 2 — Boyle's Law
1. (C)
2. (B)
3. (C)

Drill 3 — Charles' Law
1. (E)
2. (C)
3. (A)

Drill 4 — The Law of Gay-Lussac
1. (B)
2. (B)
3. (B)

Drill 5 — The Combined Gas Law
1. (D)
2. (C)
3. (A)

Drill 6 — Avogadro's Law and the Mole Concept
1. (A)
2. (E)
3. (B)

Drill 7 — The Ideal Gas Law
1. (E)   4. (B)
2. (C)   5. (D)
3. (A)

Drill 8 — Density of Gases
1. (C)
2. (D)
3. (C)

Drill 9 — Real Gases
1. (C)
2. (E)
3. (A)

GLOSSARY: GASES

Atmospheric Pressure

The pressure exerted by the air in the earth's atmosphere. At sea level atmospheric pressure is 1 atmosphere. This is equal to the pressure exerted by a column of mercury 760 mm high.

Atom

Subunit of which all matter is composed.

Avogadro's Law

The notion that at constant volume, pressure, and temperature the number of moles of a gas will be constant as well.

Avogadro's Number

The number of entities in one mole of a substance. It is 6.023×10^{23}.

Barometer

An instrument used to measure pressure.

Boyle's Law

A relation that states that when a gas sample is compressed at a constant temperature, the product of the pressure and the volume remains constant.

Charles' Law

Law which states that the volume of a quantity of gas varies directly with temperature, at a constant pressure.

Gas

A substance in the vapor phase whose atoms or molecules are kept apart by thermal motion.

Gas Constant

A constant that relates volume, pressure, amount, and temperature. Its value is 0.082056 L atm mol^{-1} °K^{-1} or 8.31441 J mol^{-1} °K^{-1}.

Gay-Lussac's Law

Law which states that at constant volume, the pressure exerted by a given mass of gas varies directly with the temperature.

Ideal Gas

An imaginary gas whose behavior can be completely explained by ideal gas laws, such as Boyle's law, Charles' law, and the law of Gay Lussac.

Ideal Gas Law

An equation which states the relationship between pressure, volume, temperature, and amount for any ideal gas at moderate pressures; $PV = nRT$.

Kelvin Temperature Scale

An absolute temperature scale in which zero is absolute zero and a degree is 1/273.16 of the temperature of the triple point of water. It also places 100 degrees between the boiling point and freezing point of water.

Kinetic Energy

Energy that is associated with motion.

Manometer

A device used to measure pressure.

Mass

A property that reflects the amount of matter in a substance.

Mole

The SI unit of amount. It is defined as the amount of a particular substance that contains 6.023×10^{23} (or Avogadro's number) atoms, molecules, or other divisions of that substance. The mass of a mole of a substance is equal to the gram molecular weight of that substance.

Molecule

Two or more atoms bonded together.

Pressure

The force exerted on a unit area. May be expressed in atmospheres, mm Hg, or torr.

STP

Standard temperature and pressure. 0 degrees C and 760 mm Hg.

Van der Waals Equation

Equation similar to the Ideal gas equation that takes into account the molecular volume and intermolecular attractions of a real gas.

CHAPTER 3

Physical Properties of Gases

➤ Diagnostic Test
➤ Physical Properties of Gases
   Review & Drills
➤ Glossary

PHYSICAL PROPERTIES OF GASES DIAGNOSTIC TEST

1. Ⓐ Ⓑ Ⓒ Ⓓ Ⓔ
2. Ⓐ Ⓑ Ⓒ Ⓓ Ⓔ
3. Ⓐ Ⓑ Ⓒ Ⓓ Ⓔ
4. Ⓐ Ⓑ Ⓒ Ⓓ Ⓔ
5. Ⓐ Ⓑ Ⓒ Ⓓ Ⓔ
6. Ⓐ Ⓑ Ⓒ Ⓓ Ⓔ
7. Ⓐ Ⓑ Ⓒ Ⓓ Ⓔ
8. Ⓐ Ⓑ Ⓒ Ⓓ Ⓔ
9. Ⓐ Ⓑ Ⓒ Ⓓ Ⓔ
10. Ⓐ Ⓑ Ⓒ Ⓓ Ⓔ
11. Ⓐ Ⓑ Ⓒ Ⓓ Ⓔ
12. Ⓐ Ⓑ Ⓒ Ⓓ Ⓔ
13. Ⓐ Ⓑ Ⓒ Ⓓ Ⓔ
14. Ⓐ Ⓑ Ⓒ Ⓓ Ⓔ
15. Ⓐ Ⓑ Ⓒ Ⓓ Ⓔ

PHYSICAL PROPERTIES OF GASES
DIAGNOSTIC TEST

This diagnostic test is designed to help you determine your strengths and your weaknesses with the physical properties of gases. Follow the directions and check your answers.

Study this chapter for the following tests:
AP Chemistry, ASVAB, CLEP General Chemistry, GRE Chemistry, MCAT, PRAXIS II Subject Assessment: Chemistry, SAT II: Chemistry

15 Questions

DIRECTIONS: Choose the correct answer for each of the following problems. Fill in each answer on the answer sheet.

1. What is the mole fraction of ethanol, C_2H_5OH, in a solution made by dissolving 9.29 g, of alcohol in 18 g H_2O? The M of $H_2O = 18$ and of $C_2H_5OH = 46$.

   (A) .2

   (B) 1

   (C) .17

   (D) .36

   (E) .18

2. Which gas has a rate of diffusion 0.25 times that of hydrogen at the same temperature and pressure?

   (A) CH_4

   (B) PH_3

   (C) Argon

   (D) N_2

   (E) O_2

3. If 200 ml of N_2 at 25°C and pressure 400 mm Hg, and 200 ml of O_2 at 25°C and pressure 300 mm Hg are placed in a vessel with volume 700 ml, what is the total pressure of the mixture?

   (A) 100 mm Hg

   (B) 200 mm Hg

   (C) 500 mm Hg

   (D) 700 mm Hg

   (E) 800 mm Hg

4. What is the partial pressure of $N_2(g)$ in a mixture which contains 40 g $He(g)$, 56 g $N_2(g)$, and 16 g $O_2(g)$, if the total pressure of the mixture is 5 atmospheres?

    (A) 4 atm

    (D) 0.2 atm

    (B) 0.8 atm

    (E) 0.4 atm

    (C) 8 atm

5. The rate of diffusion of hydrogen gas as compared to that of oxygen gas is

    (A) half as fast.

    (D) four times as fast.

    (B) identical.

    (E) eight times as fast.

    (C) twice as fast.

6. Gas A diffuses 10 times faster than gas B. From this information, the molecular weight of A as compared to B (A:B) is

    (A) 100:1.

    (D) 1:10.

    (B) 1:100.

    (E) 20:2.

    (C) 10:1.

7. Methane is burned in oxygen to produce carbon dioxide and water vapor according to the following reaction

    $$CH_4(g) + 2O_2(g) \rightarrow CO_2(g) + 2H_2O(g)$$

    The final pressure of the gaseous products was 6.25 torr. What is the partial pressure of the water vapor?

    (A) 8.32 torr

    (D) 1.75 torr

    (B) 4.16 torr

    (E) 2.00 torr

    (C) 3.01 torr

8. If 100 ml of oxygen (O_2) were collected over water in the laboratory at a pressure of 700 torr and a temperature of 20°C, what would be the volume of the dry oxygen gas at STP? (The partial pressure of H_2O at 20°C = 17.5 torr.)

    (A) 17.5 ml

    (D) 83.7 ml

    (B) 68.3 ml

    (E) 1 ml

    (C) 20 ml

9. The ratio of the rates of diffusion of oxygen to hydrogen is

   (A) 1:2.

   (D) 1:16.

   (B) 1:4.

   (E) 1:32.

   (C) 1:8.

10. It has been estimated that each square meter of the earth's surface supports 1×10^7 g of air above it. If air is 20% oxygen (O_2, molecular weight = 32 g/mole) by weight, approximately how many moles of O_2 are there above each square meter of the earth?

    (A) 2×10^6

    (D) 6.02×10^{23}

    (B) 32×10^3

    (E) 6×10^4

    (C) 1×10^7

11. Calculate the molecular weight of an unknown gas X *if* the ratio of its effusion rate to that of He is 0.378 (AW of He = 4.00 g/mole).

    (A) 9.47

    (D) 28.0

    (B) 42.3

    (E) 32.0

    (C) 10.6

12. Each of the following is a statement of Dalton's law except

    (A) any gas in a mixture exerts its partial pressure.

    (B) atoms are permanent and cannot be decomposed.

    (C) each gas's pressure depends on other gases in a mixture.

    (D) gases can exist in a mixture.

    (E) substances are composed of atoms.

13. At standard conditions 45 liters of oxygen gas weighs about 64 g, whereas 45 liters of hydrogen weighs only about 4 g. Which gas diffuses faster? Calculate how much faster.

    (A) Hydrogen, at 4 times the rate of oxygen

    (B) Hydrogen, at 2 times the rate of oxygen

    (C) Oxygen, at 8 times the rate of hydrogen

    (D) Oxygen, at 3 times the rate of hydrogen

    (E) None of the above.

14. A container holds 40 g He, 56 g N_2, and 40 g Ar. If the total pressure of the mixture is 10 atm, what is the partial pressure of He?

(A) 0.8 atm   (D) 10 atm

(B) 0.15 atm   (E) 2 atm

(C) 7.7 atm

15. How many times faster will hydrogen effuse from the same effusion apparatus than nitrogen at the same temperature?

(A) 14.0   (D) 3.8

(B) 0.3   (E) 2.7

(C) 5.3

PHYSICAL PROPERTIES OF GASES DIAGNOSTIC TEST

ANSWER KEY

1.	(C)	5.	(D)	9.	(B)	13.	(A)
2.	(E)	6.	(B)	10.	(E)	14.	(C)
3.	(B)	7.	(B)	11.	(D)	15.	(D)
4.	(B)	8.	(D)	12.	(C)		

DETAILED EXPLANATIONS
OF ANSWERS

1.   **(C)**   Mole fraction problems are similar to % composition problems. A mole fraction of a compound tells us what fraction of 1 mole of solution is due to that particular compound. Hence,

$$\text{mole fraction of solute} = \frac{\text{moles of solute}}{\text{moles of solute} + \text{moles of solvent}}$$

The solute is the substance being dissolved into or added to the solution. The solvent is the solution to which the solute is added.

The equation for finding mole fractions is:

$$\frac{\text{moles } A}{\text{moles } A + \text{moles } B} = \text{mole fraction } A$$

Moles are defined as grams/molecular weight (M). Therefore, first find the number of moles of each compound present and then use the above equation.

$$\text{moles of } C_2H_5OH = \frac{9.2 \text{ g}}{46.0 \text{ g/mole}} = .2 \text{ mole}$$

$$\text{moles of } H_2O = \frac{18 \text{ g}}{18 \text{ g/mole}} = 1 \text{ mole}$$

$$\text{mole fraction of } C_2H_5OH = \frac{.2}{1 + .2} = .17$$

2.   **(E)**   Graham's law of diffusion relates the diffusion rate and molecular weight of gases as

$$\frac{\text{rate } (A)}{\text{rate } (B)} = \sqrt{\frac{M_A}{M_B}}$$

$$\frac{\text{rate } (A)}{\text{rate } (H_2)} = 0.25$$

$$0.25 = \sqrt{\frac{2}{M_A}}$$

$$0.0625 = \frac{2}{M_A}$$

$$M_A = \frac{2}{0.0625} = 2 \times 216 = 32$$

The gas is O_2.

3. **(B)** According to Dalton's law, "the total pressure of a mixture of gases is the summation of the pressures of the individual gases if they occupied alone the volume of the mixture."

$$\left.\frac{P_1 V_1}{T_1} = \frac{PV}{T}\right\} \quad \begin{aligned} T &= T_1 \\ V &= 700 \\ P &= ? \end{aligned}$$

$$400 \times 200 = P \times 700$$

$$P = \frac{400 \times 200}{700} = \frac{80,000}{700}$$

$$\left.\frac{P_2 V_2}{T_2} = \frac{P' V'}{T}\right\} \quad \begin{aligned} T' &= T_2 \\ V &= 700 \\ P &= ? \end{aligned}$$

$$300 \times 200 = P \times 700$$

$$\frac{300 \times 200}{700} = P_1 = \frac{600}{7}$$

$$P_{total} = P + P_1 = \frac{800 + 600}{7} = \frac{1,400}{7} = 200 \text{ mm}$$

4. **(B)** In a mixture of gases, the partial pressure of each gas is proportional to its mole fraction. Here, to calculate the partial pressures of the various gases contained in this system, one must first calculate the mole fraction of each component. This is done by calculating the number of moles present of each component and dividing that by the total number of moles present in the system. To calculate the partial pressure of each component, the mole fraction must then be multiplied by the total pressure of the system.

To calculate the number of moles present of each component, divide the number of grams present by the molecular weight of the element.

$$\text{number of moles} = \frac{\text{number of grams present}}{\text{molecular weight}}$$

$$\text{Moles of He} = \frac{40 \text{ g}}{4 \text{ g/mole}} = 10 \text{ moles}$$

$$\text{Moles of N}_2 = \frac{56 \text{ g}}{28 \text{ g/mole}} = 2 \text{ moles}$$

$$\text{Moles of O}_2 = \frac{16 \text{ g}}{32 \text{ g/mole}} = 0.5 \text{ mole}$$

To calculate the total number of moles present, add the number of moles of all components together.

$$\text{total number of moles} = \text{moles He} + \text{moles N}_2 + \text{moles O}_2$$

$$= 10 + 2 + 0.5$$

$$= 12.5 \text{ moles}$$

To calculate the mole fraction of N_2, divide the number of moles present of N_2 by the total number of moles in the system.

$$\text{mole fraction N}_2 = \frac{2}{12.5} = .16$$

To find the partial pressure of N_2, multiply the mole fraction by the total pressure in the system.

$$\text{partial pressure N}_2 = .16 \times 5 \text{ atm} = 0.8 \text{ atm}$$

5. **(D)** Graham's law states that the relative rate of diffusion of a gas is inversely proportional to the square root of its molecular weight. Expressed as a ratio we have

$$\frac{\text{diffusion rate of H}_2}{\text{diffusion rate of O}_2} = \sqrt{\frac{\text{molecular weight of O}_2}{\text{molecular weight of H}_2}}$$

$$= \sqrt{\frac{32}{2}} = \sqrt{16} = 4$$

Thus, the rate of H_2 diffusion is 4 times as fast as the rate of O_2 diffusion.

6. **(B)** Graham's law of diffusion states that the rate of diffusion of a gas is inversely proportional to the square root of its molecular weight. In other terms

$$\frac{\text{rate } A}{\text{rate } B} = \sqrt{\frac{\text{molecular weight of } B}{\text{molecular weight of } A}}$$

Substituting given values we obtain

$$\frac{10}{1} = \sqrt{\frac{\text{molecular weight of } B}{\text{molecular weight of } A}}$$

$$\frac{100}{1} = \frac{\text{molecular weight of } B}{\text{molecular weight of } A}$$

$$A : B = 1 : 100$$

7. **(B)** The partial pressure of each of the components in a mixture of gaseous substances will be equal to the product of the mole fraction of that component in the gas and the total pressure of the system. Given the total vapor

pressure of the gases, one can calculate the mole fraction of the water vapor.

Let P_A = partial pressure of the water vapor, N_T = total number of moles of gases produced, N_A = number of moles of water vapor, and P_T = total pressure. This law can now be expressed in these terms:

$$P_A = \frac{N_A}{N_T} P_T$$

From the stoichiometry of the equation, the total number of moles of products is 3. Out of these, 2 moles are water vapor. Thus, substituting,

$$P_A = \frac{2}{3} (6.25 \text{ torr}) = 4.16 \text{ torr of water vapor}$$

8.   **(D)**   STP means Standard Temperature and Pressure, which is 0°C and 760 torr. In this problem oxygen is gathered over water, therefore, Dalton's law of partial pressure (each of the gases in a gaseous mixture behaves independently of the other gases and exerts its own pressure, the total pressure of the mixture being the sum of the partial pressures exerted by each gas present; that is

$$P_{total} = p_1 + p_2 + p_3 \cdots p_n)$$

is used to calculate the original pressure of the oxygen. There is both water vapor and oxygen gas present.

After you obtain the original pressure of the oxygen, you can use the combined gas law to calculate the final volume of the oxygen. The combined gas law states that for a given mass of gas, the volume is inversely proportional to the pressure and directly proportional to the absolute temperature. It can be written as follows:

$$\frac{P_1 V_1}{T_1} = \frac{P_2 V_2}{T_2}$$

where P_1 is the original pressure, V_1 is the original volume, T_1 is the original absolute temperature, P_2 is the final temperature, V_2 is the final volume, and T_2 is the final absolute temperature. You are given the temperature in °C, so it must be converted to the absolute temperature by adding 273.

To find the original pressure of oxygen using 17.5 torr as the partial pressure of water, use the following formula:

$$P_{total} = 700 \text{ torr} = p_{O_2} + p_{H_2O}$$

$$p_{O_2} = 700 - 17.5$$

$$= 682.5 \text{ torr}$$

Convert the temperature to the absolute scale.

$$T_1 = 20 + 273 = 293°K$$

$$T_2 = 0 + 273 = 273°K$$

Use the combined law.

$$\frac{P_1V_1}{T_1} = \frac{P_2V_2}{T_2}$$

$P_1 = 682.5$ torr

$V_1 = 100$ ml

$T_1 = 293°K$

$P_2 = 760$ torr

$V_2 = ?$

$T_2 = 273°K$

$$\frac{(682.5 \text{ torr})(100 \text{ ml})}{293° K} = \frac{(760 \text{ torr})V_2}{273° K}$$

$$V_2 = \frac{(682.5 \text{ torr})(100 \text{ ml})(273° K)}{(760 \text{ torr})(293° K)} = 83.7 \text{ ml}$$

9. **(B)** Graham's law of diffusion states that the rate of diffusion of a gas is inversely proportional to the square root of its molecular weight. Thus we have

$$\frac{\text{rate } O_2}{\text{rate } H_2} = \sqrt{\frac{\text{mol. wt. of } H_2}{\text{mol. wt. of } O_2}}$$

$$= \sqrt{\frac{2}{32}} = \sqrt{\frac{1}{16}} = \sqrt{\frac{1}{4}}$$

10. **(E)** This problem is solved by first calculating what weight of oxygen is present in 1×10^7 g of air and then dividing by the molecular weight of oxygen to convert this mass to moles.

Using the definition of weight percent,

$$\text{weight \% of } O_2 = \frac{\text{weight of } O_2}{\text{total weight}} \times 100\%$$

$$20\% = \frac{\text{weight of } O_2}{1 \times 10^7 \text{ g}} \times 100\%$$

$$20 = \frac{\text{weight of } O_2}{1 \times 10^7 \text{ g}} \times 100$$

Solving for the weight of O_2,

$$\text{weight of } O_2 = \frac{20}{100} \times 1 \times 10^7 \text{ g} = 2 \times 10^6 \text{ g}.$$

The number of moles of O_2 is equal to the weight of O_2 divided by the molecular weight, or

$$\text{moles of } O_2 = \frac{\text{weight of } O_2}{\text{molecular weight}} = \frac{2 \times 10^6}{32 \text{ g/mole}}$$

$$\cong 6 \times 10^4 \text{ moles}$$

Therefore, each square meter of the earth's surface supports 6×10^4 moles of O_2.

11. **(D)** Effusion rates of gases are related as

$$\frac{r_A}{r_B} = \left(\frac{M_B}{M_A}\right)^{\frac{1}{2}}$$

Thus,

$$\frac{r_x}{r_{He}} = .378 = \left(\frac{4}{x}\right)\frac{1}{2} \, x = 28.0.$$

12. **(C)** Dalton's laws of partial pressures state that gases in a mixture exert their individual pressures *independently*.

13. **(A)** Analysis:

Rate   Density (g/45 liters)

$$r_{H_2} \qquad d_{H_2} = 4$$

$$r_{O_2} \qquad d_{O_2} = 64$$

According to Graham's law, a logical statement is: The hydrogen will diffuse faster because it has the lower density. The factor is

$$\frac{r_{H_2}}{r_{O_2}} \frac{\sqrt{d_{O_2}}}{\sqrt{d_{H_2}}} \Rightarrow r_{H_2} = r_{O_2} \times \frac{\sqrt{64}}{\sqrt{4}} = r_{H_2} \times \frac{8}{2}$$

$$= 4r_{O_2}$$

14. **(C)** The mole fraction of a component in a system is defined as the number of moles of that component divided by the sum of all the moles present in the system. Here, one must first calculate the number of moles present of each

component, then the mole fractions can be found. The number of moles of each gas can be found by dividing the number of grams present of the gas by the molecular weight. (M He = 4, M N$_2$ = 28, M Ar = 40.)

$$\text{number of moles} = \frac{\text{number of grams}}{M}$$

$$\text{Moles of He} = \frac{40 \text{ g}}{4 \text{ g/mole}} = 10 \text{ moles}$$

$$\text{Moles of N}_2 = \frac{56 \text{ g}}{28 \text{ g/mole}} = 2 \text{ moles}$$

$$\text{Moles of Ar} = \frac{40 \text{ g}}{40 \text{ g/mole}} = 1 \text{ mole}$$

The total number of moles of gas in the system is the sum of the number of moles of the three gases. Thus, there are 13 moles of gas in the system. The mole fraction can now be found for each gas by dividing the number of moles of each gas by 13, the total number of moles.

$$\text{mole fraction} = \frac{\text{no. of moles}}{\text{total no. of moles in system}}$$

$$\text{mole fraction of He} = \frac{10 \text{ moles}}{13 \text{ moles}} = .77$$

In a system where various gases are present, the partial pressure of each gas is proportional to the mole fraction of the gas. The relationship between the partial pressure of a particular gas and the total pressure is

$$\text{partial pressure} = \text{total pressure} \times \text{mole fraction}$$

In this problem, one is given the total pressure of the system and one has found the mole fraction of He. One can now find the partial pressure of He.

$$\text{partial pressure of He} = 10 \text{ atm} \times .77 = 7.7 \text{ atm}$$

15.   **(D)**   Graham's law of effusion states that the rates of effusion of gases are inversely proportional to the square roots of their molecular weights or densities.

$$\text{rate of effusion} \propto \frac{1}{\sqrt{M}}$$

or, $\text{rate of effusion} \times \sqrt{M} = t_0$ a constant

Hence, when two gases effuse from the same apparatus at the same conditions:

$$\frac{\text{rate}_A}{\text{rate}_B} = \frac{\sqrt{M_B}}{\sqrt{M_A}}$$

Substituting the molecular weight for N_2 and H_2:

$$\frac{\text{rate } H_2}{\text{rate } N_2} = \frac{\sqrt{28}}{\sqrt{2}} = \frac{5.29}{1.41} = 3.8$$

PHYSICAL PROPERTIES OF GASES REVIEW

1. Mole Fractions

A useful expression for representing the number of moles present is the mole fraction. The mole fraction of component A is generally designated as X_A and is given by the expression

$$X_A = \frac{\text{number of moles of } A}{\text{total number of moles of all components}}$$

It should be noted that the mole fraction represents a fraction of molecules; since unlike molecules normally have different masses, the mole fraction and mass fraction are not the same.

PROBLEM

Calculate the mole fractions of methanol, CH_3OH, and water in a solution made by dissolving 4.5 g of alcohol in 40 g of H_2O. M of $H_2O = 18$, M of $CH_3OH = 32$.

SOLUTION

Mole fraction problems are similar to % composition problems. A mole fraction of a compound tells us what fraction of 1 mole of solution is due to that particular compound. Hence,

$$\text{mole fraction of solute} = \frac{\text{moles of solute}}{\text{moles of solute + moles of solvent}}$$

The solute is the substance being dissolved into or added to the solution. The solvent is the solution to which the solute is added.

The equation for finding mole fractions is:

$$\frac{\text{moles } A}{\text{moles } A + \text{moles } B} = \text{mole fraction } A$$

Moles are defined as grams/molecular weight (M). Therefore, first find the number of moles of each compound present and then use the above equation.

$$\text{moles of } C_2H_5OH = \frac{4.5\,g}{32.0\,g/mole} = .14 \text{ mole}$$

$$\text{moles of } H_2O = \frac{40\,g}{.18\,g/mole} = 2.2 \text{ moles}$$

$$\text{mole fraction of } C_2H_5OH = \frac{.14}{2.2 + .14} = 0.061$$

PROBLEM

Of the many compounds present in cigarette smoke, such as the carcinogen 3, 4-benzo[a]pyrene, some of the more abundant are listed in the following table along with their mole fractions:

Component	Mole Fraction
H_2	0.016
O_2	0.12
CO	0.036
CO_2	0.079

What is the mass of carbon monoxide in a 35.0 ml puff of smoke at standard temperature and pressure (STP)?

SOLUTION

Assuming ideal gas behavior, we will calculate the number of moles of ideal gas in a 35.0 ml volume. Using the mole fraction of CO in smoke, we will then obtain the number of moles of CO in a 35.0 ml volume and finally convert this to a mass.

The molar volume of an ideal gas is 22.4 liter/mole when STP conditions exist, i.e., when temp. = 0°C and pressure = 1 atm. Hence, the number of moles of ideal gas in a 35.0 ml volume is obtained by dividing this volume by the molar volume of gas, or

$$\frac{35.0 \text{ ml}}{22.4 \text{ }l/\text{mole}} = \frac{0.035 \text{ } l}{22.4 \text{ }l/\text{mole}}$$
$$= 1.56 \times 10^{-3}$$

The mole fraction is defined by the equation

$$\text{mole fraction CO} = \frac{\text{moles CO}}{\text{total number of moles}}.$$

Solving for the number of moles of CO,

$$\text{moles CO} = \text{mole fraction CO} \times \text{total number of moles}$$

$$= 0.036 \times 1.56 \times 10^{-3} \text{ mole}$$

$$\equiv 5.6 \times 10^{-5} \text{ mole}$$

This is converted to a mass by multiplying by the molecular weight of CO (28 g/mole). Hence,

$$\text{mass of CO} = \text{moles of CO} \times \text{molecular weight of CO}$$

$$= 5.6 \times 10^{-5} \text{ mole} \times 28 \text{ g/mole}$$

$$= 1.6 \times 10^{-3} \text{ g}$$

Drill 1: Mole Fractions

1. What is the mole fraction of benzene in a solution made by adding 4.5 g benzene to 12.0 g H_2O? (M H_2O) = 18, M benzene = 78)

    (A) 0.60   (D) 1.31

    (B) 0.080   (E) 0.54

    (C) 0.91

2. What is the mole fraction of solute in a solution prepared by dissolving 98 g H_2SO_4 (M 98) in 162 g H_2O (M 18)?

    (A) 0.1   (D) 0.5

    (B) 0.3   (E) 0.2

    (C) 0.4

3. A 9.00 g sample of gas is composed of 6.00 g of ethane (C_2H_6) and 3.00 g of methane (CH_4). What is the mole fraction of ethane in the sample? (AW C = 12.011, H = 1.0079)

    (A) 0.200   (D) 0.517

    (B) 0.187   (E) 0.483

    (C) 0.387

2. Dalton's Law of Partial Pressures

The pressure exerted by each gas in a mixture is called its partial pressure. The total pressure exerted by a mixture of gases is equal to the sum of the partial pressures of the gases in the mixture. This statement, known as Dalton's law of partial pressures, can be expressed

$$P_T = P_a + P_b + P_c + \dots$$

When a gas is collected over water (a typical laboratory method), some water vapor mixes with the gas. The total gas pressure then is given by

$$P_T = P_{gas} + P_{H_2O}$$

where P_{gas} = pressure of dry gas and P_{H_2O} = vapor pressure of water at the temperature of the system.

PROBLEM

The composition of dry air by volume is 78.1% N_2, 20.9% O_2, and 1% other gases. Calculate the partial pressures, in atmospheres, in a tank of dry air compressed to 10.0 atmospheres.

SOLUTION

A partial pressure is the individual pressure caused by one gas in a mixture of several gases. The total pressure, P_{total}, according to Dalton's laws, is the sum of these individual partial pressures, p_1, p_2 and p_3.

The term dry air indicates that no water vapor is present.

The partial pressure is found by multiplying the percent of each gas in the volume by the total pressure.

partial pressure = proportion by volume × total pressure

P_{N_2} = .781 × 10 atm = 7.81 atm

P_{O_2} = .209 × 10 atm = 2.09 atm

$P_{other\ gases}$ = .010 × 10 atm = .1 atm

The total pressure, as required, is 10 atm.

Drill 2: Dalton's Law of Partial Pressures

1. 200 ml of oxygen is collected over water at 25°C and 750 torr. If the oxygen thus obtained is dried at a constant temperature of 25°C and 750 torr, what volume will it occupy? Assume that the equilibrium vapor pressure of water at 25°C is 28.3 torr.

    (A) 721.7 ml   (D) 750 ml

    (B) 192.45 ml   (E) 200 ml

    (C) 28.3 ml

2. If 40 liters of N_2 gas are collected at 22°C over water at a pressure of .957 atm, what is the volume of dry nitrogen at STP (standard temperature and pressure)? The partial pressure of water at 22°C is 19.8 torr.

    (A) 26.2 *l*   (B) 29.5 *l*

(C) 30.3 *l*   (D) 34.4 *l*

(E) 14.2 *l*

3.   A 20 g chunk of dry ice (CO_2) is placed in an "empty" 0.75 liter wine bottle and tightly corked. What would be the final pressure in the bottle after all the CO_2 has evaporated and the temperature has reached 25°C?

(A) 14.81 atm   (D) 13.81 atm

(B) 44 atm   (E) 16.81 atm

(C) 15.81 atm

3. Graham's Law of Gaseous Diffusion

Effusion is the process by which a gas escapes from one chamber of a vessel to another by passing through a very small opening or orifice.

Graham's law of effusion states that the rate of effusion is inversely proportional to the square root of the density of the gas.

$$\text{rate of effusion } \alpha \sqrt{\frac{1}{d}},$$

and
$$\frac{\text{rate of effusion } (A)}{\text{rate of effusion } (B)} = \sqrt{\frac{d_B}{d_A}} = \sqrt{\frac{M_B}{M_A}}$$

where M is the molecular weight of each gas, and where the temperature is the same for both gases.

Mixing of molecules of different gases by random motion and collision until the mixture becomes homogeneous is called diffusion.

Graham's law of diffusion states that the relative rates at which gases will diffuse will be inversely proportional to the square roots of their respective densities or molecular weights:

$$\text{rate } \alpha \frac{1}{\sqrt{\text{mass}}}$$

(where, again, $T_1 = T_2$) and

$$\frac{\text{rate 1}}{\text{rate 2}} = \frac{\sqrt{M_2}}{\sqrt{M_1}} \left(\text{or } \frac{r_1}{r_2} = \frac{\sqrt{d_2}}{\sqrt{d_1}} \right)$$

PROBLEM

Under standard temperature and pressure conditions, compare the relative rates at which inert gases, Ar, He, and Kr, diffuse through a common orifice.

SOLUTION

This problem involves the application of Graham's law of diffusion. It states that the relative rates at which gases will diffuse will be inversely proportional to the square roots of their respective densities or molecular weights. That is, rate

$$\propto \frac{1}{\sqrt{mass}}$$

Thus, to compare the rates of diffusion of Ar, He, and Kr, look up their weights in the Periodic Table of Elements and substitute this value into

$$\frac{1}{\sqrt{M}} \alpha \mu_m$$

where M = mass of that element and μ_m = rate. Therefore,

$$\mu_{Ar} : \mu_{He} : \mu_{Kr}$$

$$= \frac{1}{\sqrt{M_{Ar}}} : \frac{1}{\sqrt{M_{He}}} : \frac{1}{\sqrt{M_{Kr}}}$$

$$= \frac{1}{\sqrt{39.95}} : \frac{1}{\sqrt{4.003}} : \frac{1}{\sqrt{83.80}}$$

$$= .1582 : .4998 : .1092$$

PROBLEM

Two gases, HBr and CH$_4$, have molecular weights 81 and 16, respectively. The HBr effuses through a certain small opening at the rate of 4 ml/sec. At what rate will the CH$_4$ effuse through the same opening?

SOLUTION

The comparative rates or speeds of effusion of gases are inversely proportional to the square roots of their molecular weights. This is written

$$\frac{rate_1}{rate_2} = \frac{\sqrt{M_2}}{\sqrt{M_1}}$$

For this case

$$\frac{rate_{HBr}}{rate_{CH_4}} = \frac{\sqrt{M_{CH_4}}}{\sqrt{M_{HBr}}}$$

One is given the $rate_{HBr}$, M_{CH_4}, and M_{HBr} and asked to find $rate_{CH_4}$.

Solving for $rate_{CH_4}$:

$$rate_{HBr} = 4 \text{ ml/sec}$$

$$rate_{CH_4} = ?$$

$$M_{CH_4} = 16$$

$$M_{HBr} = 81$$

$$\frac{rate_{HBr}}{rate_{CH_4}} = \frac{\sqrt{M_{CH_4}}}{\sqrt{M_{HBr}}}$$

$$\frac{4 \text{ ml/sec}}{rate_{CH_4}} = \frac{\sqrt{16}}{\sqrt{81}}$$

$$rate_{CH_4} = \frac{4 \text{ ml/sec} \times \sqrt{81}}{\sqrt{16}}$$

$$= \frac{4 \text{ ml/sec} \times 9}{4} = 9 \text{ ml/sec}$$

Drill 3: Graham's Law of Gaseous Diffusion

1. The time required for a volume of gas, X, to effuse through a small hole was 112.2 sec. The time required for the same volume of oxygen was 84.7 sec. Calculate the molecular weight of gas X.

    (A) 45.3 amu   (D) 63.2 amu

    (B) 56.2 amu   (E) 19.6 amu

    (C) 37.1 amu

2. At standard conditions, 1 liter of oxygen gas weighs almost 1.44 g, whereas 1 liter of hydrogen weighs only .09 g. Which gas diffuses faster?

    (A) O_2, 3 times   (D) O_2, 4 times

    (B) H_2, 3 times   (E) O_2, 2 times

    (C) H_2, 4 times

3. Given two gases A and B that have a ratio of rates of effusion of gas A to gas B is 0.866 and the molecular weight of A is 24.00. What is the molecular weight of B?

(A) 2.00

(B) 32.00

(C) 17.00

(D) 6.00

(E) 20.00

PHYSICAL PROPERTIES OF GASES DRILLS

ANSWER KEY

Drill 1 — Mole Fractions

1. (B)
2. (A)
3. (D)

Drill 2 — Dalton's Law of Partial Pressures

1. (B)
2. (D)
3. (C)

Drill 3 — Graham's Law of Gaseous Diffusion

1. (B)
2. (C)
3. (B)

GLOSSARY: PHYSICAL PROPERTIES OF GASES

Dalton's Law of Partial Pressures

The notion that in a mixture of gases, the total pressure exerted is equal to the sum of the partial pressures of each of the gases in the mixture.

Diffusion

The process by which a substance, because of its kinetic motion, will spread through or mix with another substance.

Effusion

The process by which a gas moves through a capillary, porous solid, or other small hole in its container into another gaseous region or vacuum.

Graham's Law of Effusion and Diffusion

Law which states that the rate at which a gas effuses or diffuses is inversely proportional to the square roots of the gases' respective densities or molecular weights.

Homogeneous

Having uniform composition.

Mole Fraction

A concentration unit that is defined as the number of moles of a component divided by the total number of moles of all components.

Partial Pressure

The portion of the total pressure in a gaseous mixture that can be attributed to a particular component.

Vapor Pressure

The pressure exerted by vapor above a liquid when the vapor and liquid are in equilibrium.

CHAPTER 4

Avogadro's Hypothesis

➤ Diagnostic Test
➤ Avogadro's Hypothesis
  Review & Drills
➤ Glossary

AVOGADRO'S HYPOTHESIS
DIAGNOSTIC TEST

1. Ⓐ Ⓑ Ⓒ Ⓓ Ⓔ
2. Ⓐ Ⓑ Ⓒ Ⓓ Ⓔ
3. Ⓐ Ⓑ Ⓒ Ⓓ Ⓔ
4. Ⓐ Ⓑ Ⓒ Ⓓ Ⓔ
5. Ⓐ Ⓑ Ⓒ Ⓓ Ⓔ
6. Ⓐ Ⓑ Ⓒ Ⓓ Ⓔ
7. Ⓐ Ⓑ Ⓒ Ⓓ Ⓔ
8. Ⓐ Ⓑ Ⓒ Ⓓ Ⓔ
9. Ⓐ Ⓑ Ⓒ Ⓓ Ⓔ
10. Ⓐ Ⓑ Ⓒ Ⓓ Ⓔ
11. Ⓐ Ⓑ Ⓒ Ⓓ Ⓔ
12. Ⓐ Ⓑ Ⓒ Ⓓ Ⓔ
13. Ⓐ Ⓑ Ⓒ Ⓓ Ⓔ
14. Ⓐ Ⓑ Ⓒ Ⓓ Ⓔ
15. Ⓐ Ⓑ Ⓒ Ⓓ Ⓔ
16. Ⓐ Ⓑ Ⓒ Ⓓ Ⓔ
17. Ⓐ Ⓑ Ⓒ Ⓓ Ⓔ
18. Ⓐ Ⓑ Ⓒ Ⓓ Ⓔ

19. Ⓐ Ⓑ Ⓒ Ⓓ Ⓔ
20. Ⓐ Ⓑ Ⓒ Ⓓ Ⓔ
21. Ⓐ Ⓑ Ⓒ Ⓓ Ⓔ
22. Ⓐ Ⓑ Ⓒ Ⓓ Ⓔ
23. Ⓐ Ⓑ Ⓒ Ⓓ Ⓔ
24. Ⓐ Ⓑ Ⓒ Ⓓ Ⓔ
25. Ⓐ Ⓑ Ⓒ Ⓓ Ⓔ
26. Ⓐ Ⓑ Ⓒ Ⓓ Ⓔ
27. Ⓐ Ⓑ Ⓒ Ⓓ Ⓔ
28. Ⓐ Ⓑ Ⓒ Ⓓ Ⓔ
29. Ⓐ Ⓑ Ⓒ Ⓓ Ⓔ
30. Ⓐ Ⓑ Ⓒ Ⓓ Ⓔ
31. Ⓐ Ⓑ Ⓒ Ⓓ Ⓔ
32. Ⓐ Ⓑ Ⓒ Ⓓ Ⓔ
33. Ⓐ Ⓑ Ⓒ Ⓓ Ⓔ
34. Ⓐ Ⓑ Ⓒ Ⓓ Ⓔ
35. Ⓐ Ⓑ Ⓒ Ⓓ Ⓔ

AVOGADRO'S HYPOTHESIS
DIAGNOSTIC TEST

This diagnostic test is designed to help you determine your strengths and your weaknesses with Avogadro's hypothesis. Follow the directions and check your answers.

Study this chapter for the following tests:
AP Chemistry, ASVAB, CLEP General
Chemistry, GRE Chemistry, MCAT, PRAXIS II
Subject Assessment: Chemistry, SAT II: Chemistry

---

35 Questions

DIRECTIONS: Choose the correct answer for each of the following problems. Fill in each answer on the answer sheet.

1. Among the choices given, the atom with the largest size is

    (A) bromine.

    (B) chlorine.

    (C) fluorine.

    (D) helium.

    (E) iodine.

2. Which of the following is not a member of the transition metals?

    (A) The scandium family

    (B) The titanium family

    (C) The vanadium family

    (D) The iron family

    (E) The beryllium family

3. Which of the following has the smallest ionic radius?

    (A) Li^+

    (B) Na^+

    (C) K^+

    (D) Rb^+

    (E) Cs^+

4. The number of atoms in a molecule of $Pb(Cr_2O_7)_2$ is

    (A) 3.

    (B) 10.

(C) 14.   (D) 17.

(E) 19.

5.   What is the gram-formula weight of $Na_2Cr_2O_7$?

(A) 174 g   (D) 262 g

(B) 196 g   (E) 286 g

(C) 240 g

6.   The element with atomic number 32 describes

(A) a metal.   (D) a halogen.

(B) a nonmetal.   (E) a noble gas.

(C) a metalloid.

7.   Which region of the periodic table represents the element with the largest atomic radius?

(A) Upper left   (D) Lower right

(B) Upper right   (E) Middle

(C) Lower left

8.   Arrange the following neutral gaseous atoms in order of decreasing atomic radius.

S, Mg, F, Cl

(A) Mg>S>Cl>F   (D) S>Mg>Cl>F

(B) F>Cl>S>Mg   (E) Cl>S>F>Mg

(C) Cl>F>S>Mg

9.   Which of the following is a compound?

(A) Bleach   (D) Baking soda

(B) Air   (E) Graphite

(C) Oxygen gas

10.   A compound that is 17% sulfur (atomic weight 32.1) reacted completely to yield 1.8 grams of $H_2S_2O_7$ (molecular weight 178.1). Which of the choices should be used to find the total weight of the original compound?

(A) $\dfrac{1.8}{178.1} \times \dfrac{2}{32.1} \times \dfrac{17}{100}$ (D) $\dfrac{1.8}{178.1} \times \dfrac{2(32.1)}{1} \times \dfrac{100}{17}$

(B) $\dfrac{1.8}{178.1} \times \dfrac{32.1}{1} \times \dfrac{100}{17}$ (E) $\dfrac{1.8}{178.1} \times \dfrac{2(32.1)}{1}$

(C) $\dfrac{1.8}{178.1} \times \dfrac{1}{2(32.1)} \times \dfrac{17}{100}$

11. A sample containing aluminum weighing 10.0 grams yielded 2.0 grams of aluminum sulfide. What is the percentage of aluminum (atomic weight 26.98) in the sample?

(A) $\dfrac{2.0}{10.0} \times 100$ (D) $\dfrac{2.0}{10.0} \times \dfrac{2 \times 26.98}{150.14} \times 100$

(B) $\dfrac{2.0}{10.0} \times \dfrac{26.98}{150.14} \times 100$ (E) None of the above.

(C) $\dfrac{2.0}{10.0} \times \dfrac{150.14}{3 \times 26.98} \times 100$

12. If a copper bearing material weighing 40 grams yielded 5 grams of CuO (MW 79.55), the percentage of copper (atomic weight 63.55) in the sample is

(A) $\dfrac{5}{40} \times 100$. (D) $\dfrac{40}{5} \times \dfrac{79.55}{63.55} \times 100$.

(B) $\dfrac{5}{40} \times \dfrac{79.55}{63.55} \times 100$. (E) None of the above.

(C) $\dfrac{5}{40} \times \dfrac{63.55}{79.55} \times 100$.

13. What is the molecular weight of $HClO_4$?

(A) 52.5 (D) 100.5

(B) 73.5 (E) 116.5

(C) 96.5

14. The formula $Cr(NH_3)_5SO_4Br$ represents

(A) 4 atoms. (B) 8 atoms

(C) 12 atoms.   (D) 23 atoms.

(E) 27 atoms.

15. Which of the following is not true for metalloids?

(A) They are borderline elements that exhibit both metallic and nonmetallic properties to some extent.

(B) They usually act as electron donors with nonmetals and as electron acceptors with metals.

(C) Some of these elements are boron, silicon, and germanium.

(D) They are all solids at room temperature.

(E) They are good conductors of heat and electricity.

16. The weight of a naturally occurring sample of chlorine was measured and resulted to be 35.5 g mole. An atom, if chosen at random, from such a sample is most likely to weigh

(A) 34 amu.   (D) 36 amu.

(B) 35 amu.   (E) 37 amu.

(C) 35.5 amu.

17. Natural chlorine occurs as a mixture of isotopes. If a mixture contains

$$75\% \ Cl_{17}^{35} \ \text{and} \ 25\% \ Cl_{17}^{37},$$

determine its molecular weight.

(A) 34.50   (D) 70.00

(B) 35.50   (E) None of the above.

(C) 72.00

18. The atomic weight of an element is calculated by considering

(A) a weighted average.

(B) all naturally occurring isotopes.

(C) electronic energy level populations.

(D) Both (A) and (B).

(E) All of the above.

19. Which is the empirical formula of a compound consisting of 70% iron and 30% oxygen?

    (A) FeO   (D) Fe_2O_3

    (B) FeO_2 (E) Fe_3O_5

    (C) Fe_2O

20. The molecular weight of nicotine, a colorless oil, is 162.1 and it contains 74.0% carbon, 8.7% hydrogen, and 17.3% nitrogen. Calculate the molecular formula of nicotine.

    (A) C_5H_7N (D) $C_4H_{14}N$

    (B) $C_{10}H_{14}N_2$ (E) $C_8H_{28}N_2$

    (C) $C_{20}H_{28}N_4$

21. Which of the following are true statements?

    (A) The atomic radius of Li is greater than that of F.

    (B) The ionic radius of Li^+ is smaller than that of Na^+.

    (C) The ionic radius of Na^+ is smaller than that of Cl^-.

    (D) Both (A) and (B).

    (E) All of the above.

22. Members of a common horizontal row of the periodic table should have the same

    (A) atomic number.

    (B) atomic mass.

    (C) electron number in the outer shell.

    (D) number of energy shells.

    (E) valence.

23. The most common isotope of hydrogen has an atomic number and mass, respectively, of

    (A) 1,0.   (D) 2,1.

    (B) 1,1.   (E) 2,2.

    (C) 1,2.

24. Which of the following statements is true of isotopes?

    (A) They have different atomic numbers and the same atomic masses.

    (B) Their electronic configurations differ.

    (C) Slight differences in chemical behavior of isotopes (called isotope effects) influence the kind of reaction rather than the rate of reaction itself.

    (D) They have identical chemical behaviors.

    (E) The first ones discovered were those of neon.

25. What is the gram-molecular weight of $CH_3 COOH$?

    (A) 42   (D) 60

    (B) 44   (E) 72

    (C) 48

26. Which of the following compounds has an approximate molecular weight of 120?

    (A) $Ca(OH)_2$ (D) $AlCl_3$

    (B) KNO_3 (E) $BeCl_2$

    (C) $MgSO_4$

27. 0.585 g of a detergent was burnt to destroy certain organics. The residue was washed in hot HCl, which converted the phosphorous (P) present to H_3PO_4. After further filtering and washing, the precipitate was converted to $Mg_2P_2O_7$ (MW = 222.6). This final residue weighed 0.432 g. Calculate the percentage of P(MW = 30.97) in the sample of detergent.

    (A) $\%P = \dfrac{0.432 \times 2 \times 30.97 \times 100}{(222.6)(0.585)}$

    (B) $\%P = \dfrac{222.6(0.432)(2)(100)}{(0.585)(30.97)}$

    (C) $\%P = \left(\dfrac{0.432}{30.97}\right)\left(\dfrac{0.585}{222.6}\right) \times \left(\dfrac{1}{2}\right) \times 100$

    (D) $\%P = \left(\dfrac{0.432}{30.97}\right)\left(\dfrac{222.6}{0.585}\right) \times \left(\dfrac{1}{2}\right) \times 100$

    (E) Not enough information is given to determine the percentage of P.

28. Hydrogen, nitrogen, and oxygen combine in the following amounts to form a compound:

    $$H = 3.18 \text{ g} \quad O = 152.64 \text{ g} \quad N = 44.52 \text{ g}$$

    The probable formula for the compound is

    (A) HNO_2.

    (D) H_2N_2O.

    (B) HNO_3.

    (E) H_3NO_2.

    (C) H_2NO_3.

29. An alkaline earth metal may be described by the atomic number

    (A) 11.

    (D) 32.

    (B) 12.

    (E) 52.

    (C) 24.

30. What is the molecular formula of a hydrocarbon if it has a molecular weight of 42?

    (A) C_2H_6

    (D) C_3H_6

    (B) C_2H_4

    (E) CH_2

    (C) C_3H_8

31. Which of the following occurs naturally as a diatomic element?

    (A) I_2

    (D) S

    (B) O_3

    (E) He_2

    (C) NO

32. Lithium (AW = 6.941) exists as two naturally occurring isotopes, ^6_3Li and ^7_3Li, with relative atomic masses of 6.015 and 7.016. Find the percent abundances of the two isotopes.

    (A) 23.1, 46.2

    (D) 92.51, 7.49

    (B) 74.30, 25.70

    (E) 94.63, 5.37

    (C) 90, 10

33. How many grams of Na are present in 30 grams of NaOH?

    (A) 10 g

    (D) 20 g

    (B) 15 g

    (E) 22 g

    (C) 17 g

34. What is the percent of carbon in sucrose, $C_{12}H_{22}O_{11}$?

    (A) 42.1

    (B) 3.5

    (C) 12.0

    (D) 6.0

    (E) 26.6

35. A compound was found to contain only carbon, hydrogen, and oxygen. The percent composition was determined as 40.0% C, 6.7% H, and 53.3% O. The empirical formula of this compound is

    (A) C_2H_4O.

    (B) C_6HO_8.

    (C) CHO.

    (D) CH_2O.

    (E) C_3H_6O.

AVOGADRO'S HYPOTHESIS DIAGNOSTIC TEST

ANSWER KEY

1.	(E)	10.	(D)	19.	(D)	28.	(B)
2.	(E)	11.	(D)	20.	(B)	29.	(B)
3.	(A)	12.	(C)	21.	(E)	30.	(D)
4.	(E)	13.	(D)	22.	(D)	31.	(A)
5.	(D)	14.	(E)	23.	(B)	32.	(D)
6.	(C)	15.	(E)	24.	(E)	33.	(C)
7.	(C)	16.	(B)	25.	(D)	34.	(A)
8.	(A)	17.	(B)	26.	(C)	35.	(D)
9.	(D)	18.	(D)	27.	(A)		

DETAILED EXPLANATIONS
OF ANSWERS

1.   **(E)**   Iodine is at the bottom of the halogens in family seven of the periodic table. Elements that appear at the bottom of any family have energy levels and are larger in size.

2.   **(E)**   Transition elements occur between calcium and gallium. They are all metals, their ions are mostly colored, and they mostly form complexes e.g.,

$$CuCl_2^-, \ NiCl_4^{2-}, \ Co(NH_3)_6^{3+}$$

Beryllium is not a transition element, but a neighbor of the transition elements.

3.   **(A)**   In the alkali metal family the smallest ion has the smallest atomic number, $_3Li$.

4.   **(E)**   The number of atoms in a molecule of lead (IV) dichromate is given by $1Pb + 4Cr + 14O = 19$ atoms.

5.   **(D)**   The gram-formula weight of a compound is obtained by multiplying the atomic weight of each constituent by the respective subscript and summing the products. For $Na_2Cr_2O_7$ we have:

$$(2 \times 23) + (2 \times 52) + (7 \times 16) = 262 \text{ g/mol}.$$

6.   **(C)**   Referring to the periodic table we see that element 32 is germanium. Germanium is a metalloid as are boron, silicon, arsenic, antimony, tellurium, polonium, and astatine. Chemically, metalloids exhibit both positive and negative oxidation states and combine with metals and nonmetals. They are characterized by approximately half-filled outer electron shells and electronegativity values between those of the metals and the nonmetals.

7.   **(C)**   The atomic radii of the elements decreases as one proceeds from left to right across a row of the periodic table and increases as one proceeds from top to bottom along a column. Thus, the element with the largest atomic radius would be found in the lower left corner of the periodic table.

8.   **(A)**   Atomic radius increases as the principal quantum number, n, increases, thus Cl>F. Within a row (or same principal quantum) number size decreases as the atomic number increases, thus Mg>S>Cl.

9.   **(D)**   Baking soda, or sodium bicarbonate, is the only compound given.

Bleach is a dilute solution of sodium hypochlorite, air is a mixture of gases, oxygen is an element, and graphite is one of the several naturally occurring forms of carbon.

10.   **(D)**   The number of moles of $H_2S_2O_7$ is the first thing that should be calculated. It is equal to

$$1.87 \text{ g } H_2S_2O_7 / 178.13 \text{ g/mol } H_2S_2O_7.$$

Next, the weight of sulfur in $H_2S_2O_7$ is calculated by multiplying the moles of $H_2S_2O_7$ by $2(32.06)$. The "2" is required because in one mole of $H_2S_2O_7$ there are 2 moles of sulfur. Once the weight of sulfur in $H_2S_2O_7$ is known, multiplication by 100/17 yields the weight of the original compound, since the weight of sulfur in $H_2S_2O_7$ represented 17% of the total weight of the original compound.

11.   **(D)**   The weight fraction of aluminum sulfide in the sample is given by:

$$\frac{2.0 \text{ g of } Al_2S_3}{10.0 \text{ g of sample}}$$

To determine the weight fraction of aluminum in aluminum sulfide, we calculate:

$$\frac{\text{weight of Al in one mole of } Al_2S_3}{\text{weight of one mole of } Al_2S_3}$$

Recalling that the molar weight of sulfur is 32.06 and using the given weight of aluminum and the atomic subscripts, we obtain:

$$\frac{2 \times 26.98}{2 \times 26.98 + 3 \times 32.06} = \frac{2 \times 26.98}{150.14}$$

Multiplying the result by 100 (to change from a fraction to a percent scale):

$$\frac{2.0}{10.0} \times \frac{2 \times 26.98}{150.14} \times 100$$

12.   **(C)**

$$\frac{5 \text{ g CuO}}{40 \text{ g material}} \times \frac{63.55 \text{ g Cu}}{79.55 \text{ g CuO}} \times \frac{100}{1}$$

13.   **(D)**   The molecular weight of a compound is the sum of its constituents atomic weights. Elements or groups followed by a subscript have their atomic weight multiplied by that subscript. Thus, the molecular weight of perchloric acid ($HClO_4$) is

atomic weight of H + atomic weight of Cl + 4 × atomic weight of O or $1 + 35.5 + 4(16) = 100.5$.

14. **(E)** $Cr(NH_3)_5 SO_4 Br$ represents 27 atoms. They are:

$1 \times Cr$, $5 \times N$, $15 \times H$, $1 \times S$, $4 \times O$, and $1 \times Br$.

15. **(E)** Metalloids are compounds that exhibit both metallic and nonmetallic properties to some extent. They also act as electron donors with nonmetals and as electron acceptors with metals. They are all solids at room temperature. They are rather poor conductors of heat and electricity.

16. **(B)** Naturally occurring samples of Cl are mixtures of two isotopes: 75% Cl-35 and 25% Cl-37. Although the observed weight is 35.5 g/mol, we are most likely to pick an atom of Cl-35.

17. **(B)** 1 mole of chlorine $= 0.75$ mole of $Cl_{17}^{35} + 0.25$ mole of Cl_{17}^{37}. Therefore, molar weight $= .75(35) + .25(37) = 35.50$.

18. **(D)** The atomic weight of an element is the weighted average of the masses of all naturally occurring isotopes. This explains the occurrence of non-integral atomic weights. The mass of an atom's electrons are negligible and are thus omitted from the atomic weight calculation.

19. **(D)** A 100 g sample of this compound would contain 70 g of iron and 30 g of oxygen. Converting these weights to moles

$$70 \text{ g of Fe} \times \frac{1 \text{ mole of Fe}}{56 \text{ g of Fe}} = 1.25 \text{ moles of Fe}$$

$$30 \text{ g of O} \times \frac{1 \text{ mole of O}}{16 \text{ g of O}} = 1.9 \text{ moles of O}$$

This gives an empirical formula of $Fe_{1.25} O_{1.9}$. We convert to whole numbers by dividing each subscript by the smallest subscript:

$$Fe_{\frac{1.25}{1.25}} O_{\frac{1.9}{1.25}} = FeO_{1.5}$$

Multiplying each subscript by 2 to obtain integer values, we obtain Fe_2O_3.

20. **(B)** Atomic weights are C = 12 g/mol, H = 1.01 g/mol, N = 14.0 g/mol.

$$\text{moles of C} = \frac{74.0 \text{ g}}{12 \text{ g/mol}} = 6.17 \text{ moles}$$

$$\text{moles of H} = \frac{8.7\,\text{g}}{1.01\,\text{g/mol}} = 8.61 \text{ moles}$$

$$\text{moles of N} = \frac{17.3\,\text{g}}{14.0\,\text{g/mol}} = 1.24 \text{ moles}$$

Ratio C : H : N is 6.17 : 8.61 : 1.24 which is the same as

$$\frac{6.17}{1.24} : \frac{8.61}{1.24} : \frac{1.24}{1.24} \cong 5 : 7 : 1$$

Thus, the empirical formula is $C_5H_7N = 81.05$, but the molecular weight of nicotine is 162.1. Therefore, the whole number is

$$\frac{162.1}{81.05} = 2.$$

Thus, the molecular formula equals

$$(C_5H_7N) \times 2 = C_{10}H_{14}N_2.$$

21. **(E)**   Periodically, we find that the atomic radius decreases as one goes from left to right across a period. Thus, the atomic radius of Li is greater than that of F. It is also seen that both the atomic and ionic radius increase as one goes from top to bottom along a group. Therefore, the ionic radius of Li^+ is smaller than that of Na^+. It may also be observed that the ionic radius as compared to the atomic radius is smaller for cations and larger for anions. This may be rationalized since the loss of an electron results in a stronger attraction between the nucleus and the remaining electrons, hence a smaller ionic radius, as contrasted to the gain of an electron. In this case, the mutual attraction between the nucleus and the electrons is reduced, resulting in an increased ionic radius. Thus, the ionic radius of Na^+ is smaller than that of Cl^-.

22. **(D)**   The number of energy shells stays constant with the valence electrons tending to increase from left to right. Atomic numbers and mass change for each element.

23. **(B)**   Hydrogen's most common atomic form has one proton and one electron.

24. **(E)**   Isotopes of a particular element are characterized by the same atomic number and different atomic masses. The only difference in composition between them is in the number of neutrons in the nucleus. Thomson and Aston in 1912-1913, discovered the first isotopes (neon). All the other choices are inaccurate and therefore should be ruled out.

25. **(D)** Consulting the periodic table we find that

$$2C + 4H + 20 =$$

$$2(12) + 4(1) + 2(16) = 60 \text{ g/mol}$$

26. **(C)** Consulting the periodic table, we find that the atomic weights of magnesium, sulfur, and oxygen are approximately 24.3, 32, and 16, respectively. Therefore, the molecular weight of magnesium sulfate is given by:

$$24.3 + 32 + 4(16) = 120.3 \approx 120$$

In a likewise manner, the molecular weights of $Ca(OH)_2$, KNO_3, $AlCl_3$, and $BeCl_2$ are 74, 101, 133.5, and 80, respectively.

27. **(A)** This problem concerns gravimetric analysis. We have a formula which is as follows:

$$\%A = \frac{m_{precipatate} \times \dfrac{a \times M_A}{b \times M_{prec.}} \times 100}{m_{sample}}$$

In our case,

$$\%P = \frac{m_{Mg_2P_3O_7} \times \dfrac{a \times M_{W.P.}}{b \times M_{Mg_2P_3O_7.}} \times 100}{0.585}$$

where

$$\frac{a}{b} = \frac{2}{1} \frac{\text{moles P in precip.}}{\text{1 mole P in sample}}$$

$$\%P = \frac{0.432 \times 2 \times 30.97 \times 100}{(0.585)(222.6)}$$

28. **(B)** First calculate the number of moles of each element present.

1 mole H = 1.0 g

1 mole O = 16.0 g

1 mole N = 14.0 g

Therefore, the number of moles of H is

$$\frac{3.18 \text{ g}}{1 \text{ g/mol}} = 3.18 \text{ moles H}$$

number of moles of O is

$$\frac{152.64 \text{ g}}{16 \text{ g/mol}} = 9.54 \text{ moles O}$$

number of moles of N is

$$\frac{44.52 \text{ g}}{14 \text{ g/mol}} = 3.18 \text{ moles N}$$

The smallest number is used to find the simplest ratio in which the elements combine as

$$H: \frac{3.18}{3.18} = 1 \quad O: \frac{9.54}{3.18} = 3 \quad N: \frac{3.18}{3.18} = 1$$

Therefore, the simplest molecular formula is HNO_3.

29. **(B)** The alkali earth metals are found in Group IIA of the periodic table. Atomic number 11 designates an alkali metal (Group IA), 24 designates a transition metal (B-groups) and 32 and 52 are metalloids.

30. **(D)** This problem may be solved without knowing the empirical formula by recalling that the formulas for alkanes, alkenes, and alkynes are C_nH_{2n+2}, C_nH_{2n}, and C_nH_{2n-2}, respectively, where n is the number of each atom in the compound. We determine the mass of each compound by multiplying n by the respective atomic mass of each element (12 for carbon and 1 for hydrogen). Assuming the unknown compound to be an alkane (C_nH_{2n+2}), we have

$$12(n) + 1(2n+2) = 14n + 2 = 42$$

and $\qquad\qquad 14n = 40$

We do not obtain an integral value of n with this calculation so the compound cannot be an alkane.

For the case of an alkene (C_nH_{2n}), we have

$$12(n) + 1(2n) = 14n = 42$$

and $\qquad\qquad n = 3$

Therefore, the molecular formula for the unknown hydrocarbon is $C_3H_{2\times3}$ or C_3H_6. It is left as an exercise to show that the alkyne formula does not yield an integral value of n.

31. **(A)** Iodine occurs naturally as a diatomic element. Ozone, O_3, an allotrope of diatomic oxygen, is a triatomic element. Nitric oxide, NO, is not an element but rather a compound. Fluorine and helium exist as monoatomic elements.

32. **(D)**   Let x = fraction of ^6_3Li and $1 - x$ = fraction of ^7_3Li.

$$6.015x = 7.016$$

$$(1 - x) = 6.941$$

$$x = .0749 \text{ or } 7.49\%$$

$$1 - x = .9251 \text{ or } 92.51\%.$$

33. **(C)**   We find that there are 23 g of sodium (its atomic weight) in 40 g of NaOH (its molecular weight). Using this as a conversion factor,

$$30 \text{ g of NaOH} \times \frac{23 \text{ g of Na}}{40 \text{ g of NaOH}} = 17 \text{ g of Na}.$$

Another method for determining this quantity is to convert 30 g of NaOH to moles,

$$30 \text{ g of NaOH} \times \frac{1 \text{ mole of NaOH}}{40 \text{ g of NaOH}} = 0.75 \text{ mole of NaOH},$$

and calculating the weight of sodium corresponding to 0.75 mole,

$$0.75 \text{ mole of Na} \times \frac{23 \text{ g of Na}}{1 \text{ mole of Na}} = 17 \text{ g of Na},$$

since 1 mole of Na is present in 1 mole of NaOH.

34. **(A)**   The percent of an element in a compound is equal to the weight of that element divided by the total weight of the compound times 100. Therefore, the percent carbon in sucrose is equal to

$$\%C = \frac{12 \times \text{atomic weight of C}}{\text{molecular weight of sucrose}} \times 100$$

$$= \frac{12 \times 12.0 \text{ g}}{(12 \times 12)_C + (22 \times 1)_H + (16 \times 11)_O} \times 100$$

$$= \frac{144}{342} \times 100 = 42.1\% \text{ carbon}$$

35. **(D)**   The empirical formula of a compound tells us the relative number of moles or atoms of each element it contains. From the data given above it is possible to calculate the mole ratio of each element. In a 100 g sample there would be 40 g of carbon, 6.7 g of hydrogen, and 53.3 g of oxygen. Therefore, the number of moles of each element would be:

$$\text{moles of C} = \frac{m_c \text{ of C}}{\text{AW of C}} = \frac{40.0 \text{ g}}{12.0 \text{ g}} = 3.3$$

$$\text{moles of H} = \frac{m_H \text{ of H}}{\text{AW of H}} = \frac{6.7 \text{ g}}{1.0 \text{ g}} = 6.7$$

$$\text{moles of O} = \frac{m_O \text{ of O}}{\text{AW of O}} = \frac{53.3 \text{ g}}{16.0 \text{ g}} = 3.3$$

The molar ratio is then

$C_{3.3}H_{6.7}O_{3.3}$.

The relative whole number ratio of each can be found by dividing each by the smallest number, which in this case is 3.3:

$$C: \frac{3.3}{3.3} = 1 \quad H: \frac{6.7}{3.3} = 2 \quad O: \frac{3.3}{3.3} = 1$$

Therefore, the empirical formula would be CH_2O.

AVOGADRO'S HYPOTHESIS
REVIEW

1. Avogadro's Law

Avogadro's hypothesis, first introduced by the Italian Renaissance physicist Amedeo Avogadro, states that for two gases at the same temperature and pressure, equal volumes will contain equal numbers of molecules. This number will be the same for all gases.

Avogadro's hypothesis has been so extensively verified that it is now widely known as Avogadro's law. It follows, from the ideal gas law, that a mole of any gas will contain the same number of molecules. This number, 6.02×10^{23}, is called Avogadro's number and applies to all matter, not just to gases. One mole of any species will contain Avogadro's number of molecules. One mole of any species is the amount of that species whose mass in grams is numerically equal to the molecular or atomic weight.

In order to calculate the number of molecules in a given mass of material, it is necessary to first compute the number of moles. To calculate the number of moles of a substance, divide the mass in grams by the molecular weight as shown in the following equation:

$$n \, (\text{moles}) = \frac{m \, \text{g}}{M \, \text{g/mole}}$$

The number of molecules is simply the number of moles times Avogadro's number (the molecules per mole for any substance).

$$\text{Number of molecules} = n \times N_A$$

PROBLEM

If the dot under a question mark has a mass of 1×10^{-6} g, and you assume it is carbon, how many atoms are required to make such a dot?

SOLUTION

Two facts must be known to answer this question. You must determine the number of moles in the carbon dot. You must also remember the number of atoms in a mole of any substance, 6.02×10^{23} atoms/mol (Avogadro's number).

A mole is defined as the weight in grams of a substance divided by the atomic weight (or molecular weight).

The atomic weight of carbon is 12 g/mol. Therefore, in the dot you have

$$\frac{1 \times 10^{-6}\,g}{12\,g/mol}\ \text{moles of carbon.}$$

Therefore, the number of atoms in such a dot is the number of moles × Avogadro's number,

$$\left(\frac{1 \times 10^{-6}\,g}{12\,g/mol}\right) 6.02 \times 10^{23}\ \text{atoms/mol} = 5 \times 10^{16}\ \text{atoms.}$$

PROBLEM

What is the approximate number of molecules in a drop of water which weighs 0.09 g?

SOLUTION

The number of molecules in a mole is defined to be 6.02×10^{23} molecules. Thus, to find the number of molecules in a drop of water, one must know the number of moles making up the drop. This is done by dividing the weight of the drop by the molecular weight of H_2O. (MW of $H_2O = 18$)

$$\text{number of moles} = \frac{0.09\,g}{18\,g/mol} = .005\ \text{moles}$$

The number of molecules present is now found by multiplying the number of moles by Avogadro's number (6.02×10^{23}).

$$\text{no. of molecules} = .005\ \text{moles} \times 6.02 \times 10^{23}\ \text{molecules/mol}$$

$$= 3.01 \times 10^{21}\ \text{molecules}$$

Drill 1: Avogadro's Law

1. 6.02×10^{23} molecules of H_2O is identical to

    (A) 1 mole of hydrogen atoms.   (D) Both (B) and (C).

    (B) 1 mole of H_2O.   (E) All of the above.

    (C) 18 g of H_2O.

2. How many moles of CO_2 are represented by 1.8×10^{24} atoms?

    (A) 1   (D) 4

    (B) 2   (E) 5

    (C) 3

3. How many atoms are there in 0.65 g of gold? (Atomic weight of gold = 196.9665)

    (A) 196.9665(0.65)

    (D) $\dfrac{0.65}{196.9665}(6.02 \times 10^{23})$

    (B) $\dfrac{0.65}{196.9665}$

    (E) $\dfrac{196.9665\,(0.65)}{6.02 \times 10^{23}}$

    (C) 196.9665(0.65)(6.02 × 10²³)

4. How many molecules are there in 22 g of CO_2? The molecular weight of CO_2 is 44.

    (A) 3

    (D) 9.03 × 10²³

    (B) 6.02 × 10²³

    (E) 3.01 × 10²³

    (C) 44

5. How many grams of Na are present in 30 grams of NaOH?

    (A) 10 g

    (D) 20 g

    (B) 15 g

    (E) 22 g

    (C) 17 g

---

2. Atomic and Molecular Weights

The atoms that make up matter can be subdivided into outer layers containing negatively charged electrons that are a great distance from a central nucleus, which is quite dense. The nucleus is made up of two types of particles—protons, which are positively charged, and neutrons which have no charge. The general term for a particle in a nucleus is "nucleon." The atomic weight tells how many nucleons are in the nucleus of that type of atom. For example, oxygen's atomic weight of 16 tells us that oxygen has 16 nucleons in its nucleus. The atomic weight of an atom is the mass of the atom compared to carbon-12 which is defined as 12 amu or 12 atomic mass units. Remember the atomic weight also tells us that 12 g of carbon would contain 1 mole (6.02 × 10²³) of atoms. Atomic weights are tabulated for all elements. Unless reference is made to a specific isotope, the tabulated values of atomic weights are average values for the mixture of isotopes that occur naturally.

For example, by looking at the periodic table, lead has an atomic weight of 207.2. Can a nucleus have two-tenths of a nucleon? Obviously, it cannot. All atoms of lead have 82 protons; however, in nature, lead atoms can exist with 122, 124, 125, and 126 neutrons, resulting in atoms with atomic weights of 204, 206, 207, and 208. These atoms, with the same number of protons, but a different number of neutrons are called "isotopes." Hence, we could say that in nature we find four isotopes of lead. To determine a value to place on a chart of atomic weights, we take a weighted average, based upon the natural abundance of each lead isotope. The abundances are as follows:

Pb-204	1.42%	abundant
Pb-206	24.1%	
Pb-207	22.1%	
Pb-208	52.4%	

Using this data we can calculate the average atomic weight of lead as:

$$(0.0142)(204) + (0.241)(206) + (0.221)(207) + (0.524)(208) = 207.2$$

Some elements, such as gold, exist as one isotope. All gold has 79 protons and 118 neutrons, and hence an atomic weight of 197. Other elements, such as lead, can have two or more isotopes of different abundances. (Tin is the "winner," with 10 different isotopes, with abundances ranging from 0.38% up to 32.4%, giving an average atomic weight of 118.69.) As can be seen, atomic weights need not be whole numbers.

The number of protons found in a nucleus is the atomic number, and it is the number of protons found in a nucleus that determines what element we have. For example, the atomic number of hydrogen is 1, and it will always be 1 (1 proton); if it were 2, the atom would be helium, not hydrogen. Likewise, lead will always have atomic number 82; if it were 83 it would be bismuth; if it were 81 it would be thallium, and not lead in either case. Atomic numbers will always be whole numbers, and never involve fractions.

To find the number of neutrons in a nucleus, one simply subtracts the atomic number from the atomic weight. For example, carbon is atomic number 6, and atomic weight 12. How many protons and neutrons are found in its nucleus? The atomic number tells us right away how many protons we have: 6. The number of neutrons is not much more difficult to determine: $12 - 6 = 6$. Therefore, we have 6 protons (which make it carbon) and 6 neutrons (which give it a total weight of 12).

Molecular weight is defined as the sum of the atomic weights of all the atoms which form the molecule. For example, the molecular weight of carbon dioxide (CO_2) is equal to the sum of the component atomic weights or

$$12 \text{ g/mol} + 16 \text{ g/mol} + 16 \text{ g/mol} = 44 \text{ g/mol}.$$

PROBLEM

What is the difference between the number of carbon atoms in 1.00 g of C-12 isotope (atomic mass = 12.000 g/mol) and 1.00 g of C-13 isotope (atomic mass = 13.003 g/mol)?

SOLUTION

The difference in the number of carbon atoms in each sample is equal to the difference in the number of moles times Avogadro's number 6.02×10^{23}. Hence, we must begin by calculating the number of moles of C-12 and of C-13 in 1.00 g samples of each.

The number of moles is equal to the mass divided by the atomic weight. Therefore,

$$\text{moles C-12} = \frac{\text{moles C-12}}{\text{atomic mass C-12}} = \frac{1.00 \text{ g}}{12.000 \text{ g/mol}} = 0.083 \text{ mole}$$

and $\quad$ $$\text{moles C-13} = \frac{\text{moles C-13}}{\text{atomic mass C-13}} = \frac{1.00 \text{ g}}{13.003 \text{ g/mol}} = 0.077 \text{ mole}$$

The difference in the number of moles between the two samples is (moles C-12) – (moles C-13) = 0.083 – 0.077 = 0.006 mole. Multiplying by Avogadro's number gives the difference in the number of carbon atoms in the two samples:

$$\text{number of carbon atoms} = 0.006 \text{ mole} \times 6.02 \times 10^{23} \text{ molecules/mol}$$

$$= 3.61 \times 10^{21} \text{ molecules}$$

PROBLEM

Determine the relative abundance of each isotope in naturally occurring gallium from the following data: AW Ga = 69.72. Masses of isotopes ^{69}Ga = 68.926, ^{72}Ga = 70.925.

SOLUTION

The relative abundance of the various isotopes of gallium can be found by using the following equation:

$$\text{atomic weight of Ga} = \% \text{ of } ^{69}\text{Ga} \times \text{AW of } ^{69}\text{Ga} +$$

$$\% \text{ of } ^{71}\text{Ga} \times \text{AW of } ^{71}\text{Ga}$$

It is given that Ga consists only of the two isotopes ^{69}Ga and ^{71}Ga. Thus, if one lets x = fraction of ^{69}Ga, then $1 - x$ = fraction of ^{71}Ga. Using the above equation one can solve for x.

$$69.72 = (x \times 68.926) + (1 - x) \times 70.925$$

$$69.72 = 68.926x + 70.925 - 70.925x$$

$$-1.205 = -1.999x$$

$$x = 1.2/2.0 = 0.60$$

This means that the ^{69}Ga makes up 60% of Ga. ^{71}Ga makes up 1 – .60 or 40% of Ga.

Drill 2: Atomic and Molecular Weights

1. Which of the following is true for isotopes of an element?

    (A) They are atoms of the same atomic number with different masses.

    (B) The only difference in composition between isotopes of the same element is in the number of neutrons in the nucleus.

    (C) The atomic weight of an element is an average of the weights of the isotopes of the element in the proportions in which they normally occur in nature.

    (D) (A) and (C) only.

    (E) (A), (B), and (C).

2. One atom of hydrogen weighs

    (A) 1.00 g.   (D) 2.00 g.

    (B) 22.4 g.   (E) 6.02×10^{23} g.

    (C) 1.66×10^{-24} g.

3. How many atoms are represented by the formula $K_3Fe(CN)_6$?

    (A) 6   (D) 18

    (B) 10   (E) 20

    (C) 16

4. The atomic weights of most elements are not whole numbers because of

    (A) the First law of thermodynamics.

    (B) the Second law of thermodynamics.

    (C) the Presence of isotopes.

    (D) the Mass of the electrons.

    (E) the Mass defect.

5. How many atoms are described by the formula $Na_2CO_3 \times 10H_2O$?

    (A) 4   (D) 60

    (B) 16   (E) 96

    (C) 36

3. Equivalent Weights

The equivalent weight of a substance is defined as the molecular weight divided by the valence—i.e., the number of electrons/molecule which enter the reaction in question. Valence, for the purposes of equivalent weights, can also be considered a measure of the number of hydrogen atoms that might react with one atom of the element in question or one molecule of a substance in question. For example, the equivalent weight of sulfuric acid H_2SO_4 would be half its molecular weight because it is capable of donating 2 protons. Equivalent weight depends on the reaction, and therefore, a substance may have different equivalent weights for different reactions.

PROBLEM

If the atomic weight of oxygen was 50, what would its equivalent weight be?

SOLUTION

The relationship of the atomic weight and the equivalent weight is

$$\frac{\text{Atomic weight}}{\text{Equivalent weight}} = \text{valence number}.$$

The valence number is a measure of the number of atoms of hydrogen that will combine with one atom of the element. Two hydrogen atoms combine with one oxygen. The valence number of oxygen is therefore 2. One can now solve for the equivalent weight of oxygen when the atomic weight is taken as 50.

$$\frac{50}{\text{equivalent weight}} = 2$$

$$\text{equivalent weight} = \frac{50}{2} = 25$$

PROBLEM

A chemist forms magnesium oxide by burning magnesium in oxygen. The oxide obtained weighed 1.2096 grams. It was formed from .7296 g of magnesium. Determine the mass equivalent of magnesium in this reaction.

SOLUTION

An equivalent is defined as that mass of oxidizing or reducing agent that picks up or releases the Avogadro number of electrons in a particular reaction. One equivalent of any reducing agent reacts with one equivalent of any oxidizing agent. In this problem, the key is to determine the number of equivalents of oxygen involved. Once this is known, you also know the number of equivalents of magnesium.

Since the oxide weighed 1.2096 g and the magnesium weighed .7296 g, the mass of the combined oxygen must be $1.2096 - .7296 = .4800$ g. Before the oxygen reacted, its oxidation state was zero. After the reaction, however, it was -2. As such, each oxygen atom gained 2 electrons. Therefore, the Avogadro number of electrons will be taken up by one-half of a mole of O atoms.

It follows, therefore, that there are 8.000 g per equivalent for oxygen, since 1 mole of oxygen atoms weighs 16 grams. It was found, however, that there were .4800 g of oxygen. As such

$$\frac{.4800 \text{ g}}{8.00 \text{ g/equiv}} = .0600 \text{ equiv. of oxygen.}$$

This means that magnesium also has .06 equiv. 0.7296 g of Mg participated in the reaction. Therefore, the grams per equivalent of Mg equals

$$\frac{.7296 \text{ g Mg}}{.060 \text{ equiv}} = 12.16 \text{ g/equiv.}$$

Drill 3: Equivalent Weights

1. It was found that a magnesium oxide contained .833 g of oxygen and 1.266 g of magnesium. Calculate the gram-equivalent weight of magnesium.

    (A) 14.15   (D) 12.16

    (B) 9.32   (E) 15.13

    (C) 6.12

2. 2.0 g of molybdenum (Mo) combines with oxygen to form 3.0 g of a molybdenum oxide. Calculate the equivalent weight of Mo in this compound.

    (A) 14   (D) 18

    (B) 12   (E) 8

    (C) 16

3. For the oxidation of VO by Fe_2O_3 to form V_2O_5 and FeO, what is the weight of one equivalent of VO and of Fe_2O_3?

    (A) 159.7 g/equiv   (D) 43.29 g/equiv

    (B) 66.94 g/equiv   (E) 75.95 g/equiv

    (C) 22.31 g/equiv

4. In acting as a reducing agent a piece of metal M, weighing 16.00 g, gives up 2.25×10^{23} electrons. What is the weight of one equivalent of the metal?

    (A) 37.45 g/equiv.   (D) 55.43 g/equiv.

    (B) 16.00 g/equiv.   (E) 12.92 g/equiv.

    (C) 42.78 g/equiv.

5. A compound of vanadium and oxygen is analyzed and found to contain 56.0% vanadium. What is the equivalent weight of vanadium in this compound?

    (A) 10.2   (D) 20.4

    (B) 56.0   (E) 30.6

    (C) 44.2

---

4. Chemical Compounds: Weight and Volume Percent

The percentage composition of a compound is the percentage of the total mass contributed by each element:

$$\% \text{ composition} = \frac{\text{mass of element in compound}}{\text{mass of compound}} \times 100\%$$

The key to determining a chemical formula from weight percent data of the elemental constituents is the reduction of the amount of each element to moles. Then, from the molar ratio, the empirical, or simplest formula will usually become obvious. When amounts of gases are expressed as volumes, the number of moles can be determined using the ideal gas law.

PROBLEM

> What is the simplest formula of a compound that is composed of 72.4% iron and 27.6% oxygen by weight?

SOLUTION

For purposes of calculation, let us assume that there is 100 g of this compound present. This means that there are 72.4 g of Fe and 27.6 g O. The simplest formula for this compound is Fe_nO_m, where n is the number of moles of Fe present and m is the number of moles of O. One finds the number of moles by dividing the number of grams by the molecular weight.

$$\text{number of moles} = \frac{\text{number of grams}}{MW}$$

For Fe: n = number of moles present. MW = 55.8.

$$n = \frac{72.4\,g}{55.8\,g/mole} = 1.30 \text{ moles}$$

For O: m = number of moles present. MW = 16.0.

$$m = \frac{27.6\,g}{16.0\,g/mole} = 1.73 \text{ moles}$$

One solves for the simplest formula by finding the ratio of Fe : O.

$$\frac{Fe}{O} = \frac{1.30}{1.73} = .75 = \frac{3}{4}$$

Therefore, $n = 3$ and $m = 4$.
The simplest formula is Fe_3O_4.

PROBLEM

An unknown compound consists of 82.98% potassium and 17.02% oxygen. What is the empirical formula of the compound?

SOLUTION

The empirical formula of any compound is the ratio of the atoms that make up the compound by weight. It is the simplest formula of a material that can be derived solely from its components. Therefore, we must determine the ratio of gram-atoms of potassium (K) to the number of gram-atoms of oxygen (O).

The number of gram-atoms of a substance equals the weight of the substance in grams divided by the weight per gram-atom of the substance. In other words,

$$\text{number of gram-atoms} = \frac{\text{weight in grams}}{\text{weight per gram-atom}}$$

In this problem, we are given the percentages of the elements that make up the compound. These percentages are, in reality, the weight in grams, since in the definition of weight, we imply percentage. The weight per gram-atom is the atomic weight of the element which can be found in the periodic table of elements.

Therefore, the number of gram-atoms for potassium is

$$\frac{82.98\,g\,(m\text{ of K})}{39.10\,(AW\text{ of K})} = 2.120 \text{ moles.}$$

For oxygen, the number of gram-atoms is

$$\frac{17.02 \text{ g } (m \text{ of O})}{16.00 \text{ (AW of O)}} = 1.062 \text{ moles.}$$

Recall, the empirical formula is the ratio of the elements by weight. Consequently, the ratio of potassium to oxygen is 2 : 1, since the gram-atom ratios are respectively 2.120 : 1.062. Therefore, the empirical formula is K_2O.

Drill 4: Chemical Compounds: Weight and Volume Percent

1. What is the molecular formula of a compound composed of 25.9% nitrogen and 74.1% oxygen?

    (A) NO

    (B) NO_2

    (C) N_2O

    (D) N_2O_4

    (E) N_2O_5

2. A certain hydrate analyzes as follows: 29.7% copper, 15.0% sulfur, 2.8% hydrogen, and 52.5% oxygen. Determine the empirical formula of this hydrate from these percentages.

    (A) $Cu_2 S_4 HO_2$

    (B) $Cu S_3 H_5O$

    (C) $Cu SH_6 O_7$

    (D) $Cu_2 S_2 H_9O_5$

    (E) $Cu_3 S_4 H_2 O_9$

3. A chemist finds that an unknown compound contains 50.05% S and 49.95% O by weight. Calculate its simplest formula.

    (A) S_2O

    (B) SO_2

    (C) S_4O_5

    (D) SO

    (E) S_3O_9

---

5. The Periodic Table

Periodic law states that chemical and physical properties of the elements are periodic functions of their atomic numbers.

Vertical columns are called groups, each containing a family of elements possessing similar chemical properties.

The horizontal rows in the periodic table are called periods.

The elements lying in two rows just below the main part of the table are called the inner transition elements.

In the first of these rows are elements 58 through 71, called the lanthanides or rare earths.

The second row consists of elements 90 through 103, the actinides.

Group IA elements are called the alkali metals.

Group IIA elements are called the alkaline earth metals.

Group VIIA elements are called the halogens, and the Group O elements, the noble gases.

The metals in the first two groups are the light metals, and those toward the center are the heavy metals. The elements found along the dark line in the chart are called metalloids. They have characteristics of both metals and nonmetals. Some examples of metalloids are boron and silicon.

Properties Related to the Periodic Table

The most active metals are found in the lower left corner. The most active nonmetals are found in the upper right corner.

Metallic properties include high electrical conductivity, luster, generally high melting points, ductility (ability to be drawn into wires), and malleability (ability to be hammered into thin sheets). Nonmetals are uniformly very poor conductors of electricity, do not possess the luster of metals and form brittle solids. Metalloids have properties intermediate between those of metals and non-metals.

Atomic Radii

The atomic radius generally decreases across a period from left to right. The atomic radius increases down a group.

Chemistry of the Main Groups and Transition Elements and Representatives of Each

The group IA elements are the most reactive of the metals. They react violently with cold water, and are so reactive that they are never found in nature in an uncombined state. Due to their high reactivity, they were not isolated and identified until the nineteenth century, when electric current became available for refining chemicals. Sodium is a good representative of this family. It reacts violently in water to form its hydroxide; in air it will oxidize to form its oxide. The further down the table the group IA elements are, the stronger bases they form, and the more reactive the element is.

The group IIA elements are also quite reactive, although not as reactive as those of group IA. Due to their high reactivity, they are not found in the free state in nature. They will slowly react with the oxygen in the air to form oxides, and will react slowly with hot water for hydroxide. They form moderate bases, with the strength of the base (and reactivity of the element) increasing as one goes downward.

Magnesium is a good representative of the group IIA metals. It will not react with cold water, but in boiling water, it produces its hydroxide. At normal temperatures it will react only very, very slowly with the oxygen in the air, but at an elevated temperature (e.g., a match flame), it will burst into flame.

The transition metals all react very slowly with oxygen or water, and some not at all. Because of their low reactivity, these metals are relatively easy to isolate, and were among the first metals to be discovered by the ancients (e.g., copper and iron). Some of these metals are so unreactive (e.g., gold) that they can be found in an uncombined state in nature.

The lightest of the group IIIA elements possess nonmetallic properties, while the remaining are all metals. The group IIIA elements are all mildly reactive. For example, aluminum resists combining with oxygen or reacting with water at normal temperature. It will react with hydrochloric acid.

The first element of the group IVA elements (carbon) is a nonmetal; it is followed by two metalloids (silicon and germanium), and then by two metals (tin and lead). The first three elements are semi-conductors and are fairly reactive. Were it not for the fact that life on this planet is based upon carbon, it would never have been found in the uncombined state in nature (e.g., graphite). The rest are never found free in nature, although the lower reactivities of the last two (tin and lead) allowed them to be separated and used by the ancients.

As one moves horizontally along the periodic table, the number of valence electrons increase. Each elements has one more valence electron than the one to its left. The exception to this rule is the transition elements, which still contain one more electron than the element to the left of them; however, it is a non-valence electron. Each element possesses one more proton than the element to its left. To sum it up: As one moves from left to right, each atom on the periodic table increases by one electron (usually a valence electron) and one proton.

The left side of the periodic table contains the metals. These metals have a tendency to lose electrons, have high melting points, are malleable and ductile, possess a silvery luster, and are good conductors of heat and electricity. The right-hand side of the table contains the nonmetals. Their properties are just the opposite of the metals. They tend to gain electrons, have varied melting points, are not malleable nor ductile, present a "dull," non-reflective appearance, and are poor conductors of heat and electricity.

If one does not include the inert gases (group VIII), reactivity decreases as one goes toward the center of the chart, and increases as one travels to the edges. The elements in rows IA and VIIA are more reactive than those in IIA and VIA; those in IIA and VIA are more reactive than those in IIIA, VA, or in the transition metals, etc.

As one travels down on the table, reactivity increases for metals and decreases for nonmetals. Cesium is more reactive than lithium, and chlorine is more reactive than astatine.

The atomic radii increases as one moves down on the table. It decreases as one moves from left to right, with the exception of group VIII being the inert gases. In this case the atomic radius increases.

The group VA elements consists of two nonmetals, two metalloids, and one metal. The more reactive (higher on the table) of these elements combine readily with oxygen and some of the more reactive metals.

The group VIA elements are quite reactive, starting with oxygen, which combines with almost all other elements. Were it not for biological processes occurring on earth, there would be no free oxygen. In fact, it is thought that due to its extreme reactivity, free oxygen was a deadly poison to early life-forms on earth. The reactivity of the group VIA elements decreases as one moves down the table. Most of the VIA elements are nonmetals, the last two, however, are metalloids.

The group VIIIA elements (also known as the Halogen Family) are quite reactive; hence, they are never found in the free state in nature. Fluorine is so reactive that it can combine with *all* of the other elements except helium. Reactivity decreases as one goes down the row. The first four elements have been used for killing life-forms from bacteria (germicide) to humans (war gas), due to their high degree of reactivity. Astatine, the last element in the row, is a metalloid and only exists in radioactive forms.

The last row, group VIIA (or O), contains the inert gases. These elements all have a filled valence shell, and hence, do not react at all in nature. They exist as monatomic gases. When Ramsey discovered argon, the first of these gases, it was a great surprise, as their existence had not been predicted, and there was no position for them on the periodic table of the time (1895). He named it from the Greek word *argos,* meaning "lazy," since the gas was too lazy to combine with any other element. Radon, the furthest down on the table, only exists in radioactive forms. Since the 1960s, chemists have been able to make compounds with all of these elements except helium (the *only* element that has no compounds), mostly fluoride salts. These compounds are unstable, and release a lot of energy upon their decomposition. Due to the fact that they can combine with other elements (albeit under extreme circumstances), it has been suggested that they be called "noble gases," rather than "inert gases."

PROBLEM

What is the correct ranking of alkali metals from most reactive to least?

SOLUTION

The alkali metals are in family one of the period table. They are progressively more active as metals, losing their one valence electron as they descend through the vertical array. So the correct ordering would be:

Cs – Rb – K – Na – Li

PROBLEM

Which of the elements Na, N, K, O, F has the smallest atomic radius?

SOLUTION

Atomic radius decreases as one moves from left to right across a period and increases from top to bottom along a group. Therefore, the elements with the smallest and largest atomic radii will be found in the upper right and lower left corners of the periodic table, respectively. So, in this group F will have the smallest atomic radius.

Drill 5: The Periodic Table

1. Which of the following pairs of elements have almost the same atomic size?

    (A) Sc and Y   (D) Mg and Ca

    (B) B and Al   (E) Be and Mg

    (C) Al and Ga

2. The correct ranking of halogens, from most reactive to least reactive, is

    (A) Cl – F – Br – I.   (D) I – Br – Cl – F.

    (B) F – Cl – Br – I.   (E) Ne – Cl – Br – I.

    (C) He – I – Br – Cl.

3. In the periodic table, metals with low melting points appear just to the left and below the nonmetals. A metal with a probable low melting point is

    (A) cadmium.   (D) selenium.

    (B) iron.   (E) rubidium.

    (C) lithium.

4. The Russian chemist Mendeleev first arranged 63 known elements in order of their increasing

    (A) atomic number.   (D) electron number.

    (B) atomic weight.   (E) silicon.

    (C) boiling point.

5. The atomic number 20 describes

    (A) an alkali metal.   (D) an inert gas.

    (B) an alkaline earth metal.   (E) a transition metal.

    (C) a halogen.

AVOGADRO'S HYPOTHESES DRILLS

ANSWER KEY

Drill 1 — Avogadro's Law

1. (D)
2. (A)
3. (D)
4. (E)
5. (C)

Drill 2 — Atomic and Molecular Weights

1. (E)
2. (C)
3. (C)
4. (C)
5. (C)

Drill 3 — Equivalent Weights

1. (D)
2. (C)
3. (E)
4. (C)
5. (A)

Drill 4 — Chemical Compounds: Weight and Volume Percent

1. (E)
2. (C)
3. (B)

Drill 5 — The Periodic Table

1. (C)
2. (B)
3. (A)
4. (B)
5. (B)

GLOSSARY:
AVOGADRO'S HYPOTHESIS

Alkali Metal

Any member of group IA in the periodic table.

Alkaline Earth Metal

Any member of group IIA in the periodic table.

Atomic Number

The number of electrons in an atom of an element or the number of protons in the nucleus of that atom.

Atomic Radius

The radius of the atom. It is generally defined as half of the distance between the centers of two nuclei in the elementary form of the substance.

Atomic Weight

The average weight of an element based on a weighted average of its isotopes.

Avogadro's Number

The number of entities in one mole of a substance. It is 6.023×10^{23}.

Electron

An entity that carries one unit of negative charge. It is generally found in the outer layers of an atom.

Empirical Formula

The form of a chemical formula that expresses its elements with the lowest set of integers.

Equivalent Weight

The molecular weight of a substance divided by its valence.

Halogens

Any element in group VIIA of the periodic table.

Isotope

An atom that has the same number of protons as another atom, but a different number of neutrons.

Metal

A substance that has the properties of high electrical conductivity, luster, generally high melting points, ductility, and malleability.

Metalloid

A substance having properties intermediate between those of metals and nonmetals.

Mole

The SI unit of amount. It is defined as the amount of a particular substance that contains 6.023×10^{23} (or Avogadro's number) atoms, molecules, or other divisions of that substance. The mass of a mole of a substance is equal to the gram molecular weight of that substance.

Molecular Weight

The sum of the atomic weights of all the atoms which form a molecule.

Neutron

Particles having no charge that are contained within the nucleus of an atom.

Noble Gases

Any element of group VIIIA of the periodic table. These elements are generally very unreactive.

Nonmetals

Elements that are uniformly very poor conductors of electricity, do not possess the luster of metals, and form brittle solids.

Nucleon

A general term for a particle in a nucleus.

Nucleus

The dense center of an atom that contains protons and neutrons.

Periodic Table

A periodic arrangement of elements by atomic number. Elements in the same column tend to have similar properties.

Proton

A positively charged entity that is found in the nucleus of an atom.

Transition Metal

Any element found in groups IB to VIIIB of the periodic table.

Valence Electrons

Electrons in the outer shell of an atom.

CHAPTER 5

Stoichiometry

➤ Diagnostic Test
➤ Stoichiometry Review & Drills
➤ Glossary

STOICHIOMETRY DIAGNOSTIC TEST

1. Ⓐ Ⓑ Ⓒ Ⓓ Ⓔ
2. Ⓐ Ⓑ Ⓒ Ⓓ Ⓔ
3. Ⓐ Ⓑ Ⓒ Ⓓ Ⓔ
4. Ⓐ Ⓑ Ⓒ Ⓓ Ⓔ
5. Ⓐ Ⓑ Ⓒ Ⓓ Ⓔ
6. Ⓐ Ⓑ Ⓒ Ⓓ Ⓔ
7. Ⓐ Ⓑ Ⓒ Ⓓ Ⓔ
8. Ⓐ Ⓑ Ⓒ Ⓓ Ⓔ
9. Ⓐ Ⓑ Ⓒ Ⓓ Ⓔ
10. Ⓐ Ⓑ Ⓒ Ⓓ Ⓔ
11. Ⓐ Ⓑ Ⓒ Ⓓ Ⓔ
12. Ⓐ Ⓑ Ⓒ Ⓓ Ⓔ
13. Ⓐ Ⓑ Ⓒ Ⓓ Ⓔ
14. Ⓐ Ⓑ Ⓒ Ⓓ Ⓔ
15. Ⓐ Ⓑ Ⓒ Ⓓ Ⓔ
16. Ⓐ Ⓑ Ⓒ Ⓓ Ⓔ
17. Ⓐ Ⓑ Ⓒ Ⓓ Ⓔ
18. Ⓐ Ⓑ Ⓒ Ⓓ Ⓔ
19. Ⓐ Ⓑ Ⓒ Ⓓ Ⓔ
20. Ⓐ Ⓑ Ⓒ Ⓓ Ⓔ

21. Ⓐ Ⓑ Ⓒ Ⓓ Ⓔ
22. Ⓐ Ⓑ Ⓒ Ⓓ Ⓔ
23. Ⓐ Ⓑ Ⓒ Ⓓ Ⓔ
24. Ⓐ Ⓑ Ⓒ Ⓓ Ⓔ
25. Ⓐ Ⓑ Ⓒ Ⓓ Ⓔ
26. Ⓐ Ⓑ Ⓒ Ⓓ Ⓔ
27. Ⓐ Ⓑ Ⓒ Ⓓ Ⓔ
28. Ⓐ Ⓑ Ⓒ Ⓓ Ⓔ
29. Ⓐ Ⓑ Ⓒ Ⓓ Ⓔ
30. Ⓐ Ⓑ Ⓒ Ⓓ Ⓔ
31. Ⓐ Ⓑ Ⓒ Ⓓ Ⓔ
32. Ⓐ Ⓑ Ⓒ Ⓓ Ⓔ
33. Ⓐ Ⓑ Ⓒ Ⓓ Ⓔ
34. Ⓐ Ⓑ Ⓒ Ⓓ Ⓔ
35. Ⓐ Ⓑ Ⓒ Ⓓ Ⓔ
36. Ⓐ Ⓑ Ⓒ Ⓓ Ⓔ
37. Ⓐ Ⓑ Ⓒ Ⓓ Ⓔ
38. Ⓐ Ⓑ Ⓒ Ⓓ Ⓔ
39. Ⓐ Ⓑ Ⓒ Ⓓ Ⓔ
40. Ⓐ Ⓑ Ⓒ Ⓓ Ⓔ

STOICHIOMETRY
DIAGNOSTIC TEST

This diagnostic test is designed to help you determine your strengths and weaknesses in stoichiometry calculations. Follow the directions and check your answers.

Study this chapter for the following tests:
AP Chemistry, ASVAB, CLEP General
Chemistry, GRE Chemistry, MCAT, PRAXIS II
Subject Assessment: Chemistry, SAT II: Chemistry

---

40 Questions

DIRECTIONS: Choose the correct answer for each of the following problems. Fill in each answer on the answer sheet.

1. Select the correct rule about solubility of substances.

    (A) All ammonium salts are insoluble.

    (B) All nitrates are insoluble.

    (C) All silver salts, except $AgNO_3$, are insoluble.

    (D) All sodium salts are insoluble.

    (E) Sulfides of sodium, potassium, and magnesium are insoluble.

2. Consider the following balanced equation. How many moles of hydrogen sulfide react with one mole of oxygen?

    $$2H_2S + 3O_2 \rightarrow 2SO_2 + 2H_2O$$

    (A) $\dfrac{2}{3}$ (D) 2

    (B) $\dfrac{3}{2}$ (E) 3

    (C) 1

3. What is the product gas if Cu metal and HNO_3 are the reactants?

    (A) NO   (B) NO_2

(C) O_2 (D) N_2

(E) H_2

4. Consider this reaction under standard lab conditions:

    $$FeS + 2HCl \rightarrow FeCl_2 + H_2S$$

    If 22 grams of iron sulfide are completely reacted to form products, the volume of hydrogen sulfide gas produced is

    (A) 5.6.   (D) 44.4.

    (B) 11.2.   (E) 88.0.

    (C) 22.4.

5. How many liters of H_2 can be produced by the decomposition of 3 moles of NH_3?

    (A) 4.5 liters   (D) 96 liters

    (B) 27 liters   (E) 101 liters

    (C) 67.2 liters

6. Consider the balanced equation:

    $$Fe + S \rightarrow FeS$$

    Approximately what amount of sulfur in grams must be available for 28 grams of iron to react to form the compound?

    (A) 1   (D) 32

    (B) 8   (E) 64

    (C) 16

7. 15 g of ethane (MW = 30 g/mol) reacts with chlorine to yield 15 g of ethyl chloride (MW = 64.5 g/mol). The percent yield of ethyl chloride is

    (A) 46.5%.   (D) 93.0%.

    (B) 50.0%.   (E) 23.3%.

    (C) 100.0%.

8. How many grams of CO must be reacted with excess O_2 to produce 33 g of CO_2?

    (A) 7 g   (D) 28 g

    (B) 14 g   (E) 44 g

    (C) 21 g

9. How many moles of Al_2O_3 can be formed when a mixture of 0.36 moles of aluminum and 0.36 moles of oxygen gas is ignited?

   (A) 0.72 moles
   (D) 0.12 moles

   (B) 0.28 moles
   (E) 0.46 moles

   (C) 0.18 moles

10. How many grams of oxygen are needed for the complete combustion of 39.0 g of C_6H_6? The molecular weight of C_6H_6 is 78.0.

   $$2C_6H_6 + 15O_2 \rightarrow 12CO_2 + 6H_2O$$

   (A) 3.75 g
   (D) 60.0 g

   (B) 120.0 g
   (E) 292.5 g

   (C) 32.0 g

11. How many milliliters of ammonia gas could be produced from the reaction of 9 ml of N_2 with 9 ml of H_2?

   (A) 1 ml
   (D) 9 ml

   (B) 3 ml
   (E) 12 ml

   (C) 6 ml

12. How many milliliters of the reactants of question 11 are left unreacted?

   (A) 3 ml of N_2
   (D) 6 ml of H_2

   (B) 6 ml of N_2
   (E) None of either

   (C) 3 ml of H_2

13. Consider this reaction:

   $$C_8H_{18} + O_2 \rightarrow CO_2 + H_2O$$

   Under standard conditions, the volume of air in liters required for *complete* combustion of 228 grams of octane is about

   (A) 22.4.
   (D) 1,560.

   (B) 2.50.
   (E) 2,800.

   (C) 560.

Questions 14–16 refer to the following:

Balance each of the following equations and reduce all coefficients to smallest whole numbers. Select from the following choices the one which is the coefficient of the last product of each of the reactions.

    (A) 1   (D) 4

    (B) 2   (E) 5

    (C) 3

14. $Na_2O_2 + H_2O \rightarrow O_2 +$ ___$NaOH$

15. $K + H_2O \rightarrow H_2 +$ ___KOH

16. $BaCl_2 + Na_2SO_4 \rightarrow NaCl +$ ___$BaSO_4$

17. What is the maximum weight of SO_3 that could be made from 40.0 g of SO_2 and 8.0 g of O_2 by the reaction below?

          $2SO_2 + O_2 \rightarrow 2SO_3$

    (A) 40.0 g   (D) 20.0 g

    (B) 31.3 g   (E) 52.5 g

    (C) 25.0 g

18. How many grams of sodium sulfate can be produced by reacting 98 g of H_2SO_4 with 40 g of $NaOH$?

    (A) 18 g   (D) 142 g

    (B) 36 g   (E) 150 g

    (C) 71 g

19. How many grams of water can be produced when 8 g of hydrogen reacts with 8 g of oxygen?

    (A) 8 g   (D) 27 g

    (B) 9 g   (E) 30 g

    (C) 18 g

Questions 20–23 are based on the following information: A student is given a soluble sodium salt containing one of eight possible anions: acetate, chloride, bromide, iodide, sulfide, sulfate, and phosphate.

20. What cations other than sodium will likely form soluble salts with all the anions?

    (A) Zn^{2+}, Pb^{2+}, K^+

    (B) Pb^{2+}, Hg^{2+}, Fe^{2+}

    (C) Ca^{2+}, Mg^{2+}

    (D) K^+, NH_4^+

    (E) Fe^{2+}

21. 3–4 drops of $AgNO_3$ are added to a solution of the salt. No precipitate results. Which anions are absent?

    (A) Cl^-, Br^-, I^-

    (B) Cl^-, Br^-, CH_3COO^-, I^-

    (C) Cl^-, Br^-, I^-, S^{2-}, PO_4^{3-}

    (D) Cl^- only

    (E) None of the above

22. Suppose that a precipitate did form in the reaction in question 21 and it is then treated with a few drops of nitric acid. No change is observed. Which anions can be present?

    (A) Cl^-, Br^-, I^-, SO_4^{2-}

    (B) CH_3COO^-, NO_3^-

    (C) Cl^-, Br^-, I^-, SO_4^{2-}

    (D) Cl^-, Br^-, I^-, only

    (E) None of the above

23. The student treats some fresh solution of the salt with $BaCl_2$. A precipitate forms. Which anions could have been present?

    (A) SO_4^{2-}, PO_4^{3-}

    (B) Br^-, Cl^-, I^-

    (C) CH_3COO^- only

    (D) Br^- only

    (E) NO_3^- only

24. The coefficients for the reaction:

    $$Na_3PO_4 + Pb(NO_3)_2 \rightarrow NaNO_3 + Pb_3(PO_4)_2$$

    if it were balanced are

    (A) 12, 18, 36, 6.

    (B) 3, 6, 9, 2.

    (C) 9, 12, 24, 6.

    (D) 6, 9, 18, 3.

    (E) 10, 12, 6, 8.

25. How much reactant remains if 92 g of HNO_3 is reacted with 24 g of LiOH assuming the reaction to be complete?

   (A) 46 g of HNO_3 (D) 2 g of LiOH

   (B) 29 g of HNO_3 (E) 12 g of LiOH

   (C) 12 g of HNO_3

26. Consider the following unbalanced equation. Coefficients are missing

   $$\_\_\_NH_3 + \_\_\_O_2 \rightarrow \_\_\_NO + \_\_\_H_2O$$

   To balance the equation, the four consecutive coefficients from left to right are

   (A) 4, 5, 4, 6.   (D) 5, 5, 4, 6.

   (B) 4, 4, 5, 6.   (E) 6, 5, 4, 4.

   (C) 5, 4, 5, 6.

27. What is the coefficient of HNO_3 necessary for the following to be a balanced equation?

   $$Pb(NO_3)_2 + H_2S \rightarrow PbS + \_\_\_HNO_3$$

   (A) 1   (D) 4

   (B) 2   (E) 5

   (C) 3

28. 20 liters of NO gas react with excess oxygen. How many liters of NO_2 gas are produced if the NO gas reacts completely?

   (A) 5 liters   (D) 40 liters

   (B) 10 liters   (E) 50 liters

   (C) 20 liters

29. The sum of the coefficients of the reaction

   $$\_\_\_C_6H_6 + \_\_\_O_2 \rightarrow \_\_\_CO_2 + \_\_\_H_2O$$

   when it is balanced is

   (A) 7.   (D) 35.

   (B) 14.   (E) 42.

   (C) 28.

30. The balanced equation of the reaction of Ba with water has the following coefficients:

    (A) 1.

    (B) 2.

    (C) 3.

    (D) Both (A) and (B)

    (E) All of the above

31. How many liters of water can be produced by the reaction of 5 liters of hydrogen with 5 liters of oxygen?

    (A) 1.0 liters

    (B) 2.5 liters

    (C) 5.0 liters

    (D) 10.0 liters

    (E) 15.0 liters

32. How many grams of sulfuric acid can be produced by the reaction SO_3 (g) + $H_2O \rightarrow H_2SO_4$ from 33.6 liters of SO_3 and excess water at STP?

    (A) 49

    (B) 98

    (C) 147

    (D) 196

    (E) 245

Question 33 refers to the table below on salt solubility.

Solubility of salts at varying temperatures (g/100 g of H_2O)

T(°C)	KNO_3	NaCl	KCl	Na_2SO_4	$NaNO_3$
0	13	35.7	28	4.8	73
10	21	35.8	30	9.0	80
20	31	36.0	34	19.5	85
30	45	36.3	37	40.9	92

33. The salt showing the least solubility change over temperature variation is

    (A) potassium chloride.

    (B) potassium nitrate.

    (C) sodium chloride.

    (D) sodium nitrate.

    (E) sodium sulfate.

34. Consider the balanced equation:

    $$2KClO_3 \rightarrow 2KCl + 3O_2$$

    If 72 grams of oxygen gas are produced, the amount of potassium chlorate required in grams is

    (A) 112.

    (B) 224.

(C) 336.

(D) 448.

(E) 1,020.

35. How many grams of calcium carbonate must be decomposed to produce 44.8 liters of CO_2?

(A) 50 g

(D) 200 g

(B) 100 g

(E) 250 g

(C) 150 g

36. How many grams of N_2 are required to produce 34 g of NH_3 with excess H_2 by the Haber process?

(A) 7 g

(D) 28 g

(B) 9 g

(E) 56 g

(C) 14 g

37. How many grams of sodium must be reacted with water to produce 22.4 liters of H_2?

$$(Na + H_2O \rightarrow NaOH + H_2)$$

(A) 11.5

(D) 46.0

(B) 23.0

(E) 57.5

(C) 34.5

38. What are the lowest possible whole number coefficients for the reaction of the previous question?

(A) 1, 1, 1, 1

(D) 2, 2, 2, 1

(B) 1, 2, 2, 1

(E) 1, 2, 1, 2

(C) 2, 1, 2, 1

39. Which of the following salts are soluble?

I. $(NH_4)_2CO_3$

II. $CaSO_4$

III. $PbCl_2$

IV. $AgClO_4$

(A) I and III only

(B) I and IV only

(C) I, II, and IV

(D) II, III, and IV

(E) II and IV only

40. The incomplete combustion of hexane (C_6H_{14}) produces CO and H_2O gases. Use a coefficient of 1 for hexane to balance the incomplete combustion reaction. What is the resulting coefficient for CO?

(A) 2

(D) 8

(B) 4

(E) 12

(C) 6

STOICHIOMETRY
DIAGNOSTIC TEST

ANSWER KEY

1. (C)	11. (C)	21. (C)	31. (C)
2. (A)	12. (B)	22. (E)	32. (C)
3. (A)	13. (E)	23. (A)	33. (C)
4. (A)	14. (D)	24. (D)	34. (C)
5. (E)	15. (B)	25. (B)	35. (D)
6. (C)	16. (A)	26. (A)	36. (D)
7. (A)	17. (A)	27. (B)	37. (D)
8. (C)	18. (C)	28. (C)	38. (D)
9. (C)	19. (B)	29. (D)	39. (B)
10. (B)	20. (D)	30. (D)	40. (C)

DETAILED EXPLANATIONS
OF ANSWERS

1. **(C)** The other statements are the opposite of what is correct due to substances' ability to ionize (soluble) or not ionize (insoluble) among the molecules of the solvent water. Silver salts form insoluble precipitates except for the ionizing silver nitrate, $AgNO_3$.

2. **(A)** If 2 moles of H_2S react with 3 moles of oxygen, $2/3$ moles react with 1 mole:

$$\frac{2}{3} = \frac{X}{1}$$
$$X = \frac{2}{3}$$

3. **(A)** The reaction is

$$3Cu + 8HNO_3 \rightarrow 3Cu(NO_3)_2 + 4H_2O + 2NO \text{ (g)}.$$

4. **(A)** One mole of FeS, 88 grams (56 + 32), forms one mole of H_2S gas in the balanced equation. By simple proportion, one-quarter mole of the given 22 grams, thus yields one-quarter mole of H_2S. One mole of a gas fills 22.4 liters. .25 mole × 22.4 liters = 5.6 liters.

5. **(E)** The equation for the reaction is

$$2NH_3 \rightarrow N_2 + 3H_2.$$

Multiplying each coefficient by $3/2$ gives

$$3NH_3 \rightarrow \frac{3}{2}N_2 + \frac{9}{2}H_2.$$

Thus, 3 moles of NH_3 decompose to produce 4.5 moles of H_2. Converting to liters (since 1 mole = 22.4 l):

$$4.5 \text{ moles } H_2 \times \frac{22.4\, l \text{ of } H_2}{1 \text{ mole } H_2} \approx 101\, l \text{ of } H_2.$$

6. **(C)** By consulting the periodic table, iron atoms have an atomic weight of about 56 while sulfur is 32. They combine in an atom-to-atom ratio, or a unitary gram equivalent to a unitary gram equivalent ratio. Thus, 56 grams of

iron combine with 32 grams of sulfur. By proportion, if iron is halved from 56 to 28, sulfur is halved from 32 to 16.

7.   **(A)**   The reaction is

$$Cl_2 + CH_3CH_3 \rightarrow CH_3CH_2Cl + HCl$$

1 mole of ethane is $(2 \times 12) + 6 = 30$ g.

From the equation, 1 mole of C_2H_6 yields 1 mole of C_2H_5Cl.

1 mole of $C_2H_5 - Cl$ weighs $(2 \times 12) + 5 + 35.5 = 64.5$ g

30 g of ethane theoretically should give 64.5 g of ethyl chloride. The amount of ethyl chloride obtained from 15 g of ethane is

$$\text{amount of } C_2H_5Cl = \frac{15}{30} \times 64.5 \text{ g} = 32.25 \text{ g}.$$

$\therefore$ The theoretical yield is 32.25 g.

But the actual yield is 15 g.

$$\therefore \text{ percent yield} = \frac{15}{32.25} \times 100 = 46.5\%$$

8.   **(C)**   Writing and balancing the reaction equation

$$CO + \frac{1}{2}O_2 \rightarrow CO_2$$

we find that CO reacts to produce an equimolar amount of CO_2. Converting to moles of CO_2,

$$33 \text{ g of } CO_2 \times \frac{1 \text{ mole of } CO_2}{44 \text{ g of } CO_2} = 0.75 \text{ mole of } CO_2.$$

Therefore, 0.75 mole of CO is required. Converting to grams of CO,

$$0.75 \text{ mole of } CO \times \frac{28 \text{ g of } CO}{1 \text{ mole of } CO} = 21 \text{ g of } CO.$$

Thus, 21 g of CO is required to produce 33 g of CO_2.

9.   **(C)**

$$4Al + 3O_2 \rightarrow 2Al_2O_3$$

$$\frac{4 \text{ moles Al}}{3 \text{ moles } O_2} = \frac{0.36 \text{ moles Al}}{x \text{ mole of } O_2}$$

$$x \text{ mole of } O_2 = \frac{3(0.36)}{4} = 0.27 \text{ moles } O_2$$

Thus, there is an excess of 0.09 mole O_2.

$$\frac{4 \text{ moles Al}}{2 \text{ moles Al}_2O_3} = \frac{0.36 \text{ moles Al}}{x}$$

$$\therefore x = \frac{2 \times 0.36}{4} = 0.18 \text{ mole Al}_2Cl_3$$

Note, since Al is the limiting reagent, we calculate the amount of Al_2O_3 produced from the amount of Al available.

10. **(B)** According to the equation:

$$2C_6H_6 + 15O_2 \rightarrow 12CO_2 + 6H_2O$$

for every 2 moles of C_6H_6, 15 moles of oxygen are required for the complete combustion of the C_6H_6. Therefore, it is necessary to find the number of moles of C_6H_6 in 39.0 g. Since the number of moles is always equal to

$$\text{moles} = \frac{g}{MW} = \frac{39.0}{78.0} = 0.5 \text{ mole of } C_6H_6.$$

According to the balanced equation, for every 2 moles of C_6H_6, 15 moles of oxygen are required. Therefore,

$$\frac{2 \text{ moles } C_6H_6}{.5 \text{ moles } C_6H_6} \quad \frac{15 \text{ moles } O_2}{x \text{ moles } O_2}$$

$x = 3.75$ moles of oxygen are needed for .5 moles of C_6H_6. However, the question asked for the answer expressed in grams, so we must now convert 3.75 moles of O_2 to grams of O_2. Since moles $= m/MW$ the number of grams can be calculated by

$$m = \text{moles} \times MW$$

$$m = 3.75 \text{ moles} \times 32 \text{ g/mol}$$

$$m = 120 \text{ g of oxygen}$$

11. **(C)** Taking the reaction equation:

$$N_2 + 3H_2 \rightarrow 2NH_3$$

and multiplying each coefficient by 3, we obtain

$$3N_2 + 9H_2 \rightarrow 6NH_3.$$

The coefficients give the proportional relationship in volume reactions so 6 ml of NH_3 can be produced from 3 ml of N_2 and 9 ml of H_2. H_2 is the limiting reactant.

12. **(B)** 9 − 3 or 6 ml of N_2 remain unreacted since there is insufficient H_2 present to continue the reaction.

13. **(E)** After balancing the reaction, we obtain

$$2C_8H_{18} + 25O_2 \rightarrow 16CO_2 + 18H_2O$$

Octane's molecular weight is 114: C(8 × 12) + H(18 × 1). Thus, one mole weighs 114 grams. From the balanced equation, two moles (228 grams) react with 25 moles of oxygen gas. At STP, one mole of a gas occupies 22.4 liters. By proportion, the 25 reacting moles occupy 560 liters (22.4 × 25). Since oxygen is about one-fifth of the atmosphere, 560 is multiplied by five for the air volume required.

14. **(D)**

$$2Na_2O_2 + 2H_2O \rightarrow 4NaOH + O_2$$

15. **(B)**

$$2K + 2H_2O \rightarrow 2KOH + H_2$$

16. **(A)**

$$BaCl_2 + Na_2SO_4 \rightarrow 2NaCl + BaSO_4$$

17. **(A)**

$$2SO_2 + O_2 \rightarrow 2SO_3$$

$$\frac{8.0\,g\,O_2}{32.0\,g/mol\,O_2} = 0.25\ \text{mole}\ O_2$$

$$\frac{40.0\,g\,SO_2}{64.0\,g/mol\,SO_2} = 0.625\ \text{mole}\ SO_2$$

O_2 is the limiting reagent. Since two moles of SO_3 are produced for every mole of O_2 consumed, the maximum yield of SO_3 is 0.50 mole. This weighs (.50 mole SO_3) (80 g/mol SO_3) = 40.0 g SO_3.

18. **(C)** Converting the given quantities to moles:

$$98\ \text{g of}\ H_2SO_4 \times \frac{1\ \text{mole of}\ H_2SO_4}{98\ \text{g of}\ H_2SO_4} = 1\ \text{mole of}\ H_2SO_4$$

$$40 \text{ g of NaOH} \times \frac{1 \text{ mole of NaOH}}{40 \text{ g of NaOH}} = 1 \text{ mole of NaOH}$$

The reaction in question is

$$H_2SO_4 + 2NaOH \rightarrow Na_2SO_4 + 2H_2O.$$

This shows that one mole of H_2SO_4 reacts with two moles of sodium hydroxide to produce one mole of sodium sulfate. Since we only have one mole of sodium hydroxide, 0.5 mole of sulfuric acid reacts with it to produce 0.5 mole of sodium sulfate. Converting to grams:

$$0.5 \text{ mole of sodium sulfate} \times \frac{142 \text{ g of sodium sulfate}}{1 \text{ mole of sodium sulfate}}$$

$$= 71 \text{ g of sodium sulfate}$$

19. **(B)**   The reaction in question is

$$2H_2 + O_2 \rightarrow 2H_2O.$$

Converting the given quantities to moles:

$$8 \text{ g of } H_2 \times \frac{1 \text{ mole of } H_2}{2 \text{ g of } H_2} = 4 \text{ moles of } H_2$$

$$8 \text{ g of } O_2 \times \frac{1 \text{ mole of } H_2}{2 \text{ g of } H_2} = 0.25 \text{ mole of } O_2$$

Oxygen is the limiting reactant in this reaction. Multiplying all coefficients by 0.25 in order to obtain 0.25 O_2, we have

$$0.5H_2 + 0.25O_2 \rightarrow 0.5H_2O.$$

Converting to grams:

$$0.5 \text{ mole of } H_2O \times \frac{18 \text{ g of } H_2O}{1 \text{ mole of } H_2O} = 9 \text{ g of } H_2O$$

20. **(D)**   Salts with the cations Na^+, K^+, or NH_4^+ are usually very soluble.

21. **(C)**   These ions form insoluble precipitates with silver.

22. **(E)**   The silver salts of Cl^-, Br^-, I^-, and S^{2-} remain insoluble in strongly acidic solution.

23. **(A)** Ba^{2+} will form insoluble salts with sulfate and phosphate.

24. **(D)**

$$6Na_3PO_4 + 9Pb(NO_3)_2 \rightarrow 18NaNO_3 + 3Pb_3(PO_4)_2$$

25. **(B)** The molecular weight of HNO_3 is 63 grams/mole and that of LiOH is 24 grams/mole. HNO_3 and LiOH react in a 1:1 ratio by mole as seen by

$$HNO_3 + LiOH \rightarrow H_2O + LiNO_3.$$

There is an excess of HNO_3, since only one mole of it can react with the one mole of LiOH available. Thus, there is an excess of $92 - 63$ or 29 grams of nitric acid.

26. **(A)** Four ammonia molecules offer the 12 hydrogens needed for six water molecules. Six water molecules require six oxygens. Five O_2 molecules offer these six as a reactant, plus the four additional oxygens (5×2 total) for the four NO molecules among the products.

27. **(B)**

$$Pb(NO_3)_2 + H_2S \rightarrow PbS + 2HNO_3$$

28. **(C)** The reaction in question is

$$NO + \frac{1}{2}O_2 \rightarrow NO_2$$

or using the given coefficients

$$20NO + 10O_2 \rightarrow 20NO_2.$$

Note that the unit of the coefficients used is liters, not moles. This does not affect the calculation since moles and liters are directly related in the case of gases (1 mole of a gas occupies 22.4 liters at STP).

29. **(D)** The balanced reaction is

$$C_6H_6 + \frac{15}{2}O_2 \rightarrow 6CO_2 + 3H_2O$$

Since the carbon in CO_2 can only be obtained from benzene, which has six carbons, we know that the coefficient of CO_2 will be six. In a similar fashion, the coefficient of H_2O will be three since benzene has six hydrogens. There are 12 oxygens in $6CO_2$ and three oxygens in $3H_2O$ so the coefficient of O_2 is $^{15}/_2$. Multiplying by 2 to remove the fraction, we obtain

$$2C_6H_6 + 15O_2 \rightarrow 12CO_2 + 6H_2O.$$

30. **(D)** The reaction of barium with water is illustrated as

$$Ba + 2H_2O \rightarrow Ba(OH)_2 + H_2.$$

Only 1 and 2 are used as coefficients.

31. **(C)** The reaction is

$$2H_2 + O_2 \rightarrow 2H_2O.$$

Multiplying each coefficient by 2.5 gives

$$5H_2 + 2.5O_2 \rightarrow 5H_2O.$$

Therefore, we find that 5 liters of H_2 reacts with 2.5 liters of O_2 to produce 5 liters of H_2O.

32. **(C)** Converting to moles,

$$33.6 \text{ liters} \times \frac{1 \text{ mole}}{22.4 \text{ liters}} = 1.5 \text{ moles of } SO_3.$$

1.5 moles of H_2SO_4 is produced since SO_3 reacts to produce H_2SO_4 in a 1:1 ratio. Converting to grams,

$$1.5 \text{ moles of } H_2SO_4 \times \frac{98 \text{ g of } H_2SO_4}{1 \text{ mole of } H_2SO_4} = 147 \text{ g of } H_2SO_4$$

33. **(C)** NaCl is sodium chloride. The amount dissolved changes by only .6 gram over 30 degrees. This is much less than potassium nitrate (KNO_3),–32 g, potassium chloride (KCl) –9 g, sodium sulfate (Na_2SO_4) –36.1 g or sodium nitrate $(NaNO_3)$ –19 g.

34. **(C)** By the balanced equation, two moles (224 grams) of $KClO_3$ yield three moles of O_2 (48 grams). By simple proportion:

$$\frac{(2KClO_3^-)}{3O_2} = \frac{224}{48} = \frac{X}{72}$$

so, $$X = 336$$

35. **(D)** The reaction in question is

$$CaCO_3 \rightarrow CaO + CO_2.$$

Converting to moles:

$$44.8 \text{ liters of ideal gas} \times \frac{1 \text{ mole of ideal gas}}{22.4 \text{ liters}} = 2 \text{ moles of ideal gas.}$$

Two moles of CO_2 are produced if it is assumed to be an ideal gas. Two moles of $CaCO_3-$ are required since $CaCO_3 : CO_2 = 1:1$ by the reaction equation. Converting to grams:

$$2 \text{ moles of } CaCO_3 \times \frac{100 \text{ g of } CaCO_3}{1 \text{ mole of } CaCO_3} = 200 \text{ g of } CaCO_3.$$

36. **(D)** The Haber process produces ammonia by the reaction

$$N_2 + 3H_2 \rightarrow 2NH_3.$$

To solve this problem we must first determine how many moles of ammonia are to be produced.

$$34 \text{ g of } NH_3 \times \frac{1 \text{ mole of } NH_3}{17 \text{ g of } NH_3} = 2 \text{ moles of } NH_3$$

Examining the reaction equation, we see that one mole of N_2^- is required to produce two moles of NH_3. Converting moles of N_2 to grams of N_2, we obtain

$$1 \text{ mole of } N_2 \times \frac{28 \text{ g of } N_2}{1 \text{ mole of } N_2} = 28 \text{ g of } N_2.$$

37. **(D)** The balanced reaction in question is

$$2Na + 2H_2O \rightarrow 2NaOH + H_2.$$

Recalling that 1 mole of an ideal gas occupies a volume of 22.4 liters at STP, we find that we require 2 moles of Na to produce 1 mole of H_2. Converting to grams:

$$2 \text{ moles of } Na \times \frac{23 \text{ g of } Na}{1 \text{ mole of } Na} = 46 \text{ g of } Na.$$

38. **(D)**

$$Na + H_2O \rightarrow NaOH + H_2$$

In balancing the equation, we see that there are equal amounts of Na and O on both sides but that there are 2H reacting and 3H produced. Placing a $1/2$ in front of H_2 gives

$$Na + H_2O \rightarrow NaOH + \frac{1}{2}H_2.$$

Now we multiply each term by 2 to remove the fraction, giving

$$2Na + 2H_2O \rightarrow 2NaOH + H_2$$

or 2, 2, 2, 1 as the respective coefficients.

39. **(B)**   From the solubility rules we know that

(1) all $(IA)^+$ and NH_4^+ salts are soluble, therefore I is soluble.

(2) all sulphates are soluble except sulphates of Pb^{+2}, Ca^{2+}, Sr^{2+}, and Ba^{2+}, therefore, II is not soluble.

(3) also all halides are soluble except halides of Pb^{+2}, Ag^+, Hg^{2+}, and Tl^+, therefore, $PbCl_2$ is insoluble.

(4) all salts of No_3^-, ClO_4^-, ClO_3^- and $C_2H_3O_2^-$ ions are soluble, therefore, IV is soluble.

40. **(C)**   The balanced equation is

$$1C_6H_{14} + 6\frac{1}{2} O_2 \rightarrow 6CO + 7H_2O.$$

STOICHIOMETRY
REVIEW

This chapter addresses problems of balancing chemical equations and calculating amounts, such as mass, volume, moles, etc., of species which enter chemical reactions. These kinds of calculations are called stoichiometric calculations. The key idea is to understand that chemical reactions are always written representing the number of molecules (or atoms for the elements) or moles of species which undergo reaction. Atomic or molecular ratios equal mole ratios. Conversion from moles to mass or, for gases, from moles to volume must be understood to solve these problems.

1. Balancing Equations

Chemical reactions are represented by equations showing reactants and products for a particular reaction. The key to balancing chemical reaction equations lies in understanding that elements are conserved in chemical reactions (the exception is nuclear reactions or reactions involving the atomic nucleus. These very special kinds of reactions are discussed in Chapter 13). The same number of moles (or atoms) of each element must appear in both reactants and products. For simple reactions, which include all those in this text, the molecular ratios of reactants are small integers. These molecular ratios are the coefficients in chemical equations and are called the stoichiometric coefficients; they are usually small integers and are usually reduced to the smallest possible integers by dividing each coefficient by common factors.

For example, ammonia (NH_3) can be combined with oxygen (O_2) to form nitric oxide (NO), and water (H_2O). Write the balanced equation for this reaction

$$a\ NH_3 + b\ O_2 \rightarrow c\ NO + d\ H_2O$$

Conserving elements leads to the following requirements:

For N $\quad a = c$

For H $\quad 3a = 2d$

For O $\quad 2b = c + d$

One additional relationship is needed to solve these equations for the four stoichiometric coefficients, a, b, c, and d. This relationship is arbitrary and is usually selected so that a, b, c, and d will be small integers. Simply pick a value of one coefficient—for example, let $a = 1$. It follows that $c = a = 1$, $d = 3/2a = 3/2$, and $b = (c + d)/2 = 5/4$. If a, b, c, and d are multiplied by 4 (this is equivalent to initially selecting $a = 4$), the coefficients become

$$a = 4$$

$b = 5$

$c = 4$

$d = 6$

or $\quad$ $4NH_3 5O_2 \rightarrow 4NO + 6H_2O$

PROBLEM

Balance the equations

(a) $Ag_2O \rightarrow Ag + O_2$

(b) $Zn + HCl \rightarrow ZnCl_2 + H_2$

SOLUTION

When balancing chemical equations, one must make sure that there are the same number of atoms of each element on both the left and right side of the arrow. For example, $H_2 + O_2 \rightarrow H_2O$ is not a balanced equation because there are two O's on the left side and only one on the right. $2H_2 + O_2 \rightarrow 2H_2O$ is the balanced equation for water because there are the same number of H and O atoms on each side of the equation.

(a) $Ag_2O \rightarrow Ag + O_2$ is not a balanced equation because there are two Ag on the left and only one on the right, and because there is only one O on the left and two O's on the right. To balance this equation one must first multiply the left side by 2 to have two O's on each side.

$$2Ag_2O \rightarrow Ag + O_2$$

There are now four Ag on the left and only one on the right, thus the Ag on the right must be multiplied by 4.

$$2Ag_2O \rightarrow 4Ag + O_2$$

The equation is now balanced.

(b) $\quad$ $Zn + HCl \rightarrow ZnCl_2 + H_2$

In this equation, there are two H and two Cl on the right and only one of each on the left therefore, the equation can be balanced by multiplying the HCl on the left by 2.

$$Zn + 2HCl \rightarrow ZnCl_2 + H_2$$

Because there are the same number of Zn, Cl, and H on both sides of the equation, it is balanced.

Drill 1: Balancing Equations

1. Balance the following equation. What is the missing species?

    $$NaOH + \underline{\hspace{1cm}} \rightarrow NaHSO_4 + HOH$$

    (A) HSO_4^- (D) $H_3SO_4^+$

    (B) HS_2O_3 (E) H_2SO_4

    (C) S_2O_6

2. $\underline{\hspace{1cm}} H_2S + \underline{\hspace{1cm}} O_2 \rightarrow \underline{\hspace{1cm}} H_2O + \underline{\hspace{1cm}} SO_2$

    Balancing this equation yields the following coefficients from left to right:

    (A) 1-1-2-2   (D) 3-2-2-2

    (B) 2-3-2-2   (E) 3-2-3-2

    (C) 2-2-2-3

For questions 3 and 4, balance the given equation. What is the missing species with its appropriate coefficient?

3. $PCl_3 + 3HOH \rightarrow \underline{\hspace{1cm}} + 3HCl$

    (A) $P(OH)_3$ (D) $P_2(OH)_3$

    (B) $P(OH)_4$ (E) $P_2(OH)_4$

    (C) $P(OH)_2$

4. $CH_4 + \underline{\hspace{1cm}} \rightarrow CCl_4 + 4HCl$

    (A) ClO_3 (D) CH_3Cl

    (B) $4Cl_2$ (E) C_2Cl_2

    (C) $3Cl_2$

2. Calculations Using Chemical Arithmetic

In order to solve weight/weight, volume/volume or volume/weight problems, it is critical to first change weights and/or volumes to moles.

For a typical weight/volume problem, follow the following steps:

Step 1: Write the balanced equation for the reaction.

Step 2: Write the given quantities and the unknown quantities for the appropriate substances.

Step 3: Calculate reacting weights or number of moles (or volume, if the reaction involves only gases) for the substances whose quantities are given. Make sure that the units for each substance are identical.

Step 4: Use the proportion method or the factor-label method.

Step 5: Solve for the unknown.

PROBLEM

$NaClO_3$, when heated, decomposes to $NaCl$ and O_2. What volume of O_2 at STP results from the decomposition of 42.6 grams $NaClO_3$?

SOLUTION

1. (Using reactive masses) balanced equation:

$$213 \text{ g NaClO}_3 \xrightarrow{\Delta} 117 \text{ g NaCl} + 96 \text{ g O}_2$$

$$\frac{\text{mass O}_2 \text{ produced}}{\text{mass NaClO}_3 \text{ decomposed}} = \frac{96 \text{ g}}{213 \text{ g}} = \frac{x}{42.6 \text{ g}}$$

$$x = \frac{(96)(42.6)}{213} = 19.2 \text{ g O}_2$$

or

2. (Using moles) balanced equation:

$$2NaClO_3 \xrightarrow{\Delta} 2 \text{ NaCl} + 3O_2$$

$$\frac{\text{mass O}_2 \text{ produced}}{\text{mass NaClO}_3 \text{ decomposed}} = \frac{3}{2} = \frac{y}{(42.6 \text{ g}/206.5 \text{ g/mol})}$$

$$y = \frac{(3)(42.6)}{(2)(106.5)} = 0.6 \text{ mole O}_2$$

Finally, at STP, 1 mole (32 g) O_2 occupies 22.4 l, so

$$V_{O_2(\text{STP})} = \left(\frac{19.2 \text{ g}}{32 \text{ g/mole}}\right)(22.4 \text{ } l \text{ /mol}) = 0.0268 \text{ } l$$

or $V_{O_2(\text{STP})} = (0.6 \text{ mole})(22.4 \text{ } l \text{ /mol}) = 0.0268 \text{ } l$

or $V_{O_2(\text{STP})} = 2.68 \times 10^{-2} \text{ } l$

The Ideal Gas Law, $PV = nRT$, can be used to determine the volume of gas or the number of moles of a gas at conditions other than STP.

PROBLEM

> In a chemical reaction requiring two atoms of phosphorus for five atoms of oxygen, how many grams of oxygen are required by 3.10 g of phosphorus?

SOLUTION

Because the relationship between the phosphorus and oxygen is given in atoms, the relationship also holds for moles. There must be two moles of phosphorus for every five moles of oxygen. This is true because there is a set number of atoms in any one mole. Therefore, one must first find the number of moles of phosphorus present. From this, one can find the number of moles of oxygen present. From the number of moles of oxygen, one can find the weight by multiplying by the molecular weight.

The number of moles of phosphorus present is found by dividing the number of grams by the molecular weight (MW = 31).

$$\text{no. of moles} = \frac{3.10\,g}{31\,g/mol} = .10\ \text{moles}$$

Because there must be five moles of oxygen for every two moles of phosphorus, the following ratio can be used to determine the number of moles of oxygen. Let x = number of moles of O.

$$\frac{2\ \text{moles P}}{5\ \text{moles O}} = \frac{.10\ \text{mole P}}{x}$$

$$x = \frac{5\ \text{moles O} \times .10\ \text{mole P}}{2\ \text{moles P}} = .25\ \text{mole O}$$

The weight of the oxygen is then found by multiplying the number of moles by the molecular weight (MW = 16).

$$\text{no. of grams of O} = 16\,g/mol \times .25\ \text{mole} = 4.00\ g\ O.$$

Drill 2: Calculations Using Chemical Arithmetic

1. A student isolates 35.8 g of $AgCl_2$ from a photographic emulsion. What is the maximum amount of silver metal he could recover from this? (a.w. Ag = 108, a.w. Cl = 35.5)

    (A) 17.9 g

    (B) 26.9 g

    (C) 21.6 g

    (D) 0.2 g

    (E) 0.6 g

2. How many moles of $KClO_3$ must be decomposed to produce 3.20 g of O_2?

    (A) .10   (D) .14

    (B) .30   (E) .07

    (C) .24

3. How many grams of CO_2 are produced by the complete reaction of 180 g of $CaCO_3$ with excess HCl?

    (A) 22 g   (D) 110 g

    (B) 44 g   (E) 132 g

    (C) 79 g

4. Calculate the number of moles of carbon dioxide produced during the combustion of 2 moles of ethane (C_2H_6).

    (A) 2   (D) 8

    (B) 4   (E) 10

    (C) 6

3. Reactions with Limiting Reagents

Limiting-Reactant Calculations

The reactant that is used up first in a chemical reaction is called the limiting reactant, and the amount of product is determined (or limited) by the limiting reactant. Simply stated, when one of the reactants is used up the reaction stops. Suppose a limited amount of ammonia (for example, 100 g) is burned in the atmosphere (21% O_2) which is, compared to 100 g, unlimited in supply. When 100 g of NH_3 is consumed, the reaction stops even though more oxygen is available.

Theoretical Yield and Percentage Yield

The theoretical yield of a given product is the maximum yield that can be obtained from a given reaction if the reaction goes to completion (rather than to equilibrium).

The percentage yield is a measure of the efficiency of the reaction. It is defined

$$\text{percentage yield} = \frac{\text{actual yield}}{\text{theoretical yield}} \times 100\%$$

For example, in the reaction

$$Pb^{2+}_{(aq)} + 2Cl^-_{(aq)} \rightarrow PbCl_{2(s)}$$

starting with two moles of Pb^{2+}, one would expect two moles of $PbCl_2$ product. This is the theoretical yield. If only one mole of $PbCl_2$ is obtained (the actual yield), a percentage yield of 50% can be calculated by (2 mol/1mol) × 10.

PROBLEM

Chromic oxide (Cr_2O_3) may be reduced with hydrogen according to the equation

$$Cr_2O_3 + 3H_2 \rightarrow 2Cr + 3H_2O$$

(a) What weight of hydrogen would be required to reduce 7.6 g of Cr_2O_3? (b) For each mole of metallic chromium prepared, how many moles of hydrogen will be required? (c) What weight of metallic chromium can be prepared from one ton of Cr_2O_3? 1 lb = 454 g

SOLUTION

(a) From the equation for the reaction, one knows that it takes three moles of H_2 to reduce one mole of Cr_2O_3. Thus, in solving this problem one should first determine the number of moles of Cr_2O_3 in 7.6 g, then using the ratio

$$\frac{3}{1} = \frac{\text{number of moles of } H_2}{\text{number of moles of } Cr_2O_3}$$

one can find the number of moles of H_2 necessary to reduce 7.6 g of Cr_2O_3. After finding the number of moles of H_2 needed, one can obtain the weight by multiplying the number of moles by the molecular weight of H_2.

Solve for the number of moles of Cr_2O_3 in 7.6 g. This is done by dividing 7.6 g by the molecular weight of Cr_2O_3 (MW = 152).

$$\text{number of moles} = \frac{7.6 \text{ g}}{152 \text{ g/mol}} = .05 \text{ moles}$$

Determine the number of moles of H_2 necessary. The ratio

$$\frac{3}{1} = \frac{\text{number of moles of } H_2}{\text{number of moles of } Cr_2O_3}$$

will be used. This ratio was made using the stoichiometric coefficients of the equation for the reaction

$$\frac{3}{1} = \frac{\text{number of moles of } H_2}{.05 \text{ moles}}$$

$$\text{number of moles of H} = \frac{.05 \text{ moles} \times 3}{1} = .15 \text{ mole}$$

Solve for the weight of H_2. (MW = 2) The weight of a compound is found by multiplying the number of moles present by the molecular weight.

$$\text{weight of } H_2 = .15 \text{ mole} \times 2 \text{ g/mol} = .30 \text{ g}$$

(b) From the equation for the reaction, it is seen that three moles of H_2 are needed to form two moles of Cr. This means that for every mole of Cr formed $^3/_2$ this amount of H_2 is needed. Thus, 1.5 moles of H_2 is necessary to form one mole of Cr.

(c) Using the equation for the reaction, one is told that for every mole of Cr_2O_3 reduced two moles of Cr are formed. Thus, one must determine the number of moles in 1 ton of Cr_2O_3; there will be twice as many moles of Cr formed. After one knows the number of moles of Cr formed, one can determine its weight by multiplying the number of moles by the molecular weight of Cr.

Determine the number of moles of Cr_2O_3 in one ton of the compound (1 ton = 2,000 lbs). The number of moles is found by dividing the weight of the Cr_2O_3 present by its molecular weight. Because the molecular weight is given in grams per mole, it must be converted to pounds per mole before using it to determine the number of moles present. There are 454 g in one pound, thus grams can be converted to pounds by multiplying the number of grams by the conversion factor 1 lb/454 g. (MW of Cr_2O_3 = 152)

$$\text{MW of } Cr_2O_3 \text{ in lbs} = 152 \text{ g/mol} \times 1 \text{ lb/454 g}$$

$$= .33 \text{ lbs/mol}$$

Determine the number of moles of Cr_2O_3 in one ton. The number of moles present can be found by dividing one ton (2,000 lbs) by the molecular weight in pounds.

$$\text{number of moles} = \frac{2,000 \text{ lbs}}{.33 \text{ lb/mol}} = 6,060 \text{ moles}$$

Thus, there are twice this many moles of Cr produced.

$$\text{number of moles of Cr} = 2 \times 6,060 \text{ moles} = 12,120 \text{ moles}$$

Find the weight of Cr formed. One knows that 12,120 moles of Cr are produced. To find the weight of this quantity, the number of moles is multiplied by the molecular weight. To find the weight in pounds, one must first convert the molecular weight from grams to pounds. This is done by multiplying the molecular weight by the conversion factor 1 lb/454 g. This is used because there are 454 g in 1 lb (MW of Cr = 52).

$$\text{MW of Cr in pounds} = 52 \text{ g/mol} \times 1 \text{ lb/454 g}$$

$$= 0.11 \text{ lb/mol}$$

The weight of the Cr formed is now found by multiplying this molecular weight by the number of moles present.

$$\text{weight} = 0.11 \text{ lb/mol} \times 12,120 \text{ moles} = 1,333 \text{ lbs.}$$

PROBLEM

When 10.0 g of silicon dust, Si, is exploded with 100.0 g of oxygen, O_2, forming silicon dioxide, SiO_2, how many grams of O_2 remain uncombined? The reaction equation is

$$Si + O_2 \rightarrow SiO_2.$$

SOLUTION

From the equation, it can be seen that Si and O_2 react in a 1:1 ratio. This means that one mole of Si will react with one mole of O_2. To determine the amount of O_2 left unreacted after the reaction is performed, calculate the number of moles of O_2 and Si present and then subtract the number of moles of Si from the number of moles of O_2. (MW of Si = 28, MW of O_2 = 32)

$$\text{number of moles} = \frac{\text{number of grams}}{MW}$$

$$\text{number of moles of } O_2 = \frac{100.0 \text{ g}}{32 \text{ g/mol}} = 3.12 \text{ moles}$$

$$\text{number of moles of Si} = \frac{10.0 \text{ g}}{28 \text{ g/mol}} = 0.357 \text{ moles}$$

$$\text{number of moles of excess } O_2 = 3.12 - 0.357 = 2.763 \text{ moles}$$

$$\text{weight of excess } O_2 = \text{number of moles} \times MW$$

$$= 2.763 \text{ moles} \times 32 \text{ g/mol} = 88.5 \text{ g}$$

Drill 3: Reactions with Limiting Reagents

1. How many moles of Al_2O_3 can be formed when a mixture of 0.36 mole of aluminum and 0.36 mole of oxygen is ignited?

    $$4Al + 3O_2 \rightarrow 2Al_2O_3$$

    (A) 2 moles   (D) 4 moles

    (B) 0.36 mole   (E) 0.27 mole

    (C) 0.18 mole

2. What is the maximum weight of SO_3 that could be made from 25.0 g of SO_2 and 6.00 g of O_2 by the following reaction?

    $$2SO_2 + O_2 \rightarrow 2SO_3$$

    (A) 30.1 g   (D) 37.6 g

    (B) 1.88 g   (E) 64.1 g

    (C) 3.91 g

Questions 3 and 4 use the following information: A chemist reacts ferric sulfate with barium chloride and obtains barium sulfate and ferric chloride. He writes the following balanced equation to express this reaction:

$$Fe_2(SO_4)_3 + 3BaCl_2 \rightarrow 3BaSO_4\downarrow + 2FeCl_3$$

3. How much $BaCl_2$ should be used to react with 10 grams of $Fe_2(SO_4)_3$?

    (A) 4.39   (D) 15.62

    (B) 13.95   (E) 7.43

    (C) 12.16

4. How much $Fe_2(SO_4)_3$ will be necessary to produce 100 g of $BaSO_4$?

    (A) 57.11   (D) 75.43

    (B) 43.29   (E) 49.31

    (C) 67.96

5. If 2.0 grams of $NiCl_2$ is mixed in aqueous solution with 3.0 NaOH to yield 0.019 mole of $Ni(OH)_2(s)$ what is the percentage yield?

    (A) 43%   (D) 50%

    (B) 90%   (E) 16%

    (C) 85%

4. Net Ionic Equations

A precipitation reaction is one in which soluble reactants are mixed together to form an insoluble product. For example, when a silver nitrate solution is mixed with a sodium chloride solution, silver nitrate is formed, and being insoluble, settles out of the solution. To observe this, one would see two colorless solutions poured together, immediately turning white and milky. The white silver chloride formed will settle to the bottom of the container, leaving a colorless solution of sodium nitrate above it. There are three ways to show what compounds precipitate in a gross chemical reaction:

1) $AgNO_3 + NaCl \rightarrow NaNO_3 + AGCl\downarrow$

An arrow pointing down designates a precipitate.

2) $AgNO_3 + NaCl \rightarrow NaNO_3 + \underline{AgCl}$

as does underling the precipitate.

3) $AgNO_3(aq) + NaCl(aq) \rightarrow NaNO_3(sq) + AgCl(s)$

The nomenclature (aq) stands for "aqueous," which means the material is dissolved in water. The nomenclature (s) stands for "solid," i.e., it is not dissolved. This latter method is the preferred one to use.

Another method to show a precipitation is by writing the "net ionic equation," as opposed to the gross equations above. Recall that all of the above compounds *except* silver nitrate exist as ions when in water. Since we are working with aqueous (water) solutions, we can write:

$$Ag^+ + NO^-_3 + Na^+ + Cl^- \rightarrow Na^+ + NO^-_3 + AgCl(s)$$

It will be noted that we have sodium and nitrate ions in solution on both sides of the arrow. These are called "spectator ions" and can be cancelled out to give us the net ionic equation:

$$Ag^+ + \underline{Cl^-} \rightarrow AgCl(s)$$

We could have also shown the precipitation with the arrow pointing down or by underlining the precipitation. It should be noted that not all molecular species are insoluble; hence, the product of a net ionic equation does not always yield a precipitate. Always look for one of the designations used for precipitate to identify them.

All strong acids, strong bases, and soluble salts will disassociate into ions when dissolved in water:

Sulfuric acid	$H2SO_4 \rightarrow 2H^+ + SO_4^=$
Cesium hydroxide	$CsOH \rightarrow Cs^+ + OH^-$
Ammonium phosphate	$(NH_4)_3PO_4 \rightarrow 3NH_4^+ + PO_4^=$

Weak acids and weak bases only disassociate to a very small extent. Insoluble material and covalently bonded compounds (e.g., sugars, alcohols, etc.) will not disassociate to form ions at all. A solution containing ions will conduct an electric current, whereas one without them will not.

PROBLEM

> Describe, using equations, what takes place when the following solutions are prepared:
>
> (a) A solution containing equal amounts of 0.10M lead nitrate and 0.10M potassium chromate;
>
> (b) A solution containing one mole of sodium chloride and one mole of potassium bromide.

SOLUTION

In such problems, two processes can occur: mixing of ions in solution and precipitation.

(A) Both lead nitrate $(Pb(NO_3)_2)$ and potassium chromate (K_2CrO_4) are highly soluble, so that four ions are present in solution:

$$Pb^{2+}, K^+, NO_3^-, CrO_4^{2-}$$

There are four possible combinations of these ions: $Pb(NO_3)_2$, $PbCrO_4$, KNO_3, and $K_2Cr_2O_4$. Of these, only lead chromate, $PbCrO_4$, is insoluble, so that the reaction

$$Pb^{2+} + CrO_4^{2-} \Rightarrow PbCrO_4$$

is driven to the right, and yields a precipitate, leaving K^+ and NO_3^- ions in solution (along with trace amounts of Pb^{2+} and CrO_4^{2-}, which exist in equilibrium with solid lead chromate).

(B) Both sodium chloride (NaCl) and potassium bromide (KBr) are highly soluble, giving rise to Na^+, K^+, Cl^-, and Br^- ions in solution. The four possible combinations of these ions are NaCl, NaBr, KCl, and KBr, all of which are highly soluble. Therefore, no precipitate is formed, but only mixing of the ions according to the equation

$$(Na^+, Cl^-) + (K^+, Br^-) \leftrightarrow (Na^+, Br^-) + (K^+, Cl^-).$$

Drill 4: Net Ionic Equations

1. Which of the following are examples of normal salts?

(A) KNO_3

(B) LiO_2CCH_3

(C) $NaHCO_3$

(D) Both (A) and (B)

(E) All of the above

2. Complete ionization of an aluminum hydroxide particle yields

    (A) Al^+, OH^-.   (D) $2Al^+$, $3OH^-$.

    (B) Al^+, $2OH^-$.   (E) $2Al^+$, OH^-.

    (C) Al^+, $3OH^-$.

3. What is/are the product gas/gases if NH_4Cl and $Ca(OH)_2$ are used as reactants?

    (A) N_2 (D) $NH_3 + N_2$

    (B) NH_3 (E) $NH_3 + H_2O$

    (C) H_2O

4. Complete ionization of a calcium hydroxide molecule yields

    (A) Ca^{++}, OH^-.   (D) $2Ca^{++}$, $2OH^-$.

    (B) Ca^{++}, $2OH^-$.   (E) $3Ca^{++}$, OH^-.

    (C) $2Ca^{++}$, OH^-.

5. The net ionic equation that represents the reaction of aqueous hydrochloric acid with aqueous ammonia is

    (A) $HCl(aq) + NH_3(aq) \rightarrow NH_4^+(aq)$.

    (B) $HCl(aq) + NH_3(aq) \rightarrow NH_4^+(aq) + Cl^-(aq)$.

    (C) $H^+(aq) + NH_3(aq) \rightarrow NH_4^+(aq)$.

    (D) $H^+(aq) + Cl^-(aq) + NH_4^+(aq) + OH^-(aq)$

    $\rightarrow H_2O(aq) + NH_4^+(aq) + Cl^-(aq)$.

    (E) $H^+(aq) + NH_4^+(aq) + OH^-(aq)$ fi $H_2O(aq) + NH_4^+(aq)$.

STOICHIOMETRY
DRILLS

ANSWER KEY

Drill 1 — Balancing Equations
1. (E)
2. (B)
3. (A)
4. (B)

Drill 2 — Calculations Using Chemical Arithmetic
1. (C)
2. (D)
3. (C)
4. (B)

Drill 3 — Reactions with Limiting Reagents
1. (C)
2. (A)
3. (D)
4. (A)
5. (B)

Drill 4 — Net Ionic Equations
1. (A)
2. (C)
3. (E)
4. (B)
5. (C)

GLOSSARY:
STOICHIOMETRY

Anion

A negatively charged ion.

Balancing Equations

Procedure by which one makes sure that there are the same number of atoms of each element in a chemical equation on both sides of the arrow.

Cation

A positively charged ion.

Disassociation

The breaking up of a salt or an acid into its component ions.

Insoluble

A compound or other species that cannot dissolve in water or another solvent is considered to be insoluble.

Ion

A positively or negatively charged species.

Limiting Reagent

The reactant that is used up first in a chemical reaction. This reactant determines the quantity of product that is produced.

Net Ionic Equations

A balanced equation which shows only the reactants and products that actually participate in a reaction.

Percentage Yield

A measure of the efficiency of a reaction obtained by comparing the actual yield with the theoretical yield.

Precipitate

An insoluble product of a reaction.

Salt

An ionic compound made up of a positive ion and a negative ion. Compounds containing the ions H^+, OH^-, or O^{2-} are generally not included in this definition.

Soluble

A compound or substance that dissolves in water or another solvent is considered to be soluble.

Solute

A component which is dissolved in a solvent.

Solvent

The liquid into which a solute is dissolved.

Spectator Ions

Ions which do not take part in a chemical reaction and are "crossed out" before a final net ionic equation is written.

Stoichiometry

The way in which quantities of reactants or products in a chemical reaction are determined.

Theoretical Yield

The maximum yield that can be obtained from a given reaction if the reaction goes to completion.

Weak Acid

An acid which does not disassociate completely in its solvent.

Weak Base

A base which does not disassociate completely in its solvent.

CHAPTER 6

Solids

➤ Diagnostic Test
➤ Solids Review & Drills
➤ Glossary

SOLIDS
DIAGNOSTIC TEST

1. Ⓐ Ⓑ Ⓒ Ⓓ Ⓔ		16. Ⓐ Ⓑ Ⓒ Ⓓ Ⓔ
2. Ⓐ Ⓑ Ⓒ Ⓓ Ⓔ		17. Ⓐ Ⓑ Ⓒ Ⓓ Ⓔ
3. Ⓐ Ⓑ Ⓒ Ⓓ Ⓔ		18. Ⓐ Ⓑ Ⓒ Ⓓ Ⓔ
4. Ⓐ Ⓑ Ⓒ Ⓓ Ⓔ		19. Ⓐ Ⓑ Ⓒ Ⓓ Ⓔ
5. Ⓐ Ⓑ Ⓒ Ⓓ Ⓔ		20. Ⓐ Ⓑ Ⓒ Ⓓ Ⓔ
6. Ⓐ Ⓑ Ⓒ Ⓓ Ⓔ		21. Ⓐ Ⓑ Ⓒ Ⓓ Ⓔ
7. Ⓐ Ⓑ Ⓒ Ⓓ Ⓔ		22. Ⓐ Ⓑ Ⓒ Ⓓ Ⓔ
8. Ⓐ Ⓑ Ⓒ Ⓓ Ⓔ		23. Ⓐ Ⓑ Ⓒ Ⓓ Ⓔ
9. Ⓐ Ⓑ Ⓒ Ⓓ Ⓔ		24. Ⓐ Ⓑ Ⓒ Ⓓ Ⓔ
10. Ⓐ Ⓑ Ⓒ Ⓓ Ⓔ		25. Ⓐ Ⓑ Ⓒ Ⓓ Ⓔ
11. Ⓐ Ⓑ Ⓒ Ⓓ Ⓔ		26. Ⓐ Ⓑ Ⓒ Ⓓ Ⓔ
12. Ⓐ Ⓑ Ⓒ Ⓓ Ⓔ		27. Ⓐ Ⓑ Ⓒ Ⓓ Ⓔ
13. Ⓐ Ⓑ Ⓒ Ⓓ Ⓔ		28. Ⓐ Ⓑ Ⓒ Ⓓ Ⓔ
14. Ⓐ Ⓑ Ⓒ Ⓓ Ⓔ		29. Ⓐ Ⓑ Ⓒ Ⓓ Ⓔ
15. Ⓐ Ⓑ Ⓒ Ⓓ Ⓔ		30. Ⓐ Ⓑ Ⓒ Ⓓ Ⓔ

SOLIDS

DIAGNOSTIC TEST

This diagnostic test is designed to help you determine your strengths and your weaknesses in solids. Follow the directions and check your answers.

> **Study this chapter for the following tests:**
> **ACT, AP Chemistry, CLEP General Chemistry,**
> **GED, MCAT, MSAT, PRAXIS II Subject Assessment:**
> **Chemistry, SAT II: Chemistry, GRE Chemistry**

30 Questions

DIRECTIONS: Choose the correct answer for each of the following problems. Fill in each answer on the answer sheet.

1. The unit cell cube edge length for LiCl (NaCl-like structure, face-centered cubic) is 5.14 Å. Assuming anion-anion contact, what is the ionic radius for chloride ion?

(A) $\sqrt{2}\,(5.14\text{Å})$

(D) $\dfrac{\sqrt{2}}{2}(2.57\text{Å})$

(B) $\dfrac{\sqrt{2}}{2}(5.14\text{Å})$

(E) $\sqrt{2}\,(2.57\text{Å})$

(C) $\dfrac{1}{2}(5.14\text{Å})$

Questions 2 and 3 use the following answer choices

(A) Molecules are moving least rapidly and are closest together.

(B) Water is in this state at 12°C.

(C) Mercury is in this state at room temperature.

(D) Molecules are moving most rapidly.

(E) Molecules maintain a definite volume, but shape depends upon the contours of the container holding them.

2. Gas

3. Solid

4. Water could be made to boil at 90°C by

    (A) increasing the pressure with volume constant.

    (B) decreasing the pressure with volume constant.

    (C) increasing the volume with pressure constant.

    (D) decreasing the volume with pressure constant.

    (E) increasing the pressure while decreasing the volume.

5. The temperature above which a liquid cannot exist is indicated by

    (A) the triple point.   (D) the boiling point.

    (B) the critical point.   (E) the sublimation point.

    (C) the eutectic point.

6. Which of the following does *not* show a tetrahedral structure?

    (A) Diamond   (D) Zinc blend (ZnS)

    (B) LiF   (E) Wurtzite (ZnS)

    (C) Ice

Answer questions 7–9 using the phase diagram below.

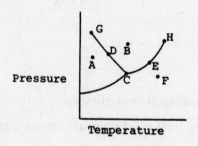

7. At which point can all three phases coexist at equilibrium?

    (A) *C*   (D) *G*

    (B) *D*   (E) *H*

    (C) *E*

8. At which point can only the solid phase exist?

(A) A   (D) E

(B) B   (E) F

(C) C

9. Which is the critical point?

(A) B   (D) E

(B) F   (E) G

(C) H

10. The triple-point pressure for H_2O is 4.58 mm Hg and the triple-point temperature is 273.16°K. From this we can conclude

(A) steam cannot exist at temperatures below 273.16°K.

(B) the vapor pressure of ice is 4.58 mm Hg for temperatures below 273.16°K.

(C) the vapor pressure of water is less than 4.58 mm Hg in most cases.

(D) water cannot exist at pressures below 4.58 mm Hg.

(E) ice cannot exist at pressures below 4.58 mm Hg.

11. The critical pressure of a substance is a value that is necessary to

(A) convert a gas to a solid at its critical temperature.

(B) convert a liquid to a solid at its critical temperature.

(C) freeze a liquid at its critical temperature.

(D) liquefy a gas at its critical temperature.

(E) vaporize a liquid at its critical temperature.

12. The solid-liquid line in the phase diagram shown below for H_2O has a negative slope because

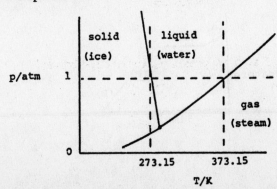

(A) the freezing process is exothermic.

(B) ice contracts on melting.

(C) H_2O has a high entropy of fusion.

(D) ice can exist in several forms.

(E) the triple-point pressure is very low.

13. All of the following share the same crystal structure except

(A) LiCl.   (D) RbCl.

(B) NaCl.   (E) CsCl.

(C) KCl.

14. How many lattice points are found in the face-centered unit cell illustrated below?

(A) 2

(B) 3

(C) 4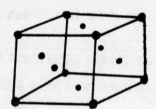

(D) 5

(E) 14

15. Arrange the following ion pairs in order of increasing lattice energy

    LiCl, $BaCl_2$, LiBr, LiI

(A) LiI, LiBr, LiCl, $BaCl_2$ (D) LiCl, LiBr, LiI, $BaCl_2$

(B) $BaCl_2$, LiI, LiBr, LiCl   (E) LiCl, $BaCl_2$, LiI, LiBr

(C) $BaCl_2$, LiCl, LiBr, LiI

For questions 16–18 use the labeled graph below.

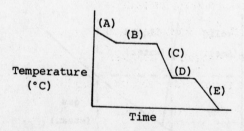

16. Where is the state of the substance a mixture of vapor and liquid?

17. Where is the heat required to change the state of the compound the least (assume constant heating)?

18. Where is the specific heat of the compound the greatest (assume constant heating)?

19. The name of the compound $HClO_2$ is

    (A) hydrochloric acid.   (D) chloric acid.

    (B) hypochlorous acid.   (E) perchloric acid.

    (C) chlorous acid.

20. The following phase diagram was obtained for compound X:

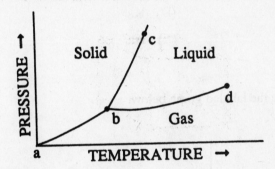

    The melting point of X would

    (A) be independent of pressure.

    (B) increase as pressure increases.

    (C) decrease as pressure increases.

    (D) depends on the amount of X present.

    (E) Both (C) and (D).

21. A substance will boil at a lower temperature if

    (A) the volume of the system is decreased.

    (B) the pressure of the system is decreased.

    (C) an additional amount of the substance is added.

    (D) the substance is dissolved in water.

    (E) Both (A) and (B).

22. The melting point of a substance is

    (A) 0°C.

    (B) very high for solids.

    (C) the same as its freezing point.

    (D) very low for liquids.

    (E) different from the freezing point.

23. The *least* symmetrical unit cell, where there are unequal sides and angles, is called

    (A) monoclinic.

    (B) triclinic.

    (C) orthorhombic.

    (D) cubic.

    (E) tetragonal.

24. An example of a metal which is ductile and malleable is

    (A) Au.

    (B) Cd.

    (C) Hg.

    (D) K.

    (E) Na.

For questions 25–28 use the labeled graph below.

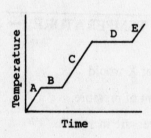

25. Which region indicates a solid?

26. Which region indicates a liquid?

27. Which region indicates a gas?

28. Which region indicates a liquid and a gas?

29. A decrease in kinetic energy of a molecule occurs when

    (A) two different compounds are mixed together.

    (B) the volume is decreased.

(C) the pressure is increased.

(D) the substance sublimes.

(E) the substance condenses.

30. Which one of the following processes indicates sublimation?

(A) gas → liquid   (D) solid → gas

(B) gas → solid   (E) liquid → gas

(C) solid → liquid

SOLIDS DIAGNOSTIC TEST

ANSWER KEY

1.	(D)	9.	(E)	17.	(D)	25.	(A)
2.	(D)	10.	(D)	18.	(A)	26.	(C)
3.	(A)	11.	(D)	19.	(C)	27.	(E)
4.	(B)	12.	(B)	20.	(B)	28.	(D)
5.	(B)	13.	(E)	21.	(B)	29.	(E)
6.	(C)	14.	(C)	22.	(C)	30.	(D)
7.	(A)	15.	(A)	23.	(B)		
8.	(A)	16.	(B)	24.	(A)		

DETAILED EXPLANATIONS
OF ANSWERS

1.   **(D)**   For a face-centered cubic structure, the distance between the center of a lithium ion is one-half the edge length of the cubic unit cell.

$$a = 5.14 \frac{\text{Å}}{2} = 2.57 \text{ Å}$$

$$b = 5.14 \frac{\text{Å}}{2} = 2.57 \text{ Å}$$

$$c = \sqrt{2} \, (2.57 \text{Å})$$

since we assume $Cl^- - Cl^-$ contact, the ionic radius for Cl^- is

$$\frac{\sqrt{2}}{2} (2.57 \text{Å}).$$

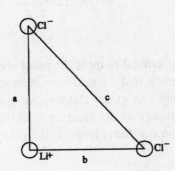

2.   **(D)**   3.   **(A)**
The two choices are general principles comparing solids and liquids for rapidity of particle movement and intermolecular distance. Liquid water does not freeze and become a solid until 0°C. Mercury, unlike most metals, is a liquid at room temperature.

4.   **(B)**   Boiling occurs when the vapor pressure of a liquid is equal to the atmospheric pressure. Therefore, decreasing the pressure would lower the boiling point of water.

5.   **(B)**   The critical point on a phase diagram indicates the temperature above which a liquid cannot exist regardless of the pressure. The triple point specifies the temperature and pressure at which gas, liquid, and solid can coexist in equilibrium.

6.   **(C)**   LiF has the octahedral NaCl structure.

7. **(A)** All three phases (solid, liquid, and gas) may coexist at a single pressure/temperature combination known as the triple point. This point occurs at the intersection of the solid-liquid, solid-gas, and liquid-gas equilibrium curves as illustrated by point *C*.

8. **(A)** Examining a labeled phase diagram, we see that the solid phase can only exist at point *A*.

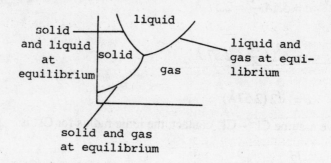

9. **(E)** The critical point is the point above which a gas cannot change into a liquid. This means that a liquid cannot exist above this point, but at and below this point a liquid can exist. The temperature at the critical point is called the critical temperature, and the pressure is called the critical pressure. The critical point in the phase diagram shown is the point *H*, since above it a gas cannot be liquefied.

10. **(D)** 4.58 mm Hg is the lowest pressure at which water can be observed. Below this point ice sublimes directly into steam.

11. **(D)** For each gas, a temperature is reached where the kinetic energy of the molecules is so great that no pressure, however large, can liquefy the gas. Any pressure is insufficient to compress gas molecules back to the liquid state where molecules are closer together. This temperature is the critical temperature. The accompanying pressure is the critical pressure.

12. **(B)** Since ice takes up more volume than an equal molar amount of water (at the freezing point) higher pressures favor the water phase.

13. **(E)** LiCl, NaCl, KCl, and RbCl all have the 6PO(NaCl) structure. Because of the large ionic radius of Cs, CsCl takes on the 3·2PTOT(CsCl) structure.

14. **(C)** The lattice points (*LP*) "belonging" to a unit cell are calculated as follows:

$$\frac{1}{8} \text{ (no. of corner } LP) + \frac{1}{4} \text{ (no. of edge } LP) + \frac{1}{2} \text{ (no. of face } LP)$$

$$+ \, 1 \text{ (no. of body } LP).$$

Applied to the face-centered cell we have

$$\frac{1}{8}(8) + \frac{1}{2}(6) = 4LP.$$

15. **(A)** Lattice energy depends on (i) ionic charge (a greater charge results in a greater lattice energy), (ii) ionic radius (a smaller sum of the ionic radii results in a greater lattice energy), and (iii) lattice geometry. Here, ionic charge (Ba^{2+} versus Li^+) and ionic radius ($Cl^- < Br^- < I^-$) are the important factors.

16. **(B)** The states present on this cooling curve are as follows:

    (A) vapor phase being cooled to the boiling point

    (B) mixture of vapor and liquid at a constant temperature (the boiling point)

    (C) all the vapor has been converted to the liquid which is being cooled to the freezing point

    (D) mixture of liquid and solid at a constant temperature (the freezing point)

    (E) all the liquid has been converted to the solid which continues to cool

17. **(D)** The phase changes occur at a constant temperature, thus they are depicted as horizontal lines on the cooling curve. Assuming constant heating, the phase change requiring the least energy will be the one that requires the shortest time interval to occur. This is depicted to be region *D*.

18. **(A)** A large value for the specific heat of a substance indicates a small change in temperature results for every unit of heat introduced or removed to or from the system. Assuming constant heating, a longer period of time will be required to change the temperature a given amount if there is a large specific heat. This appears as the smallest non-zero slope on the cooling curve. In our example, region *A* gives the phase with the largest specific heat.

19. **(C)** The structure $HClO_2$ is chlorous acid. Binary acids, such as HCl, are given the prefix hydro- in front of the stem of the nonmetallic element and the ending -ic.

    Ternary acids (composed of three elements—usually hydrogen, a nonmetal, and oxygen) usually have a variable oxygen content so the most common member of the series has the ending -ic. The acid with one less oxygen than the -ic acid has the ending -ous. The acid with one less oxygen than the -ous acid has the prefix hypo- and the ending -ous. The acid containing one more oxygen than the -ic acid has the prefix per- and the ending -ic. For example,

HCl hydrochloric acid

HClO hypochlorous acid

$HClO_2$ chlorous acid

$HClO_3$ chloric acid

$HClO_4$ perchloric acid

20. **(B)** A phase diagram relates the pressure and the temperatures at which the gaseous, liquid, and solid states of a material can exist. In the question above we are interested in the relationship between the temperature at which the material changes from a solid to a liquid, which is the melting point, and the pressure. This corresponds to the line indicated by *bc*. It is apparent that this line slants to the right which shows that an increase in the pressure will cause an increase in the temperature at which this material melts.

Note: The point T_C is called the triple point, and at that pressure and temperature the solid, liquid, and gaseous states are in equilibrium with each other.

21. **(B)** The boiling point of a substance is defined as the temperature at which its vapor pressure equals the external pressure. Since the vapor pressure of a substance increases with temperature, the boiling point may be depressed by decreasing the external pressure.

22. **(C)** The melting point of a substance is the same as its freezing point. It is 0°C only for water. Liquids do not melt. The melting point is different for different solids.

23. **(B)** The following are the properties for each of the unit cells in the question:

Unit cell	Edge Length	Angles
monoclinic	$a \neq b \neq c$	$\alpha = \beta = 90° \neq \gamma$
triclinic	$a \neq b \neq c$	$\alpha \neq \beta \neq \gamma$
orthorhombic	$a \neq b \neq c$	$\alpha = \beta = \gamma = 90°$
cubic	$a = b = c$	$\alpha = \beta = \gamma = 90°$
tetragonal	$a = b = c$	$\alpha = \beta = \gamma = 90°$

24. **(A)** "Au" is the symbol for gold, which is malleable (capable of being hammered into a desired shape) and ductile (capable of being drawn into a wire). Metals such as Fe, Cu, Ni, and Pt in the middle of the periodic table have these

properties. Cadmium (Cd), mercury (Hg), potassium (K), and sodium (Na) lack this table location and these properties.

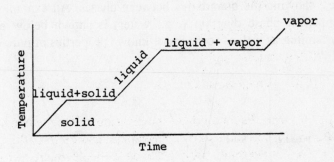

25. **(A)** Region *A* indicates a solid being heated.

26. **(C)** Region *C* indicates a liquid being heated.

27. **(E)** Region *E* indicates a gas being heated.

28. **(D)** Region *D* indicates a liquid undergoing a phase change to a gas at a constant temperature.

29. **(E)** The kinetic energy of a system is directly related to the temperature. Thus, by decreasing the temperature and allowing the substance to condense, we would observe a decrease in kinetic energy. This shows us that the kinetic energy of a solid is less than that of a liquid which is less than that of a gas.

30. **(D)** Sublimation is the phase change from solid to gas without passing through a liquid phase. Other changes of phase are

            solid → liquid   fusion

            liquid → gas   vaporization

            gas → liquid   condensation

            liquid → solid   freezing

SOLIDS REVIEW

1. Phase Diagrams

Phase diagrams are simply graphs, usually on pressure and temperature coordinates, showing the boundaries between phases. An example for a simple single-component phase diagram (e.g., water) is shown below and illustrates some very familiar as well as some lesser known properties of pure compounds.

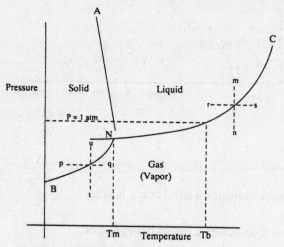

Figure 1 — Phase diagram for single component system

Figure 1.1

The scales on the axes of the sample phase diagram shown above have deliberately been left off so that the diagram applies to any single-component system which has only one solid phase.* If the diagram above were to apply for water, then the normal boiling point, T_b (the temperature where the liquid and gas phases are in equilibrium at 1 atmosphere total pressure), would be 100°C. The melting point, T_m, is 0°C, and at a pressure of one atmosphere, water is a liquid between 0°C and 100°C. At temperatures and pressures represented by points along the lines separating the phases, both phases can exist in equilibrium with each other.

The line *BN* represents the temperatures and pressures where the solid and gas phases are in equilibrium. Similarly, the line *NA* represents the points where the liquid and solid phases are in equilibrium; for water and many other species,

---

* Many single-component systems have more than one solid phase. These are typically different crystalline forms of the component. Sulfur, for example, has a monoclinic and orthorhombic solid crystalline phase. In a phase diagram for such a material, areas of the diagram exist which represent the temperatures and pressures where each of the solid phases is stable.

it slants slightly to the left of vertical as shown in this figure illustrating the effect of pressure on lowering the melting point. The solid-liquid equilibrium boundary between the solid and liquid phases may instead have a positive slope. The line *NC* represents the points where the liquid and gas phases are in equilibrium and this line ends at point *C,* the critical point. Above the critical point (218 atm and 375°C for water), the liquid and gas phases are indistinguishable. In this part of the phase diagram, the material is sometimes called a supercritical fluid since it is not quite correct to refer to it as either a liquid or gas. There is normally no corresponding end point at *A* where the solid and liquid phases become indistinguishable (except, perhaps, at extreme pressures).

Note that at point *N* (0.0098°C and 4.58 torr for water) the solid, liquid, and gas phases meet. At this point, called the triple point, all three phases can exist in equilibrium. The triple point is unique; that is, there is only one point where this can occur, and both the temperature and pressure must be exactly equal to those represented by the coordinates of the triple point. On the other hand, there are many values of temperature and pressure where both liquid and gas or liquid and solid or even gas and solid phases exist in equilibrium. When a solid vaporizes directly without passing through the liquid phase it is called sublimation.

If the temperature is raised at constant pressure along the line *rs,* the material will change from a liquid to a gas. The same thing will happen if the pressure is lowered at constant temperature along the line *mn.* similarly, the solid phase will sublime (change to a gas) if the appropriate state variable (temperature/pressure) changes along the lines *pq* or *ut.*

Phase Equilibrium

In a closed system, when the rates of evaporation and condensation are equal, the system is in phase equilibrium.

In a closed system, when opposing changes are taking place at equal rates, the system is said to be in dynamic equilibrium. Virtually all of the equilibria considered in this review are dynamic equilibria.

PROBLEM

The diagram below is an example of a phase diagram for a pure substance. To what phases do the regions *A, B,* and *C* correspond?

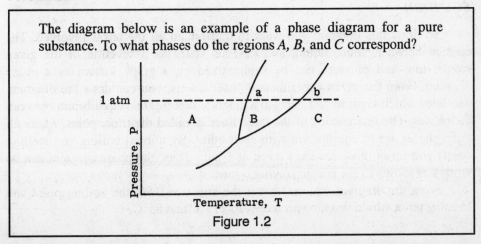

Figure 1.2

SOLUTION

Following the 1 atm constant pressure line from left to right, we are proceeding from low values of the temperature to high values. Therefore, we will intersect the three regions in the order solid-liquid-vapor. The regions *A, B,* and *C* hence correspond to the solid, liquid, and vapor phases, respectively. Point *a* denotes the normal freezing (melting) point of the substance and point *b* denotes the normal boiling point.

PROBLEM

Draw a labelled phase diagram for a substance Z which has the following properties; normal boiling point = 220°C, normal freezing point 80°C, and triple point 60°C and 0.20 atm. Predict the freezing and boiling, if the pressure were 0.80 atm.

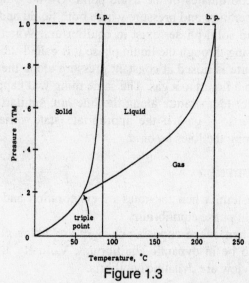

Figure 1.3

SOLUTION

To draw this diagram you want to understand all the terms involved. The relation between solid, liquid, and gaseous states as a function of the given temperature and pressure can be summarized on a graph known as a phase diagram. From the given experimental observations, you can draw the diagram. The lines which separate the states in a diagram represent an equilibrium between the phases. The intersection of the three lines is called the triple point, where all three phases are in equilibrium with each other. By normal boiling and melting points you mean those readings taken at 1 atm. Thus, the phase diagram can be written as shown in the accompanying figure.

From the diagram you see that if the atm was 0.80, the boiling point and freezing point would drop, respectively, to 215°C and 85°C.

Drill 1: Phase Diagrams

1. At a certain temperature and pressure, ice, water, and steam are found to coexist at equilibrium. This pressure and temperature corresponds to

   (A) the critical temperature.   (D) the triple point.

   (B) the critical pressure.   (E) two of the above.

   (C) the sublimation point.

2. The graph below is a cooling curve for compound Y, beginning with gaseous Y. What is the melting point of Y?

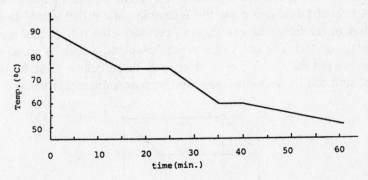

   (A) 90°C   (D) 50°C

   (B) 75°C   (E) None of the above

   (C) 60°C

3. Using the graph for the problem above, what is the boiling temperature for Y?

   (A) 90°C   (D) 45°C

   (B) 60°C   (E) 75°C

   (C) 50°C

2. Crystal Structure: Ionic and Molecular Substances

In general, solids are characterized by their tendency to retain their shape when transferred from one container to another, their virtual incompressibility, as well as their slow rates of diffusion.

In a solid, the attractive forces between the atom's molecules or ions are relatively strong. The particles are held in a rigid structural array, wherein they exhibit only vibrational (as opposed to rotational or translational) motion.

There are two types of solids, amorphous and crystalline. Amorphous substances do not display geometric regularity in the solid; glass is an example of an amorphous solid. Amorphous substances have no sharp melting point, but melt over a wide range of temperatures.

Crystalline solids on the other hand form very ordered, geometrically regular structures. In a crystal structure, the constituent atoms or molecules are in a repeating three-dimensional array. Metals, simple salts, and semiconducting solids like silicon are all crystalline. The most common non-crystalline or amorphous solid is glass. In this solid, the molecules are in a random orientation with respect to each other. Glasses do not have a sharp melting point, rather they have a melting range, thus leading some to classify glass as a supercooled or extremely viscous liquid.

A crystal may be defined as a homogeneous body having the natural shape of a polyhedron. A representative portion of a crystal is called a unit cell. Just as a small swatch of fabric can show the repeating pattern that would be seen on many meters of the fabric, so can the unit cell show the pattern that is repeated throughout the crystal. The unit cell is a parallelepiped, which by variation in its dimensions we get the seven groups of crystals listed below. First, however, let us look at a unit cell for a simple compound such as sodium chloride.

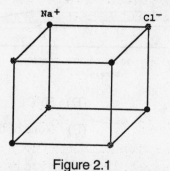

Figure 2.1

Notice the distribution of the sodium and chloride ions. If we were to graph this structure on a rectangular solids coordinate grid, we would notice that the dimensions of the unit cell in the *X, Y,* and *Z* planes are all equal. These are known as sides *a, b,* and *c*. The angles these three sides form with one another (designated as α, β, and γ) are also equal in this instance, and have a value of 90°.

A crystal is said to be body centered if each unit cell contains at its center the same type of atoms found at its corners. A crystal is said to be face centered if the unit cell contains at the midpoints of its faces atoms of the same type that are found at its corners. Crystals that only have atoms at the corners of their lattices (e.g., wall) are called simple.

Different compounds that crystallize with the same structure are said to be isomorphisms.

When one substance can occur in two or more crystalline forms, it is said to be polymorphic.

The principles of stoichiometry, along with information on the size of unit cells, calculate the densities, volumes, and many other properties of crystalline solids. For example, the theoretical density of a crystalline solid can be obtained by:

$$\rho = \frac{\text{mass of unit cell}}{\text{volume of unit cell}}$$

This is the maximum value of a pure crystal; because of imperfections in the crystalline lattice, the measure density is usually slightly less than that calculated by this procedure.

A key solution in calculating the stoichiometric properties of unit cells is to ascertain how many cells share a particular atom or molecule. For cubic lattices, a corner species is shared by eight cells, an edge species by four cells, a face centered species by two cells, and a body-centered species by only one cell.

In crystals, the atoms or molecules remain in a fixed orientation with respect to each other. The distances between atoms can be determined from X-ray diffraction patterns because the distances are comparable in magnitude to the wavelengths of X-rays. Hence, X-ray diffraction patterns are determined by the regular spacings between the atoms or molecules.

Crystals may be broken up into two types: ionic and molecular. Following is a listing of various properties of the two types.

Ionic Crystals

1. Ionic crystals have large lattice energies because the electrostatic forces between them are strong.

2. In the solid phase, they are poor electrical conductors.

3. In the liquid phase, they are relatively good conductors of electric current; the mobile charges are the ions (in contrast to metallic conduction, where the electrons constitute the mobile charges).

4. They have relatively high melting and boiling points.

5. They are relatively non-volatile and have low vapor pressure.

6. They are brittle.

7. Those that are soluble in water form electrolytic solutions that are good conductors of electricity.

Molecular Crystals and/or Liquids

1. Molecular crystals tend to have small lattice energies and are easily deformed because their constituent molecules have relatively weak forces between them.

2. Both the solids and liquids are poor electrical conductors.

3. Many exist as gases at room temperature and atmospheric pressure; those that are solid or liquid at room temperature are relatively volatile.

4. Both the solids and liquids have low melting and boiling points.

5. The solids are generally soft and have a waxy consistency.

6. A large amount of energy is often required to chemically decompose the solids and liquids into simpler substances.

PROBLEM

Distinguish between crystalline and amorphous solid substances, using some specific examples. To what extent is the distinction useful?

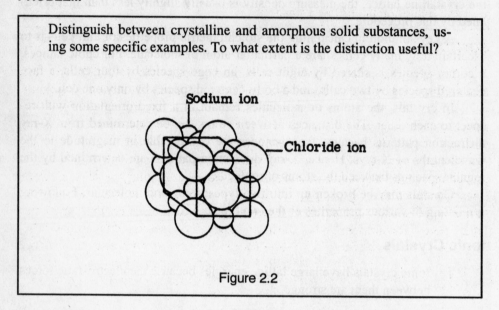

Figure 2.2

SOLUTION

Crystalline substances can be generally thought of as species composed of structural units with specific geometric patterns. The accompanying drawing of sodium chloride would be an example of such a pattern. The important point is that there exists a regularity in the arrangement of structural units. Structures with regularity generally show a sharp and characteristic melting point, which is the case with crystalline substances.

Amorphous substances, however, tend to be shapeless and without definite order. That is, you have a randomness. For example, glassy or glasslike materials such as Plexiglas and silicate glasses. In substances with a general lack of order, the melting points vary over a range or temperature interval. For amorphous substances, this is exactly what you find. It would, however, be *incorrect* to state categorically that amorphous substances are without ANY order. For they do tend to have short range order even though they do contain long-range randomness.

PROBLEM

Excluding hexagonal unit cells, when counting the number of points inside a cell, a point on an edge is $1/4$ inside the cell, and a lattice point at a corner is $1/8$ inside the cell. Justify these fractions. Calculate, also, the net number of lattice points in the following unit cells: simple cubic, body-centered cubic, face-centered cubic, and tetragonal. (See the figures on the following page.)

SOLUTION

To solve this problem properly you need to know the definition of a unit cell and lattice point. You need, also, to know the actual structures of the unit cells given. You proceed as follows: A unit cell is that small fraction of a space lattice, which sets the pattern for the whole lattice. In other words, it is the smallest portion of the space lattice (which is just the pattern of points which describes the arrangement of atoms or molecules in the crystal), which, when moved repeatedly a distance equal to its own dimensions along the various directions, generates the whole space lattice equivalent to the original lattice. The lattice is, of course, just points that denote atoms or molecules; thus, the term lattice point. The accompanying illustrations give you the structures in question. From this, you can obtain justification for the fractions and find the net number of lattice points. For example, take the simple cubic structure. In a space lattice, each atoms shared with three other unit structures as seen in figure 2.3(D).

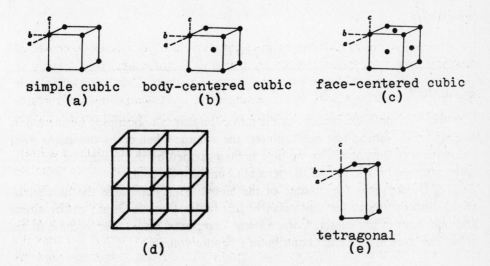

simple cubic   body-centered cubic   face-centered cubic
(a)   (b)   (c)

(d)   tetragonal
(e)

Figure 2.3

Note: This point is shared by eight unit cells, if you place another 4 cubes on the side. Thus, each point is only $1/8$ in the cell. This same sort of procedure and reasoning can be used to justify all fractions in all unit cells given. Calculations now become easy.

Simple cubic: Have eight lattice points. Each is only $1/8$ inside the cell. Thus, you have a net of $(1/8)(8) = 1$.

Face-centered cubic: Eight lattice points at corners, each only $1/8$ inside cell. Have a point on each face. Each point is only $1/2$ inside cell. Have six points. Thus, net = $(8)(1/8) + (6)(1/2) = 4$.

Body-centered: Again, have $(8)(1/8)$ from lattice points at corners. One point in center, which is entirely in cell. Thus, net becomes $(8)(1/8) + (1)(1) = 2$.

Tetragonal: Only have a net of $(8)(1/8)$ from lattice points at corners.

PROBLEM

What fraction of the total space in a body-centered cubic unit cell is unoccupied? Assume that the central atom touches each of the eight corner atoms of the cube.

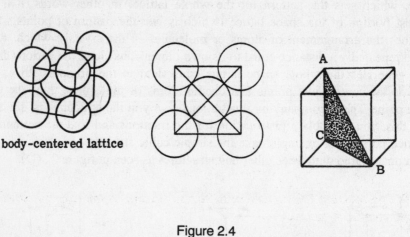

body-centered lattice

Figure 2.4

SOLUTION

To determine the percent of the unit cube that is unoccupied by the atoms making up the lattice, one must subtract the volume taken up by the atoms from the volume of the cube. This volume is then divided by the volume of the cube and multiplied by 100 to find the percent of unoccupied space.

1) Determining the volume of the atoms: First determine the number of atoms contributing to the unit cube. In this lattice there are eight corner atoms and one atom in the center. Corner atoms contribute $1/8$ of their volume to the cube. The atom in the center contributes its entire volume.

$$\text{no. of atoms in unit cube} = (\frac{1}{8} \text{ atom/corner} \times 8 \text{ corners})$$

$$+ 1 \text{ atom (in center)}$$

$$= 2 \text{ atoms}$$

Thus, the volume in the unit cube taken up by the atoms is equal to the volume of two atoms. The radius of these atoms is taken to be 1. The volume of a sphere is $^4/_3\,\pi r^3$.

$$\text{volume of 2 atoms} = 2 \times \frac{4}{3}\,\pi(1)^3 = 8.38$$

2) Volume of the cube: The corner atoms are assumed to be touching the central atom. The diagonal of the cube can be visualized as shown in Figure 2.4.

Because $\overline{AB}$ is shown to be $4r$ or 4 the side of the cube $\overline{AC}$ can be found using the Pythagorean Theorem. Once $\overline{AC}$ is known the volume of the cube can be found. From geometry, it is known that

$$\overline{CB} = \overline{AC} \times \sqrt{2} \qquad\qquad 4^2 = \overline{AC}^2 + 2(\overline{AC})^2$$

$$\overline{AB}^2 = \overline{AC}^2 + \overline{CB}^2 \qquad\qquad 4^2 = 3\overline{AC}^2$$

$$\overline{AB} = 4 \qquad\qquad\qquad\quad \frac{4}{\sqrt{3}} = \overline{AC}$$

$$CB = \overline{AC} \times \sqrt{2}$$

The volume of the cube is equal to the length of the side cubed.

$$\text{volume of cube} = \left(\frac{4}{\sqrt{3}}\right)^3 = 12.32$$

3) The space in the cube is equal to the volume of the spheres subtracted from the volume of the cube.

$$\text{vol. of space} = \text{vol. of cube} - \text{vol. of spheres}$$

$$\text{vol. of space} = 12.32 - 8.38 = 3.94$$

$$\text{percentage of cube taken up by space} = \frac{3.94}{12.32} \times 100 = 32\%$$

Drill 2: Crystal Structure: Ionic and Molecular Substances

Questions 1–3 refer to the accompanying figure. Determine the number of unit particles (atoms) in a unit cell of each of the three type lattices.

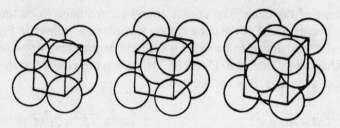

Unit cells in cubic lattices; (1) simple cubic lattice; (2) body-centered lattice; (3) face-centered lattice.

(A) 1 atom

(D) 4 atoms

(B) 2 atoms

(E) 5 atoms

(C) 3 atoms

4. Iron may crystallize in the face-centered cubic system. If the radius of an Fe atom is 1.26Å, determine the length of the unit cell.

(A) 5.04Å

(D) 10.08Å

(B) 3.55Å

(E) 2.12Å

(C) 1.26Å

5. Given the situation in the above question, calculate the density of Fe if its atomic weight is 55.85.

(A) 8.30 g/cm³

(D) 1.92 g/cm³

(B) 5.85 g/cm³

(E) 2.12 g/cm³

(C) 4.12 g/cm³

SOLIDS DRILLS

ANSWER KEY

Drill 1 — Phase Diagrams

1. (D)
2. (C)
3. (E)

Drill 2 — Crystal Structure: Ionic and Molecular Substances

1. (A)
2. (B)
3. (D)
4. (B)
5. (A)

GLOSSARY: SOLIDS

Amorphous Solid

A solid which does not display geometric regularity. They melt over a wide range of temperatures rather than sharply at one temperature.

Body-Centered Crystal

A crystal in which the center of the unit cell contains the same type of atom that is found at the corners.

Critical Pressure

The pressure at which a substance changes from liquid to vapor or from vapor to liquid and the two are no longer in equilibrium.

Critical Temperature

The temperature at which a substance changes from liquid to vapor or from vapor to liquid and the two are no longer in equilibrium.

Crystalline Solid

A solid which is characterized by structural units that are bounded by regular geometric patterns. They have sharp melting points.

Equilibrium

A situation in which the rate of a forward reaction is equal to that of the backward reaction. There is then no net change in the quantity of either products or reactants.

Face-Centered Crystal

A crystal in which the midpoints of the faces of the unit cell are occupied by the same type of atom that is found at the corner of the unit cell.

Gas

A substance in the vapor phase whose atoms or molecules are kept apart by thermal motion.

Isomorphous

Different compounds that crystallize with the same structure are termed isomorphous.

Phase Diagram

A plot of temperature verses pressure which displays the three phases of a substance (gas, liquid, solid) and the places of equilibria between these phases.

Phase Equilibrium

When rates of evaporation and condensation are equal in a closed system.

Sublimation

The conversion of a solid directly to the gas phase without going through the liquid phase.

Triple Point

The point at which all three phases (gas, liquid, and solid) are all in equilibrium.

Unit Cell

A representative portion of a crystal.

CHAPTER 7

Properties of Liquids

➤ Diagnostic Test
➤ Properties of Liquids
   Review & Drills
➤ Glossary

PROPERTIES OF LIQUIDS DIAGNOSTIC TEST

1. Ⓐ Ⓑ Ⓒ Ⓓ Ⓔ
2. Ⓐ Ⓑ Ⓒ Ⓓ Ⓔ
3. Ⓐ Ⓑ Ⓒ Ⓓ Ⓔ
4. Ⓐ Ⓑ Ⓒ Ⓓ Ⓔ
5. Ⓐ Ⓑ Ⓒ Ⓓ Ⓔ
6. Ⓐ Ⓑ Ⓒ Ⓓ Ⓔ
7. Ⓐ Ⓑ Ⓒ Ⓓ Ⓔ
8. Ⓐ Ⓑ Ⓒ Ⓓ Ⓔ
9. Ⓐ Ⓑ Ⓒ Ⓓ Ⓔ
10. Ⓐ Ⓑ Ⓒ Ⓓ Ⓔ
11. Ⓐ Ⓑ Ⓒ Ⓓ Ⓔ
12. Ⓐ Ⓑ Ⓒ Ⓓ Ⓔ
13. Ⓐ Ⓑ Ⓒ Ⓓ Ⓔ
14. Ⓐ Ⓑ Ⓒ Ⓓ Ⓔ
15. Ⓐ Ⓑ Ⓒ Ⓓ Ⓔ
16. Ⓐ Ⓑ Ⓒ Ⓓ Ⓔ
17. Ⓐ Ⓑ Ⓒ Ⓓ Ⓔ
18. Ⓐ Ⓑ Ⓒ Ⓓ Ⓔ
19. Ⓐ Ⓑ Ⓒ Ⓓ Ⓔ
20. Ⓐ Ⓑ Ⓒ Ⓓ Ⓔ
21. Ⓐ Ⓑ Ⓒ Ⓓ Ⓔ
22. Ⓐ Ⓑ Ⓒ Ⓓ Ⓔ
23. Ⓐ Ⓑ Ⓒ Ⓓ Ⓔ
24. Ⓐ Ⓑ Ⓒ Ⓓ Ⓔ
25. Ⓐ Ⓑ Ⓒ Ⓓ Ⓔ

PROPERTIES OF LIQUIDS
DIAGNOSTIC TEST

This diagnostic test is designed to help you determine your strengths and your weaknesses in properties of liquids. Follow the directions and check your answers.

Study this chapter for the following tests:
ACT, AP Chemistry, CLEP General Chemistry,
GED, MCAT, MSAT, PRAXIS II Subject Assessment:
Chemistry, SAT II: Chemistry, GRE Chemistry

---

25 Questions

DIRECTIONS: Choose the correct answer for each of the following problems. Fill in each answer on the answer sheet.

1. All of the following are colligative properties except

   (A) osmotic pressure.
   (B) pH of buffer solutions.
   (C) vapor pressure.
   (D) boiling point elevation.
   (E) freezing point depression.

2. Which of the following is not a colligative property of a solution?

   (A) Freezing point
   (B) Vapor pressure over the solution
   (C) Molecular weight of the solute
   (D) Boiling point
   (E) None of the above.

3. Suppose that solutions are made up in 1 molal concentrations for five substances. Which of these solutions would have the lowest freezing point?

   (A) NaBr
   (B) $C_6H_{12}O_6$
   (C) NaCl
   (D) Na_2SO_4
   (E) CH_3COOH

4. Which of the following liquid solutions show positive deviation from Raoult's law?

    (A) Benzene and toluene

    (B) Water and ethanol

    (C) Pentane and hexane

    (D) Acetone and carbon disulfide

    (E) Acetone and chloroform

5. Which of the following salts will have the greatest freezing point depression for a .1 M solution of the salt?

    (A) Na_2SO_4

    (B) NaCl

    (C) KCl

    (D) KNO_3

    (E) NH_4Cl

6. How many moles of sodium chloride should be dissolved in one liter of water so that it freezes at $-1.86°C$? Molal freezing constant for water is $1.86°C$ kg/mol.

    (A) 0.25

    (B) 0.50

    (C) 1.00

    (D) 1.50

    (E) 2.00

7. Which of the following methods is best suited to separate a 500 ml sample of two miscible liquids whose boiling points differ by approximately 60°?

    (A) Distillation

    (B) Fractional distillation

    (C) Paper chromatography

    (D) Use of a separatory funnel

    (E) None of the above.

8. Benzene and toluene form nearly ideal solutions. The vapor pressure of pure toluene is 22 torr at 20°C. For an equimolar mixture of benzene and toluene at 20°C the vapor pressure of toluene is

    (A) 5.5 torr.

    (B) 7.3 torr.

    (C) 11 torr.

    (D) 22 torr.

    (E) 44 torr.

9. A solution is made by combining 1 mole of ethanol and 2 moles of water. What is the total vapor pressure above the solution? (At the same temperature, the vapor pressure of pure ethanol is .53 atm and the vapor pressure of water is .24 atm.)

    (A) .34 atm

    (B) .41 atm

(C) .43 atm   (D) .56 atm

(E) .77 atm

10. Consider a mixture of two liquids A and B in a non-ideal solution. Under which conditions will Henry's law accurately describe the vapor pressure of component B?

(A) The vapor pressures of pure A and pure B are very similar.

(B) The vapor pressure of pure A is very low.

(C) The vapor pressure of pure B is very low.

(D) The mole fraction of B in the mixture is close to zero.

(E) The mole fraction of B in the mixture is close to unity.

11. Water has a vapor pressure of 23.76 torr at 25°C. What is the vapor pressure of a solution of sucrose, if the mole fraction of sucrose is 0.250?

(A) 15.2 torr   (D) 5.9 torr

(B) 23.8 torr   (E) 17.8 torr

(C) 29.7 torr

12. A certain substance with molecular weight 62 g × mole^{-1} will cause an aqueous solution to have a change of freezing point of 1.86°/m, where m stands for molal. What would be the change in the freezing point of an aqueous solution containing 200 g of H_2O, if we added 8 g of this substance?

(A) 2.4°C   (D) 0.6°C

(B) 1.2°C   (E) Cannot be determined.

(C) 3.6°C

13. The average osmotic pressure π, of blood is 12.3 atm at 27°C. What concentration of glucose, $C_6H_{12}O_6$, will be isotonic with blood?

(A) 2.0 M   (D) 0.025 M

(B) 0.5 M   (E) 1.5 M

(C) 1.0 M

14. An important application of colligative properties of solutions is

(A) the determination of boiling points.

(B) the determination of heat of fusion.

(C) the determination of molecular weight.

(D) the determination of atomic weight.

(E) the evaluation of electronegativities.

15. The height to which a liquid will rise in an open capillary tube is inversely proportional to:

(A) temperature of liquid.   (D) surface tension.

(B) density of liquid.   (E) viscosity of the liquid.

(C) air pressure.

16. The vapor pressure of pure acetone is 347 mm Hg. A mixture of 58.0 g of acetone and 2.0 g of water is made. According to Raoult's law, what is the partial vapor pressure of the acetone in this mixture?

(A) 382 mm Hg   (D) 298 mm Hg

(B) 335 mm Hg   (E) 242 mm Hg

(C) 312 mm Hg

17. If oxygen is collected over water at 25°C and at a pressure of 760 torr, the pressure due to just the oxygen is: (The vapor pressure of H_2O at 25°C is 19.0 torr.)

(A) 779 torr.   (D) 741 torr.

(B) 760 torr.   (E) Insufficient data to solve.

(C) 19 torr.

18. One mole of a substance dissolved in 1,000 grams of water elevates the boiling point by .52°C and depresses the freezing point by 1.86°C. Twenty-three grams of an alcohol, C_2H_5OH, is dissolved in a kilogram of water. At standard conditions, water's new boiling and freezing points are respectively (in °C):

(A) 100.26°, –.93°   (D) 100.52°, –1.86°

(B) .26°, –1.86°   (E) 101.04°, –1.86°

(C) .52°, –.93°

19. What is the boiling point of an aqueous solution containing 117 g of NaCl in 1,000 g of H_2O?

k_b (H_2O) = 0.52°C kg/mol

(A) 98.96°C   (B) 99.48°C

(C) 100.52°C   (D) 101.04°C

(E) 102.08°C

20. The vapor pressure of the water in a 1 M NaCl solution is _____ the vapor pressure of pure water.

(A) less than

(B) greater than

(C) the same as

(D) not determinable in this question by

(E) always four times

21. Ethanol boils at 78.5°C. If 34.2 g of sucrose (MW = 342) is dissolved in 200 g of ethanol, at what temperature will the solution boil? (Assume k_b = 1.20°C/m for the alcohol.)

(A) 79.1°C   (D) 78.56°C

(B) 77.9°C   (E) 84.5°C

(C) 0.60°C

22. How many grams of Na_2CO_3 must be dissolved in 1,000 g of water to produce a solution that boils at 103.06°C? (Molal boiling point constant for water is 0.51°C kg/mol; MW Na_2CO_3 = 106 g/mol)

(A) 27 g   (D) 159 g

(B) 53 g   (E) 212 g

(C) 106 g

23. How many grams of HCl must be added to 500 ml of water to produce a solution that freezes at –1.86°C? (Molal freezing constant = 1.86°C kg/mol.).

(A) 4.6   (D) 36.5

(B) 9.1   (E) 73.0

(C) 18.3

24. What would be the freezing point of a 2 liter solution containing 58.5 g NaCl in water? (The molal freezing point coefficient for water is = 1.86°C kg/mol.)

(A) 0°C   (B) – 0.47°C

(C) −0.93°C   (D) −1.86°C

(E) −3.72°C

25. What would be the freezing point of a solution containing 8.0 g of ethylene glycol ($C_2H_6O_2$) in 100 g of H_2O? (The K_f for H_2O is 1.86°C/m and the freezing point of pure water is 0° C.)

(A) 0.0°C   (D) +0.2°C

(B) −1.8°C   (E) −2.4°C

(C) +2.4°C

PROPERTIES OF LIQUIDS
DIAGNOSTIC TEST

ANSWER KEY

1.	(B)	8.	(C)	15.	(B)	22.	(E)
2.	(C)	9.	(A)	16.	(C)	23.	(B)
3.	(D)	10.	(D)	17.	(D)	24.	(D)
4.	(D)	11.	(E)	18.	(A)	25.	(E)
5.	(A)	12.	(B)	19.	(E)		
6.	(B)	13.	(B)	20.	(A)		
7.	(A)	14.	(C)	21.	(A)		

DETAILED EXPLANATIONS
OF ANSWERS

1.   **(B)**   pH is not a colligative property as it depends on specific properties of acids and bases (k_a, k_b) and not just their concentration.

2.   **(C)**   Physical properties of solutions that only depend on the number of particles that are present, and not the kind of particles that are present, are referred to as colligative properties.

It has been shown that a non-volatile solute in a solution will retard the rate of escape of solvent molecules in the solution. The total number of solvent molecules per unit area on the surface is reduced because of the presence of the solute. This results in a lowering of the vapor pressure of the solution. The lowering of the vapor pressure depresses the freezing point of the solution from that of only the pure solvent. The vapor pressure lowering is Raoult's law, which states that the vapor pressure of a solvent in a solution decreases as the mole fraction of the solution decreases. The molecular weight of a solute does not effect the total number of particles in a solution. For example, 1 mole of NaCl and 1 mole of KCl will both produce the same number of particles when dissolved in a solvent, but NaCl and KCl have different molecular weights.

3.   **(D)**   Freezing point depression is a colligative property—it depends on the number of moles of particles (ions, molecules, etc.) in solution. The 1 molal Na_2SO_4 ionizes into a 3 molal concentration of ions ($1Na_2SO_4 \rightarrow 2Na^+ + SO_4^{2-}$). This is greater than any of the other choices and will result in the solution with the lowest freezing point.

4.   **(D)**   Raoult's law states that $P_A = X_A P_{A\ partial}$ where P^A_A is the partial vapor pressure of a component in solution, X_A is the mole fraction of component A, and P_A is the vapor pressure of pure substance A. Solutions that obey Raoult's law are called ideal. Ideal solutions form between substances in which solute-solute, solute-solvent, and solvent-solvent interactions are the same

If unlike molecules interact less strongly than like molecules do, the partial vapor pressures of the components will be greater than predicated by Raoult's law. This is known as positive deviation. Of the responses listed, only the solution of acetone and carbon disulfide would show positive deviation.

Choices (A) and (C) are examples of ideal solutions; choices (B) and (E) are examples of negative deviation.

5.   **(A)**   The freezing point depression or boiling point elevation of a solution depends on the number of particles present in the solution. A solution that contains a strong electrolyte, such as the salts listed above, will depress the freezing point according to the number of particles that are produced when the

salt dissolves and ionizes completely. When Na_2SO_4 ionizes, it forms $2Na^+$ and $1SO_4^{2-}$ ions:

$$Na_2SO_4 \rightarrow 2NA^+ + 1SO_4^{2-}$$

So the effective molality of the solution will be three times the molality of the undissociated Na_2SO_4. The other salts will only produce an effective molality two times the undissociated form of the salt.

$$NaCl \rightarrow Na^+ + Cl^-$$

Therefore, the freezing point depression would be the greatest for Na_2SO_4.

6.   **(B)**   The freezing point depression observed for a 1 molal aqueous solution is $1.86°C$. Molality is defined as the number of moles of particles dissolved in 1 kg of solvent. One liter of water is approximately 1 kg since the density of water is approximately 1 g/cm^3. We require 0.5 mole of sodium chloride since NaCl dissociates into two particles in solution and one mole of particles is required to produce an aqueous solution freezing at $-1.86°C$.

This can also be demonstrated by substitution in

$$\Delta T_o = m \times k_f \times \#particles$$

$$1.86°\ C = m \times 1.86\ (°\ C\ kg/mol) \times 2$$

$$\therefore\ m = \text{molality of NaCl} = 0.5\ \text{mol/kg}$$

Since we have 1 kg of water then,

$$0.5\ (\text{mol/kg}) \times 1\ \text{kg} = 0.25\ \text{moles NaCl}$$

7.   **(A)**   A simple distillation is sufficient to separate the mixture into its components. Fractional distillation would be the method of choice if the boiling points of the components were similar. Paper chromatography separates compounds by virtue of their differing amounts of interaction with the solvent used and has no use in separating this mixture. In addition, only small quantities of testing material are reasonable for paper chromatography. A separatory phase is used to extract a compound from one immiscible liquid to another and is not suited for this purpose.

8.   **(C)**   For ideal solutions, the vapor pressure of each component is given by Raoult's law:

$$P = XP^*$$

where X is the mole fraction of the component and P^* is the vapor pressure of the pure component. In this case $X = 0.5$ and $P^* = 22$ torr, therefore:

$$P = XP^* = (0.5)\ (22\ \text{torr}) = 11\ \text{torr}$$

9. **(A)** Applying Raoult's law, we can solve the problem.

$$P_A = P_A°X_A$$

P_A = vapor pressure of compound A above the solution

P_Ao = vapor pressure of pure compound A,

X_A = mole fraction of compound A.

Let A = ethanol, B = water.

$$P_{total} = P_A + P_B = P_A°X_A + P_B°X_B$$

$$= (.53)(.33) + (.24)(.67)$$

$$= .34 \text{ atm}$$

10. **(D)**

The graph shows the vapor pressure of component B for non-ideal mixtures of A and B, where X_B is the mole fraction of B. For X_B close to unity, Raoult's law applies:

$$P_B = X_B P^*_B$$

where P_B^* is the vapor pressure of pure B. For X_B close to zero, Henry's law applies:

$$P_B = X_B k_B$$

where k_B is an experimentally determined constant.

11. **(E)** The vapor pressure of a solution containing a volatile solvent and a nonvolatile solute is given by:

$$P_{solution} = P_1^0(1 - x_2)$$

So $P_{solution} = (23.76 \text{ torr})(1 - 0.250) = 17.8 \text{ torr}$

12. **(B)** The first step consists of determining the molarity of the solution, which is obtained by dividing the mass of the solute by its molecular weight:

$$\text{molarity} = \frac{8}{62} = 0.13 \text{ mole}$$

To obtain the molality, it is necessary to divide the molarity by the mass of solvent as below:

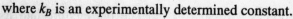
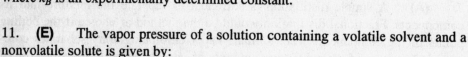

$$\text{molality} = \frac{\text{molarity}}{k_p \text{ of solvent}} = \frac{0.13}{0.2} = 0.645 \text{ m}$$

The change in freezing point Δt_f is:

$$\frac{1.86°C}{M} \times 0.645M = 1.2°C$$

13.   **(B)**   Osmotic pressure (π) where $\pi = MRT$ and M is the molarity. R is the ideal gas constant and T is the temperature in degrees Kelvin. The concentration of glucose that will be isotonic (i.e., of the same osmotic pressure) with blood can be determined as follows:

$$\pi = MRT$$

$$M = \frac{\pi}{RT}$$

$$= \frac{12.3 \text{ atm}}{(0.082 \frac{1\text{-atm}}{\text{k-mol}})(300° K)}$$

$$= 0.50M$$

14.   **(C)**   Properties of a solution that depend only on the concentration of the solute and not on its nature, are called colligative properties. Examples are the freezing-point depression, boiling point elevation, osmotic pressure, and vapor pressure depression. Each of these can be used to find the molecular weight of an unknown solute.

15.   **(B)**

$$h = \frac{2\gamma}{\rho \text{ gr}}$$

where

γ = surface tension

r = radius

ρ = density

g = gravity

Thus, height is inversely proportional to density.

16.   **(C)**   According to Raoult's law, $P_A = x_A P_A{}^*$, where P_A is the partial vapor pressure of component A, x is the mole fraction of A, and $P_A{}^*$ is the vapor pressure of pure A. Raoult's law is fairly accurate when x_A is close to unity. We are given that $P_A{}^* = 347$ mm Hg. We must now calculate x_A:

$$\text{moles of acetone} = \frac{58.0\,\text{g}}{58.0\,\text{g/mol}} = 1\,\text{mole}$$

$$\text{moles of water} = \frac{2.0\,\text{g}}{18.0\,\text{g/mol}} = \frac{1}{9}\,\text{mole}$$

$$\text{mole fraction of acetone} = \frac{1.0\,\text{mole}}{(1+\frac{1}{9})\,\text{mole}} = \frac{9}{10}$$

Our answer, P_A, is thus

$$P_A = x_A P^*_A = (.9)\,(347\,\text{mm Hg}) = 312\,\text{mm Hg}$$

17. **(D)**   According to Dalton's law of partial pressures, the total pressure is equal to the sum of the partial pressures.

$$P_T = P_a + P_b + P_c + \dots$$

In this case there are two gases present in the container, the oxygen and the water vapor. Since the vapor pressure of water is 19 torr at 25°C, the partial pressure of the water vapor is 19 torr, and the total pressure in the container is 760 torr. From Dalton's law, we can calculate the pressure due to just the oxygen.

$$P_{\text{total}} = P_{O_2} + P_{H_2O\ \text{vapor}}$$

$$760\,\text{torr} = P_{O_2} + 19\,\text{torr}$$

$$P_{O_2} = 760\,\text{torr} - 19\,\text{torr}$$

$$= 741\,\text{torr}$$

18. **(A)**   The alcohol's molecular weight is 46: $24(C_2) + 6(6H) + 16(O)$. Twenty-three grams is one-half mole. It, therefore, changes boiling and freezing points by one-half the stated increments.

19. **(E)**   Converting to moles,

$$117\,\text{g of NaCl} \times \frac{1\,\text{mole of NaCl}}{58.5\,\text{g of NaCl}} = 2\,\text{moles of NaCl}.$$

The molality of a solution is defined as the number of moles of dissolved solute per 1,000 g of solvent. Therefore, the molality of the solution is 2 m in NaCl. However, since NaCl dissociates completely to Na^+ and Cl^-, the molality of the solution is 4 m in particles. It has been found that a 1 m aqueous solution freezes at −1.86°C and boils at 100.52°C, a change of −1.86°C and +0.52°C, respectively. Thus, the boiling point increase for a 4 m solution (since boiling point elevation is a colligative property) is

$$4 \text{ m} \times \frac{0.52°\text{C}}{1 \text{ m}} = 2.08°\text{C}.$$

Therefore, the boiling point of the solution is

$$100°\text{C} + 2.08°\text{C} = 102.08°\text{C}.$$

20. **(A)** The vapor pressure of a solvent is lowered when a solute is added.

21. **(A)**

$$\Delta T_b = K_b n$$

MW of sucrose: MW = 342

$$\text{moles of solute} = \frac{\text{g of solute}}{\text{MW of solute}} = \frac{34.2 \text{ g}}{342 \text{ g/mol}}$$

$$= 0.10 \text{ moles}$$

$$200 \text{ g of solvent} = \frac{200 \text{ g}}{1,000 \text{ g}} \times 1 \text{ kg} = 0.20 \text{ kg}$$

$$\text{Molality} = \frac{\text{moles solute}}{1 \text{ kg solvent}} = \frac{0.10 \text{ moles}}{0.20 \text{ kg}} = 0.5 \text{ m}$$

Therefore, the elevation of the boiling point is

$$\Delta T_b = K_b n = (1.20°\text{C/m})(0.5 \text{ m}) = .60°\text{C}$$

Thus, the boiling point of the solution is $78.5°\text{C} + 0.6°\text{C} = 79.1°\text{C}$.

22. **(E)** If m = molality, k_b is the boiling point constant, and ΔT is the change in the boiling point, then

$$\Delta T = \text{m} \times k_b \times \text{number of effective particles}.$$

Since Na_2CO_3 in water dissociates to 2, Na + $1CO_3^{-2}$, then we would have 3 effective particles. Substituting

$$3.06 \text{ (°C)} = \text{m (mol/kg)} \times 0.51 \text{ (°C kg/mol)} \times 3$$

$$\therefore \text{ m} = 2 \text{ moles } Na_2CO_3/\text{kg water}$$

Since we have 1 kg of solvent then,

$$2 \text{ (mole/kg)} \times 1 \text{ kg} = 2 \text{ moles } Na_2CO_3$$

Finally, converting moles to grams:

$$2 \text{ moles} \times 106 \text{ g/mol} = 212 \text{ g } Na_2CO_3$$

23. **(B)** If ΔT is the change in the melting point, k_m is the melting or freezing point constant, and m = molality, then

$$\Delta T = m \times k_m \times \text{number of effective particles.}$$

Since HCl dissociates into H^+ and Cl^- then substituting we obtain:

$$1.86 \ (°C) = m \ (\text{mol/kg}) \times 1.86 \ (°C \ \text{kg/mol}) \times 2$$

∴ $m = 0.5$ = moles of HCl (solute)/kg of water (solvent). Since we have 0.5 kg of water then, 0.5 (mol/kg) × 0.5 kg = 0.25 mole HCl. Finally, converting 0.25 mole to grams, 0.25 mole × 36.5 g/mol = 9.1 g HCl.

24. **(D)** Converting to moles of NaCl:

$$58.5 \text{ g NaCl} \times \frac{1 \text{ mole NaCl}}{58.5 \text{ g NaCl}} = 1 \text{ mole NaCl}$$

$$\Delta T = k_f \times m \times \text{\# ions or particles per mole}$$

Since $NaCl \rightarrow Na^+ + Cl^-$ then # of effective particles is 2.

$$\Delta T = (18.6°C \text{ kg/mol}) \times (1 \text{ mol/2kg}) \times 2$$

∴ $\Delta T = 1.86°C$, and

$$T = -1.86°C$$

25. **(E)** The presence of a non-volatile solute such as $C_2H_6O_2$ will depress the freezing point of the solution according to Raoult's law. The freezing point depression is given by the expression:

$$\Delta T = k_f \times m$$

where k_f is the freezing point depression constant, ΔT is the change in the freezing point, and m is the molality of the solution, or

$$m = \frac{\text{moles of solute}}{\text{kg of solvent}}$$

and since the number of moles equals

$$\frac{\text{g of solute}}{\text{MW of solute}}$$

we now have

$$m = \frac{\frac{\text{g of solute}}{\text{MW of solute}}}{\text{kg of solvent}}$$

Substituting this expression for molality in the first equation we have,

$$\Delta T = k_f \ \frac{\frac{\text{g of solute}}{\text{MW of solute}}}{\text{kg of solvent}}$$

$$\Delta T = 1.86 \left(\frac{\frac{8.0}{62.0}}{.1} \right)$$

$$\Delta T = 2.4$$

Since the normal freezing point of water is 0.0°C, the freezing point of this solution will be 2.4°C lower than pure H_2O or −2.4°C.

PROPERTIES OF LIQUIDS
REVIEW

Liquids differ from solids in that the molecular structure is random, and the liquid will normally assume the shape of the vessel or container in which it is placed.

In a liquid, the attractive forces hold the molecules close together, so that increasing the pressure has little effect on the volume. Therefore, liquids are incompressible. Changes in temperature cause only small volume changes.

Liquids diffuse much more slowly than gases because of the constant interruptions in the short mean free paths between molecules. (The rates of diffusion in liquids are more rapid at higher temperatures.)

The strength of the inward forces of a liquid is called the liquid's surface tension. Surface tension decreases as the temperature is raised.

Increases in temperature increase the average kinetic energy of molecules and the rapidity of their movement. If a particular molecule gains enough kinetic energy when it is near the surface of a liquid, it can overcome the attractive forces of the liquid phase and escape into the gaseous phase. This is called a change of phase (specifically, evaporation).

1. Density

The mass per unit volume occupied by a liquid is an important property used to characterize a liquid. Liquids are nearly incompressible and require extreme pressures to make a significant difference in the density. Also, only small differences in the density occur with moderate changes in the temperature; hence the density of a liquid is frequently considered to be a constant, except when very precise values are required. Water has a density of 1 gram per cubic centimeter; common hydrocarbons, such as those in gasoline, have densities of approximately 0.7 g/cc and float on water, while chloroform ($CHCl_3$), a common organic solvent, has a density of 1.5 g/cc and sinks in water.

PROBLEM

What volume of a block of wood (density = 0.80 g/cm^3), which weighs 1.0 kg, will be above the water surface when the block is afloat? (Note: A floating object displaces its own weight of water.)

SOLUTION

Since a floating object displaces its own weight in water, 1 kg of water is displaced by this block of wood. One can find the volume of the block of wood above the water by solving for the volume of the block and subtracting the volume of 1 kg of water from it. One uses the density to solve for the volume.

$$\text{density} = \frac{\text{weight}}{\text{volume}} = \frac{\text{g}}{\text{cm}^3}$$

Therefore:

$$\text{volume} = \frac{\text{weight}}{\text{density}}$$

Solving for the volume of the wood:

$$1 \text{ kg} = 1,000 \text{ g}$$

$$\text{volume} = \frac{1,000 \text{ g}}{.80 \text{ g/gm}^3} = 1.25 \times 10^3 \text{ cm}^3$$

Solving for the volume of the water displaced:

By definition the density of water is 1.0 g/cm^3.

$$\text{volume} = \frac{1,000 \text{ g}}{1.0 \text{ g/gm}^3} = 1.00 \times 10^3 \text{ cm}^3$$

volume of wood above water = volume of wood – volume of water

volume of wood above water = $1,250 \text{ cm}^3 - 1,000 \text{ cm}^3$

$$= 250 \text{ cm}^3$$

PROBLEM

A chemist dropped a 200 g object into a tank of water. It displaced 60 ml of the water when it sunk to the bottom of the tank. In a similar experiment, it displaced only 50 g of an oil into which it was dropped. Calculate the density of the object and the oil.

SOLUTION

The density (ρ) of a substance is defined as its mass divided by its volume.

$$\rho = \frac{\text{mass}}{\text{volume}}$$

Thus, to solve for the densities of the object and the oil, one must first calculate their respective masses and volumes. The mass of the object is 200 g, but the volume is not given. An object dropped in any liquid displaces a volume of liquid equal to the volume of the object. The object displaces 60 ml of water; therefore, the volume of the object is 60 ml. Solving for the density of the object:

$$\rho = \frac{\text{mass}}{\text{volume}} = \frac{200 \text{ g}}{60 \text{ ml}} = 3.33 \text{ g/ml.}$$

One is given that the object displaces 50 g of the oil, and from the water experiment, it is known that the volume of the object is 60 ml. Since the object displaces the same volume of liquid as it occupies, 60 ml of the oil weighs 50 g. Solving for the density of the oil:

$$\rho = \frac{\text{mass}}{\text{volume}} = \frac{50 \text{ g}}{60 \text{ ml}} = .833 \text{ g/ml.}$$

Drill 1: Density

1. Assuming that the density of water is .9971 g/cm^3 at 25°C and that of ice at 0° is .917 g/cm^3, what percent of a water jug at 25°C should be left empty so that, if the water freezes, it will just fill the jug?

    (A) 10%   (D) 6%

    (B) 12%   (E) 2%

    (C) 8%

2. The molecular diameter of an N_2 molecule, as deduced from the Van der Waals b parameter, is 3.15×10^{-8} cm. The density of liquid nitrogen is 0.8081 g/cm^3. On a hard-sphere model, what fraction of the liquid volume appears to be empty space?

    (A) 69.25   (D) 40.16

    (B) 38.12   (E) 90.12

    (C) 71.3

3. If acetic acid has a density of 1.11 g/ml, how many kilograms will 2.34 liters of acetic acid weigh?

    (A) 4.74   (D) 2.88

    (B) 2.11   (E) 2.60

    (C) 4.15

2. Freezing Point Depression and Boiling Point Elevation

The boiling point of a liquid is the temperature at which the pressure of vapor escaping from the liquid equals atmospheric pressure. The normal boiling point of a liquid is the temperature at which its vapor pressure is 760 mm Hg, that is, standard atmospheric pressure.

Liquids relatively strong with attractive forces have high boiling points. The melting point of a substance is the temperature at which its solid and liquid phases are in equilibrium.

Colligative property law: The freezing point, boiling point, and vapor pressure of a solution differ from those of the pure solvent by amounts which are directly proportional to the molal concentration of the solute.

The vapor pressure of an aqueous solution is always lowered by the addition of more solute, which causes the boiling point to be raised (boiling point elevation).

The freezing point is always lowered by addition of solute (freezing point depression). The freezing point depression, ΔT_f, equals the negative of the molal freezing point depression constant, K_f, times molality (m):

$$\Delta T_f = - K_f(m)$$

The boiling point elevation, ΔT_b, equals the molal boiling point elevation constant, K_b, times molality (m):

$$\Delta T_b = K_b (m)$$

$$\Delta T = T_{\text{solution}} - T_{\text{pure solvent}}$$

PROBLEM

> The freezing point constant of toluene is 3.33°C per mole per 1,000 g. Calculate the freezing point of a solution prepared by dissolving 0.4 mole of solute in 500 g of toluene. The freezing point of toluene is −95.0°C.

SOLUTION

The freezing point constant is defined as the number of degrees the freezing point will be lowered per 1,000 g of solvent per mole of solute present. The freezing point depression is related to this constant by the following equation.

freezing pt depression = molality of solute × freezing pt constant

The molality is defined as the number of moles per 1,000 g of solvent. Here, one is given that 0.4 moles of solute are added to 500 g of solvent, therefore, there will be 0.8 moles in 1,000 g.

$$\frac{0.4 \text{ moles}}{500 \text{ g}} = \frac{0.8 \text{ moles}}{1,000 \text{ g}}$$

The molality of the solute is thus 0.8 m. One can now find the freezing point depression. The freezing point constant for toluene is 3.33°.

$$\text{freezing point depression} = \text{molality} \times 3.33°$$

$$= 0.8 \times 3.33° = 2.66°$$

The freezing point of toluene is thus lowered by 2.66°.

$$\text{freezing point of solution} = (-95°C) - 2.66° = -97.66°C.$$

PROBLEM

By how much will 50 grams of water have its freezing point depressed if you add 30 grams (molecular weight 80) of glucose to it?

SOLUTION

The addition of any substance to water will alter its boiling or freezing point. To determine the amount of change, you must know the concentration of solute (the substance dissolved) in the solvent (water). This information is required because the freezing point depression, ΔT_f, equals the molal freezing point constant, k_f, times the molality (m).

$$\Delta T_f = k_f \text{ (m)}$$

The concept of molality refers to the number of moles of solute per 1 kilogram of solvent.

The solute, glucose, weighs 30 grams. Therefore, the number of moles of glucose is 30 g/180 g/mol = 1/6 moles. You have 50 grams of water. However, molality refers to moles per thousand grams. As such, a conversion is required; namely 50/1,000. Hence, the molality is

$$\frac{\frac{1}{6} \text{ moles}}{\frac{50}{1,000} \text{ g}}$$

Recall that the amount of depression of the freezing point is defined as $\Delta T_f = (k_f \text{ m})$. k_f is given for water as $-1.86°$ mole^{-1}. You calculated m. The temperature depression is thus $\Delta T_f = k_f \times \text{m} = -1.86° \times 3.33 \text{ m} = -6.2°$.

Drill 2: Freezing Point Depression and Boiling Point Elevation

1. What is the approximate melting point of 0.2 liters of water containing 6.20 g of ethylene glycol ($C_2H_6O_2$)?

    (A) −1.86°C

    (B) −0.93°C

    (C) 0°C

    (D) 0.93°C

    (E) 1.86°C

2. What is the freezing point of a solution of 92 g of alcohol (C_2H_5OH) and 500 g of H_2O?

    (A) 8.29°

    (B) 7.4°

    (C) 4.12°

    (D) 6.15°

    (E) 11.32°

3. Calculate the composition (molality) of an alcohol—water mixture which will not freeze above a temperature of −10°C (+14°F). (MW of alcohol = 46; freezing point constant for water (k_f) = 1.86°.)

    (A) 6.2 m

    (B) 5.4 m

    (C) 25 m

    (D) 7.2 m

    (E) 65 m

4. The molal freezing point constant for a certain liquid is 0.500°C. 26.4 g of a solute dissolved in 250 g of this liquid yields a solution which has a freezing point 0.125° below that of the pure liquid. Calculate the molecular weight of this solute.

    (A) 105.6

    (B) 250

    (C) 211.2

    (D) 63.2

    (E) 422.4

5. Liquid naphthalene normally freezes at 80.2°C. When 1 mole of solute is dissolved in 1,000 g of naphthalene, the freezing point of the solution is 73.2°C. When 6.0 g of sulfur is dissolved in 250 g of naphthalene, the freezing point is 79.5°C. What is the molecular weight of sulfur?

    (A) 24 g/mol

    (B) 19 g/mol

    (C) 300 g/mol

    (D) 240 g/mol

    (E) 64 g/mol

3. Raoult's Law and Vapor Pressure

Liquids are in equilibrium with the gas or vapor above the liquid. The pressure of the gas above a pure liquid is called the vapor pressure or sometimes, for emphasis, the pure component vapor pressure.

The vapor pressure is a characteristic of a liquid and is a strong function of temperature. It is precisely defined as the equilibrium gas (or vapor) pressure above a pure liquid at a specified temperature with no other species present. Except at extreme pressures, it is an excellent approximation to assume that the vapor pressure is independent of the total pressure, i.e., unaffected by the presence of other species in the gas phase.

There are empirical equations that permit the vapor pressure to be calculated, but these equations contain approximations and, for precise work, measurements should be used.

Raoult's law predicts the partial pressure above a liquid which contains a mixture of compounds. It is correct only for ideal liquid mixtures, i.e., liquid mixtures where the attractive forces are the same between like and unlike molecules. It is a good approximation for mixtures of isomers (such as normal pentane, isopentane, and neopentane) or adjacent members of homologous series (such as benzene, toluene, and xylene).

Raoult's law is written as follows:

$$p_i = x_i P^\circ_i$$

where p is the partial pressure in the gas phase, x is the mole fraction in the liquid phase, and P° is the pure component vapor pressure (or just vapor pressure). The subscript i indicates that this equation can be applied to each species in the solution, i.e., $i = 1, 2, 3,....$ If two of the terms in the above equation are known, the third can be easily calculated. If Raoult's law holds for one species in a solution, it will hold for all the species. Departures from Raoult's Law are common in nature, but are not discussed in this text.

One key to working the problems involving Raoult's law is to observe that the total pressure is the sum of the partial pressures. From the Ideal gas law, the mole fractions in the vapor phase are simply the respective partial pressures divided by the total pressure (sum of the partial pressures). As well, the mole fractions of the various solutes must add up to 1.

$$YI = P_i/P$$

where Y is the vapor phase mole fraction and P is the total pressure.

PROBLEM

> You have two 1-liter containers connected to each other by a valve
> which is closed. In one container, you have liquid water in equilibrium
> with water vapor at 25°C. The other container contains a vacuum.
> Suddenly, you open the valve. Discuss the changes that take place,
> assuming temperature is constant with regard to (a) the vapor pressure,
> (b) the concentration of the water molecules in the vapor, and (c) the
> number of molecules in the vapor state.

SOLUTION

The vapor pressure is the pressure exerted by the gas molecules when they
are in equilibrium with the liquid. When the valve is opened some of the gas
molecules will move to the empty container. At this point, the pressure will be
less than the equilibrium pressure because the concentration of the gas molecules
will be lowered. Very quickly, though, the equilibrium will be attained again by
the action of more liquid molecules vaporizing. Therefore, the vapor pressure and
the concentration of the gaseous molecules of the system remains essentially
unchanged.

Since the concentration of the gaseous molecules remains unchanged when
the volume of the system is doubled, the number of molecules must also be
doubled. This is true because concentration is an expression of the number of
molecules per unit volume.

PROBLEM

> A chemist decides to find the vapor pressure of water by the gas
> saturation method. One hundred liters of N_2 gas is passed through
> 65.44 g of water. After passage of the gas, 63.13 g remained. The
> temperature of the H_2O (water) is 25°C. Find the vapor pressure of
> water at this temperature.

SOLUTION

In the gas saturation method, a dry, unreactive gas, such as nitrogen or air,
is bubbled through a specific amount of liquid maintained at constant temperature. After the gas has been bubbled away, the loss in weight of the liquid is
determined. The weight loss is the number of grams of liquid in the vapor state.
There exists an equation that relates the volume, pressure, weight loss, and molecular weight of the liquid.

$$P = \frac{m}{MW \times V} RT,$$

where P = vapor pressure, m = grams of vapor, MW = molecular weight of liquid,

R = universal gas constant (.0821 liter atm/mol °K) V = volume, and T = temperature in Kelvin (Celsius plus 273°).

The pressure will be expressed in mm, so that you use the conversion factor of 760 mm/atm. $m = 65.44 - 63.13 = 2.31$ grams or the weight of the liquid in the vapor state.

The molecular weight of water = 18.02 g/mol. Thus,

$$P = \frac{m}{MW \times V} = \frac{(2.31 \text{ grams})(.0821 \frac{\text{liter atm}}{\text{mole °K}})(298°K)\frac{760 \text{ mm}}{1 \text{ atm}}}{(18.02 \text{ g/mole})(200 \text{ liters})}$$

$= 23.8$ mm = vapor pressure of H_2O.

Drill 3: Raoult's Law and Vapor Pressure

Questions 1–3 refer to the following information. The vapor pressures of pure benzene and toluene at 60°C are 385 and 139 torr, respectively. The mole fraction of toluene in the solution is 0.60.

1. Calculate the partial pressure of the toluene in torr.

    (A) 61.2   (D) 94.2

    (B) 76.9   (E) 83.4

    (C) 35.3

2. Calculate the total vapor pressure of the solution.

    (A) 237.4 torr   (D) 61.2 torr

    (B) 83.4 torr   (E) 435.2 torr

    (C) 76.9 torr

3. Calculate the mole fraction of toluene in the vapor above the solution.

    (A) 0.621   (D) 0.165

    (B) 0.351   (E) 0.213

    (C) 0.124

4. Osmotic Pressure

Osmosis is the diffusion of a solvent through a semipermeable membrane into a more concentrated solution.

The osmotic pressure of a solution is the minimum pressure that must be applied to the solution to prevent the flow of solvent from pure solvent into the solution.

The osmotic pressure for a solution is:

$$\pi = CRT$$

where π is the osmotic pressure, C is the concentration in molality or molarity, R is the gas constant, and T is the temperature (in Kelvin). This equation can be used to calculate the molecular weight of a species by measuring the osmotic pressure produced by a known mass of solute. It should be noted that this equation applies only to very dilute solutions. (Note the formal similarity of the osmotic pressure equation to the Ideal gas law, $C = n/V$.)

Solutions that have the same osmotic pressure are called isotonic solutions.

Reverse osmosis is a method for recovering pure solvent from a solution.

PROBLEM

A sugar solution was prepared by dissolving 9.0 g of sugar in 500 g of water. At 27°C, the osmotic pressure was measured at 2.46 atm. Determine the molecular weight of the sugar.

SOLUTION

The molecular weight of the sugar is found by determining the concentration, C, of sugar from the equation for osmotic pressure,

$$\pi = CRT,$$

where π is the osmotic pressure, R = universal gas constant = 0.08206 liter atm/mole °K, and T is the absolute temperature.

The osmotic pressure is measured as $\pi = 2.46$ atm and the absolute temperature is $T = 27°C + 273 = 300°K$, hence

$$\pi = CRT,$$

$$2.46 \text{ atm} = C \times 0.08206 \text{ liter atm/mol °K} \times 300°K,$$

or
$$C = \frac{2.46 \text{ atm}}{0.08206 \text{ liter atm/mol °K} \times 200°K}$$

$$= 0.10 \text{ mole/liter}$$

If we assume that the volume occupied by the sugar molecules in so small a concentration can be neglected, then in 1 liter of solution there is approximately 1 liter of water, or 1,000 g of water, and

$$C = 0.10 \text{ mole}/1,000 \text{ g.}$$

Therefore, there is 0.10 mole of sugar dissolved in 1,000 g of water.

9.0 g of sugar dissolved in 500 g of water is equivalent to 18.0 g of sugar dissolved in 1,000 g of water (9.0 g/500 g = 18.0 g/1,000 g). But since $C = 0.10$ mole/1000 g, the 18.0 g of sugar must correspond to 0.10 mole of sugar. Therefore, the molecular weight of the sugar is

$$18 \text{ g}/0.1 \text{ mole} = 180 \text{ g/mol.}$$

Drill 4: Osmotic Pressure

1. What is the osmotic pressure of a 2.0 m solution of a non-electrolyte at 298°K?

    (A) 60 atm   (D) 61 atm

    (B) 49 atm   (E) 53 atm

    (C) 30 atm

2. A chemist dissolves 10 g of an unknown protein in a liter of water at 25°C. The osmotic pressure is found to be 9.25 mmHg. What is the protein's molecular weight? Based upon the number of moles in 10 g of protein, what would the freezing point depression be? Assume R = Universal gas constant = .0821 liter atm/mol°K, $k_f = 1.86$°C/m.

    (A) 9.3×10^{-4}°C   (D) 8.4×10^{-3}°C

    (B) 6.2×10^{-2}°C   (E) 10.4×10^{-6}°C

    (C) 4.1×10^{-6}°C

3. A solution is made by dissolving 4.35 g of a powdered substance in 375 ml of water. The osmotic pressure was found to be 1.41 atm at 28.3°C. What is the molecular weight of this powered substance?

    (A) 301 g/mol   (D) 203 g/mol

    (B) 821 g/mol   (E) 19.1 g/mol

    (C) 435 g/mol

5. Van't Hoff Factor and Electrolytes

Electrolytes are compounds which form charged ions when dissolved. Most salts are electrolytes, as are most common inorganic acids and bases. Strong electrolytes disassociate completely when dissolved, weak electrolytes on the other hand, dissociate only partially.

This dissociation has an effect on the boiling point and freezing point as it serves to introduce more species into a solution than one might initially expect. The Van't Hoff factor takes this into account and is in part a measure of an electrolytes dissociation.

The Van't Hoff "factor," i, is defined as the ratio of the observed freezing point depression produced by a solute in solution, to the freezing point that the solution would exhibit if the solute were a nonelectrolyte. For example, since NaCl yields 2 moles of dissolved particles (Na^+ and Cl^-) in water, its Van't Hoff factor is 2, and a 1 molal solution of NaCl (aq) yields a freezing point depression which is twice as large as that produced by sucrose, a nonelectrolyte.

$$i = \frac{(\Delta T_f) \text{ measured}}{(\Delta T_f) \text{ calculated as nonelectrolyte}}$$

PROBLEM

Explain the following phenomena: a) Liquid HCl is a nonelectrolyte, but aqueous HCl is a strong electrolyte and b) Liquid HCN is a non-electrolyte as is aqueous HCN.

SOLUTION

An electrolyte is a substance which exists as ions. Both liquid HCN and HCl are not electrolytes because they consist of neutral atoms. No ions exist. The case with an aqueous solution requires a little more investigation. In water, HCl is ionized via the reaction

$$HCl + H_2O \leftrightarrow H_3O^+ + Cl^-.$$

Thus, you have hydronium ions and chloride ions in solution. Thus, aqueous HCl is an electrolyte. HCN also dissociates when placed in water. But, to what extent? H_3O^+ and CN^- ions are only to a very slight extent produced, since HCN's dissociation constant is so small. Thus, while HCN may be called an electrolyte, it is an extremely weak one.

Drill 5: Van't Hoff Factor and Electrolytes

1. Select the compound that is *not* a conductor in aqueous solution.

   (A) CH_3OH (D) NaCl

   (B) $CuSO_4$ (E) NaOH

   (C) HCl

2. The weak electrolyte is

   (A) HNO_3.   (B) KI.

(C) HCl.   (D) NaCl.

(E) NH₄OH.

3. What is the Van't Hoff factor, *i*, for NaCl?

(A) 1   (D) 4

(B) 2   (E) None of the above.

(C) 3

PROPERTIES OF LIQUIDS
DRILLS

ANSWER KEY

Drill 1 — Density

1. (C)
2. (C)
3. (E)

Drill 2 — Freezing Point Depression and Boiling Point Elevation

1. (B)
2. (B)
3. (B)
4. (E)
5. (D)

Drill 3 — Raoult's Law and Vapor Pressure

1. (E)
2. (A)
3. (B)

Drill 4 — Osmotic Pressure

1. (B)
2. (A)
3. (D)

Drill 5 — Van't Hoff Factor and Electrolytes

1. (A)
2. (E)
3. (B)

GLOSSARY:
PROPERTIES OF LIQUIDS

Boiling Point

The temperature at which the pressure of vapor escaping from the liquid equals atmospheric pressure.

Boiling Point Elevation

The raising of the boiling point by the addition of a solute to a liquid.

Colligative Properties

Any physical property that is dependent on the concentration of a solute, but not the identity of that solute.

Density

Mass per unit volume.

Electrolyte

Any substance that splits up into ions when placed in solution.

Freezing Point

The temperature at which the solid and liquid phases of a substance are in equilibrium.

Freezing Point Depression

The lowering of the freezing point of a substance resulting from the addition of a solute.

Molality

A measure of concentration. Number of moles of solute per liter of solution.

Molarity

A measure of concentration. Number of moles of solute per kilogram of solvent.

Mole Fraction

A concentration unit that is defined as the number of moles of a component divided by the total number of moles of all components.

Osmosis

The diffusion of a solvent through a semipermeable membrane into a more concentrated solution.

Osmotic Pressure

The minimum pressure that must be applied to a solution to prevent the flow of solvent from pure solvent into the solution through a semipermeable membrane.

Raoult's Law

A law which states that the vapor pressure of a solution at a particular temperature is equal to the mole fraction of the solvent in the liquid phase multiplied by the vapor pressure of the pure solvent at the same temperature.

Solute

A component which is dissolved in a solvent.

Solvent

The liquid into which a solute is dissolved.

Surface Tension

The strength of the inward forces of a liquid.

Vapor Pressure

The pressure exerted by vapor above a liquid when the vapor and liquid are in equilibrium.

Volatile

The nature of a substance to easily enter the vapor phase.

Van't Hoff Factor

The ratio of the observed freezing point depression produced by a solute in solution, to the freezing point that the solution would exhibit if the solute were a nonelectrolyte.

CHAPTER 8

Solution Chemistry

➤ Diagnostic Test
➤ Solution Chemistry
  Review & Drills
➤ Glossary

SOLUTION CHEMISTRY DIAGNOSTIC TEST

1. Ⓐ Ⓑ Ⓒ Ⓓ Ⓔ   19. Ⓐ Ⓑ Ⓒ Ⓓ Ⓔ
2. Ⓐ Ⓑ Ⓒ Ⓓ Ⓔ   20. Ⓐ Ⓑ Ⓒ Ⓓ Ⓔ
3. Ⓐ Ⓑ Ⓒ Ⓓ Ⓔ   21. Ⓐ Ⓑ Ⓒ Ⓓ Ⓔ
4. Ⓐ Ⓑ Ⓒ Ⓓ Ⓔ   22. Ⓐ Ⓑ Ⓒ Ⓓ Ⓔ
5. Ⓐ Ⓑ Ⓒ Ⓓ Ⓔ   23. Ⓐ Ⓑ Ⓒ Ⓓ Ⓔ
6. Ⓐ Ⓑ Ⓒ Ⓓ Ⓔ   24. Ⓐ Ⓑ Ⓒ Ⓓ Ⓔ
7. Ⓐ Ⓑ Ⓒ Ⓓ Ⓔ   25. Ⓐ Ⓑ Ⓒ Ⓓ Ⓔ
8. Ⓐ Ⓑ Ⓒ Ⓓ Ⓔ   26. Ⓐ Ⓑ Ⓒ Ⓓ Ⓔ
9. Ⓐ Ⓑ Ⓒ Ⓓ Ⓔ   27. Ⓐ Ⓑ Ⓒ Ⓓ Ⓔ
10. Ⓐ Ⓑ Ⓒ Ⓓ Ⓔ   28. Ⓐ Ⓑ Ⓒ Ⓓ Ⓔ
11. Ⓐ Ⓑ Ⓒ Ⓓ Ⓔ   29. Ⓐ Ⓑ Ⓒ Ⓓ Ⓔ
12. Ⓐ Ⓑ Ⓒ Ⓓ Ⓔ   30. Ⓐ Ⓑ Ⓒ Ⓓ Ⓔ
13. Ⓐ Ⓑ Ⓒ Ⓓ Ⓔ   31. Ⓐ Ⓑ Ⓒ Ⓓ Ⓔ
14. Ⓐ Ⓑ Ⓒ Ⓓ Ⓔ   32. Ⓐ Ⓑ Ⓒ Ⓓ Ⓔ
15. Ⓐ Ⓑ Ⓒ Ⓓ Ⓔ   33. Ⓐ Ⓑ Ⓒ Ⓓ Ⓔ
16. Ⓐ Ⓑ Ⓒ Ⓓ Ⓔ   34. Ⓐ Ⓑ Ⓒ Ⓓ Ⓔ
17. Ⓐ Ⓑ Ⓒ Ⓓ Ⓔ   35. Ⓐ Ⓑ Ⓒ Ⓓ Ⓔ
18. Ⓐ Ⓑ Ⓒ Ⓓ Ⓔ

SOLUTION CHEMISTRY
DIAGNOSTIC TEST

This diagnostic test is designed to help you determine your strengths and your weaknesses in solution chemistry. Follow the directions and check your answers.

> ## Study this chapter for the following tests:
> ### ACT, ASVAB, AP Chemistry, CLEP General Chemistry, GED, MCAT, MSAT, PRAXIS II Subject Assessment: Chemistry, SAT II: Chemistry, GRE Chemistry

35 Questions

DIRECTIONS: Choose the correct answer for each of the following problems. Fill in each answer on the answer sheet.

1. How many grams of $NaHCO_3$ must be added to water to produce 200 ml of 0.5 M solution?

   (A) 8.4 g

   (B) 21 g

   (C) 42 g

   (D) 66 g

   (E) 84 g

2. 58.5 g of NaCl in

   (A) 1 liter of solution is 2 molar.

   (B) 2 liters of solution is 0.75 molar.

   (C) 1 liter of solution is 1 molal.

   (D) 2 kilograms of solvent is 0.5 molal.

   (E) 1 kilogram of solvent is 1 molar.

3. To what final volume should 100 ml of 4 N H_2SO_4 be diluted to produce a 1 M solution?

   (A) 100 ml

   (B) 300 ml

   (C) 800 ml

   (D) 700 ml

   (E) 900 ml

4. 2.3 g of ethanol (C_2H_5OH, MW = 46 g/mol) is added to 500 g of water. Determine the molality of the resulting solution.

    (A) 0.01 m   (D) 10.0 m

    (B) 0.1 m   (E) 1.1 m

    (C) 1.0 m

5. The solubility of a gas in a liquid is increased by

    (A) increased pressure.

    (B) increased temperature.

    (C) decreased pressure and increased temperature.

    (D) both (A) and (B).

    (E) None of the above.

6. What is the molality of a solution made by dissolving 18.4 grams of toluene (C_7H_8) in 200 grams of benzene (C_6H_6)?

    (A) 1.00 m   (D) .001 m

    (B) 0.5 m   (E) 2.00 m

    (C) .0001 m

7. A saturated solution of KNO_3 contains 63 g KNO_3 at 40°C. If a solution at the same temperature is found to contain more than 63 g of KNO_3, but with no precipitation, then the solution is probably

    (A) dilute.   (D) saturated.

    (B) concentrated.   (E) supersaturated.

    (C) unsaturated.

8. If 20 ml of 0.5 N salt solution is diluted to 1 liter, what is the new concentration?

    (A) 0.01 N   (D) 10 N

    (B) 0.001 N   (E) 5 N

    (C) 1 N

9. A student weighs an amount of NaCl placed on a piece of filter paper. The paper weighs 0.455 g and the total weight is 11.085 g. He dissolves the salt in distilled water in a 200 ml volumetric flask and then adds water to the line. What is the concentration of NaCl in moles/liter? (A.W. Na = 22.99,

A.W. Cl = 35.45)

(A) $\dfrac{(11.085 - .455)\,(22.99 + 35.45)}{200}$

(B) $\dfrac{(11.085 - .455)\,(22.99 + 35.45)}{200} \times 1,000$

(C) $\dfrac{(11.085 - .455)}{(22.99 + 35.45)} \times \dfrac{1}{200}$

(D) $\dfrac{(11.085 - .455)}{(22.99 + 35.45)} \times \dfrac{1,000}{200}$

(E) $\dfrac{(11.085 - .455)}{(22.99 + 35.45)} \times \dfrac{200}{1,000}$

10. A 10% solution of HNO_3 would be produced by dissolving 63 g of HNO_3 in _____ ml of water.

    (A) 100
    (B) 300
    (C) 567
    (D) 630
    (E) 1,000

11. Which of the following sequences lists the relative sizes of particles in a water mixture from smallest to largest?

    (A) Solutions, suspensions, colloids

    (B) Solutions, colloids, suspensions

    (C) Colloids, solutions, suspensions

    (D) Colloids, suspensions, solutions

    (E) Suspensions, colloids, solutions

12. How many moles of sulfate ions are in 200 ml of a 2 M sodium sulfate solution?

    (A) 0.2 mole
    (B) 0.4 mole
    (C) 0.6 mole
    (D) 0.8 mole
    (E) 1.0 mole

13. How much barium nitrate is required to prepare 250.0 ml of a 0.1000 M solution? (The molecular weight of barium nitrate, MW = 199.344.)

(A) $\dfrac{(250)(199.344)}{(1,000)(0.1000)}$

(D) $\dfrac{(250)}{(0.1000)(199.344)(1,000)}$

(B) $\dfrac{(250)(0.1000)}{(199.344)(1,000)}$

(E) None of the above.

(C) $\dfrac{(250)(0.1000)(199.344)}{(1,000)}$

14. What is the molarity of a 10 ml solution in which 3.7 g of KCl are dissolved?

(A) 0.05 M

(D) 5 M

(B) 0.1 M

(E) 10 M

(C) 1 M

15. A small crystal of NaCl is added to a sodium chloride solution resulting in the precipitation of more than 1 gram of sodium chloride. This solution had been

(A) unsaturated.

(D) dilute.

(B) saturated.

(E) concentrated.

(C) supersaturated.

16. 2.4 liter of HNO_3 solution reacts with 63 ml of 1.9 N $Ba(OH)_2$ to produce a neutral solution. What is the molarity (M) of the original HNO_3 solution?

(A) $\dfrac{63(1.9)}{1,000(2.4)}$ M

(D) $\dfrac{63(1.9)}{1,000(2)(2.4)}$ M

(B) $\dfrac{63(1.9)}{1,000}$ M

(E) $\dfrac{63(1.9)}{100(2.4)}$ M

(C) $\dfrac{2(63)(1.9)}{1,000}$ M

17. How many moles of sulfate ions are there in 500 ml of a 5 M sulfuric acid solution?

(A) 0.5

(D) 5.0

(B) 1.0

(E) 10.0

(C) 2.5

18. What is the normality, N, of a 1 M solution of H_2SO_4, given the following reaction?

$$H_2SO_4 + 2KOH \rightarrow K_2SO_4 + 2H_2O$$

(A) 4 N

(D) 1 N

(B) 3 N

(E) None of the above.

(C) 2 N

19. How much water must be evaporated from 500 ml of 1M NaOH to make it 5M?

(A) 100 ml

(D) 300 ml

(B) 200 ml

(E) 400 ml

(C) 250 ml

20. Compute the quantity in grams of sucrose ($C_{12}H_{22}O_{11}$) required to make a 1 M strength solution of 500 ml.

(A) 85.5

(D) 684

(B) 171

(E) 982

(C) 342

21. In a methanol-ethanol-propanol solution (consisting of a mixture of 42.0 g methanol, 35.0 g ethanol and 50.0 g propanol), the partial molar volumes are respectively 16.0 ml, 20.0 ml and 50.0 ml. The volume of 1.00 mole of the solution is: (Take the mole fraction of CH_3OH, C_2H_5OH and C_3H_7OH to be 0.452, 0.260, and 0.287 respectively.)

(A) $(.452)(16.0) + (0.26)(20.0) + (0.287)(50.0)$.

(B) $(0.26)(16.0) + (0.452)(20.0) + (0.287)(50.0)$.

(C) $\dfrac{0.26}{16.0} + \dfrac{0.452}{20.0} + \dfrac{0.287}{50.0}$.

(D) $\dfrac{(0.452)(16.0) + (0.26)(20.0) + (0.287)(50.0)}{(0.452)(0.26)(0.287)}$.

(E) $\dfrac{0.26}{20.0} + \dfrac{0.452}{16.0} + \dfrac{0.287}{50.0}$.

22. A liquid containing some microscopic particles just like particles from tobacco smoke can be detected not by the naked eye, but by the fact that they reflect light. These particles could probably be described as

(A) a saturated solution. (D) colloidal.

(B) an unsaturated solution. (E) a suspension.

(C) a supersaturated solution.

23. A chemist adds 0.1000 mole of KCl to 1.000 liter of distilled water at 25°C. The concentration of the resulting solution is

(A) 7.455 M. (D) > 0.1000 M and < 1.000 M.

(B) 1.000 M. (E) < 0.1000 M.

(C) 0.1000 M.

24. A 20% solution of NaOH will be produced by dissolving one mole of NaOH in

(A) 160 ml of acetone. (D) 160 ml of carbon tetrachloride.

(B) 160 ml of water. (E) None of the above.

(C) 160 ml of ammonia.

25. A 1 molal solution of NaCl results when 58.5 g of sodium chloride is dissolved in

(A) one liter of water. (D) 100 g of water.

(B) 100 ml of water. (E) one cubic meter of water.

(C) one kilogram of water.

26. A suspension of particles in solution will

(A) not settle upon standing. (D) have a cloudy or opaque color.

(B) exhibit Brownian movement. (E) not be visible with a microscope.

(C) pass through filter paper.

27. What is the molarity of a solution that is prepared by dissolving 32.0 g of KCl in enough water to make 425 ml of solution?

(A) 1.0 (D) .425

(B) 2.3 (E) .0075

(C) 1.0×10^{-3}

28. Calculate the volume of 4 M HCl needed to prepare 1 liter of a 0.5 M solution.

(A) 0.125 l (B) 0.0125 l

(C) 0.875 l (D) 0.0875 l

(E) 12.5 l

29. Which of the following is *not* a homogeneous mixture?

(A) Sugar in water   (D) Gasoline

(B) Salt in water   (E) Soft drinks

(C) Sand in water

30. What is the mole fraction of CH_3OH in a solution that contains 53.0 g of water, 20.3 g of CH_3OH and 15.0 g of CH_3CH_2OH?

(A) .33   (D) .96

(B) .48   (E) .16

(C) 1.22

31. What is the normality of a solution that contains 23.2 g of H_2SO_4 and enough water to make 400 ml of solution?

(A) 2.4   (D) 6.0

(B) .60   (E) .50

(C) 1.2

32. What volume of a 1.3 M solution of NaCl contains 2.3 g of NaCl.

(A) 130 ml   (D) 177 ml

(B) 30.2 ml   (E) None of the above.

(C) 3.9 ml

33. How many moles of ions are present in one liter of a 2 M solution of NaCl?

(A) 0.2   (D) 4.0

(B) 1.0   (E) 8.0

(C) 2.0

34. A stock solution of 10 M NaOH was used to prepare 2 liters of 0.5 M NaOH. How many milliliters of sodium hydroxide stock solution were used?

(A) 10 ml   (D) 200 ml

(B) 100 ml   (E) 2,000 ml

(C) 1,000 ml

35. A 0.5 molal solution could be prepared by dissolving 20 g of NaOH in

    (A) 0.5 liter of water.

    (B) 0.5 kg of water.

    (C) 1 liter of water.

    (D) 1 kg of water.

    (E) 2 liters of water.

SOLUTION CHEMISTRY DIAGNOSTIC TEST

ANSWER KEY

1.	(A)	10.	(C)	19.	(E)	28.	(A)
2.	(D)	11.	(B)	20.	(B)	29.	(C)
3.	(C)	12.	(B)	21.	(A)	30.	(E)
4.	(B)	13.	(C)	22.	(D)	31.	(C)
5.	(A)	14.	(D)	23.	(E)	32.	(B)
6.	(A)	15.	(C)	24.	(B)	33.	(D)
7.	(E)	16.	(A)	25.	(C)	34.	(B)
8.	(A)	17.	(C)	26.	(D)	35.	(D)
9.	(D)	18.	(C)	27.	(A)		

DETAILED EXPLANATIONS
OF ANSWERS

1. **(A)**   The molecular weight of sodium bicarbonate is determined to be 84 from the sum of its constituent atomic weights. Thus, one mole of $NaHCO_3$ weighs 84 g. A 1 M solution of sodium bicarbonate is composed of 84 g of $NaHCO_3$ in one liter of solution. Thus, one liter of a 0.5M solution contains 42 g of $NaHCO_3$. Hence 0.2 liters of a 0.5 M solution contains 0.2×42 grams or 8.4 grams of $NaHCO_3$.

2. **(D)**   One mole of NaCl weighs 58.5 g as obtained by: atomic weight of Na + atomic weight of Cl from the periodic table. Thus, 58.5 g of NaCl in one liter of solution is 1 molar and 58.5 g of NaCl in one kilogram of solvent is 1 molal. By simple proportions, 58.5 g of NaCl in 2 kilograms of solvent is 0.5 molal.

3. **(C)**   From the expression $M_1V_1 = M_2V_2$ we have,

    $$(8 \text{ M*}) (100 \text{ ml}) = (1 \text{ M}) (V_2) : V_2 = 800 \text{ ml}$$

(*) Note that we are required to change normality to molarity before we substitute. With 4 N H_2SO_4 being equal to 8 M H_2SO_4.

4. **(B)**   In this case, solute is ethanol, solvent is water.

$$\text{molality} = \frac{\text{moles of solute (ethanol)}}{\text{kg of solvent (water)}}$$

$$\text{mole of ethanol} = \frac{2.3 \text{ g}}{46 \text{ g/mol}} = 0.05 \text{ mole}$$

$$\therefore \text{molality} = \frac{0.05 \text{ mole}}{0.5 \text{ kg}} = 0.1 \text{ molal}$$

If we need the molarity,

$$\text{molarity} = \frac{\text{moles of solute}}{\text{liter of solvent}} = \frac{0.05 \text{ mole}}{0.5 \text{ kg/}l \text{ kg/m}^3}$$

$$= \frac{0.05 \text{ mole}}{0.5 \, l}$$

$$= 0.1 \text{ molar}$$

5. **(A)** The solubility of a gas in a liquid is increased by increased pressure and decreased temperature and decreased by the opposite. On the other hand, the solubility of a solid in a liquid is increased by increased temperature and only negligibly affected by changes in pressure.

6. **(A)** The molality of a solution is defined as the number of moles of solute in one kilogram of solvent. To solve this problem therefore, the moles of C H (solute) and the kilograms of C H (solvent) need to be determined as follows:

$$(18.4 \text{ grams } C_7H_8) \left(\frac{1 \text{ mole } C_7H_8}{92 \text{ grams } C_7H_8} \right) = 0.200 \text{ moles } C_7H_8$$

$$(200 \text{ grams } C_6H_6) \left(\frac{1 \text{ kilogram } C_6H_6}{1,000 \text{ grams } C_6H_6} \right) = 0.200 \text{ kilograms } C_6H_6$$

$$\text{molarity} = \frac{0.200 \text{ moles } C_7H_8}{0.200 \text{ kilograms } C_6H_6} = 1.00 \text{ M}$$

7. **(E)** Converting to grams,

$$1.3 \text{ moles of } KNO_3 \times \frac{101 \text{ g of } KNO_3}{1 \text{ mole of } KNO_3} = 131 \text{ g of } KNO_3.$$

A solution that contains more solute than necessary for saturation but without precipitation is an example of supersaturation.

8. **(A)**

$$N_A V_A = N_B V_B$$

$$0.5 \, N \times 20\text{ml} = N_B \times 1,000 \text{ ml}$$

$$N_B = \frac{0.5 \times 20 \text{ ml}}{1,000 \text{ ml}} = 0.01 \text{ N}$$

9. **(D)** The net weight of NaCl is

11.085 g − .445 g.

The number of moles of NaCl is:

$$\frac{(11.085 \text{ g} - .445 \text{ g})}{MW_{NaCl} \, (\text{moles/g})} = \frac{(11.085 \text{ g} - .455 \text{g})}{(22.99 + 35.45) \text{ moles/g}}.$$

The concentration is the number of moles of solute divided by the volume of the solution in liters.

$$\frac{(11.085 \text{ g} - .455\text{g})}{(22.99 + 35.45) \text{ moles/g}} = \frac{1{,}000 \text{ ml } l}{200 \text{ ml}} = .91 \text{ moles/}l$$

10. **(C)**   Percent solutions are based on the mass of the solute rather than the number of moles. Since a 10% solution is produced with 63 g of HNO_3, we know that $10(10\%) = 100\%$ of the solution has a mass of $10(63 \text{ g}) = 630$ g. This gives the mass of the water as $630 - 63 = 567$ g. Using the density of water: $\rho = 1$ g/cm^3 and the conversion factor

$$1 \text{ ml} = 1 \text{ cm}^3$$

we have

$$567 \text{ g of } H_2O \times \frac{1 \text{ cm}^3}{1 \text{ g}} \times \frac{1 \text{ ml}}{1 \text{ cm}^3} = 567 \text{ ml of } H_2O$$

11. **(B)**   These terms describe the relative sizes of the particles in a water mixture. Solutions involve the smallest particles, which are invisible and do not settle on standing. The particles of a colloid are visible with an ultramicroscope, exhibit Brownian motion, and do not settle on standing. Suspensions may be visible with the naked eye, show no Brownian motion, and settle upon standing.

12. **(B)**   A 1 M sodium sulfate (Na_2SO_4) solution contains one mole of sulfate ion per liter of solution. Thus, 0.2 l of a 23 M solution contains 0.2 mole of sulfate ion. 0.2 l of a 2 M solution would then contain 0.4 mole of sulfate ion.

13. **(C)**

grams of materials present = (moles) (MW)

$$\text{moles} = \text{volume} \times \text{molarity} = 250 \text{ ml} \times \frac{0.1 \text{ mole}}{\text{liter}}$$

$$= 250 \text{ ml} \times \frac{0.1000 \text{ moles}}{1} \times \frac{11}{1{,}000 \text{ ml}}$$

grams barium nitrate present = $MWX \times$ moles

$$= (199.344) (250) (0.1000) / (1{,}000)$$

14. **(D)**   Converting to moles

$$3.7 \text{ g of KCl} \times \frac{1 \text{ mole of KCl}}{74 \text{ g of KCl}} = 0.05 \text{ mole of KCl}$$

Converting to liters of solution

$$10 \text{ ml} \times \frac{1 \text{ liter}}{1,000 \text{ ml}} = 0.01 \text{ of solution}$$

Molarity is defined as the number of moles of solute dissolved in one liter of solution. Thus

$$M = \frac{0.05 \text{ mole of KCl}}{0.01 \text{ liter of solution}} = 5$$

15. **(C)** The solution in question had been supersaturated as is seen by the precipitation of more solute than what had been added. The same amount of solute would have precipitated if the solution was saturated and no precipitation would have occurred if the solution was unsaturated. The terms dilute and concentrated cannot be used in this context since a dilute solution may be saturated if the solute is only slightly soluble while a concentrated solution may be unsaturated if the solute is exceptionally soluble.

16. **(A)** Molarity is by definition the number of moles of solute divided by the liters of solution. HNO_3 reacts with OH^- on a one-to-one molar basis. The moles of OH^- are equal to

$$63 \text{ ml} \times \frac{1 \text{ liter}}{1,000 \text{ ml}} \times \frac{1.9 \text{ moles}}{1 \text{ liter}} = 0.12 \text{ moles}$$

The original HNO_3 solution had to contain 0.12 moles of HNO_3 in 2.4 liters of solution. Thus, its molarity is

$$\frac{0.12 \text{ moles } H^+}{2.4 \text{ liters}} = 0.05 \text{ M}$$

17. **(C)** Using the following conversion factors:

$$500 \times \frac{1 \text{ liter}}{1,000 \text{ ml}} \times \frac{5 \text{ moles}}{\text{liter}} \times \frac{1 \text{ mole of } SO_4^{2-}}{1 \text{ mole of } H_2SO_4}$$

we find that there are 2.5 moles of sulfate ion present.

18. **(C)**

$$N = \frac{\text{equivalents of solute}}{\text{liters of solution}}$$

while $\quad M = \dfrac{\text{moles of solute}}{\text{liters of solution}}$

It should be evident that 1 M solution of HCl is also 1 N because a gram-equivalent weight of HCl is the same as the gram-molecular weight. However, a 1 M solution of H_2SO_4 is $2N$, because one mole of H_2SO_4 is equal to two equivalents of H_2SO_4, if both hydrogens react.

e.g., $\qquad H_2SO_4 + 2KOH \rightarrow K_2SO_4 + 2H_2O$

$$\therefore N = \frac{\text{moles of solute}}{\text{liters of solution}} \times \frac{\text{equivalents of solute}}{\text{moles of solute}}$$

$$= M \times \frac{\text{equivalents}}{\text{moles of solute}}$$

$$= \frac{1 \text{ mole of solute}}{1 \text{ liter of solution}} \times \frac{2 \text{ equivalents } H_2SO_4}{1 \text{ mole of solute}}$$

$$= \frac{2 \text{ equivalents}}{\text{liter of solution}}$$

19. **(E)** Using the relationship

$$M_1V_1 = M_2V_2$$

where M is the molarity and V is the volume we have

$$(1 \text{ M}) (500 \text{ ml}) = (5 \text{ M}) (x)$$

and $\qquad x = 100 \text{ ml}$

This value gives us the volume of the final solution obtained by evaporating $500 - 100$ or 400 ml of water from the initial solution.

Note that the product MV gives the number of moles of solute

$$MV = (\text{moles / liter}) \times \text{liter} = \text{moles}$$

20. **(B)** One mole of sucrose is 342 grams.

$$\text{C} \quad 12 \times 12 \text{ (atomic weight)} = 144$$

$$\text{H} \quad 22 \times 1 \qquad\qquad\quad = 22$$

$$\text{O} \quad 11 \times 16 \qquad\qquad\quad = 176$$

These total to 342. 342 grams in one liter makes a 1 M strength. In one-half liter, 500 ml, this measured amount is also halved.

21. **(A)**

$$\text{mole fraction } X_{CH_3OH} = 0.452$$

$$\text{mole fraction } X_{C_2H_5OH} = 0.260$$

$$\text{mole fraction } X_{C_3H_7OH} = 0.287$$

Therefore, the volume of 1.00 mole of this solution is

$$0.452\ (16.0) + 0.260\ (20) + 0.287\ (50)$$

22. **(D)** These are examples of colloids. Usually colloids are not visible to the naked eye (diameter in the range of 1 to 500 nm). Colloids, however, are large enough to reflect light from their surface (Tyndall effect). The scattered light can be viewed at right angles to the beam. True solutions do not display this effect due to the smaller solute particles.

23. **(E)** The molarity of a solution is the number of moles of solute per liter of solution, not solvent. Adding the salt to the solvent increases its volume. Since the final volume is greater than 1.000 liter, the final concentration is:

$$\frac{0.1000\ \text{mole}}{> 1.000\ \text{liter}} = < 0.1000\ \text{M}$$

24. **(B)** One mole of NaOH has a weight of 40 g. We may determine that 40 g is 20% of 200 g from:

$$40 = (0.2)\ x$$

$$x = 200$$

Thus, we require $200 - 40 = 160$ g of solvent. This condition is satisfied by water since it has a density of 1 g/cm^3 but not by acetone, ammonia, or carbon tetrachloride.

25. **(C)** Converting 58.5 g of sodium chloride to moles of sodium chloride:

$$58.5\ \text{g} \times \frac{1\ \text{mole}}{58.5\ \text{g}} = 1\ \text{mole of NaCl}$$

The molality of a solution, m, is defined as the number of moles of solute dissolved in one kilogram of solvent. Therefore we have

$$m = 1 = \frac{1\ \text{mole}}{x\ \text{kg}}$$

and $\quad x = 1$ kg of water

26. **(D)** A suspension will have a cloudy or opaque color. It will settle upon standing, will not exhibit Brownian motion, will not pass through filter paper, and the particles will be visible with a microscope.

27. **(A)** The molarity of a solution is defined as the number of moles of solute in a liter of solution.

$$\text{molarity} = \frac{\text{moles of solute}}{\text{liters of solution}}$$

To solve this problem, we must first determine the number of moles of KCl in 32.0 g of KCl, which is:

$$\text{moles} = \frac{m_{KCl}}{MW_{KCl}} = \frac{32.0\ g}{74.6\ g/mol} = .428\ \text{moles}$$

Substituting into this expression for molarity and expressing the volume in liters,

$$M = \frac{.428\ \text{moles}}{.425\ \text{liters}} = 1.0$$

28.　**(A)**

$$V_A \times M_A = V_B \times M_B$$

$$V_A = \frac{11 \times 0.5\ M}{4\ M} = 0.125\ l$$

29.　**(C)**　A homogeneous mixture is one of uniform composition. All of the choices are homogeneous except for the heterogeneous mixture of sand in water, in which case you can see the separate bits of sand dispersed in the water medium.

30.　**(E)**　The mole fraction of a component in a solution is the number of moles of that component divided by the total number of moles,

$$X_{CH_3OH} = \frac{\text{moles of } CH_3OH}{\text{total number of moles present in solution}}$$

The number of moles of each component:

$$\text{moles } H_2O\ \frac{53.0}{18.0} = 2.90$$

$$\text{moles } CH_3OH = \frac{20.3}{32.0} = .63$$

$$\text{moles } CH_3CH_2OH = \frac{15.0}{46.0} = .32$$

Therefore, the mole fraction of CH_3OH is:

$$X_{CH_3OH} = \frac{.63}{21.90 + .63 + .32}$$

$$= \frac{.63}{3.85} = .16$$

31. **(C)** Normality is defined as the number of equivalent weights of solute per liter of solution.

$$N = \frac{\text{Number of equivalent weights}}{1}$$

To solve this problem we must first determine the number of equivalent weights there are in 23.2 g of H_2SO_4. For acids the equivalent weight is defined as the mass of the acid that will furnish 1 mole of hydrogen ions. Since 1 mole of H_2SO_4 (98.0 g) produces 2 moles of H^+ ions,

$$H_2SO_4 \rightarrow 2H^+ + SO_4^-,$$

its equivalent weight is 1/2 the molecular weight or 49.0 g. The number of equivalent weights in 23.2 g of H_2SO_4 will be:

$$\text{The number of equivalent weights} = \frac{\text{g present}}{\text{equivalent weight of } H_2SO_4}$$

$$= \frac{23.2 \text{ g}}{49.0 \text{ g/eq wt}} = .47$$

and the normality is then calculated by

$$N = \frac{\text{number of eq wt}}{\text{liter}}$$

$$N = \frac{.47 \text{ eq wt}}{.40 \, l} = 1.2 \text{ N}$$

32. **(B)** The molarity, M, is defined as the number of moles of solute per liter of solution.

$$M = \frac{\text{moles}}{V_1}$$

Rearranging and solving for volume, we have

$$V = \frac{\text{moles}}{M}$$

Since the problem is asking for the number of grams, it is necessary to substitute m/MW for the number of moles.

$$V = \frac{m/MW}{M} = \frac{2.3 \text{ g}/58.5 \text{ g/mol}}{1.3 \text{ mol}/l}$$

$$V = .302 \ l \text{ or } 30.2 \text{ ml}$$

33. **(D)**   One liter of a 2 M solution contains two moles of solute. Sodium chloride dissociates completely to Na^+ and Cl^- ions, so a 2 M sodium chloride solution contains two moles of Na^+ and two moles of Cl^- for a total of four moles of ions.

34. **(B)**   Since $M_1V_1 = M_2V_2$, we have:

$$(0.5)\,(2,000) = (10)\ V_2$$

Upon rearranging the equation:

$$V_2 = \frac{(0.5)\,(2,000)}{10} = 100 \text{ ml}$$

Note that 2 liters was converted to 2,000 ml.

35. **(D)**   The molality of a solution (m) is defined as the number of moles of solute dissolved in one kilogram of solvent. The number of moles of NaOH to be used is determined to be:

$$20 \text{ g of NaOH} \times \frac{1 \text{ mole of NaOH}}{40 \text{ g of NaOH}} = 0.5 \text{ mole of NaOH}$$

Thus,

$$0.5 \text{ m} = \frac{0.5 \text{ mole of NaOH}}{x \text{ kilograms of water}}$$

Rearranging,

$$x = \frac{0.5}{0.5} = 1 \text{ kg of water}$$

SOLUTION CHEMISTRY REVIEW

Formality (F) of a solution is the number of gram formula weights per liter of solution.

Normality (N) of a solution is the number of gram equivalent weights per liter of solution.

The differences in molarity, formality, and normality are sufficiently subtle to justify additional explanation. The "formula weight" may differ from the "molecular weight" if hydration is present in the solute but, of course, loses its identity when dissolved in an aqueous solution. For example, the formula weight of $Mg(NO_3)_2 \times 6H_2O$ is 256.43 but, in an aqueous solution, the water of hydration loses its identity and only magnesium (MG^{+2}) and nitrate ($NO_3)^-$ ions can be traced to the solute. Formula weights and molecular weights also differ for complicated molecules, such as polymers, where the formula is known but the number of monomer units in an individual molecule, and hence the size of the molecule, are uncertain.

Normality is related to the reactivity of a species. In the example given above, if 76.9 g of $Mg(NO_3)_2 \times 6H_2O$ is dissolved in water to form one liter of solution, the formality and molarity are both 0.3 (76.9/256.43). The salt completely ionizes to form magnesium (Mg^{+2}) ions and nitrate ($NO_3)^-$ ions. The concentration of magnesium ions (Mg^{+2}) is 0.3 M, and of nitrate ions ($NO_3)^-$ is 0.6 M. However, the normality of both the magnesium ions (Mg^{+2}) and the nitrate ions ($NO_3)^-$ is 0.6 N. For magnesium, the normality is twice the molarity because each ion has a charge of +2. While it is generally true, for ions, that the normality is equal to the molarity times the charge on the ion, it is important to examine the reaction context before specifying the normality.

In calculating the volumes of solutions required to mix a new solution of intermediate concentration or to neutralize species from separate solutions, it is important to calculate the amount (i.e., moles) of solute in each solution. In the problems in this text, non-ideal volume effects are, in general, neglected, and you may assume the volumes are additive. The procedure is best illustrated by an example.

Determine the volume of 0.1 N H_2SO_4 and 0.1 M H_2SO_4 needed to neutralize 100 ml of 0.5 M NaOH. It is important to understand the meaning of molarity and normality.

In 100 ml of 0.5 M NaOH, there exists 0.05 moles of OH^- ions. Therefore, 0.05 moles of H^+ ions will be required for neutralization. In 0.1 M H_2SO_4, there are 0.2 moles of H^+ per liter available for reaction, while in 0.1 N H_2SO_4, there are only 0.1 moles of H^+ per liter available for reaction. For the 0.1 N solution, 500 ml (0.5 liters) will contain 0.05 moles of H^+, while for the 0.1 M (0.2 N) solution, 250 ml (0.25 liters) will contain the required 0.05 moles of H^+.

There are three types of solutions: gaseous, liquid, and solid.

The most common type of solution consists of a solute dissolved in a liquid.

The atmosphere is an example of a gaseous solution.

Solid solutions, of which many alloys (mixtures of metals) are examples, are of two types:

Substitutional solid solutions in which atoms, molecules, or ions of one substance take the place of particles of another substance in its crystalline lattice.

Interstitial solid solutions are formed by placing atoms of one kind into voids, or interstices, that exist between atoms in the host lattice.

1. Ways of Expressing Concentration in Solution

Mole fraction is the number of moles of a particular component of a solution divided by the total number of moles of all of the substances present in the solution:

$$X_A = \frac{n_A}{n_A + n_B + n_C + \ldots}$$

$$\sum_{i=1}^{N} X_i = 1$$

Mole percent is equal to $100\% \times$ mole fraction. Weight fraction specifies the fraction of the total weight of a solution that is contributed by a particular component. Weight percent is equal to $100\% \times$ weight fraction.

Molarity (M) of a solution is the number of moles of solute per liter of solution.

$$\text{molarity (M)} = \frac{\text{moles of solute}}{\text{liters of solution}}$$

Molality of a solution is the number of moles of solute per kilogram (1,000g) of solvent.

$$\text{molality (m)} = \frac{\text{moles of solute}}{\text{kilograms of solvent}}$$

When using water as a solvent, the molality (m) of a dilute solution is equal to the molarity (M).

PROBLEM

In over 90 percent of cases, the concentration of ethanol (C_2H_5OH, density = 0.80 g/ml;) in blood necessary to produce intoxication is 0.0030 g/ml. A concentration of 0.0070 g/ml is fatal. What volume of 80 proof (40% ethanol by volume) Scotch whiskey must an intoxicated person consume before the concentration of ethanol in his blood reaches a fatal level? Assume that all the alcohol goes directly to the blood and that the blood volume of a person is 7.0 liters.

SOLUTION

The difference between a fatal concentration of ethanol and an intoxicating concentration of ethanol is 0.0070 g/ml – 0.0030 g/ml = 0.0040 g/ml. We must determine the total volume of ethanol in the blood which corresponds to a concentration of 0.0040 g/ml and then calculate the volume of Scotch whiskey which will provide this concentration.

The total mass of ethanol in blood needed to raise the concentration from an intoxicating to a fatal level is equal to concentration of ethanol × volume of blood = 0.0040 g/ml × 7 liters = 0.0040 g/ml × 7000 ml = 28 g of ethanol. Dividing this mass by the density, we obtain the corresponding volume of ethanol, or 28 g/0.80 g/ml = 35 ml of ethanol.

The amount of Scotch whiskey that must be consumed must provide 35 ml of ethanol. But the Scotch whiskey is only 40% ethanol, or 0.40 ml ethanol/ml Scotch. Let v denote the volume of Scotch in ml. Then

(ratio of ethanol to Scotch) × volume of Scotch = volume of ethanol

0.40 ml ethanol/ml Scotch × v ml Scotch = 35 ml ethanol

or, $$v = \frac{35 \text{ ml ethanol}}{6.40 \text{ ml ethanol/Scotch}} = 88 \text{ ml Scotch.}$$

Thus, under our assumptions, an intoxicated person must drink 88 ml of 80 proof Scotch whiskey (about 3 ounces) before ethanol reaches a fatal level in his blood.

PROBLEM

What is the molality of a solution in which 49 g of H_2SO_4 (MW 98) is dissolved in 250 grams of water?

SOLUTION

Molality is defined as the number of moles of solute per kilogram of solvent.

$$\text{molality} = \frac{\text{moles of solute}}{\text{no. of kg of solvent}}$$

Here, the solute is H_2SO_4 and the solvent is water. In this problem, one is given the number of grams of solute and the number of grams of solvent. One must calculate the number of moles of solute and the number of kilograms of solvent.

The number of moles of solute (H_2SO_4) is found by dividing the number of grams available by the molecular weight.

$$\text{no. of moles} = \frac{\text{no. of grams}}{\text{MW}}$$

$$\text{no. of moles of } H_2SO_4 = \frac{49\text{ g}}{98\text{ g/mol}} = 0.5\text{ moles}$$

Grams can be converted to kilograms by multiplying the number of grams by the conversion factor 1 kg/1,000 g. For the water, 250 × 1 kg/1,000 g = .250 kg. The molality can now be found.

$$\text{molality} = \frac{\text{no. of moles of } H_2SO_4}{\text{no. of kg of } H_2O}$$

$$\text{molality} = \frac{0.5\text{ moles}}{.250\text{ kg}} = 2.0\text{ moles/kg}$$

PROBLEM

Calculate the molarity of a solution containing 10.0 grams of sulfuric acid in 500 ml of solution. (MW of H_2SO_4 = 98.1.)

SOLUTION

The molarity of a compound in a solution is defined as the number of moles of the compound in one liter of the solution. In this problem, one is told that there are 10.0 grams of H_2SO_4 present. One should first calculate the number of moles that 10.0 g represents. This can be done by dividing 10.0 g by the molecular weight of H_2SO_4.

$$\text{number of moles} = \frac{\text{amount present in grams}}{\text{molecular weight}}$$

$$\text{number of moles of } H_2SO_4 = \frac{10.0\text{ g}}{98.1\text{ g/mol}} = 0.102\text{ moles}$$

Since molarity is defined as the number of moles in one liter of solution, and since, one is told that there is 0.102 mole in 500 ml ($^1/_2$ of a liter), one should multiply the number of moles present by 2. This determines the number of moles in H_2SO_4 present in 1,000 ml.

Number of moles in 1,000 ml = 2 × 0.102 = 0.204

Because molarity is defined as the number of moles in 1 liter, the molarity (M) here is 0.204 M.

PROBLEM

Calculate the normality of a solution containing 2.45 g of sulfuric acid in 2.00 liters of solution. (MW of H_2SO_4 = 98.1.)

SOLUTION

Normality is defined by the following equation.

$$normality = \frac{grams\ of\ solute}{equivalent\ weight \times liters\ of\ solution}$$

In this problem, one is given the grams of solute (H_2SO_4) present and the number of liters of solution it is dissolved in. One equivalent of a substance is the weight in which the acid contains one gram atom of replaceable hydrogen. This means that when the acid is dissolved in a solution, and it ionizes, that in one equivalent weight, one hydrogen atom is released. The equivalent weight for acids is defined as:

$$equivalent\ weight = \frac{MW}{no.\ of\ replaceable\ H}.$$

When H_2SO_4 is dissolved in a solution, there are two replaceable hydrogen atoms as shown in the following equation:

$$H_2SO_4 \leftrightarrow 2H^+ + SO_4^=$$

This means that in calculating the equivalent weight for H_2SO_4 the molecular weight is divided by 2.

$$equivalent\ weight\ of\ H_2SO_4 = \frac{98.1\ g}{2\ equivalents}$$

$$= 49\ g/equiv$$

The normality can now be calculated.

$$normality = \frac{grams\ of\ solute}{equivalent\ weight \times liters\ of\ solution}$$

$$normality = \frac{2.45\ g}{49\ g/equiv \times 2.0\ liters}$$

$$= 0.025\ equiv/liter$$

Drill 1: Ways of Expressing Concentration in Solution

1. A one liter of a solution of 2 M NaOH can be prepared with

   (A) 20 g of NaOH.   (D) 80 g of NaOH.

   (B) 40 g of NaOH.   (E) 100 g of NaOH.

   (C) 60 g of NaOH.

2. A solution has a density of 1.2 in grams/ml. Pure water is added to it. A probable new solution density is

   (A) .9.   (D) 1.2.

   (B) 1.0.   (E) 1.5.

   (C) 1.1.

3. How many grams of hydrochloric acid are there in 500 ml of a 5M solution?

   (A) 36.5 g   (D) 91.25 g

   (B) 54.75 g   (E) 109.5 g

   (C) 73.0 g

4. How many grams of sulfuric acid are contained in 3.00 liters of 0.500 N solution? (MW of H_2SO_4 = 98.1.)

   (A) 73.5 g   (D) 69.2

   (B) 95.2   (E) 35.3

   (C) 42.1

5. 2.3 g of ethanol (C_2H_5OH, molecular weight = 46 g/mol) is added to 500 g of water. Determine the molality of the resulting solution.

   (A) 0.11 molal   (D) 0.05 molal

   (B) 0.001 molal   (E) 0.07 molal

   (C) 0.1 molal

SOLUTION CHEMISTRY DRILLS

ANSWER KEY

Drill 1 — Ways of Expressing Concentration in Solution

1. (D)
2. (C)
3. (D)
4. (A)
5. (C)

GLOSSARY: SOLUTION CHEMISTRY

Formality

A measure of concentration. The number of gram-formula weights of solute per liter of solution.

Molality

A measure of concentration. Number of moles of solute per kilogram of solvent.

Molarity

A measure of concentration. Number of moles of solute per liter of solution.

Mole Fraction

A concentration unit that is defined as the number of moles of a component divided by the total number of moles of all components.

Normality

A concentration unit that is defined as the number of gram-equivalent weights of a solute per liter of solution.

Solution

A homogeneous mixture of two or more substances.

CHAPTER 9

Equilibrium

➤ Diagnostic Test
➤ Equilibrium Review & Drills
➤ Glossary

EQUILIBRIUM DIAGNOSTIC TEST

1. Ⓐ Ⓑ Ⓒ Ⓓ Ⓔ
2. Ⓐ Ⓑ Ⓒ Ⓓ Ⓔ
3. Ⓐ Ⓑ Ⓒ Ⓓ Ⓔ
4. Ⓐ Ⓑ Ⓒ Ⓓ Ⓔ
5. Ⓐ Ⓑ Ⓒ Ⓓ Ⓔ
6. Ⓐ Ⓑ Ⓒ Ⓓ Ⓔ
7. Ⓐ Ⓑ Ⓒ Ⓓ Ⓔ
8. Ⓐ Ⓑ Ⓒ Ⓓ Ⓔ
9. Ⓐ Ⓑ Ⓒ Ⓓ Ⓔ
10. Ⓐ Ⓑ Ⓒ Ⓓ Ⓔ
11. Ⓐ Ⓑ Ⓒ Ⓓ Ⓔ
12. Ⓐ Ⓑ Ⓒ Ⓓ Ⓔ
13. Ⓐ Ⓑ Ⓒ Ⓓ Ⓔ
14. Ⓐ Ⓑ Ⓒ Ⓓ Ⓔ
15. Ⓐ Ⓑ Ⓒ Ⓓ Ⓔ
16. Ⓐ Ⓑ Ⓒ Ⓓ Ⓔ
17. Ⓐ Ⓑ Ⓒ Ⓓ Ⓔ
18. Ⓐ Ⓑ Ⓒ Ⓓ Ⓔ
19. Ⓐ Ⓑ Ⓒ Ⓓ Ⓔ
20. Ⓐ Ⓑ Ⓒ Ⓓ Ⓔ

21. Ⓐ Ⓑ Ⓒ Ⓓ Ⓔ
22. Ⓐ Ⓑ Ⓒ Ⓓ Ⓔ
23. Ⓐ Ⓑ Ⓒ Ⓓ Ⓔ
24. Ⓐ Ⓑ Ⓒ Ⓓ Ⓔ
25. Ⓐ Ⓑ Ⓒ Ⓓ Ⓔ
26. Ⓐ Ⓑ Ⓒ Ⓓ Ⓔ
27. Ⓐ Ⓑ Ⓒ Ⓓ Ⓔ
28. Ⓐ Ⓑ Ⓒ Ⓓ Ⓔ
29. Ⓐ Ⓑ Ⓒ Ⓓ Ⓔ
30. Ⓐ Ⓑ Ⓒ Ⓓ Ⓔ
31. Ⓐ Ⓑ Ⓒ Ⓓ Ⓔ
32. Ⓐ Ⓑ Ⓒ Ⓓ Ⓔ
33. Ⓐ Ⓑ Ⓒ Ⓓ Ⓔ
34. Ⓐ Ⓑ Ⓒ Ⓓ Ⓔ
35. Ⓐ Ⓑ Ⓒ Ⓓ Ⓔ
36. Ⓐ Ⓑ Ⓒ Ⓓ Ⓔ
37. Ⓐ Ⓑ Ⓒ Ⓓ Ⓔ
38. Ⓐ Ⓑ Ⓒ Ⓓ Ⓔ
39. Ⓐ Ⓑ Ⓒ Ⓓ Ⓔ
40. Ⓐ Ⓑ Ⓒ Ⓓ Ⓔ

EQUILIBRIUM
DIAGNOSTIC TEST

This diagnostic test is designed to help you determine your strengths and your weaknesses in equilibrium. Follow the directions and check your answers.

Study this chapter for the following tests:
ACT, AP Chemistry, CLEP General Chemistry,
GED, MCAT, MSAT, PRAXIS II Subject Assessment:
Chemistry, SAT II: Chemistry, GRE Chemistry

---

40 Questions

DIRECTIONS: Choose the correct answer for each of the following problems. Fill in each answer on the answer sheet.

1. At 25°C, the k_{sp} for $CaSO_4$ and Ag_2SO_4 are 2.4×10^{-5} and 1.2×10^{-5} respectively. Which of the following is true?

    (A) The solubility of $CaSO_4$ is twice that of Ag_2SO_4.

    (B) The solubility of Ag_2SO_4 is twice that of $CaSO_4$.

    (C) The solubility of Ag_2SO_4 is sensitive to the square of the sulphate ion concentration.

    (D) The solubilities of $CaSO_4$ and Ag_2SO_4 are equal.

    (E) The solubilities of Ag_2SO_4 and $CaSO_4$ in mol liter^{-1} differ by a factor of 2.9.

2. Which of the following reactions goes to completion because a gas is evolved?

    (A) $2H_2 + O_2 \rightarrow 2H_2O$ (D) $Zn + 2HCl \rightarrow ZnCl_2 + H_2$

    (B) $N_2 + 3H_2 \rightarrow 2NH_3$ (E) $NO + 2O_2 \rightarrow 2NO_2$

    (C) $2CO + O_2 \rightarrow 2CO_2$

3. For a certain first order reaction, the time required for half of an initial amount to decompose is 3 minutes. If the initial concentration of A is 1 molar, the time required to reduce the concentration of A to 0.25M is (Take ln2 to be 0.693 and ln4 to be 1.39):

(A) 4.5 min.   (D) 8.0 min.

(B) 6.0 min.   (E) 6.6 min.

(C) 12.0 min.

4.   The yield of AB (g)

   A (g) + B (g) $\leftrightarrow$ AB (g) + heat

   would be increased by

   (A) decreasing the pressure.

   (B) adding additional AB to the reaction mixture.

   (C) decreasing the temperature.

   (D) adding a nonreactive liquid to the reaction mixture.

   (E) decreasing the volume of the reaction container.

5.   A catalyst will increase the rate of a chemical reaction by

   (A) shifting the equilibrium to the right.

   (B) lowering the activation energy.

   (C) shifting the equilibrium to the left.

   (D) increasing the activation energy.

   (E) None of the above.

6.   What is the equilibrium constant for the following reaction?

   $aA + bB + cC \leftrightarrow dD + eE$

   (A) $[A]^a[B]^b[C]^c[D]^d[E]^e$ (D) $\dfrac{1}{[A]^a[B]^b[C]^c}$

   (B) $[D]^d[E]^e$ (E) $\dfrac{[D]^d[E]^e}{[A]^a[B]^b[C]^c}$

   (C) $[A]^a[B]^b[C]^c$

7.   What would the equilibrium constant be if D is a liquid and E is a pure solid?

   $aA + bB + cC \leftrightarrow dD + eE$

   (A) $[A]^a[B]^b[C]^c[D]^d[E]^e$ (B) $[D]^d[E]^e$

(C) $[A]^a[B]^b[C]^c$ (D) $\dfrac{1}{[A]^a[B]^b[C]^c}$

(E) $\dfrac{[D]^d[E]^e}{[A]^a[B]^b[C]^c}$

8. A solution contains 0.01 mol KI, 0.10 mol KBr, and 0.10 mol KCl per liter. $AgNO_3$ is gradually added to this solution. Which will be precipitated first, AgI, AgBr, or AgCl? (Solubility products are $K_{SP\ AgI} = 1.5 \times 10^{-16}$, $K_{SP\ AgBr} = 3.3 \times 10^{-13}$, $K_{SP\ AgCl} = 1.8 \times 10^{-10}$)

(A) AgI

(B) AgBr

(C) AgCl

(D) Both AgBr and AgCl.

(E) Cannot solve with given information.

9. Suppose that the reaction rate of an inorganic reaction mixture at 35°C is double the reaction rate at an earlier temperature setting. All other environmental factors are held constant. This earlier temperature was most likely

(A) 0°.   (D) 40°.

(B) 10°.   (E) 45°.

(C) 25°.

10. For a certain second order reaction $A + B \rightarrow C$, it was noted that when the initial concentration of A is doubled while B is held constant, the initial reaction rate doubles, and when the initial concentration of B is doubled while A is held constant, the initial reaction rate increases fourfold. What is the rate expression for this reaction?

(A) $r = k[A][B]$ (D) $r = k[A]^3[B]^1$

(B) $r = k[A]^2[B]^3$ (E) $r = k[A]^2[B]^1$

(C) $r = k[A]^1[B]^2$

11. What is the solubility of AgCl in water if $k_{sp} = 1.6 \times 10^{-10}$?

(A) 1.6×10^{-10} (D) 1.6×10^{-5}

(B) 3.2×10^{-10} (E) 3.2×10^{-5}

(C) 1.3×10^{-5}

12. Which of the following serves as a catalyst in the reaction

$$CH_2 = CH_2 + H_2 + Pt \rightarrow CH_3CH_3 + Pt$$

(A) C

(D) Pt

(B) $CH_2 = CH_2$

(E) CH_3CH_3

(C) H_2

13. The equilibrium constant for the reaction

$$CO_2(g) + O_2(g) \rightarrow CO_2(g) \text{ (not balanced)}$$

may be expressed as

(A) $K = \dfrac{[CO_2]}{[CO][O_2]}$

(D) $K = \dfrac{[CO]^2[O_2]}{[CO]}$

(B) $K = \dfrac{[CO][O_2]}{[CO_2]}$

(E) $K = \dfrac{[CO_2]^2}{[CO]^2[O_2]}$

(C) $K = [CO]^2[O_2][CO_2]^2$

14. K_c for a reaction $A(g) + 3B(g) \leftrightarrow 2C(g) + 2D(s)$ has units

(A) mole^{-2} liter^2

(D) mole^{-1} liter

(B) mole^2 liter^{-2}

(E) mole^{-1} liter^{-1}

(C) mole liter^{-2}

Questions 15–18

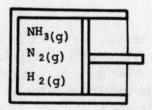

Three gases are in equilibrium in a closed chamber sealed with a piston. The equilibrium expression is

$$2NH_3(g) \leftrightarrow N_2(g) + 3H_2(g)$$

(A) The mole fraction of N_2 increases.

(B) The mole fraction of N_2 decreases.

(C) The mole fraction of N_2 remains the same.

(D) The mole fraction of N_2 increases and then decreases.

(E) The direction of change cannot be predicted with the given information.

Which of the previous choices occurs in each of the following cases?

15. The piston is pushed into the chamber.

16. More H_2 is added as the piston is adjusted to maintain constant pressure.

17. The chamber is heated while the piston is held steady.

18. A catalyst is added while the piston is held steady.

19. Which of the expressions below represents the correct rate law of the reaction?

    $2A + B \rightarrow C$

Experiment	initial conc. of A mol/l	initial conc. of B mol/l	initial rate (mole/$l \cdot$sec)
1	1	1	1.2
2	2	1	4.8
3	1	2	2.4
4	3	1	10.8
5	1	3	3.6

   (A) rate = $k[A][B]$

   (B) rate = $k[A]^2[B]$

   (C) rate = $k[A]$

   (D) rate = $k[B]$

   (E) rate = $k[A][B]^2$

20. For the previous reaction, determine the value of k

   (A) $1.2\ l^2/\text{mole}^2$ sec

   (B) $1.2\ \text{sec}^{-1}$

   (C) 1.2 sec

   (D) $1.2\ l^{-1}\,\text{mole}^{-1}$sec

   (E) $1.2\ l^{-1}\,\text{mole}^{-1}\,\text{sec}^{-1}$

21. The specific rate constant of the natural radioactive potassium isotope $^{40}_{19}K$ is 5.33×10^{-10} year^{-1}, so the half-life of $^{40}_{19}K$ is

   (A) 1.3×10^9 years.

   (B) 1.3×10^{-9} years.

   (C) $9 \times 10^{1.3}$ years.

   (D) 5.33×10^5 years.

   (E) 2.665×10^{-10} years.

22. Given the mechanism below for the oxidation of nitrogen (II) oxide; determine which expression gives the rate of formation of NO_2.

$$NO + NO \underset{k_2}{\overset{k_1}{\rightleftarrows}} N_2O_2 \text{ Fast equilibrium}$$

$$N_2O_2 + O_2 \overset{k_3}{\rightarrow} 2NO_2$$

(A) $k_3[N_2O_2][O_2]$

(D) $2\dfrac{k_2}{k_1}k_3[NO][O_2]$

(B) $\dfrac{k_1}{k_2}k_3[NO]^2[O_2]$

(E) $2\dfrac{k_1}{k_2}k_3[NO][O_2]$

(C) $\dfrac{k_2}{k_1}k_3[NO]^2[O_2]$

23. $$N_2 + 3H_2 \leftrightarrow 2NH_3 \uparrow + \text{heat}$$

In this reversible reaction, the equilibrium shifts to the right because of all the following factors *except:*

(A) adding heat.

(D) increasing pressure on reactants.

(B) adding reactant amounts.

(E) yielding an escaping gas.

(C) formation of ammonia gas.

24. For the following reaction at 500°K

$$C(s) + CO_2(g) \leftrightarrow 2CO(g)$$

the equilibrium mixture contained CO_2 and CO at partial pressures of 7.6 atm and 3.2 atm respectively. The value of the k_p is

(A) 2.4 atm.

(D) 1.0 atm.

(B) 18.1 atm.

(E) .4 atm.

(C) .6 atm.

25. $$HCOO^- H_2O \rightarrow HCOOH + OH^-$$

$$\text{Rate} = k[HCOOH][OH^-]$$

What is the order of this reaction?

(A) 1

(D) 5

(B) 2

(E) 6

(C) 4

26. For which one of the following equilibrium equations will k_p equal k_c?

    (A) $PCl_5 \leftrightarrow PCl_3 + CL_2$.

    (D) $3H_2 + N_2 \leftrightarrow 2NH_3$.

    (B) $COCl_2 \leftrightarrow CO + Cl_2$.

    (E) $2SO_3 \leftrightarrow 2SO_2 + O_2$.

    (C) $H_2 + I_2 \leftrightarrow 2HI$.

27. A chemist dissolves an excess of $BaSO_4$ in pure water at 25°C. If its $k_{sp} = 1 \times 10^{-10}$, what is the concentration of the barium in the water?

    (A) 10^{-4} M

    (D) 10^{-10} M

    (B) 10^{-5} M

    (E) 10^{-20} M

    (C) 10^{-6} M

28. The equilibrium expression, $k = [Ag^+] [Cl^-]$ describes the reaction

    (A) $AgCl \rightarrow Ag^+ + Cl^-$.

    (D) $Ag + Cl \rightarrow Ag^+ + Cl^-$.

    (B) $Ag^+ + Cl^- \rightarrow AgCl$.

    (E) none of the above.

    (C) $Ag^+ + Cl^- \rightarrow Ag + Cl$.

29. In a lab, each of the following factors will vary to affect reaction rate *except*

    (A) catalyst used.

    (D) oxygen availability.

    (B) concentration of reactants.

    (E) temperature.

    (C) identity of reactants.

30. The k_{sp} of silver chromate (Ag_2CrO_4) is 1.4×10^{-12}. What is the solubility in moles per liter of silver chromate

    (A) 7.0×10^{-5} moles/liter

    (D) 1.0×10^{-18} moles/liter

    (B) 3.5×10^{-13} moles/liter

    (E) 4.7×10^{-23} moles/liter

    (C) 6.2×10^{-15} moles/liter

31. Which of the following shifts the equilibrium of the following reaction to the right?

    $$A(g) + B(g) + C(g) \leftrightarrow A(g) + BC(g)$$

    (A) Addition of more A

    (D) Decreasing the temperature

    (B) Removal of B

    (E) Increasing the temperature

    (C) Increasing the pressure

32. The equilibrium expression, $K_e = [CO_2]$ represents the reaction:

    (A) $C(s) + O_2(g) \leftrightarrow CO_2(g)$

    (D) $CO_2(g) \leftrightarrow C(s) + O_2(g)$

    (B) $CO(g) + \dfrac{1}{2}O_2(g) \leftrightarrow CO_2(g)$

    (E) $CaO(s) + CO_2(g) \leftrightarrow CaCO_3(s)$

    (C) $CaCO_3(s) \leftrightarrow CaO(s) + CO_2(g)$

33. The half life of $^{224}_{88}Rn$ is 3.64 days. How many μg of $^{224}_{88}Rn$ would be left after 18.2 days if the starting amount weighed 2.0 μg?

    (A) .06

    (D) .69

    (B) .40

    (E) None

    (C) .19

34. Consider the following reversible reaction:

    $$H_2 + 3N_2 \leftrightarrow 2NH_3$$

    Its equilibrium constant "K" is expressed as:

    (A) $\dfrac{[NH_3]}{[N_2][H_2]^3}$

    (D) $[NH^3]$

    (B) $\dfrac{[NH_3]^2}{[N_2]^3[H_2]}$

    (E) $[N_2]^2[H_2]^3$

    (C) $\dfrac{[NH_3]}{[N_2][H_2]}$

35. The k_{sp} for PbI_2 is 8.7×10^{-9}. What is the molar solubility of PbI_2?

    (A) 1.3×10^{-3}

    (D) 1×10^{-9}

    (B) 8.7×10^{-3}

    (E) 0

    (C) 9.3×10^{-5}

36. The reaction $2H_2 + NO \rightarrow H_2O + \frac{1}{2}N_2$ has a rate law of the form: rate = $k[H_2]^x[NO]^y$.

    Find the sum of x and y from the given data.

experiment	$[H_2]$, M	$[NO]$, M	rate, M s^{-1}
a	1.0×10^{-3}	4.6×10^{-2}	3.1×10^{-4}
b	2.0×10^{-3}	4.6×10^{-2}	6.2×10^{-4}
c	1.0×10^{-3}	1.84×10^{-1}	5.0×10^{-3}

(A) $\dfrac{1}{2}$ (D) 4

(B) 2   (E) 5

(C) 3

37. Which statement is true for a liquid/gas mixture in equilibrium?

    (A) The equilibrium constant is dependent on temperature.

    (B) The amount of the gas present at equilibrium is independent of pressure.

    (C) All interchange between the liquid and gas phases has ceased.

    (D) All of the above.

    (E) None of the above.

38. Equilibrium reactions are characterized by

    (A) going to completion.

    (B) being nonspontaneous.

    (C) the presence of both reactants and products in a definite proportion.

    (D) (A) and (B)

    (E) (A), (B), and (C)

39. How long will it take for a sample of radioactive material to disintegrate to the extent that only 2% of the original concentration remains if the material has a half-life of 5.2 years? Note that

    $\ln(C_o/C) = kt.$

(A) $\dfrac{\ln 2}{\ln(5.2)}$ (D) $\dfrac{(\log 50)(5.2)}{(\log 2)(2.303)}$

(B) $\left(\dfrac{\ln 50}{\ln 2}\right) 5.2$ (E) $\left(\dfrac{\ln(0.02)}{\ln 2}\right) 5.2$

(C) $(\ln 0.02)(5.2)$

40. For the reaction

    $2A + 1B \rightarrow 2C$

the following rate data was obtained:

Experiment	initial conc. of A mol/l	initial conc. of B mol/l	initial rate mol/l sec
1	.5	.5	10
2	.5	1.0	20
3	.5	1.5	30
4	1.0	.5	40

Which expression below states the rate law for the reaction above?

(A) Rate $= k\,[A]^2\,[B]$

(B) Rate $= k\,[A]\,[B]$

(C) Rate $= k\,[A]^2\,[B]^2$

(D) Rate $= k\,[A]^3$

(E) Rate $= k\,[A]^2\,[B]^3$

EQUILIBRIUM
DIAGNOSTIC TEST

ANSWER KEY

1.	(E)	11.	(C)	21.	(A)	31.	(C)
2.	(D)	12.	(D)	22.	(B)	32	(C)
3.	(B)	13.	(E)	23.	(A)	33.	(A)
4.	(C)	14.	(A)	24.	(B)	34.	(B)
5.	(B)	15.	(B)	25.	(B)	35.	(A)
6.	(E)	16.	(B)	26.	(C)	36.	(C)
7.	(D)	17.	(E)	27.	(B)	37.	(A)
8.	(A)	18.	(C)	28.	(A)	38.	(C)
9.	(C)	19.	(B)	29.	(D)	39.	(B)
10.	(C)	20.	(A)	30.	(A)	40.	(A)

DETAILED EXPLANATIONS
OF ANSWERS

1. **(E)** k_{sp} is the solubility product of a slightly soluble salt.

For $CaSO_4$ $k_{sp} = [Ca^{2+}] [SO_4^{2-}]$

If we let X represent the concentration of SO_4^{2-} anions in mol liter^{-1}, $CaSO_4$ dissociates in equimolar proportions so

$$k_{sp} = [Ca^{2+}] [SO_4^{2-}] = [X] [X] = X_1^2 = 2.4 \times 10^{-5}$$

$$\therefore X_1 \approx 5. \times 10^{-3} \text{ mol liter}^{-1}$$

For Ag_2SO_4 the k_{sp} expression is different because $Ag_2SO_4 \leftrightarrow 2 \, Ag^+ + SO_4^{2-}$ i.e., each mole of Ag_2SO_4 produces 2 moles of Ag^+ and one mole of SO_4^{2-} ions.

$$[Ag^+] = 2X_2 \text{ and } [SO_4^{2-}] = X_2$$

$$k_{sp} = [Ag^+]^2 [SO_4^{2-}] = (2X_2)^2 (X_2)$$

$$k_{sp} = 4X^3 = 1.2 \times 10^{-5}$$

$$X_2 = 0.0144$$

$$X_2 = 14.4 \times 10^{-3}$$

$$\frac{X_2}{X_1} = \frac{14.4 \times 10^{-3}}{5 \times 10^{-3}} = 2.9$$

2. **(D)**   The addition of hydrochloric acid to zinc is a reaction which can be made to go to completion if the product (which in this case happens to be the only gas involved in the reaction) is removed or allowed to escape as the reaction proceeds (LeChatelier).

In the other cases such as in the formation of ammonia, the reaction will also go close to completion but by the removal specifically of NH_3, since the reactants nitrogen and hydrogen are also in the gas phase. The same is true for the other reactions, they will require the removal of the product not just the gas because there are gases both as reactants and products.

3. **(B)**

$$t_{\frac{1}{2}} = \frac{\ln 2}{k} = \frac{0.693}{k} = 3 \text{ min.,}$$

therefore $k = \dfrac{0.693}{3} \text{ min.}^{-1}$

$$t = \ln\left(\frac{[A]_0}{[A]}\right) \times \frac{1}{k} = \ln\left(\frac{1\,\text{mole}}{.25\,\text{mole}}\right) \times \frac{3}{0.693}$$

$$t = \frac{1.39 \times 3}{0.693} \cong 2(3) = 6\,\text{min.}$$

4.   **(C)**   According to Le Chatelier's Principle, if a stress is placed on an equilibrium system, the equilibrium is shifted in the direction which reduces the effect of that stress. This stress may be in the form of changes in pressure, temperature, concentrations, etc. By decreasing the pressure on the system, the system shifts in a direction so as to increase the pressure. For our reaction, the reverse reaction rate would increase since a larger volume (hence a greater pressure) results. This is due to the fact that two moles of gaseous product will result as opposed to one mole if the forward reaction were favored. Note that only gaseous products are accounted for when considering pressure effects. Adding AB to the reaction mixture would also serve to favor the reverse reaction since the system reacts to this stress by producing more A and B. Decreasing the temperature favors the forward reaction since heat (which will counteract the stress by increasing the temperature) is liberated in this process. The addition of a nonreactive liquid to the reaction mixture has no effect on the reaction rates (assuming pressure and temperature to remain constant). Decreasing the volume of the reaction mixture has the same effect as increasing the pressure (Boyle's Law) if temperature is constant.

5.   **(B)**   A catalyst increases the rate of a chemical reaction by lowering the activation energy. The activation energy of a reaction is the amount of energy the molecules or atoms must have, so that when they collide the collision will result in product formation. A lower activation energy means there will be a greater number of molecules possessing sufficient kinetic energy to form products. This will result in a greater reaction rate.

6.   **(E)**   The equilibrium constant (K) includes terms for gases and substances in solution (excluding terms for pure liquids or solids since at a given temperature the concentrations of these are essentially constant). In this case:

$$aA + bB + cC \leftrightarrow dD + eE$$

is $$K = \frac{[D]^d\,[E]^e}{[A]^a\,[B]^b\,[C]^c}$$

7.   **(D)**   The concentrations of pure liquids and pure solids are omitted from the equilibrium expression since they remain essentially constant throughout the reaction. Thus, we have

$$K = \frac{1}{[A]^a \, [B]^b \, [C]^c}$$

8.   **(A)**   The best approach to solve this problem is to first write out the equations of the reactions.

$$AgNO_3 + KI \rightarrow KNO_3 \; AgI \downarrow$$

$$AgNO_3 + KBr \rightarrow KNO_3 + AgBr \downarrow$$

$$AgNO_3 + KCl \rightarrow KNO_3 + AgCl \downarrow$$

So $k_{sp} = [Ag^+] \, [x^-]$

$\therefore k_{AgI} = [Ag^+] \, [I^-] = 1.5 \times 10^{-16}$

where   0.01mol/liter = 0.01M,

so $$[Ag^+] = \frac{1.5 \times 10^{-16}}{[0.01]}$$

$$= 1.5 \times 10^{-14} \text{ mol/liter}$$

This is the minimum concentration of Ag^+ necessary for precipitation of AgI.

$$k_{AgBr} = [Ag^+] \, [Br^-] = 3.3 \times 10^{-13}$$

where   0.1mol/liter = 0.1M

So $[Ag^+] = 3.3 \times 10^{-13}/[0.1] = 3.3 \times 10^{-12}$ mol/liter

$$k_{AgCl} = [Ag^+] \, [Cl^-] = 1.8 \times 10^{-10}$$

where   0.1 mol/liter = 0.1M

So $$[Ag^+] = \frac{1.8 \times 10^{-10}}{[0.1]} = 1.8 \times 10^{-9} \text{ mol/liter}$$

From the above results, it can be noted that a greater concentration of Ag^+ is necessary to cause the precipitation of AgBr and AgCl, so AgI will precipitate first.

9.   **(C)**   Reaction rates of inorganic substances usually double with every 10° increase in temperature. Therefore, if 35°C represents the new, doubled rate, then the original temperature must be 10° less: 35° – 10° = 25°C.

10.   **(C)**   For the second order reaction A + B → C, when [B] is held constant while [A] is doubled, the reaction rate doubles. This means that the rate is directly proportional to [A], so $a = 1$. Then when [A] is held constant and [B] is doubled, the reaction rate increases four times. This means that the rate is propor-

tional to $[B]^2$.

$$\therefore \text{ rate} = k[A]^a [B]^b = k[A]^1 [B]^2$$

11. **(C)** The solubility product of AgCl is

$$k_{sp} = [Ag^+] [Cl^-] = 1.6 \times 10^{-10}$$

Letting $x = [Ag^+]$ and since $[Ag^+] = [Cl^-]$, we have

$$x^2 = 1.6 \times 10^{-10}$$

and $x = 1.3 \times 10^{-5}$

Since $[Ag^+] = [Cl^-] = [AgCl]$, the solubility of silver chloride is 1.3×10^{-5} mole/liter.

12. **(D)** One characteristic of a catalyst is that it remains unchanged by the reaction process. It may now be seen that the platinum, Pt, is the catalyst for this reaction.

13. **(E)** The equilibrium expression of a reaction is specified as the concentrations of the products divided by the concentrations of the reactants each being raised to the power of the corresponding stoichiometric coefficient. For the general reaction

$$aA + bB \rightarrow cC + dD$$

the equilibrium expression is

$$K = \frac{[C]^c[D]^d}{[A]^a[B]^b}$$

The concentrations of pure liquids and solids are omitted from the equilibrium expression since their concentrations change negligibly during the reaction. For the reaction in question, we have

$$K = \frac{[CO_2]^2}{[CO]^2 [O_2]}$$

since the balanced reaction is

$$2CO + O_2 \rightarrow 2CO_2$$

14. **(A)** At equilibrium $A(g) + 3B(g) \, 2C(g) + 2D(s)$ has equilibrium constant as shown below:

$$K_c = \frac{[C]^2}{[A][B]^3} \frac{[\text{mole}/l]^2}{\left[\frac{\text{mole}}{1}\right]\left[\frac{\text{mole}}{1}\right]^3}$$

$$[K_c] = \text{mole}^{-2} \text{ liter}^2$$

15. **(B)** Le Chatelier's Principle applies to both questions 14 and 15. The principle states that a system perturbed from equilibrium will react in the opposite direction so as to restore equilibrium. In the first case, pushing the piston into the chamber increases the pressure of the reaction mixture. The system reacts so as to bring the pressure back down to a new equilibrium point. In the equilibrium expression for the reaction there are two moles of gas on the left side for every four moles on the right. In order to reduce the pressure, then, the equilibrium must shift to the left. Thus, the mole fraction of N_2 decreases.

16. **(B)** In this problem the equilibrium is perturbed by an increase in the mole fraction of H_2. In order to correct this, the reaction must shift to the left until a new equilibrium point is reached. This results in a decrease in the amount of N_2 present.

17. **(E)** Several competing effects may arise from heating the chamber. One important unknown is whether the reaction is exothermic or endothermic. For example, if the reaction is endothermic, we might write:

$$2NH_{3(g)} + \text{heat} \leftrightarrow N_{2(g)} + 3H_{2(g)}$$

Adding heat would drive the reaction forward in the same way that adding more NH_3 would. Since we do not know the heat of reaction we cannot predict the effect of heating the chamber.

18. **(C)** Catalysts affect the rates of reactions by reducing their activation energies. They do *not* affect equilibrium constants or the energy changes (free energy, enthalpy, etc.) associated with reactions.

19. **(B)** Let the rate be expressed as follows:

$$\text{rate} = k[A]^a[B]^m$$

To find n, examine the data and find the ratios of rates for reactions where $[B]$ was kept constant as:

$$\frac{\text{rate}_2}{\text{rate}_1} = \frac{k[2A]^n [B]^m}{k[A]^n [B]^m} = \frac{4.8}{1.2} = 4$$

Simplifying we have

$$2^n = 4 = 2^2$$

$$\rightarrow \quad n = 2$$

or $\quad \dfrac{\text{rate}_4}{\text{rate}_1} = \dfrac{k[3A]^n [B]^m}{k[A] [B]^m} = \dfrac{10.8}{1.2} = 9$

$\longrightarrow \qquad 3^n = 9 = 3^2$

$n = 2$

Similarly for m, consider reactions where $[A]$ = constant.

$$\frac{\text{rate}_3}{\text{rate}_1} = \frac{k[A]^n [2B]^m}{k[A]^n [B]} = \frac{2.4}{1.2} = 2$$

or $\qquad 2^m = 2 = 2^1$

$\longrightarrow \qquad m = 1$

or $\qquad \dfrac{\text{rate}_5}{\text{rate}_1} = \dfrac{k[A]^n [3B]^m}{k[A]^n [B]^m} = \dfrac{3.6}{1.2} = 3$

$3^m = 3$

$m = 1$

$\therefore \text{rate} = k[A]^2 [B]$

20. **(A)** k is given by the expression below:

$\text{rate} = k[A]^2[B]$

$$\frac{1.2 \text{ mole}}{l \times s} = k \left[\frac{1 \text{ mole}}{l}\right]^2 \left[\frac{1 \text{ mole}}{l}\right]$$

$k = 1.2 \text{ liter}^2/\text{mole}^2 \text{ sec.}$

21. **(A)** Radioactive decay is governed by a first-order rate law $R \rightarrow P$

$$\frac{-d[R]}{dt} = k[R]$$

The minus sign indicates a rate of decrease.

$[R]$ – amount present at time t

k – specific rate constant

$$\frac{d[R]}{[R]} = -kdt$$

integrating gives

$\ln[R] = -kt + c$

at $t = 0$, Let $[R] = [R]_o$

$\rightarrow \qquad c = \ln [R]_o$

$\ln[R] = -kt + \ln[R]_o$

or $\qquad \ln[R] - \ln[R]_o = -kt$

$$\frac{\ln[R]}{[R]_0} = -kt$$

The half-life $t_{1/2}$ is the time at which

$$[R] = \frac{[R]_0}{2}$$

$$\frac{\ln[R]_0 / 2}{[R]_0} = -kt_{1/2}$$

$$\ln_{1/2} = -kt_{1/2}$$

$$t_{1/2} = -\frac{(-\ln 2)}{k} = \frac{\ln 2}{k}$$

$$t_{1/2} = \frac{0.693}{5.33 \times 10^{-10}} = 1.3 \times 10^9 \text{ years.}$$

22. **(B)** Given the mechanism as:

$$NO + NO \underset{k_2}{\overset{k_1}{\rightleftarrows}} N_2O_2$$

$$N_2O_2 + O_2 \overset{k_3}{\rightarrow} 2NO_2$$

Then the rate of formation of NO_2 is

$$\frac{d[NO_2]}{dt} = k_3 [N_2O_2] [O_2]$$

The concentration of N_2O_2 is found from the equilibrium relation

$$k_1[NO] [NO] = k_2 [N_2O_2]$$

$$[N_2O_2] = \frac{k_1}{k_2} [NO]^2$$

the rate $= \dfrac{k_1}{k_2} k_3 [NO]^2 [O_2]$

23. **(A)**   Adding reactant amounts increases frequency of collision for more product formation. Removing ammonia, an escaping product gas, creates a void filled by an equilibrium shift to the right to form more NH_3. In the equation, 4 gas volumes form 2. By Le Chatelier's Principle, an altered equilibrium reacts to a stress to relieve the stress. Pressurizing the high volume reactants forces the reaction to the right to relieve this stress. The reaction, however, is exothermic, producing heat, and therefore the *right-to-left* direction absorbs heat. Adding heat throws it in this direction.

24. **(B)** K_p is the notation for the equilibrium constant when the concentrations of both the reactants and products are expressed in terms of their partial pressures. The K_p is defined for this reaction as

$$K_p = \frac{P^2_{CO_2}}{P_{CO}}$$

By convention, concentrations and partial pressures for pure liquids and solids are omitted from the equilibrium constant expression. Therefore,

$$K_p = \frac{7.6^2}{3.2} = 18.1 \text{ atm}$$

25. **(B)**   Two reactants are involved in this reaction, each with a coefficient of one. The order is found by adding their understood coefficients, one plus one.

26. **(C)** K_c is the equilibrium constant when the concentrations are expressed in moles per liter. K_p is the equilibrium constant where the partial pressures of the gases is used in place of the molar concentrations. The relationship between K_p and K_c is:

$$K_p = K_c(RT)^{\Delta n}$$

where Δn is the change in the number of moles of the gas upon going from reactants to products. The correct answer is (C), since there is no change in the number of moles in going from reactants to products. Thus $\Delta n = 0$ and K_p will then be equal to K_c.

27. **(B)**

$$BaSO_4 \leftrightarrow Ba^{2+} + SO_4^{2-}$$

$$K_{sp} = [Ba^{2+}] [SO_4^{2-}] = 1 \times 10^{-10}$$

Since both ions Ba^{2+} and SO_4^{2-} exist in equimolar amounts,

$$K = [Ba^{2+}] [SO_4^{2-}] = X^2$$

$$\therefore X^2 = 10^{-10}$$

$$X = \sqrt{10^{-10}} = 10^{-5} \text{ M}$$

28. **(A)** The equilibrium constant is given by the product of the concentrations of the products divided by the product of the reactant concentrations. Recall that the concentrations of solid products or reactants and water are omitted since they are assumed to be constant. This gives:

$$K = [Ag^+] [Cl^-] \text{ for } AgCl \rightarrow Ag^+ + Cl^-$$

$$K = \frac{1}{[Ag^+][Cl^-]} \text{ for } Ag^+ + Cl^- \rightarrow Ag$$

Atomic chlorine does not exist in nature, so the reactions proposed for it are irrelevant.

29. **(D)** Oxygen availability does not vary in the lab, and is normally not manipulated in an experiment. Increased temperature or reactant concentrations increase the probability of reactant molecule collisions. A catalyst can alter reaction rate, and substances vary in reactivity.

30. **(A)** A saturated solution of a substance is defined as an equilibrium between the substance and its dissociated ions. The substance concentration is incorporated into the equilibrium constant to give the k_{sp} since its concentration remains constant. Thus, for silver chromate

$$k = \frac{[Ag^+]^2 [CrO_4^{2-}]}{[Ag_2CrO_4]}$$

and $k_{sp} = [Ag^+]^2 [CrO_4^{2-}]$

Solving for the solubility of Ag_2CrO_4 we let

$$x = [Ag^+] = [CrO_4^{2-}]$$

so $k_{sp} = (2x)^2(x)$

since two moles of Ag^+ are present for every mole of CrO_4^{2-}. Thus,

$$4x^3 = 1.4 \times 10^{-12}$$

and $x = 7.0 \times 10^{-5}$ moles/liter

31. **(C)** Addition of more A does not affect the equilibrium because A appears both as a reactant and a product. Removal of B causes the equilibrium to shift to the left. Increasing the pressure causes the system to move to the right in an effort to reduce the number of moles of gas present. Changes in temperature will not affect the equilibrium since there is no heat released to or absorbed from the environment during the reaction. The previous explanations are all based on Le Chatelier's Principle: a system when subjected to a stress will shift in a direction so as to minimize that stress.

32. **(C)** The equilibrium constant is defined as the product of the concentrations of the gaseous products raised to the power of their coefficients divided by the product of the gaseous reactant concentrations raised to the power of their coefficients. Only gaseous reactants and products are included in K_e since the concentrations of liquids and solids participating in the reaction are assumed to be large (as compared to those of the gases) and relatively constant. The expressions of the equilibrium constants for the reactions given are:

(A) $\quad K_e = \dfrac{[CO_2]}{[O_2]}$

(B) $\quad K_e = \dfrac{[CO_2]}{[CO][O_2]^{\frac{1}{2}}}$

(C) $\quad K_e = [CO_2]$

(D) $\quad K_e = \dfrac{[O_2]}{[CO_2]}$

(E) $\quad K_e = \dfrac{1}{[CO_2]}$

33. **(A)** For first order kinetics we can evaluate the rate constant from

$$t_{1/2} = \frac{.693}{k}$$

$$k = \frac{.693}{t_{1/2}} = \frac{.693}{3.64} = .190 \text{ day}^{-1}$$

The ratio of initial concentration of the Rn, A_0, to the concentration, A, after time t is given by

$$\log \frac{A_0}{A} = \frac{kt}{2.303}$$

Substituting the values of k and t

$$\log \frac{A_0}{A} = \frac{.190 \text{ days}^{-1} \, 18.2 \text{ days}}{2.303} = \frac{.345}{2.303}$$

$$\log \frac{A_0}{A} = 1.50$$

Taking the anti log of both sides

$$\frac{A_0}{A} = 31.7$$

and since the initial concentration, A_0 was 2.0 μg we can calculate the final concentration:

$$A = \frac{A_0}{31.7} = \frac{2.0\,\mu g}{31.7} = .06\,\mu g$$

34. **(B)** "K," the proportionality for reaction rate, is derived by the multiplication of product molar amounts divided by the multiplication of reactant's molar amounts. Coefficients in the balanced equation translate into exponents outside the bracketed molar amounts of the molecules.

35. **(A)** $PbI_2(a)$ dissolves in water to form a saturated solution according to the equation:

$$PbI_2(a) \leftrightarrow Pb^{2+} + 2I^-$$

The expression for the k_{sp} is

$$k_{sp} = [Pb^{2+}] + [I-]^2 = 8.7 \times 10^{-9}$$

The molar concentration of Pb^{2+} will be equal to the molar concentration of PbI_2 since each mole of PbI_2 that dissolves produces the same number of moles of Pb^{2+} ions. Hence, the concentration of Pb^{2+} will also equal the molar solubility of PbI_2. If x equals the Pb^{2+} concentration, the I^- concentration is then $2x$. Substituting into the k_{sp} expression:

$$[x]\,[2x]^2 = 8.7 \times 10^{-9}$$

$$4x^3 = 8.7 \times 10^{-9}$$

$$x = 1.3 \times 10^{-3}$$

36. **(C)** To determine x, use the data from experiment a and b, where the H_2 concentration varies but the NO concentration is the same. Thus,

$$\text{rate (b)/rate (a)} = 2 = k[H_2]\frac{x}{b}[NO]\frac{y}{b} / k[H_2]\frac{x}{a}[NO]\frac{y}{a}$$

$$= (2.0 \times 10^{-3} / 1.0 \times 10^{-3})^x = 2^x;$$

$$x = 1$$

In a similar way, use experiment a and c to determine y; $y = 2$.

37. **(A)** The equilibrium constant is dependent only on temperature but the amount of each substance present at equilibrium is dependent on pressure, volume, and temperature. There is still an interchange between the phases, but the same number of molecules leave and enter both phases so the equilibrium concentrations and equilibrium constant are the same for a given pressure, volume, and temperature.

38. **(C)** Equilibrium reactions are characterized by the presence of both the reactants and products in a definite ratio in the final reaction mixture. Reactions which go to completion are the result of one of the reactants being depleted. An equilibrium reaction may or may not be spontaneous.

39. **(B)**

$$\ln\left(\frac{1}{.5}\right) = kt_{1/2}$$

therefore $k = \dfrac{\ln 2}{t_{1/2}}$

$$\ln\left(\frac{1}{.02}\right) = \left(\frac{\ln 2}{t_{1/2}}\right) \times t$$

$$t = \left(\frac{\ln 50}{\ln 2}\right) \times t_{1/2} = \frac{\ln 50}{\ln 2} \times 5.2 \text{ yrs.}$$

40. **(A)** The generalized rate for this reaction can be expressed as

rate $= k\,[A]^x\,[B]^y$

The values of x and y are determined from the experimental data. In experiment 1 and 2 the concentration of A is constant but the concentration of B has been doubled. The effect of doubling B increased the rate by a factor of 2. Also, in experiment 1 and 3, a tripling of B triples the rate. Hence, the concentration of B is to the first power and the exponent y must be equal to 1. In experiment 1 and 4 the concentration of B is constant but the concentration of A is doubled. We see that by doubling the concentration of A the rate increases by a factor of 4. Therefore, the value of the exponent x must be to the second power, $x = 2$. The overall rate is then

rate $= k\,[A]^2\,[B]$

EQUILIBRIUM REVIEW

1. The Equilibrium Constant

Most physical and chemical phenomena encountered in chemistry (e.g., reactions, solubility, ionization, and dissociation) proceed to some equilibrium state where concentrations and other properties do not change with time. The equilibrium state, which might well appear to be a state where nothing is happening, is a dynamic state where reactions or other changes occur in opposite directions at equal rates. (i.e., the forward and backward reactions take place at the same rate.)

The key to working all equilibrium problems lies in the definition of the equilibrium constant and conversion of this definition to an equation with only one unknown variable. For the general reaction

$$bB + dD \leftrightarrow rR + sS$$

(which can represent a classical reaction, ionization, solubility, dissociation, or any other equilibrium phenomenon) the equilibrium constant is defined as follows:

$$K_{eq} = (a_R{}^r \, a_S{}^s)/(a_B{}^b \, a_D{}^d)$$

where $a_I{}^i$ is the activity of the species, I, raised to the stoichiometric coefficient i. For gases at low pressures, the activity is equal to the partial pressure in atmospheres; for liquids and solutes in dilute solution, the activity is equal to the concentration in moles/liter. The activity of pure species is one.

The above concept with concentration as the activity leads to the following:

The law of mass action states that the rate of an elementary chemical reaction is proportional to the product of the concentrations of the reacting substances, each raised to its respective stoichiometric coefficient.

For the reaction $aA + bB \leftrightarrow eE + fF$, at constant temperature,

$$K_c = \frac{[E]^e[F]^f}{[A]^a[B]^b},$$

where the [...] denotes equilibrium molar concentrations, and K_c is the equilibrium constant.

The entire relationship is known as the law of mass action.

$$\frac{[E]^e[F]^f}{[A]^a[B]^b}$$

is known as the mass action expression. Note that if any of the species (A, B, E, F) is a pure solid or pure liquid, it does not appear in the expression for K_c. With pressure as the activity for the reaction

$$N_2(g) + 3H_2(g) \leftrightarrow 2\,NH_3(g),$$

$$K_p = \frac{(P_{NH_3})^2}{P_{N_2}(P_{H_2})^3}$$

where K_p is the equilibrium constant derived from partial pressures. K_p and K_c can be related in the following manner.

$$K_p = \frac{P_E^e P_F^f}{P_A^a P_B^b} = \frac{[E]^e (RT)^e [F]^f (RT)^f}{[A]^a (RT)^a [B]^b (RT)^b}$$

and

$$K_p = \frac{[E]^e [F]^f}{[A]^a [B]^b} = (RT)^{(e+f)-(a+b)}$$

Therefore,

$$K_p = K_c (RT)^{\Delta ng}$$

where Δng is the change in the number of moles of gas upon going from reactants to products.

For heterogeneous reactions, the equilibrium constant expression does not include the concentrations of pure solids and liquids.

For the equation

$$2NaHCO_3(s) \leftrightarrow Na_2CO_3(s) + CO_2(g) + H_2O(g),$$

$$K_p = P_{CO_2(g)} P_{H_2O\,(g)},$$

$$K_c = [CO_2(g)]\,[H_2O(g)]$$

and

$$K_p = K_c (RT)^{\Delta ng},$$

where $\Delta ng = +2$ for the reaction.

An equilibrium constant for a solubility reaction is given by a solubility product constant or ion product constant. If ions of several salts are present, the equilibrium solubility of each possible reaction must simultaneously be satisfied. Two examples will illustrate the application of these principles to solubility equilibrium.

First, the solution of a relatively insoluble salt such as silver chloride:

$$AgCl \leftrightarrow Ag^+ + Cl^-$$ Equation 1

$$K_{sp} = (AG^+Cl^-)/(AgCl) = 1.1 \times 10^{-10}$$ Equation 2

Second, let us consider the concentration of silver and chloride ions in a solution initially containing 0.1M silver nitrate, a strong electrolyte, and 0.2M sodium chloride, also a strong electrolyte. Several possible ionic equilibrium reactions might be considered. The equilibrium of each must be satisfied.

$$AgNO_3 \leftrightarrow Ag^+ + NO_3^-$$ Equation 3

$$NaCl \leftrightarrow Na^+ + Cl^-$$ Equation 4

$$AgCl \leftrightarrow Ag^+ Cl^-$$ Equation 5

The equilibrium constant k_{sp}, for strong electrolytes, is a large number and so the concentration of $AgNO_3$ and $NaCl$ can be considered to be zero. The 0.2M Cl^- from Equation 4 will react with essentially all of the 0.1M Ag^+ from Equation 5 leaving a Cl^- concentration of approximately 0.1M. To calculate the (Ag^+) concentration in this case, Equation 2 can be employed. Note that (Cl^-) is very nearly 0.1 and the AgCl activity is one.

$$1.1 \times 10^{-10} = (Ag^+)(Cl^-)/(AgCl) = (Ag^+)(0.1)/(1)$$ Equation 6

$$(Ag+) = 1.1 \times 10^{-9}$$ Equation 7

PROBLEM

One of the two most important uses of ammonia is as a reagent in the first step of the Osfwald process, a synthetic route for the production of nitric acid. This first step proceeds according to the equation

$$4NH_3 \text{ (g)} + 5O_2 \text{ (g)} \leftrightarrow 4NO(g) + 6H_2O(g).$$

What is the expression for the equilibrium constant of this reaction?

SOLUTION

This problem is an exercise in writing the equilibrium constant of a reaction. In general, for a reaction in which reactants A, B, C, ... go to products W, X, Y, ... according to the equation

$$aA + bB + cC + \ldots \leftrightarrow wW + xX + yY + \ldots,$$

where a, b, c, w, x, y, ... are the stoichiometric coefficients, the equilibrium constant is given by

$$K = \frac{[W]^w [X]^x [Y]^y \ldots}{[A]^a [B]^b [C]^c \ldots}.$$

Hence, for the reaction

$$4NH_3 + 5O_2 \leftrightarrow 4NO + 6H_2O,$$

the equilibrium constant is given by

$$K = \frac{[NO]^4 [H_2O]^6}{[NH_3]^4 [O_2]^5}.$$

PROBLEM

Given that k_{sp} for $Mg(OH)_2$ is 1.2×10^{-11}, calculate the solubility of this compound in grams per 100ml of solution. The reaction equation is

$$Mg(OH)_2 \leftrightarrow Mg^{+2} + 2OH^-.$$

SOLUTION

k_{sp} is the solubility product constant; it measures the equilibrium established between the ions in the saturated solution and the excess solid phase. Knowing the k_{sp}, we can calculate the solubility of the compound. The k_{sp} equation for general compound A_xB_y is

$$k_{sp} = [A]^x[B]^y.$$

From the equation, it can be seen that if x moles per liter of Mg $(OH)_2$ dissolves, x moles of Mg^{+2} and $2x$ moles of OH form per liter,

$$k_{sp} = [Mg^{+2}] [OH^-]^2$$

$$1.2 \times 10^{-11} = x(2x)^2 = 4x^3.$$

Solving,

$$x = 1.4 \times 10^{-4} \text{ mole/liter of Mg } (OH)_2 \text{ dissolved.}$$

One is asked, however, for grams per 100ml of solution. 1.4×10^{-4} mole/liter can be converted to grams/100ml by the following method: 100ml = .1 liters since there exists 1,000ml = 1 liter. If one has 1.4×10^{-4} moles in 1 liter, then in .1 liters, there are $(.1)(1.4 \times 10) = 1.4 \times 10^{-5}$ moles. The molecular weight of Mg $(OH)_2 = 58.312$g/mol. Therefore, 1.4×10^{-5} moles/100ml translates into

$$(1.4 \times 10^{-5} \text{ moles/100ml}) (58.312\text{g/mol}) = 8.16 \times 10^{-4} \text{ grams/100ml.}$$

PROBLEM

The solubility product constant of magnesium hydroxide is 8.9×10^{-12}, calculate its solubility in (a) water and (b) .05M NaOH.

SOLUTION

Whenever an ionic solid is placed in water, an equilibrium is established between its ions and the excess solid phase. The solubility constant, k_{sp}, measures this equilibrium. For the general reaction $A_xB_y \leftrightarrow xA^+ + yB^-$, the k_{sp} is defined as being equal to $[A^+]^x[B^-]^y$. The concentration of solid is always constant, no matter how much is in contact with the ions. This means the solid phase will not appear in the equilibrium constant expression. For this problem, part (a), you have

$$Mg(OH)_2(s) \leftrightarrow Mg^{2+} + 2OH^-.$$

The $\quad k_{sp} = [Mg^{2+}] [OH^-]^2 = 8.9 \times 10^{-12}.$

You are asked to find these concentrations. From the chemical equation, you find 2 moles of OH^- will be generated per mole of Mg^{2+}. Thus, at equilibrium, if the concentration of $Mg^{2+} = x$ (mol/liter), then the concentration of OH^- $= 2x$ (mol/liter). (Note: The dissociation of water contributes some OH^-, but this amount is very small, and can be ignored.) Thus, you have

$$[Mg^{2+}] [OH^-]^2 = x(2x)^2 = 8.9 \times 10^{-12} = k_{sp}.$$

Solving for x, you obtain $x = 1.3 \times 10^{-4}$ mol/liter. Thus, 1.3×10^{-4} mole of $Mg(OH)_2$ dissolves per liter of water, producing a solution of $1.3 \times 10^{-4}M$ Mg^{2+} and $2.6 \times 10^{-4}M$ OH^-.

To find the solubility in .05M NaOH, part (b), perform the same process, except you must realize that the NaOH supplies OH^- in addition to the OH^- that dissolves from the salt. This means, as such,

$$[Mg^{2+}] [OH^-]^2 = k_{sp} = (x) (2x + .050)^2 = 8.9 \times 10^{-12}.$$

If you solve for x, you obtain

3.6×10^{-9} mol/liter.

Drill 1: The Equilibrium Constant

1. What is the equilibrium constant for the gaseous reaction shown below?

$$1/2N_2(g) + 3/2 H_2(g) \leftrightarrow NH_3(g)$$

(A) $K_{eq} = \dfrac{a_{NH_3}}{a_{N_2} a_{H_2}}$

(D) $K_{eq} = \dfrac{a_{NH_3}}{\frac{1}{2} a_{N_2} \frac{3}{2} a_{H_2}}$

(B) $K_{eq} = \dfrac{a_{NH_3}^{1/2}}{a_{N_2} a_{H_2}}$

(E) $K_{eq} = \dfrac{\frac{1}{2} a_{N_2} a_{H_2}}{a_{HN_3}}$

(C) $K_{eq} = \dfrac{a_{NH_3}}{a_{N_2}^{1/2} a_{H_2}^{3/2}}$

(Note "a" represents activity)

2. Given the reaction $A + B \leftrightarrow C + D$, find the equilibrium constant for this reaction if .7 moles of C are formed when 1 mole of A and 1 mole of B are initially present.

(A) 5.44

(D) 1.23

(B) 6.22

(E) 2.06

(C) 9.67

3. At a certain temperature, K_{eq} for the reaction $3C_2H_2 \leftrightarrow C_6H_6$ is 4. If the equilibrium concentration of C_2H_2 is 0.5 mole/liter, what is the concentration of C_6H_6?

(A) 1.2 M   (D) 0.5 M

(B) 0.1 M   (E) 0.003 M

(C) 2.3 M

2. Equilibrium Calculations

In order to solve for the equilibrium composition, it is necessary to convert the activity (the partial pressure or concentration) of each of the species to an expression in a single variable. This can almost always be done by referring to the stoichiometry of the reaction and writing the concentration of each species as a function of the fractional or absolute conversion of one species. An example will best illustrate the technique.

Consider the reaction in

$$H_2(g) + I_2(g) \leftrightarrow 2HI(g). \hspace{3cm} \text{Equation 1}$$

Since all species in this reaction are gases at moderate temperatures, the equilibrium constant is

$$k_{eq} = P_H^2/(P_{H2}P_{I2}) = n_{HI}^2/(n_{H2}n_{I2})\,(P/n_t)^0 \hspace{1.5cm} \text{Equation 2}$$

where p represents the partial pressure in atm, P the total pressure, and n the number of moles. (Note that for gases at low pressures, $p_i = (n_i/n_t)P$.) In order to solve Equation 2 for the equilibrium composition, n_{HI}, n_{H_2}, n_{I_2}, and n_t must be written in terms of a single variable. A convenient variable is x, the fraction of the initial H_2 consumed at equilibrium. From Equation 1, the stoichiometric equation representing the reaction, the amounts of each species can be written in terms of the variable x.

$$n_{H_2} = 1.0(1-x)$$

$$n_{I_2} = 1.0(1-x)$$

$$n_{HI} = 1.0(2x)$$

$$n_T = 2$$

Equation 2 can now be written in terms of a single variable, x, and solved as follows:

$$K = (2x)^2/(1-x)^2 \hspace{3cm} \text{Equation 3}$$

$$K^{1/2} = 2x/(1-x) \hspace{3cm} \text{Equation 4}$$

$$x = K^{1/2}/(2 + K^{1/2}) \hspace{3cm} \text{Equation 5}$$

The equilibrium constant varies with temperature, but is independent of pressure. However, the equilibrium composition will be changed by changes in the pressure if the number of moles changes during the reaction. In Equation 2, the zero exponent on pressure is a consequence of the fact that there is no change in the number of moles in the reactants and products. If the number of moles changes, the exponent on total pressure will be non-zero and the composition will change as pressure changes.

PROBLEM

At 986°C, you have the following equilibrium:

$$CO_2(g) + H_2(g) \leftrightarrow CO(g) + H_2O(g).$$

Initially, 49.3 mole percent CO_2 is mixed with 50.7 mole percent H_2. At equilibrium, you find 21.4 mole percent CO_2, 22.8 mole percent H_2, and 27.9 mole percent of CO and H_2O. Find K. If you start with a mole percent ratio of 60:40, CO_2 to H_2, find the equilibrium concentrations of both reactants and products.

SOLUTION

An equilibrium constant k_{eq} measures the ratio of the concentrations of products to reactants, each raised to the power of their respective coefficients in the chemical equation. Thus, k_{eq} for this reaction equals

$$\frac{[CO][H_2O]}{[CO_2][H_2]}.$$

If you assume each substance occupies the same volume in liters, the concentration can be expressed in moles because concentration = moles/liter, i.e., liters cancel out of the equilibrium constant expression. Thus, to find K, you need to find the number of moles of each of the products and reactants and then to substitute into the equilibrium expression. You are told the final product mole percents. The reactants, then, at equilibrium, have mole percents that equal their initial amounts minus the amount that decomposed to produce the products.

Thus, at equilibrium, $[CO_2] = 49.3 - 21.4 = 27.9$, which was given. Similarly $[H_2] = 50.7 - 22.8 = 27.9$, which was given. Thus, by substitution into

$$K = \frac{[CO][H_2O]}{[CO_2][H_2]},$$

you obtain

$$K = \frac{(.279)(.279)}{(.214)(.228)} = 1.60,$$

which is the equilibrium constant for this reaction.

The second part follows. You begin with a 60:40 ratio of CO_2:H_2, which means that initially you have .600 moles CO_2 and .400 moles H_2. To find the equilibrium concentrations, let x = moles of CO formed. Thus, x = moles H_2O formed since coefficients tell us they are formed in equimolar amounts. The fact that moles of a product form, means that x moles of a reactant must have decomposed. Thus, at equilibrium, you have $0.600 - x$ moles of CO_2 and $0.400 - x$ moles of H_2. Recalling,

$$K = \frac{[CO][H_2O]}{[CO_2][H_2]},$$

you can now substitute these values to give

$$K = \frac{x^2}{(0.6 - x)(0.4 - x)}.$$

From previous part, $K = 1.60$, therefore,

$$1.60 = \frac{x^2}{(0.6 - x)(0.4 - x)}.$$

Solving, $x = 0.267$. Thus, $[CO] = [H_2O] = x = 0.267M$, $[CO_2] = 0.6 - x = 0.333M$, and $[H_2] = 0.4 - x = 0.133M$.

PROBLEM

A chemist mixes nitric oxide and bromine to form nitrosyl bromide at 298° K, according to the equation

$2NO(g) + Br_2(g) \leftrightarrow 2NOBr(g)$.

Assuming $K = 100$, what is the quantity of nitrosyl bromide formed, if the reactants are at an initial pressure of 1 atm? $R = 0.0821$ liter-atm./mole°K.

SOLUTION

You are given the equilibrium constant for this reaction and asked to calculate the quantity of nitrosyl bromide produced. The first step is to write out the equilibrium expression and equate it with the given value. For the general reaction,

$$xA + yB \rightarrow zC,$$

K is defined

$$\frac{[C]^2}{[B]^y [A]^x},$$

where the brackets represent concentrations. For this reaction,

$$K = \frac{[NOBr]^2}{[NO]^2\,[Br_2]} = 100.$$

To find out how much NOBr is produced, you would have to know how many moles of NO and Br_2 were reacted. Once this is known, you can find the number of grams produced. You know that the equilibrium expression is based on concentration of reactants and products. Concentration is expressed in moles per liter. This means that if the volume of the NOBr and its concentration is known, you can find moles, since concentration × volume (liters) = moles. Let us represent the concentration as

$$\frac{n}{V} = \frac{moles}{Volume}.$$

Thus, the equilibrium expression becomes

$$K = \frac{(n_{NOBr}\,/\,V)^2}{(n_{NO}\,/\,V)^2\,(n_{Br_2}\,/\,V)}.$$

Let x = moles of NOBr formed. Then, x moles of NO and $x/2$ moles of Br_2 are consumed, since the coefficients of the reaction show a 2:2:1 ratio among $NOBr:NO:Br_2$. The equilibrium expression becomes

$$100 = \frac{n_{NOBr}^2\,V}{n_{NO}^2\,n_{Br_2}} = \frac{x^2 v}{(2-x)^2\,(1-.5x)}.$$

If x moles of NOBr form, and you started with 2 moles of NO, then, at equilibrium, you have left $2 - x$ moles of NO. You started with only 1 mole of Br_2 and $\frac{1}{2}x$ moles of it form NOBr; thus you have $1 - .5x$ moles left. Therefore, you need to determine only the volume to find the quantity NOBr formed. V can be found from the equation of state, $PV = nRT$, where P = pressure, V = volume, n = moles, R = universal gas constant, and T = temperature in kelvin (celsius plus 273°). You are told that the reactants are under a pressure of 1 atm at 298°K. $N = 3$, since the coefficients inform you that a relative sum of 3 moles of reagents exists. You know R. Thus,

$$V = \frac{nRT}{P} = \frac{(3)(.0821)(298)}{1} = 73.4 \text{ liters.}$$

Now that V is known, the equilibrium expression becomes

$$\frac{x^2\,(73.4)}{(2-x)^2\,(1-0.5x)} = 100.$$

Solving for x, you obtain $x = .923$ moles = moles of NOBr formed. Molecular weight = 110. Grams produced = .923 × 110 = 101.53g.

Drill 2: Equilibrium Calculations

1. The table below gives solubility product constants for a few salts of Ag. From the table, we can conclude that the saturated solution with greatest value for $[Ag^+]$ is:

Salt	Solubility Product Constants
AgI	8.3×10^{-17}
AgBr	5.3×10^{-13}
AgCl	1.8×10^{-10}
$AgIO_3$	3.0×10^{-8}
$AgBrO_3$	5.3×10^{-5}

    (A) AgI

    (B) AgBr

    (C) $AgIO_3$

    (D) AgCl

    (E) $AgBrO_3$

2. $$2SO_2(g) + O_2(g) \leftrightarrow 2SO_3(g)$$

    The equilibrium constant for the reaction above is 3.47. SO_2, O_2, and SO_3 are in equilibrium in a closed vessel. The partial pressures of SO_2 and O_2 are 2.0 atm and 1.0 atm, respectively. The partial pressure of SO_3 is:

    (A) $2\sqrt{3.47}$ atm.

    (B) $2/\sqrt{3.47}$ atm.

    (C) $4\sqrt{3.47}$ atm.

    (D) $3.47/4$ atm.

    (E) $2(3.47)$ atm.

3. Calculate the concentration of HI present in an equilibrium mixture produced by the reaction

    $$H_2(g) + I_2(g) \leftrightarrow 2HI(g)$$

    if $K_e = 3.3 \times 10^{-1}$ and the concentrations of H_2 and I_2 are 0.1M and 0.3M, respectively at equilibrium.

    (A) 0.01 M

    (B) 0.03 M

    (C) 0.05 M

    (D) 0.1 M

    (E) 1 M

4. What would be the concentration of Ag^+ at equilibrium if $k_{sp} = 1.6 \times 10^{-10}$?

    (A) 1×10^{-10}

    (B) 1.6×10^{-10}

    (C) 1×10^{-5}

    (D) 1.3×10^{-5}

    (E) 1×10^{-2}

3. Le Chatelier's Principle

Le Chatelier's principle states that when a system at equilibrium is disturbed by the application of a stress (change in temperature, pressure, or concentration) it reacts to minimize the stress and attain a new equilibrium position.

When a system at equilibrium is disturbed by adding or removing one of the substances, all the concentrations will change until a new equilibrium point is reached with the same value of K_{eq}.

Increase in the concentrations of reactants shifts the equilibrium to the right, thus increasing the amount of products formed. Decreasing the concentrations of reactants shifts the equilibrium to the left and thus decreases the concentrations of products formed.

An increase in temperature causes the position of equilibrium of an exothermic reaction to be shifted to the left, while that of an endothermic reaction is shifted to the right.

Increasing the pressure on a system at equilibrium will cause a shift in the position of equilibrium in the direction of the fewest number of moles of gaseous reactants or products.

A catalyst lowers the activation energy barrier that must be overcome in order for the reaction to proceed. A catalyst merely speeds the approach to equilibrium, but does not change K_{eg} (or $\Delta G°$) at all.

If an inert gas is introduced into a reaction vessel containing other gases at equilibrium, it will cause an increase in the total pressure within the container. However, this kind of pressure increase will not affect the position of equilibrium.

PROBLEM

A solute of formula AB is slightly dissociated into A^+ and B^-. In this system, there is a dynamic equilibrium such that $A^+ + B^- \leftrightarrow AB$. Explain what happens if more acid is introduced into this system.

SOLUTION

An acid is a species which, when added to a solvent (such as H_2O), dissociates into protons (H^+) and anions. In this particular case, the proton is represented as A^+. When more acid is added to this general solvent system, more A^+ is introduced. The increased concentration of A^+ places a stress on the equilibrium and the result is a shift in this equilibrium. According to Le Chatelier's Principle, an equilibrium system will readjust to reduce a stress if one is applied. Thus, the equilibrium $A^+ + B^- \leftrightarrow AB$ will readjust to relieve the stress of the increased A^+ concentration. The stress is relieved by the reaction of A^+ with B^- to produce more AB. The concentration of B^- will decrease as compared to its concentration prior to the addition of the acid. Also, the concentration of the product AB will increase with the addition of the acid.

PROBLEM

At 986°C, K = 1.60 for the reaction, $H_2(g) + CO_2(g) \leftrightarrow H_2O(g) + CO_2$. If you inject one mole each of H_2, CO_2, H_2O, and CO simultaneously in a 20-liter box at time $t = 0$ and allow them to equilibrate at 986°C, what will be the final concentrations of all the species? What would happen to these concentrations if additional H_2 was injected and a new equilibrium was established?

SOLUTION

Final concentrations of the species can be found by using the equilibrium constant expression. This expression equates K, the equilibrium constant, to the concentration ratio of products to reactants, each raised to the power of its coefficient in the chemical equation. Thus, for this reaction, you can say

$$K = \frac{[H_2O][CO]}{[H_2][CO_2]} = 1.60.$$

Initially, there was 1 mole of each component in the 20-liter container. Since, concentration = moles/liter, all had an initial concentration of

$$\frac{1\,mole}{20\,liter} = .05M.$$

At equilibrium, let x = the number of moles/liter of H_2 that have reacted. Thus, its concentration, at equilibrium, becomes .05 – x. If x moles/liter of H_2 react, the same number of moles/liter of CO_2 must react also, since they react in equimolar amounts; this is seen from the chemical reaction. Thus, $[CO_2]$ = .05 – x, at equilibrium. These x moles/liter have been converted to products. Thus, $[H_2O]$ = $[CO]$ = .05 + x, at equilibrium. The two products have the same concentration, since, again, the reaction shows they are formed in equimolar amounts. As such, you can substitute these values to obtain

$$\frac{(.05 + x)(.05 + x)}{(.05 - x)(.05 - x)} = 1.60$$

or $\qquad \dfrac{.05 + x}{.05 - x} = \sqrt{1.60}.$

Solving for x, you obtain x = .00585. Thus, the concentrations become $[H_2]$ = $[CO_2]$ = .0442 and $[H_2O]$ = $[CO]$ = .0558. If more H_2 is injected, the equilibrium is subjected to a stress, one component's concentration has been increased, and according to Le Chatelier's Principle, the system will act to relieve the stress by shifting the equilibrium. To do this, more H_2 reacts with CO_2, thus, decreasing their concentrations, to produce more H_2O and CO, thereby, increasing their concentrations.

Drill 3: Le Chatelier's Principle

Questions 1 and 2

    (A) No change in the position of equilibrium

    (B) Shifts position of equilibrium to the left

    (C) Shifts position of equilibrium to the right

    (D) Cannot predict from data given

    (E) Will only shift position of equilibrium if a catalyst is added

Consider the following reaction:

$$N_2(g) + 3H_2(g) \leftrightarrow 2NH_3(g)$$

In which direction will the position of equilibrium shift if:

1.   Additional NH_3 is added to the system?

2.   The pressure on the system is increased from 1 atm to 2 atm?

3. $COCl_2(g) \rightarrow CO(g) + Cl_2(g)$ $K_p = 6.7 \times 10^{-9}$ atm (100°C)

For the above reaction, equal pressures of the reagent and products ($P_{CO} = P_{COCl_2} = P_{Cl_2} = 1$ atm) are placed in a flask. At equilibrium (100°C) it is true that

    (A) the total pressure of the system must remain at 3 atm.

    (B) P_{CO} and P_{Cl_2} must decrease from their initial values.

    (C) P_{CO} and P_{Cl_2} must increase from their initial values.

    (D) P_{COCl} remains constant.

    (E) final pressures cannot be determined because a volume is not specified.

4. $H_2 + S \leftrightarrow H_2S + energy$

In this reversible reaction, select the factor that will shift the equilibrium to the right.

    (A) Adding heat

    (B) Adding H_2S

    (C) Blocking hydrogen gas reaction

    (D) Removing hydrogen sulfide gas

    (E) Removing sulfur

5. The result of adding a small crystal of sodium chloride to a saturated solution of NaCl would be

    (A) the same crystal would precipitate.

    (B) a larger amount of NaCl would precipitate.

    (C) the crystal would dissolve in solution.

    (D) the same amount of NaCl would precipitate.

    (E) the solution becomes supersaturated.

4. Kinetics: Rates and Orders of Reactions

The rate of a homogeneous chemical reaction typically depends on the temperature and the concentrations of the species which enter the reaction. For a typical general homogeneous reaction,

$$A + B \rightarrow \text{Products},$$
Equation 1

experimental results have shown that the rate of disappearance of the reactant A, defined as $(-d[A]/dt)$, can be written as

$$\text{rate} = - k[A]^n [B]^m$$
Equation 2

In this expression, k, the specific reaction rate constant, is a function of the temperature, but not of the concentrations, and n and m are called the order of the reaction with respect to species A and B respectively. It is important to understand that, while the order and the stoichiometric coefficients for a reaction may be the same, the order is *not* necessarily equal to the stoichiometric coefficient. However, the order is equal to the stoichiometric coefficient for each species in an elementary reaction—a reaction that represents the actual molecular path (or mechanism) the reaction follows.

To calculate the order of a reaction, it is important to examine the data. If rate data vs. concentrations are available (presumably at a constant temperature), simply try different orders until Equation 2 is satisfied. For example, if the rate at which A is consumed, $-r_A$, is tabulated as a function of the concentration of A, the different values of n can be tried until the correct one is determined.

$$-r_A = k [A]^n$$
Equation 3

If the half-life is measured, the data can also be used to determine the order and specific rate constant of a reaction. The half-life is defined as the time for one half of a reactant to disappear. For first-order reactions, it is

$$t_{1/2} = \frac{\ln 2}{k}$$

PROBLEM

> A group of mountain climbers set up camp at a 3 km altitude and experience a barometric pressure of 0.69 atm. They discover that pure water boils at 90°C and that it takes 300 minutes of cooking to make a "three-minute" egg. What is the ratio of the rate constant $k_{100°C}$ and $k_{90°C}$?

SOLUTION

Since we do not know the rate expression for cooking an egg, we will assume one of the form

$$\text{rate} = k\,[A]^m\,[B]^n\,...$$

where k is the rate constant, A, B, ... are the reactants, and the overall order of the reaction is $m + n + ...$. We will write the rate equations at the normal boiling point of water (100°C) and at 90°C as

$$\text{rate}_{100°C} = k_{100°C}\,[A]^m[B]^n\,...$$

and

$$\text{rate}_{90°C} = k_{90°C}\,[A]^m[B]^n\,...$$

Dividing the first of these by the second gives

$$\frac{\text{rate}_{100°C}}{\text{rate}_{90°C}} = \frac{k_{100°C}\,[A]^m\,[B]^n...}{k_{90°C}\,[A]^m\,[B]^n...} = \frac{k_{100°C}}{k_{90°C}}$$

Since the egg cooks 100 times faster at 100°C than at 90°C (300 min/3 min = 100), $\text{rate}_{100°C}/\text{rate}_{90°C} = 100$.

Hence,

$$\frac{\text{rate}_{100°C}}{\text{rate}_{90°C}} = 100 = \frac{k_{100°C}}{k_{90°C}}$$

or

$$\frac{k_{100°C}}{k_{90°C}} = 100.$$

PROBLEM

> The reaction
>
> $$A + B \rightarrow C$$
>
> was studied kinetically and the following data was obtained.
>
> | Experiment | initial conc. of A mol/l | initial conc. of B mol/l | Rate (mole/liter – min) |
> |---|---|---|---|
> | 1 | 1.0 M | 1.0 M | 0.15 |
> | 2 | 2.0 M | 1.0 M | 0.30 |
> | 3 | 1.0 M | 2.0 M | 0.15 |
>
> Determine the rate expression.

SOLUTION

The rate of reaction is equal to some rate constant, k, multiplied by the concentrations of A and B raised to the appropriate powers. That is,

$$rate = k \, [A]^m \, [B]^n$$

where the exponents m and n are to be determined.

Comparing experiments 2 and 1, we see that holding [B] constant while doubling [A] doubles the rate of reaction (from 0.15 to 0.30). Hence, the rate is directly proportional to [A] and $m = 1$. Thus, if we hold [B] constant and triple [A], the rate triples.

Comparing experiments 3 and 1, we see that holding [A] constant and changing [B] (from 1.0M to 2.0M) has no effect on the rate. Hence, the rate is independent of [B] and $n = 0$ (so that $[B]^0 = 1$ and [B] does not appear in the rate expression).

Substituting $m = 1$ and $n = 0$ into the rate expression gives

$$rate = k \, [A]^1 \, [B]^0, \text{ or, } rate = k \, [A].$$

PROBLEM

Under certain conditions, the rate equation for the formation of water

$$2H_2 + O_2 \rightarrow 2H_2O$$

is given by

$$rate = k \, [H_2]^2 \, [H_2O]$$

where k is the rate constant. What is the overall order of this rate equation?

SOLUTION

The overall order of a rate equation is equal to the sum of the exponents to which the concentrations are raised. In the equation

$$rate = k \, [H_2]^2 \, [H_2O],$$

$[H_2]$ is raised to the second power and $[H_2O]$ is raised to the first power. Hence, the rate is second order in $[H_2]$, first order in $[H_2O]$, and $2 + 1 = 3$, or third order overall.

PROBLEM

What is the half-life of an unstable substance if 75% of any given amount of the substance decomposes in one hour?

SOLUTION

The half-life is defined as the time it takes for one half of the amount of a substance to decompose. When given the time elapsed and the percent of decomposition, one can find the half-life. One knows that $1/2$ of the substance decomposes in the time equal to the half-life. This leaves $1/2$ of the substance. One-half of this decomposes in the next span of time elapsed equal to the half-time. This leaves $1/4$ of the substance. 75% of it has decomposed after two half-lives have elapsed. Thus, two half-lives equal one hour or the half-life of the substance is $1/2$ hour.

PROBLEM

In the reaction $N_2O_5 \rightarrow N_2O_4 + 1/2 O_2$, the N_2O_5 decomposes by a first-order mechanism. At 298°K, the half-life is 340 minutes. Find the value of the reaction rate constant. Calculate the number of minutes required for the reaction to proceed 70% towards completion.

SOLUTION

Half-life may be defined as the time necessary for half the particular reactant present initially, in this case N_2O_5, to decompose. For a first-order reaction, the rate constant can be expressed in terms of half-life, $t_{1/2}$. Namely,

$$t_{1/2} = \frac{.693}{k},$$

where k is the rate constant. (Caution: This expression is only true for a first order reaction.) Solving this expression for k to obtain

$$k = \frac{.693}{t_{1/2}} = \frac{.693}{340} = 2.04 \times 10^{-3}\, min^{-1}.$$

To find the amount of time required for the reaction to proceed to 70% completion, use the fact that

$$t = \frac{2.303}{k} \log \frac{C_0}{C}$$

in a first-order reaction, where t = time, k = rate constant, C_0 = initial concentration, and C = existing concentration.

Suppose the initial concentration is x, then at 70% completion, the existing concentration is $.30x$. Having calculated k, substitute to find t = minutes.

$$t = \frac{2.303}{k} \log \frac{C_0}{C} = \frac{2.303}{2.04 \times 10^{-3}} \log\left(\frac{x}{.30x}\right)$$

$$= (1,129)(.523) = 590\ min$$

Drill 4: Kinetics: Rates and Orders of Reaction

1. The gas phase decomposition of A was experimentally determined to be first order. If the initial pressure of A is 2,000 torr, what time is required for the pressure to decrease to 250 torr if it decreases to 500 torr after 7.8 minutes?

   (A) 3.9 minutes

   (B) 11.7 minutes

   (C) 15.6 minutes

   (D) 27.3 minutes

   (E) 31.2 minutes

2. For the reaction $PCl_5 \rightarrow PCl_3 + Cl_2$, the rate of the reaction is proportional by mole amounts to

   (A) $Cl_2 \times PCl_3$,

   (B) PCl_5.

   (C) $\dfrac{Cl_2 \times PCl_3}{PCl_5}$.

   (D) $\dfrac{PCl_5}{Cl_2 \times PCl_3}$.

   (E) $PCl_5 \times PCl_3 \times Cl_2$.

Questions 3–5

$$A(aq) + 2B(aq) \rightarrow C(aq)$$

The rate law for the reaction above is:

$$rate = k\,[B]^2$$

3. What is the order of the reaction with respect to B?

   (A) 0

   (B) 1

   (C) 2

   (D) 3

   (E) 4

4. What will happen to the rate of the reaction if the amount of A in the solution is doubled?

   (A) The rate will double.

   (B) The rate will halve.

   (C) The rate will be four times bigger.

   (D) The rate will be four times smaller.

   (E) No effect.

5. The rate constant k is expressed in which units?

    (A) $l/\text{mol} \times \text{sec}$ (D) sec^{-1}

    (B) $\text{mol}/l \times \text{sec}$ (E) sec^{-2}

    (C) $\text{mol}^2/l^2 \times \text{sec}$

5. Temperature Changes and Effect On Rate

There are five important factors that control the rate of a chemical reaction. These are summarized below:

1. The nature of the reactants and products, i.e., the nature of the transition state formed. Some elements and compounds, because of the bonds broken or formed, react more rapidly with each other than do others.

2. The surface area exposed. Since most reactions depend on the reactants coming into contact, increasing the surface area exposed, proportionally increases the rate of reaction.

3. The concentrations. The reaction rate usually increases with increasing concentrations of the reactants.

4. The temperature. A temperature increase of 10°C above room temperature usually causes the reaction rate to double.

5. The catalyst. Catalysts speed up the rate of a reaction but do not change the equilibrium constant (i.e., it simply speeds up the rate of approach to equilibrium).

Temperature Dependence of Reaction Rate: The Arrhenius Equation

The reaction rate increases as the temperature increases. For every 10°K increase in temperature, the reaction rate doubles. The rate constant is related to the temperature by the following Arrhenius equation:

$$k = A \exp\left(\frac{-E_a}{RT}\right)$$

$$\ln k = \ln A \, \frac{-E_a}{RT}$$

where k = the rate constant

A = the frequency factor or the pre-exponential factor

E_a = the activation energy

T = the temperature in °K

$$\ln \frac{k_2}{k_1} = -\left(\frac{\Delta E_a}{R}\right)\left(\frac{1}{T_2} - \frac{1}{T_1}\right)$$

$$\frac{d(\ln k)}{dt} = \frac{\Delta E}{RT^2}$$

Where K is the equilibrium constant and ΔE is the change in energy. k is small when the activation energy is very large or when the temperature of the reaction mixture is low. A plot of $\ln k$ versus $^i/_t$ gives a straight line whose slope is equal to $-Ea/R$ and whose intercept with the ordinate is $\ln A$.

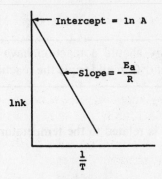

Figure 5.1

Arrhenius plot that shows the relationship between the rate constant and the temperature.

PROBLEM

Using a specific case, show that the effect of a 10°K rise in temperature will have a greater effect on the rate constant, k, at low temperatures than it does at high temperatures.

SOLUTION

The best way to demonstrate this fact is to use the natural logarithm form of Arrhenius' equation, which gives an indication of the effect of temperature on reaction rate. The expression may be written $\ln k = \ln A - E_a/2.303\, RT$, where k is the rate constant, A = the Arrhenius constant, E_a = activation energy, R = universal gas constant (8.314 J mol^{-1} deg^{-1}), and T = temperature in Kelvin (Celsius plus 273). For a low temperature, consider room temperature ($T = 300°$K). For a first-order reaction, A might be 1.0×10^{14} sec^{-1} and E might be 80 k J/mole. Thus, at $T = 300°$K, you have log k = log (1×10^{14}) − 80,000/(2.303) (8.314) (300) = .07. Thus, $k = 1.2$ sec^{-1}. If you increase the temperature 10°K to 310°K, you find by exactly the same method of calculation, that $k = 3.3$ sec^{-1}. Thus, at a temperature of 10°K higher (in a low temperature range), k was increased nearly three times.

Now consider a higher temperature range, say $T = 900°K$. For this, log $k = (1 \times 10^{-14}) - 80,000/(2.303)\,(8.314)\,(900) = 9.358$. Solving log $k = 9.358$, $k = 2.28 \times 10^9$. Again, let us increase the temperature by $10°K$ to $910°K$. Using the same type of calculations, you find that k becomes 2.56×10^9. The percent change equals

$$\frac{2.56 \times 10^9 - 2.28 \times 10^9}{2.28 \times 10^9} \times 100 = 12.28\%.$$

Thus, at the higher temperatures, the rate constant was increased only 12% as compared to 300% at the low temperatures.

PROBLEM

What activation energy should a reaction have so that raising the temperature by $10°C$ at $0°C$ would triple the reaction rate?

SOLUTION

The activation energy is related to the temperature by the Arrhenius equation which is stated

$$k = Ae^{-E/RT}$$

where A is a constant characteristic of the reaction; e is the base of natural logarithms, E is the activation energy, R is the gas constant (8.314 J mol^{-1} deg^{-1}) and T is the absolute temperature. Taking the natural log of each side:

$$\ln k = \ln A - E/RT$$

For a reaction that is 3 times as fast, the Arrhenius equation becomes

$$3\,k = Ae^{-E/R(T+10°)}$$

Taking the natural log:

$$\ln 3 + \ln k = \ln A - E/R(T + 10°)$$

Subtracting the equation for the final state from the equation for the initial state:

$$\ln k = \ln A - E/RT$$
$$\underline{-(\ln 3 + \ln k = \ln A - E/R(T + 10))}$$
$$- \ln 3 = -E/RT + E/R(T + 10)$$

Solving for E:

$$-\ln 3 = - E/RT + E/R(T + 10)$$

$$R = 8.314 \text{ J/mole } °K$$

$T = 0 + 273 = 273$

$-\ln 3 = -E/(8.314 \text{ J/mol-K}) (273\text{K}) + E/(8.314 \text{ J/mol-K}) (283\text{K})$

$-1.10 = -E/2269.72 \text{ J/mol} + E/2352.86 \text{ J/mol}$

$(2,269.72 \text{ J/mol}) (2,352.86 \text{ J/mol}) \times -1.10 =$

$\quad (-E/2,269.72 \text{ J/mol} + E/2,352.86 \text{ J/mol}) (2,269.72 \text{ J/mol})$

$\quad (2,352.86 \text{ J/mol})$

$-5.874 \times 10^6 \text{ J}^2/\text{mol}^2 = (-E) (2,352.86 \text{ J/mol}) + E (2,269.72 \text{ J/mol})$

$-5.874 \times 10^6 \text{ J}^2/\text{mol}^2 = -8.314 \times 10^1 \text{ J/mol} \times E$

$\quad 7.06 \times 10^4 \text{ J/mol} = E$

Drill 5: Temperature Changes and Effect on Rate

1. It has been suggested that a $10°$ rise in temperature results in a twofold response in the rate of a chemical reaction. What is the implied activation energy (in Kcal/mol) at $27°C$?

    (A) 113

    (B) 215

    (C) 106

    (D) 319

    (E) 226

2. A reaction proceeds five times as fast at $60°C$ as it does at $30°C$. Estimate its energy of activation in kcal/mol.

    (A) 1.61

    (B) 13.4

    (C) 10.8

    (D) 11.6

    (E) 20.1

3. For a gas phase reaction with $E_A = 40,000$ cal/mol, estimate the change in rate constant due to a temperature change from $1000°C$ to $2000°C$.

    (A) 1,050 times larger

    (B) 46 times smaller

    (C) 1,195 times smaller

    (D) 615 times larger

    (E) 1,512 times larger

EQUILIBRIUM DRILLS

ANSWER KEY

Drill 1 — The Equilibrium Constant
1. (C)
2. (A)
3. (D)

Drill 2 — Equilibrium Calculations
1. (E)
2. (A)
3. (D)
4. (D)

Drill 3 — Le Chatelier's Principle
1. (B)
2. (C)
3. (B)
4. (D)
5. (D)

Drill 4 — Kinetic's Rates and Orders of Reaction
1. (B)
2. (B)
3. (C)
4. (E)
5. (A)

Drill 5 — Temperature Changes and Effect on Rate
1. (E)
2. (C)
3. (A)

GLOSSARY: EQUILIBRIUM

Arrhenius Equation

Equation which describes how the reaction rate is related to reaction temperature.

Equilibrium

A situation in which the rate of a forward reaction is equal to that of the backward reaction. There is then no net change in the quantity of either products or reactants.

Equilibrium Constant

The constant arising from the law of mass action.

Kinetics

The study of chemical reaction rates.

Law of Mass Action

The notion that the rate of an elementary chemical reaction is proportional to the product of the concentrations of the reacting substances, each raised to its respective stoichiometric coefficient.

Le Chatelier's Principle

States that when a system at equilibrium is disturbed by the application of a stress, the system reacts to minimize the stress and attain a new equilibrium position.

Reaction Order

The exponent to which the concentration of a reactant is raised to account for the observed dependence of reaction rate on concentration.

Reaction Rate

The rate of appearance of a product or disappearance of a reactant. It is expressed as a change in concentration per unit time.

Solubility Product Constant

The equilibrium constant for the solution reaction of a relatively insoluble ionic substance. It serves to measure the solubility of a compound.

CHAPTER 10

Acid and Base Equilibria

➤ Diagnostic Test
➤ Acid and Base Equilibria
Review & Drills
➤ Glossary

ACID AND BASE EQUILIBRIUM
DIAGNOSTIC TEST

1. (A) (B) (C) (D) (E)	21. (A) (B) (C) (D) (E)
2. (A) (B) (C) (D) (E)	22. (A) (B) (C) (D) (E)
3. (A) (B) (C) (D) (E)	23. (A) (B) (C) (D) (E)
4. (A) (B) (C) (D) (E)	24. (A) (B) (C) (D) (E)
5. (A) (B) (C) (D) (E)	25. (A) (B) (C) (D) (E)
6. (A) (B) (C) (D) (E)	26. (A) (B) (C) (D) (E)
7. (A) (B) (C) (D) (E)	27. (A) (B) (C) (D) (E)
8. (A) (B) (C) (D) (E)	28. (A) (B) (C) (D) (E)
9. (A) (B) (C) (D) (E)	29. (A) (B) (C) (D) (E)
10. (A) (B) (C) (D) (E)	30. (A) (B) (C) (D) (E)
11. (A) (B) (C) (D) (E)	31. (A) (B) (C) (D) (E)
12. (A) (B) (C) (D) (E)	32. (A) (B) (C) (D) (E)
13. (A) (B) (C) (D) (E)	33. (A) (B) (C) (D) (E)
14. (A) (B) (C) (D) (E)	34. (A) (B) (C) (D) (E)
15. (A) (B) (C) (D) (E)	35. (A) (B) (C) (D) (E)
16. (A) (B) (C) (D) (E)	36. (A) (B) (C) (D) (E)
17. (A) (B) (C) (D) (E)	37. (A) (B) (C) (D) (E)
18. (A) (B) (C) (D) (E)	38. (A) (B) (C) (D) (E)
19. (A) (B) (C) (D) (E)	39. (A) (B) (C) (D) (E)
20. (A) (B) (C) (D) (E)	40. (A) (B) (C) (D) (E)

ACID AND BASE EQUILIBRIA DIAGNOSTIC TEST

This diagnostic test is designed to help you determine your strengths and your weaknesses in acid and base equilibria. Follow the directions and check your answers.

Study this chapter for the following tests:
ACT, ASVAB, AP Chemistry, CLEP General Chemistry,
GED, MCAT, MSAT, PRAXIS II Subject Assessment:
Chemistry, SAT II: Chemistry, GRE Chemistry

---

40 Questions

DIRECTIONS: Choose the correct answer for each of the following problems. Fill in each answer on the answer sheet.

1. Arrhenius would define a base as

    (1) something which yields hydroxide ions in solution.

    (2) a proton acceptor.

    (3) an electron pair donator.

    (A) 1, 2, and 3   (D) 1 only

    (B) 1 and 3   (E) 1 and 2

    (C) 2 only

2. 20 ml of NaOH is needed to titrate 30 ml of a 6 M HCl solution. The molarity of the NaOH is

    (A) 1 M.   (D) 9 M.

    (B) 3 M.   (E) 11 M.

    (C) 6 M.

3. The pH of a 1.0 M acetic acid solution, $C_2H_3O_2$, was found to be 3.5. The percent ionization of acetic acid is

    (A) 1.0.   (D) 0.031.

    (B) 1.8×10^{-5}.   (E) 35.0.

    (C) 3.5.

4. What volume of 3 M NaOH is required to neutralize a one liter solution containing 98 g of H_2SO_4?

    (A) 100 ml   (D) 1,333 ml

    (B) 333 ml   (E) 2,000 ml

    (C) 667 ml

5. An acid H_3X is classified as

    (A) monoprotic.   (D) bidentate.

    (B) diprotic.   (E) None of the above.

    (C) triprotic.

6. Which of the following indicates a basic solution?

    (A) $[H^+] > 10^{-7}$ (D) $pH = 7$

    (B) $[H^+] > 10^{-10}$ (E) $pH = 9$

    (C) $pH = 5$

7. What is the pH of a 1.0 M solution of formic acid if the k_a is 1.77×10^{-4}?

    (A) 14   (D) 2

    (B) 11   (E) 1

    (C) 4

8. Which of the following is closest to the pH of 10^{-4} M NaOH?

    (A) 11   (D) 7

    (B) 10.2   (E) 12

    (C) 9

9. What is the hydrogen ion concentration of a buffer solution that is .05 M in acetic acid and .1 M in sodium acetate? (The k_i for acetic acid is 1.8×10^{-5}.)

    (A) 9.0×10^{-6} (D) 1.8×10^{-6}

    (B) 1.8×10^{-5} (E) 1.0×10^{-7}

    (C) 1.0×10^{-14}

10. What is the concentration of $[OH^-]$ in a NH_4OH solution if it has a pH of 11?

    (A) 10^{-1} (B) 10^{-3}

(C) 10^{-5} (D) 10^{-9}

(E) 10^{-11}

11. What is the hydronium ion concentration, H_3O^+, of a solution that has a hydroxide concentration of 1.4×10^{-4} M?

(A) 7.2×10^{-11} (D) 1.8×10^{-5}

(B) 1.4×10^{-10} (E) 7.0×10^{-7}

(C) 1.0×10^{-14}

12. How many grams of HNO_3 are required to produce a one liter aqueous solution of pH 2?

(A) 0.063   (D) 1.26

(B) 0.63   (E) 12.6

(C) 6.3

13. $HA + B \rightarrow H^+ + B + A^-$

By the Bronsteäd theory, the acid in this question is

(A) A.   (D) B.

(B) HA.   (E) HB.

(C) AB.

14. Which of the following is most acidic?

(A) $HClO_4$ (D) HCN

(B) HF (E) HCl

(C) H_3PO_4

15. Hydrolysis of sodium acetate yields

(A) a strong acid and a strong base.

(B) a weak acid and a weak base.

(C) a strong acid and a weak base.

(D) a weak acid and a strong base.

(E) none of the above.

16. k_e for the reaction $CH_3COOH \rightarrow H^+ + CH_3COO^-$ is 1.8×10^{-5}. This indicates that

    (A) acetic acid is a strong acid.

    (B) the forward reaction does not occur.

    (C) a catalyst is required.

    (D) most of the acetic acid is undissociated.

    (E) the concentration of CH_3COOH is equal to the concentration of H^+.

17. What is the hydroxide ion concentration in a solution with a pH of 5?

    (A) 10^{-3} (D) 10^{-9}

    (B) 10^{-5} (E) 10^{-11}

    (C) 10^{-7}

18. Which of the following salts will hydrolyze in water to produce a neutral solution?

    (A) $BaSO_4$ (D) Na_2CO_3

    (B) $AlCl_3$ (E) Na_3PO_4

    (C) $NaNO_3$

19. What is k_b for a 0.1 M solution of NH_4OH if $[OH^-] = 1.3 \times 10^{-3}$?

    (A) 7.6×10^{-1} (D) 3.7×10^{-4}

    (B) 1.1×10^{-2} (E) 1.7×10^{-5}

    (C) 4.2×10^{-3}

20. What is the concentration of an HNO_3 acid solution with a pH of 3?

    (A) 3   (D) −3

    (B) −Antilog[3]   (E) None of the above

    (C) 10^{-3}

21. How many milliliters of 5 M NaOH are required to completely neutralize 2 liters of 3 M HCl?

    (A) 600 ml   (D) 1,500 ml

    (B) 900 ml   (E) 1,800 ml

    (C) 1,200 ml

22. A professor needs to make a buffer solution for a class demonstration. He mixes together equal volumes of various solutions. Identify the combination of solutions that would *not* produce a buffer solution.

    (A) 1 M CH_3COOH, 1 M HCl, .2 M $NaCH_3COO$

    (B) 1 M CH_3COOH, 1 M $NaCH_3COO$

    (C) 1 M CH_3COOH, .5 M NaOH

    (D) 1 M CH_3COOH, 1 M HCl, 1 M NaOH, 1 M $NaCH_3COO$

    (E) 2 M NH_3, 1 M HCl

23. A salt formed by a neutralization reaction of a strong acid and weak base is

    (A) HCl.   (D) NH_4Cl.

    (B) NaCl.   (E) NH_4CN.

    (C) Na_2CO_3.

24. Amphoteric substances are best described as

    (A) having the same number of protons and electrons but different numbers of neutrons.

    (B) having the same composition but occurring in different molecular structures.

    (C) being without definite shape.

    (D) having both acid and base properties.

    (E) having the same composition but occurring in different crystalline form.

25. What is the hydronium ion concentration of an HCl solution at pH 3?

    (A) 0.001 M   (D) 0.3 M

    (B) 0.003 M   (E) 3 M

    (C) 0.1 M

26. Which of the following solutions do not constitute a buffer pair?

    (A) .2 M NH_4OH and .1 M NH_4Cl

    (B) .2 M NaCl and .1 M HCl

    (C) .1 M acetic acid and .05 M sodium acetate

    (D) .1 M NH_4OH and .1 M $(NH_4)_2SO_4$

    (E) .1 M formic acid and .1 M sodium formate

27. What is the pH of a solution that has a hydronium ion concentration, H_3O^+ of 1.2×10^{-4}?

    (A) 3.92

    (B) 4.00

    (C) 3.80

    (D) 5.20

    (E) 7.90

28. Arrange the acids in order of increasing strength.

    (A) $HClO_4$, H_2SO_4, H_3PO_4, $HClO$

    (B) $HClO$, $HClO_4$, H_2SO_4, H_3PO_4

    (C) H_3PO_4, H_2SO_4, $HClO_4$, $HClO$

    (D) $HClO$, H_3PO_4, H_2SO_4, $HClO_4$

    (E) H_3PO_4, H_2SO_4, $HClO$, $HClO_4$

29. $H_2O + (CH_3)_3N \leftrightarrow (CH_3)_3NH^+ + OH^-$

    For the above reaction, which conjugate acid-base pair is correctly listed?

    (A) H_2O, $(CH_3)_3N$

    (B) $(CH_3)_3NH^+$, OH^-

    (C) H_2O, $(CH_3)_3NH^+$

    (D) H_2O, H^+

    (E) None of the above

30. What is the pOH of a solution with $[H^+] = 1 \times 10^{-3}$?

    (A) -3

    (B) 1

    (C) 3

    (D) 11

    (E) 14

31. $\_MnO_4^- + \_H^+ + \_Fe^{2+} \rightarrow \_Fe^{3+} + \_Mn^{2+} + \_H_2O$

    When the skeleton equation above is balanced and all coefficients reduced to their lowest whole-number terms, what is the coefficient for H^+?

    (A) 4

    (B) 6

    (C) 8

    (D) 9

    (E) 10

32. Oxalic acid will have ____ k_a's.

    (A) 0

    (B) 1

    (C) 3

    (D) 2

    (E) 4

33. Which of the following salts would hydrolyze to form a solution whose pH is below 7?

    (A) $CaCl_2$ (D) NH_4Cl

    (B) CH_3COONa (E) Both (B) and (D)

    (C) NaCl

34. Which of the following is the weakest acid?

    (A) HCl   (D) HF

    (B) HBr   (E) None of the above

    (C) HI

35. The H^+ concentration of a .01 molar solution of HCN is (The k_i for HCN is 4×10^{-10}.)

    (A) 2.5×10^{-11} M.   (D) 2×10^{-6} M.

    (B) .01 M.   (E) 2×10^{-5} M.

    (C) 4×10^{-10} M.

36. A student titrates 100 ml of acid with 5 M NaOH. Phenolphthalein indicator changes color after 50 ml of NaOH have been added. What is the molarity of the monoprotic acid?

    (A) 0.1 M   (D) 2.5 M

    (B) 1 M   (E) 25 M

    (C) 1.5 M

37. A buffer solution was prepared by mixing 100 ml of a 1.2 M NH_3 solution and 400 ml of a 0.5 M NH_4Cl solution. What is the pH of this buffer solution, assuming a final volume of 500 ml and $k_b = 1.8 \times 10^{-5}$?

    (A) 1.08   (D) 9.03

    (B) 4.96   (E) 8

    (C) 5.8

38. $H_2CO_3 \leftrightarrow H^+ + HCO_3$

    .5 moles/liter of carbonic acid dissociates hydrogen and carbonic ions at .1 mole per liter, each in a lab aqueous setting. Its ionization constant is

    (A) 1×10^{-2}.   (D) 2×10^1.

    (B) 2×10^{-2}.   (E) 2×10^2.

    (C) 1×10^2.

39. Which one of the following hydroxides is amphoteric?

    (A) $Ba(OH)_2$ (D) $Al(OH)_3$

    (B) $Ca(OH)_2$ (E) NaOH

    (C) $Mg(OH)_2$

40. What is the percent dissociation of this solution?

    (A) 0.093%   (D) 1.3%

    (B) 0.073%   (E) 13%

    (C) 0.73%

ACID AND BASE EQUILIBRIA DIAGNOSTIC TEST

ANSWER KEY

1. (D)	11. (A)	21. (C)	31. (C)
2. (D)	12. (B)	22. (A)	32. (D)
3. (D)	13. (B)	23. (D)	33. (D)
4. (C)	14. (A)	24. (D)	34. (D)
5. (C)	15. (D)	25. (A)	35. (D)
6. (E)	16. (D)	26. (B)	36. (D)
7. (D)	17. (D)	27. (A)	37. (D)
8. (B)	18. (C)	28. (D)	38. (B)
9. (A)	19. (E)	29. (E)	39. (D)
10. (B)	20. (C)	30. (D)	40. (D)

DETAILED EXPLANATIONS
OF ANSWERS

1.   **(D)**   The Arrhenius theory defines a base as a substance that gives hydroxide ions in aqueous solution. A base is defined as a proton acceptor by the Brönsted theory. This theory defines an acid as a proton donator. The Lewis theory defines an acid as an electron pair acceptor and a base as an electron pair donator.

2.   **(D)**   Using the formula $M_1V_1 = M_2V_2$ and rearranging we obtain

$$M_2 = \frac{M_2V_1}{V_2} = \frac{(6\,\text{M})\,(30\,\text{ml})}{20\,\text{ml}} = 9\,\text{M NaOH}.$$

3.   **(D)**   The percent ionization of a weak acid is the molar H_3O^+ ion concentration divided by the molar concentration of the undissociated acid. In this case:

$$\% \text{ ionization} = \frac{H_3O^+}{HC_2H_3O_2} \times 100.$$

   To solve this problem the pH of 3.5 must be converted to H_3O^+ concentration.

Since, $pH = -\log [H_3O^+]$

          $3.5 = -\log [H_3O^+]$

   $H_3O^+ = $ antilog of -3.5

and since -3.5 is equal to $-4.0 + .5$ we write,

     $\log [H_3O^+] = +.5 - 4$

          $H_3O^+ = $ antilog of $.5 - 4$

          $H_3O^+ = 3.16 \times 10^{-4}$

and to determine the percent,

$$\% \text{ ionization} = \frac{3.16 \times 10^{-4}}{1.0} \times 100 = 0.031\%$$

4.   **(C)**   The reaction in question is

$$H_2SO_4 + 2NaOH \rightarrow Na_2SO_4 + 2H_2O.$$

98 g of H_2SO_4 is equivalent to one mole of H_2SO_4 so two moles of NaOH are

required. Since we have a 3 M solution of NaOH, we require a volume of

$$2 \text{ moles of NaOH} \times \frac{1 \text{ liter of solution}}{3 \text{ moles of NaOH}} = 0.667 \text{ liter of solution}$$

or 667 ml of 3 M NaOH solution.

5.　　**(C)**　　An acid H_3X is classified as triprotic since it may "give up" three protons to a base. An example of a triprotic acid is phosphoric acid, H_3PO_4. Examples of monoprotic and diprotic acids are hydrochloric, HCl and sulfuric, H_2SO_4, respectively. The term bidentate, rather than referring to acids, is associated with ligands. Bidentate ligands have two atoms that may coordinate to a metal ion.

6.　　**(E)**　　A basic (alkaline) solution is indicated by a hydronium ion concentration less than 10^{-7} or identically, pH > 7 since pH = $-\log[H^+]$. A solution with $[H^+] > 10^{-10}$ may be basic ($10^{-10} < [H^+] < 10^{-7}$) or acidic ($[H^+] > 10^{-7}$).

7.　　**(D)**　　The acid dissociation constant, k_a, applies to this reaction:

$$HA_{(aq)} \leftrightarrow H^+_{(aq)} + A^-_{(aq)}$$

where

$$K_a = \frac{[H^+][A^-]}{[HA]}$$

We know that $k_a = 1.77 \times 10^{-4}$ and [HA] = 1.0 M (minus a negligible amount).

$$1.77 \times 10^{-4} = \frac{[H^+][A^-]}{1.0}$$

$$1.77 \times 10^{-4} = [H^+][A^-]$$

Since $[H^+] = [A^-]$,

$$1.77 \times 10^{-4} = [H^+]^2$$

$$1.33 \times 10^{-2} = [H^+]$$

Since pH = $-\log[H^+]$,

$$pH = 1.88 \cong 2.$$

8.　　**(B)**　　Because pH is defined in terms of hydrogen ion concentration, it is first necessary to find the concentration of this ion by substituting in the ion-product expression for water. The [OH⁻] from the NaOH is given as 10^{-4} M.

$$k_W = [H] [OH] \text{ because } k_W = 10^{-14}$$

$$\therefore \qquad 10^{-14} = [H^+] [10^{-4}],$$

$$\text{so} \qquad [H^+] = \frac{10^{-14}}{10^{-4}}$$

$$= 10^{-10};$$

since $\qquad pH = -\log [H^+]$

$$\therefore \qquad pH = -\log [10^{-10}] = 10.0$$

The answer closest to 10.0 is 10.2.

9. **(A)** Acetic acid ionizes,

$$HC_2H_3O_2 \leftrightarrow H^+ + C_2H_3O_2^-$$

and sodium acetate completely dissociates,

$$NaC_2H_3O_2 \rightarrow Na^+ + C_2H_3O_2^-$$

The solution is a buffer solution since it is composed of a weak acid and a salt of the weak acid. The hydrogen ion concentration of such a buffer can be calculated from the expression:

$$H^+ = \frac{[acid]}{[salt]} \times k_i$$

$$[H^+] = \frac{.05}{.1} \times 1.8 \times 10^{-5}$$

$$= .90 \times 10^{-5} \text{ or } 9.0 \times 10^{-6}$$

10. **(B)** A solution of pH 11 has $[H^+] = 1 \times 10^{-11}$ by the equation $pH = -\log [H^+]$. Recalling that $k_w = [H^+] [OH^-] = 10^{-14}$ we have

$$[OH^-] = \frac{k_w}{[H^+]} = \frac{10^{-14}}{[H^+]} = \frac{10^{-14}}{10^{-11}} = 10^{-3}.$$

11. **(A)** The hydronium ion concentration can be calculated from the ion – product constant for water, k_w,

$$k_w = [H_3O^+] [OH^-] = 1 \times 10^{-14}.$$

Due to the auto ionization of water both the OH^- and the H_3O^+ ions exist in acid or basic solutions. However, the product of the OH^- ion concentration and the H_3O^+ ion concentration will always be equal to 1×10^{-14}; therefore, substituting the OH^- concentration into the k_w expression, we can calculate the H_3O^+ ion concentration.

$$k_w = [H_3O^+][OH^-] = 1 \times 10^{-14}$$

$$H_3O^+ = \frac{1 \times 10^{-14}}{1.4 \times 10^{-4}} = 7.2 \times 10^{-11}$$

12. **(B)** The pH of a solution is defined as

$$pH = -\log[H^+].$$

Solving for $[H^+]$, we obtain

$$[H^+] = 10^{-pH} = 10^{-2}.$$

Nitric acid is assumed to dissociate completely in solution since it is a strong acid. In addition, nitric acid has only one proton so the concentration of hydronium ions is equal to the initial concentration of HNO_3. We are working with one liter of solution so the concentration is identical to the number of moles by

$$C = \frac{\text{number of moles of solute}}{\text{liters of solution}}.$$

Thus, we require 10^{-2} mole of HNO_3. Converting to grams:

$$10^{-2} \text{ mole of } HNO_3 \times \frac{63 \text{ g of } HNO_3}{1 \text{ mole of } HNO_3} = 0.63 \text{ g of } HNO_3.$$

13. **(B)** By this theory, acids are hydrogen ion (proton) donors in solution. The HA reactant yields H^+ by dissociation in solution.

14. **(A)** The greater the extent of its ionization in water, the stronger the acid. Thus,

$$HClO_4 + H_2O \rightarrow H_3O^+ + ClO_4^- \text{ (complete in dilute solution)}$$

$HClO_4$ dissociates especially well because of the stability of the ClO_4^- ion. This results from the ability of the ion to delocalize its negative charge over all four oxygen atoms.

15. **(D)** The hydrolysis reaction for sodium acetate proceeds as follows:

$$CH_3\overset{\overset{\displaystyle O}{\displaystyle \|}}{C}ONa + H_2O \rightarrow CH_3\overset{\overset{\displaystyle O}{\displaystyle \|}}{C}OH + NaOH$$

The products of the reaction are a weak acid (acetic acid) and a strong base (sodium hydroxide).

16. **(D)** Since

$$k_e = \frac{[H^+][CH_3COO^-]}{[CH_3COOH]} = 1.8 \times 10^{-5},$$

we see that a larger amount of undissociated acid is present as compared to the dissociated acid. Acetic acid is therefore a weak acid. Also, $[H^+] = [CH_3COO^-] \neq [CH_3COOH]$.

17. **(D)** Since pH = $-\log H^+$, we find that $[H^+] = 10^{-pH}$. Substituting our value gives

$$[H^+] = 10^{-5}.$$

Recalling that $[H^+][OH^-] = 10^{-14}$, we have

$$[OH^-] = \frac{10^{-14}}{[H^+]} = \frac{10^{-14}}{10^{-5}} = 10^{-9}.$$

18. **(C)** A salt hydrolyzes to produce a neutral solution if a strong acid and a strong base or a weak acid and a weak base are the hydrolysis products.

$$NaNO_3 + H_2O \rightarrow NaOH + HNO_3$$
$$\text{strong base} \quad \text{strong acid}$$

An acidic solution results when a strong acid and a weak base are produced.

$$BaSO_4 + 2H_2O \rightarrow Ba(OH)_2 + H_2SO_4$$
$$\text{weak base} \qquad \text{strong acid}$$

$$AlCl_3 + 3H_2O \rightarrow Al(OH)_3 + 3HCl$$
$$\text{weak base} \qquad \text{strong acid}$$

A basic solution is produced in cases where a weak acid and a strong base are the result:

$$Na_2CO_3 + 2H_2O \rightarrow 2NaOH + H_2CO_3$$
$$\text{strong base} \quad \text{weak acid}$$

$$Na_3PO + 3H_2O \rightarrow 3NaOH + H_3PO_4$$
$$\text{strong base} \quad \text{weak acid}$$

19. **(E)** k_b, the base dissociation constant, is defined as

$$k_b = \frac{[X^+][OH^-]}{[XOH]}.$$

Since $[OH^-] = 1.3 \times 10^{-3}$, we know that $[NH_4^+] = 1.3 \times 10^{-3}$ and that the original concentration of NH_4OH is 0.1 M. Substituting values we obtain

$$k_b = \frac{(1.3 \times 10^{-3})(1.3 \times 10^{-3})}{(0.1 - 1.3 \times 10^{-3})} = \frac{1.69 \times 10^{-6}}{0.0987}$$

$$k_b = 1.7 \times 10^{-5}$$

20. **(C)** pH = −log [H⁺], but pH of HNO_3 = 3

∴ −log [H⁺] = 3

log [H⁺] = −3

∴ [H⁺] = antilog [−3] = 10^{-3}

21. **(C)** The relationship $M_1V_1 = N_2V_2$ in neutralization problems involving a strong acid and strong base. We have

(3 M HCl) (2,000 ml) = (5 M NaOH) ($V2$ ml)

$$V2 = \frac{(3M)(2000 \ ml)}{(5M)} = 1,200 \ ml$$

22. **(A)** A buffer solution contains a weak acid (or base) and the salt of its conjugate base (or acid). Solutions (B) through (D), when mixed, yield CH_3COOH and $NaCH_3COO$. Solution (E), when mixed, yield NH_3 and NH_4Cl. These are buffer solutions. Solution (A) yields CH_3COOH, HCl, and NaCl. There is no $NaCH_3COO$ remaining; hence, we do not have a buffer.

23. **(D)** HCl is an acid. NaCl is formed from a strong acid and strong base. Na_2CO_3 is a product of a weak acid and strong base. NH_4CN is produced from two weak compounds. A strong acid, HCl, and weak base NH_4OH, react to produce NH_4Cl.

24. **(D)** An amphoteric substance has both acid and base properties. Isotopes of an element have the same number of protons and electrons but different numbers of neutrons.① Isomers of a compound are indicated by the same molecular formulas but different structures.② Amorphous substances are designated as having no definite shape.③ Allotropes of a substance have the same composition but have different crystalline structures.④

① (for example $^{12}_{6}C$ and $^{13}_{6}C$)

② (for example 1-propanol and 2-propanol)

③ (for example the product obtained when liquid sulfur is poured in water)

④ (for example rhombic and monoclinic sulfur)

25. **(A)**   The pH of a solution is defined as

$$pH = -\log [H^+],$$

where $[H^+]$ is the hydronium ion concentration. Rearranging this equation, we obtain

$$[H^+] = 10^{-pH}.$$

Substituting the given pH, we have $[H^+] = 10^{-3} = 0.001$.

26. **(B)**   Buffer solutions contain conjugate acid-base pairs. They are usually prepared by mixing either a weak acid and a salt of the weak acid or a weak base and a salt of the weak base. In the answers given, only (B), .2 M NaCl and .1 M HCl, does not meet the requirement of a buffer. HCl is a strong acid. The others given are either a weak acid and a salt of the weak acid, or a weak base and a salt of the weak base.

27. **(A)**   The pH of a solution is defined as $pH = -\log [H^+]$

Therefore, $pH = -\log (1.2 \times 10^{-4})$

$$= -\log 1.2 + \log 10^{-4}$$

The log of 1.2 is obtained either from a log table or from a calculator that has a log function key. The log of 10^{-4} is -4. Therefore:

$$pH = -(.079 - 4.00)$$

$$= -(-3.92)$$

$$= 3.92$$

28. **(D)**   The strength of oxyacids depends on the electronegativity and oxidation state of the central atom of the acid. The greater that both of these are, the greater the strength of the acid. If H_aXO_b represents a general oxyacid, then $b - a$ is proportionate to acid strength. Thus, $HClO_4$ is a strong acid and $HClO$ is weak.

29. **(E)**   Conjugate acid-base pairs consist of an acid and base related by the exchange of a proton. These are H_2O, OH^- and $(CH_3)_3N$, $(CH_3)_3NH+$.

30. **(D)**   Using the definition of pH, we find that

$$pH = -\log [H^+] = 3.$$

Since $pH + pOH = 14$, we have $pOH = 11$.

31. **(C)**   The completely balanced reaction is

$$MnO_4^- + 8H^+ + 5Fe^{2+} \rightarrow 5Fe^{3+} + Mn^{2+} + 4H_2O.$$

32. **(D)**   Oxalic acid $H_2C_2O_4$ has the structural formula

$$\underset{HO}{\overset{O}{\underset{\diagup}{\overset{\diagdown}{C}}}} - \underset{OH}{\overset{O}{\underset{\diagdown}{\overset{\diagup}{C}}}}$$

It is a diprotic acid. It will therefore have two dissociation constants:

$$k_{a_1} = 5.36 \times 10^{-2} \text{ and } k_{a_2} = 5.3 \times 10^{-5}.$$

33. **(D)**   Salts of weak acids and strong bases will hydrolyze to produce basic solutions whereas salts of weak bases and strong acids will hydrolyze to form acid solutions. Both $CaCl_2$ and $NaCl$ are salts formed from the reaction of strong acids with strong bases and will not hydrolyze, i.e.,

$$HCl + NaOH \rightarrow NaCl + H_2O.$$

Sodium acetate CH_3COONa is a salt of a weak acid and a strong base:

$$CH_3COOH + NaOH \rightarrow CH_3COO^-NA^+ + H_2O.$$

The acetate ion, CH_3COO^- will hydrolyze:

$$CH_3COO^- + HOH \leftrightarrow CH_3COOH + OH^-$$

forming OH^- ions which are basic. The salt of a weak base and a strong acid, will hydrolyze to form an acid solution. NH_4Cl is such a salt and the NH_4^+ ion will hydrolyze:

$$NH_4^+ + HOH \; H_3O^+ + NH^3$$

forming the hydronium ion resulting in an acid solution whose pH is below 7.

34. **(D)**   All of the hydrogen halides are colorless gases that dissolve in water to give acid solutions. HCl, HBr_4, and HI ionize extensively and are strong acids. However, HF only ionizes slightly in water. In order for ionization to occur, the hydrogen – halogen bond must be broken:

$$HX \leftrightarrow H^+ + X^-.$$

The $H - F$ bond is considerably stronger than either the bonds in $H - Cl$, $H - Br$ or $H - I$, and ionization is more difficult. In addition, when HF ionizes,

$$HF \leftrightarrow H^+ + F^-$$

a second ionization reaction also takes place due to hydrogen bonding. This results in the formation of the hydrogen difluoride ion:

$$F^- + HF \leftrightarrow HF_2^-.$$

This hydrogen bonding effectively ties up some of the HF molecules and inhibits their dissociation.

35. **(D)** Since a k_i is given, the acid is a weak acid and reaches a state of equilibrium. In this case, $HCN \leftrightarrow H^+ + CN^-$ and the expression for the equilibrium constant is

$$k_i = \frac{[H^+][CN^-]}{HCN} = 4 \times 10^{-10}.$$

The initial amount of HCN present is .01 molar; however, the amount at equilibrium is .01 minus the amount that has ionized $(.01 - x)$. For x amount that ionizes, there will be x amount of H^+ and CN^- ions:

$$0.01 - x \qquad\qquad x \qquad x$$
$$HCN \qquad \leftrightarrow \qquad H^+ \quad CN^-$$

To solve for the H^+ concentration, we substitute

$$\frac{(x)(x)}{0.01 - x} = 4 \times 10^{-10}.$$

Since x is very small compared to the total amount of acid present, we may neglect it in the denominator without introducing a measurable error, thus

$$\frac{x^2}{0.01} = 4 \times 10^{-10}$$

$$x^2 = 4 \times 10^{-12}$$

$$x = [H^+] = 2 \times 10^{-6}$$

36. **(D)** Since $M_1V_1 = M_2V_2$, we have

$$M_2 = \frac{M_1V_1}{V_2} = \frac{(5)(50)}{100} = 2.5$$

37. **(D)** The total amount of NH_3 added is:

$$1.2 \text{ mole/}l \times 0.11 = 0.12 \text{ mole.}$$

The total amount of NH_4^+ added is:

$$0.5 \text{ mole/}l \times 0.41 = 0.2 \text{ mole.}$$

Therefore, the "new" concentrations for a total volume of $(0.1 + 0.4)$ l are:

$$[NH_3] = \frac{0.12}{0.5} = 0.24 \text{ M}$$

$$[NH_4^+] = \frac{0.2}{0.5} = 0.4 \text{ M}$$

From the reaction below:

$$NH_3 + H_2O \leftrightarrow NH_4^+ + OH^-$$

we obtain the expression for k_b:

$$k_b = \frac{[NH^+_4][OH^-]}{[NH_3]}$$

$$[OH^-] = \frac{(1.8 \times 10^{-5}) \times (0.24)}{0.4}$$

$$[OH^-] = 1.08 \times 10^{-5}$$

$$pOH = -\log [OH^-]$$

$$pOH = -(-5 + 0.334)$$

$$pOH = 4.966 \text{ and } pH = 14 - pOH = 9.0334$$

$$pH = 9.03$$

38. **(B)** The ionization constant is computed by the multiplication of product mole amounts over the mole amount of ionizing reactant. By division:

$$\frac{.1 \times .1}{.5} = \frac{.01}{.5} = 0.02 = 2 \times 10^{-2}$$

39. **(D)** Amphoteric compounds can function as either acids or bases. Hydroxides of Group I metals form solutions that are strongly basic when dissolved in water. Hydroxides of Group II metals, although less soluble than Group I hydroxides, also form basic solutions when dissolved in water. Hydroxides of metals that are of intermediate electronegativity and in relatively high oxidation states are usually amphoteric. $Al(OH)_3$ can function either as an acid or a base:

$$Al(OH)_3 + 3OH^- \rightarrow AlO_3^{3-} + 3H_2O \qquad \text{(as an acid)}$$

or

$$Al(OH)_3 + 3H^+ \rightarrow Al^{3+} + 3H_2O \qquad \text{(as a base)}$$

The metal hydroxides that exhibit amphoteric behavior are generally found along the diagonal of the periodic table that separates the metals from the nonmetals.

40. **(D)** The percent dissociation is given by

$$\frac{[OH^-]_{equilibrium}}{[NH_4OH]_{initial}} \times 100\% = \frac{1.3 \times 10^{-3}}{0.1} \times 100\% = 1.3\%.$$

ACID AND BASE EQUILIBRIA REVIEW

1. Acids and Bases

DEFINITIONS OF ACIDS AND BASES

Arrhenius Theory

The Arrhenius theory states that acids are substances that ionize in water to give H^+ ions, and bases are substances that produce OH^- ions in water.

Bronsted-Lowry Theory

This theory defines acids as proton donors and bases as proton acceptors.

Lewis Theory

This theory defines an acid as an electron-pair acceptor and a base as an electron-pair donor.

Whenever an Arrhenius acid and base are mixed together in equal quantities, the result will be a salt (nonmetal combined with a metal) and water. This process is termed "neutralization." For example, if hydrochloric acid is mixed with sodium hydroxide, what will the product be? Since we are mixing an Arrhenius acid and base together, we know we will get water and a salt. If we look at the equation for the reaction:

$$HCl + NaOH \rightarrow H_2O + NaCl,$$

we see that the salt formed is sodium chloride.

The hydroxides of elements that border the metal/nonmetal line on the periodic table act as acids toward strong bases and as bases toward strong acids. Such compounds are called "amphoteric." For example, aluminum hydroxide is quite insoluble in water (where the number of OH^- equals the number of H^+), but is quite soluble in either strong base or acid.

$$Al(OH)_3 + H_2O \rightarrow No\ reaction \text{ — insoluble}$$

$$Al(OH)_3 + 3H^+ \rightarrow Al^{+++} + 3H_2O \text{ — soluble}$$

$$Al(OH)_3 + OH^- \rightarrow [Al(OH)_4]^- \text{ — soluble}$$

In the last equation, the combination represented as $[Al(OH)_4]^-$ is known as a "coordination complex." This term is applied to compounds and ions in which negative or neutral polar molecules are attached to metal ions or atoms.

PROBLEM

> Can I^+ (the iodine cation) be called a Lewis base? Explain your answer.

SOLUTION

A Lewis base may be defined as an electron pair donor. Writing out its electronic structure is the best way to answer this question, because it will show the existence of any available electron pairs.

The electronic structure of I^+ may be written as

$$[: \ddot{I} :]^+.$$

There are three available electron pairs. This might lead one to suspect that it is indeed a Lewis base. But note, I^+ does not have a complete octet of electrons; it does not obey the octet rule. According to this rule, atoms react to obtain an octet (8) of electrons. This confers stability.

Therefore, I^+ would certainly rather gain two more electrons than lose six. In reality, then, I^+ is an electron pair acceptor. Such substances are called Lewis acids.

PROBLEM

> Write the equations for the stepwise dissociation of pyrophosphoric acid, $H_4P_2O_7$. Identify all conjugate acid-base pairs.

SOLUTION

Pyrophosphoric acid is an example of a polyprotic acid. Polyprotic acids furnish more than one proton per molecule. From its molecular formula, $H_4P_2O_7$, one can see there exist four hydrogen atoms. This might lead one to suspect that it is tetraprotic, i.e. having 4 protons that can be donated per molecule. This is in fact the case, which means there exist four dissociation reactions. In general, the equation for a dissociation reaction is

$$HA + H_2O \rightarrow H_3O^+ + A^-.$$

Polyprotic acids follow this pattern. Thus, one can write the following equations for the stepwise dissociation of $H_4P_2O_7$.

(1) $\qquad H_4P_2O_7 + H_2O \rightarrow H_3O^+ + H_3P_2O_7^-$

(2) $\qquad H_3P_2O_7^- + H_2O \rightarrow H_3O^+ + H_2P_2O_7^{-2}$

(3) $\qquad H_2P_2O_7^{-2} + H_2O \rightarrow H_3O^+ + HP_2O_7^{-3}$

(4) $\qquad HP_2O_7^{-3} + H_2O \rightarrow H_3O^+ + P_2O_7^{-4}$

To identify all conjugate acid-base pairs, note the definition of the term. The base that results when an acid donates its proton is called the conjugate base.

The acid that results when a base accepts a proton is called the conjugate acid. From these definitions, one sees that in all cases H_3O^+ is the conjugate acid of H_2O (the base in these reactions) and $H_3P_2O_7^-$, $H_2P_2O_7^{-2}$, $HP_2O_7^{-3}$ and $P_2O_7^{-4}$ are the conjugate bases of $H_4P_2O_7$, $H_3P_2O_7^-$, $H_2P_2O_7^{-2}$ and $HP_2O_7^{-3}$, respectively.

Drill 1: Acids and Bases

1. In the reaction

    $$HCN + H_2O \leftrightarrow H_3O^+ + CN^-$$

    which species are functioning as Brönsted bases?

    (A) Only H_2O

    (B) HCN and H_3O^+

    (C) H_2O and CN^-

    (D) CN^- and HCN

    (E) H_3O^+ and CN^-

2. Given VO_4^{3-}, CrO_4^{2-}, and MnO_4^-, which of the following is true?

    (A) VO_4^{3-} (strongest base); $HMnO_4$ (strongest acid)

    (B) H_3VO_4 (strongest acid); MnO_4^- (strongest base)

    (C) H_3VO_4 (strongest acid); CrO_4^{2-} (strongest base)

    (D) H_2CrO_4 (strongest acid); MnO_4^- (strongest base)

    (E) VO_4^{3-} (weakest base);$_{22}$ MnO_4^{2-} (weaker base); CrO_4^{2-} (weak base)

3. Which of the acids below has the formula $HBrO_2$?

    (A) Bromic

    (B) Bromous

    (C) Hydrobromic

    (D) Hypobromous

    (E) Perbromic

4. Which of the following can function as a Lewis acid?

    (A) $:\underset{..}{I}:^+$

    (B) $:CN^-$

    (C) $CH_3 - \overset{..}{\underset{..}{O}} - CH_3$

    (D) $:NH_3$

    (E) $:\overset{..}{\underset{..}{Br}}:^-$

5. Which of the following is/are both a Brönsted-Lowry base and a Lewis base?

    I. NH_3

    II. BBr_3

    III. H_2O

    IV. $NaOH$

    (A) I and IV

    (B) IV only

    (C) I only

    (D) II and III

    (E) I and III

2. The Autoionization of Water

For the equation

$$H_2O + H_2O \leftrightarrow H_3O + OH^-, k_w = [H_3O^+] [OH^-]$$

$$(or\ k_w = [H^+] [OH^-]) = 1.0 \times 10^{-14} \text{ at } 25°C,$$

where $[H_3O^+] [OH^-]$ is the product of ionic concentrations, and k_w is the ion product constant for water (or simply the ionization constant or dissociation constant).

$$pH = -log [H^+]$$

$$pOH = -log [OH]$$

$$pk_w = pH + pOH = 14.0$$

In a neutral solution, pH = 7.0. In an acidic solution, pH is less than 7.0. In basic solutions, pH is greater than 7.0. The smaller the pH, the more acidic is the solution. Note that since k_w (like all equilibrium constants) varies with temperature, neutral pH is less than (or greater than) 7.0 when the temperature is higher than (or lower than) 25°C.

PROBLEM

A 0.10 M solution of HCl is prepared. What species of ions are present at equilibrium, and what will be their equilibrium concentrations?

SOLUTION

Two processes are occurring simultaneously: the reaction of HCl with H_2O (dissociation of HCl) and the autoionization of H_2O.

HCl reacts with H_2O according to the equation

$$HCl + H_2O \leftrightarrow H_3O^+ + Cl^-.$$

For every mole of HCl that dissociates, one mole of Cl^- and one mole of H_3O^+ are produced. The initial concentration of HCl is 0.10 M. Thus, if we assume that HCl dissociates completely,

$$[H_3O^+] = 0.10 \text{ M and } [Cl^-] = 0.10 \text{ M}.$$

Water autoionizes according to the equation

$$H_2O + H_2O \leftrightarrow H_3O^+ + OH^-.$$

The water constant for this process is

$$k_w = 10^{-14} \text{ moles}^2/\text{liter}^2 = [H_3O^+][OH^-].$$

Hence, $[OH^-] = \dfrac{10^{-14} \text{ moles}^2/\text{liter}^2}{[H_3O^+]}.$

$[H_3O^+]$ was determined to be $0.10 \text{ M} = 10^{-1} \text{ M} = 10^{-1} \text{ moles/liter}$,

$$[OH^-] = \frac{10^{-14} \text{ moles}^2/\text{liter}^2}{[H_3O^+]}$$

$$= \frac{10^{-14} \text{ moles}^2/\text{liter}^2}{10^{-1} \text{ moles/liter}} = 10^{-13} \text{ moles/liter} = 10^{-13} \text{ M}$$

Hence, at equilibrium, H_3O^+, OH^-, and Cl^- are present in the concentrations

$$[H_3O^+] = 0.10 \text{ M}, [OH^-] = 10^{-13} \text{ M}, [Cl^-] = 0.10 \text{ M}.$$

PROBLEM

A 0.10 M solution of NaOH is prepared. What species of ions are present at equilibrium, and what will be their equilibrium concentrations?

SOLUTION

Two processes are occurring simultaneously, the dissociation of NaOH and the autoionization of H_2O.

NaOH dissociates according to the equation

$$NaOH \leftrightarrow Na^+ + OH^-.$$

For every mole of NaOH that dissociates, one mole of Na^+ and one mole of OH^- are produced. The initial concentration of NaOH is 0.10 M. Thus, if we assume that NaOH dissociates completely $[Na^+] = 0.10 \text{ M}$ and $[OH^-] = 0.10 \text{ M}$.

Water autoionizes according to the equation

$$H_2O + H_2O \leftrightarrow H_3O^+ + OH^-.$$

The water constant for this process is

$$k_w = 10^{-14} \text{ moles}^2/\text{liter}^2 = [H_3O^+] [OH^-].$$

Hence, $[H_3O^+] = \dfrac{10^{-14} \text{ moles}^2/\text{liter}^2}{[OH^-]}.$

Since $[OH^-]$ was determined to be $0.10 \text{ M} = 10^{-1} \text{ M} = 10^{-1}$ moles/liter, by substitution, we obtain

$$[H_3O^+] = \dfrac{10^{-14} \text{ moles}^2/\text{liter}^2}{[OH^-]} = \dfrac{10^{-14} \text{ moles}^2/\text{liter}^2}{10^{-1} \text{ moles/liter}}$$

$$= 10^{-13} \text{ moles/liter} = 10^{-13} \text{ M}$$

Hence, at equilibrium, H_3O^+, OH^-, and Na^+ are present in the concentrations

$$[H_3O^+] = 10^{-13} \text{ M}, [OH^-] = 0.10 \text{ M}, [Na^+] = 0.10 \text{ M}.$$

Drill 2: The Autoionization of Water

1. What is the hydrogen ion H^+ concentration of a 0.0020 M potassium hydroxide solution?

    (A) $[H^+] = \dfrac{k_w}{0.0020}$

    (B) $[H^+] = k_w [0.0020]$

    (C) $[H^+] = \dfrac{0.0020}{k_w}$

    (D) $[H^+] = -\log_{10} \left[\dfrac{k_w}{0.0020} \right]$

    (E) None of the above

2. What is the concentration of OH^- of a solution where the concentration of $H^+ = 0.1$?

    (A) 1×10^{-13}

    (B) 1×10^{-15}

    (C) 1×10^{-2}

    (D) 1×10^{-9}

    (E) 1×10^{-6}

3. A solution of KOH is prepared by adding 6.00 g of KOH to 50.0 ml of water. What is the concentration of H_3O^+ present at equilibrium?

    (A) 2.18×10^{-29}

    (B) 4.67×10^{-15}

    (C) 1.00×10^{-14}

    (D) 2.14

    (E) 2.14×10^{-7}

3. Autoprotolysis

Autoprotolysis is defined as the donation of a proton from a molecule of one specie to another molecule of the same specie to produce positive and negative ions.

For example, the autoprotolysis of hydroxylamne (H_2NOH) can be written as

$$H_2NOH + H_2NOH \leftrightarrow H_3NOH^+ + H_2NO^-.$$

PROBLEM

> Find the equation for the autoprotolysis of water. Indicate which species is the acid, the base, the conjugate acid, and the conjugate base.

SOLUTION

One can begin by defining autoprotolysis. It may be defined as the donation of a proton from a molecule of one specie to another molecule of the same specie to produce positive and negative ions. Thus, for water, the equation is $H_2O + H_2O \rightarrow H_3O^+ + OH^-$. An acid is defined as a specie that donates protons. A base is a substance that accepts protons. From the equation, one sees that either water (H_2O) molecule can be the base or acid. A conjugate base is a specie obtained by abstracting a proton (H^+). If one abstracts a proton from water, one obtains OH^-. Thus, OH^- is the conjugate base. The conjugate acid is defined as the base plus a proton. It was stated that either H_2O molecule could be the base. If one adds a proton to one of them, one obtains H_3O^+. Thus, H_3O^+ is the conjugate acid.

Drill 3: Autoprotolysis

1. Consider the autoprotolysis of liquid ammonia:

   $$NH_3 + NH_3 \leftrightarrow NH_4^+ + NH_2^-$$

   If $KNH_3 = 10^{-22}$, how many molecules of ammonia are ionized in one mole of ammonia? Assume a density of 0.771 g/ml for ammonia.

   (A) 16.1×10^{11} (D) 6.12×10^{10}

   (B) 3.5×10^{10} (E) 2.66×10^{11} molecules

   (C) 10.1×10^{12} molecules

2. Given that the density of liquid ammonia (NH_3) is 0.771 g/ml, if the autoprotolysis constant of NH_3 is 1×10^{-22}, what would the equilibrium concentration of NH_4^+ be if one starts with 10 ml of ammonia?

   (A) 1×10^{-22} (B) 1×10^{-11}

(C) 0.0450   (D) 0.0560

(E) 1×10^{-7}

3. For a gas phase reaction with $E_a = 40,000$ cal/mol, estimate the change in rate constant due to a temperature change from 1,000°C to 2,000°C.

(A) 1,050   (D) 2,273

(B) 2,303   (E) 2,060

(C) 1,987

---

4. pH

A widely used measure of acidity, pH, is defined as the negative logarithm of the H^+ ion concentration.

$$pH = -\log_{10}(H^+) \tag{1}$$

Hence, the H^+ ion concentration is equal to $10^{(-pH)}$. When both the H^+ and OH^- ion concentrations are equal, the equation $[H^+] [OH^-] = 10^{-14} = k_w$ dictates that they both equal 10^{-7}. The neutral pH, therefore, is 7; acidic solutions have a pH less than 7, basic solutions greater than 7. A similar quantity, pOH, is defined for the OH^- ion concentration, but it is much less common than pH. The sum of pH and pOH will always equal 14 in order to satisfy the k_w equation.

PROBLEM

a) Determine the pH of a solution with a hydrogen ion concentration of 3.5×10^{-4}.

b) If a solution has a pH of 4.25, what is the hydrogen ion concentration?

SOLUTION

To determine the acidity or basicity of an aqueous solution, the hydrogen ion concentration must be measured. The pH of a solution expresses this concentration. pH is defined as the negative logarithm of the hydrogen ion concentration. In other words, $pH = (-\log [H^+])$, where the brackets around H^+ signify concentration. As such, to solve the problem, you must substitute into the equation. For part (a), you have

$$pH = -\log [3.5 \times 10^{-4}]$$

now $-\log [3.5 \times 10^{-4}] = -\log 3.5 - \log 10^{-4}$

$$= -.54 - (-4)$$

$$= -.54 + 4$$

$$= 3.46$$

It follows, then, that pH = 3.46 for a hydrogen ion concentration of 3.5×10^{-4}. Part (b) is similar, but here you are given the pH and asked to find the ion concentration. Therefore, you have $4.25 = -\log [H^+]$ or $-4.25 = \log [H^+]$. Now, logarithm numbers give only positive mantissas. As such, -4.25 must be in the form of $-5 + .75$. If you take the antilogarithm of each, .75 is 5.6 and -5 is 10^{-5}, you obtain a hydrogen ion concentration of 5.6×10^{-5} moles/liter.

PROBLEM

Determine the pH of each of the following solutions: (a) 0.20 M HCl, (b) 0.10 M NaOH.

SOLUTION

A pH scale has been devised to express the H_3O^+ concentration in solution. By definition,

$$pH = -\log [H_3O^+] \text{ or } [H_3O^+] = 10^{-pH}.$$

It has been shown that water dissociates to H_3O^+ and OH^- ions to a small degree.

$$H_2O + H_2O \rightarrow H_3O^+ + OH^-$$

The equilibrium constant is defined as k_w for this reaction and is expressed as $[H_3O^+][OH^-]$. The H_2O does not appear, since it is presumed to be a constant. From the dissociation equation, it can be seen that the concentration of H_3O^+ equals OH^-. By experimentation, k_w has been shown to equal 1.0×10^{-14}. This means that in water, therefore, H_3O^+ and OH^- each have a concentration of 1.0×10^{-7} M. With this information in mind, one can now solve the problem.

(a) The concentration of HCl is 0.20 M. Since HCl is a strong electrolyte, dissociation is complete. Therefore, the concentration of H_3O^+ is also 0.20 M = 2.0×10^{-1} M. By definition, then

$$pH = -\log (2.0 \times 10^{-1}) = 1 - 0.3 = 0.7.$$

(b) The $[OH^-]$ equals the concentration of NaOH, since it is also a strong electrolyte. One wants the pH, therefore, to employ the expression for k_w.

$$[H_3O^+] = \frac{k_w}{[OH^-]} = \frac{1 \times 10^{-14}}{0.10} = 1.0 \times 10^{-13} \text{ M}$$

Therefore, $pH = -\log (1.0 \times 10^{-13}) = 13$.

Drill 4: pH

1. What is the pH of a 10^{-13} M HCl solution?

    (A) 1   (D) 10

    (B) 5   (E) 13

    (C) 7

2. The pOH of a .0001 M KOH solution is

    (A) 1.   (D) 7.

    (B) 2.   (E) 11.

    (C) 4.

3. If the pOH of solution A is 2.5 and the pOH of B is 10.1, then which of the following is true?

    (A) Solution A has a higher concentration of protons than B.

    (B) Solution B is more basic than solution A.

    (C) Solution A is more basic than solution B.

    (D) Solution A is more acidic than solution B.

    (E) Solution B has a higher concentration of hydroxyl ions than A.

4. The pOH of a 1.0 M solution of HCl is

    (A) 1.   (D) 0.

    (B) 13.   (E) 15.

    (C) 14.

5. What is the pH of a 0.01 M NaOH solution?

    (A) 2   (D) 10

    (B) 4   (E) 12

    (C) 7

5. The Ionization Constant

The ionization constant for an acid (k_a) is defined as the concentration of H^+ ions times the concentration of the conjugate base ions of a given acid divided by the concentration of the unionized acid. For an acid HA,

$$k_a = \frac{[H^+][A^-]}{[HA]}$$

where k_a is the ionization constant, $[H^+]$ is the concentration of H^+ ions, $[A^-]$ is the concentration of the conjugate base ions, and $[HA]$ is the concentration of the unionized acid. A similar equation can be written for a weak base (BA).

$$k_b = \frac{[B^-][A^+]}{[BA]}$$

PROBLEM

The ionization constant for acetic acid is 1.8×10^{-5}.

a) Calculate the concentration of H^+ ions in a 0.10 molar solution of acetic acid.

b) Calculate the concentration of H^+ ions in a 0.10 molar solution of acetic acid in which the concentration of acetate ions has been increased to 1.0 molar by addition of sodium acetate.

SOLUTION

The ionization constant (k_a) is defined as the concentration of H^+ ions times the concentration of the conjugate base ions of a given acid divided by the concentration of unionized acid. For an acid, HA,

$$k_a = \frac{[H^+][A^-]}{[HA]},$$

where k_a is the ionization constant, $[H^+]$ is the concentration of H^+ ions, $[A^-]$ is the concentration of the conjugate base ions, and $[HA]$ is the concentration of unionized acid. The k_a for acetic acid is stated as

$$k_a = \frac{[H^+][\text{acetate ion}]}{[\text{acetic acid}]} = 1.8 \times 10^{-5}.$$

The chemical formula for acetic acid is $HC_2H_3O_2$. When it is ionized, one H^+ is formed and one $C_2H_3O^-$ (acetate) is formed, thus the concentration of H^+ equals the concentration of $C_2H_3O^-$.

$$[H^+] = [C_2H_3O^-]$$

The concentration of unionized acid is decreased when ionization occurs. The new concentration is equal to the concentration of H^+ subtracted from the concentration of unionized acid.

$$[HC_2H_3O] = 0.10 - [H^+]$$

Since $[H^+]$ is small relative to 0.10, one may assume that $0.10 - [H^+]$ is approximately equal to 0.10.

$$0.10 - [H^+] = 0.10$$

Using this assumption, and the fact that $[H^+] = [C_2H_3O^-]$, k_a can be rewritten as

$$k_a = \frac{[H^+][H^+]}{0.10} = 1.8 \times 10^{-5}.$$

Solving for the concentration of H^+:

$$[H^+]^2 = (1.0 \times 10^{-1})(1.8 \times 10^{-5}) = 1.8 \times 10^{-6}$$

$$[H^+] = \sqrt{1.8 \times 10^{-6}} = 1.3 \times 10^{-3}$$

The concentration of H^+ is thus 1.3×10^{-3} M.

b) When the acetate concentration is increased, the concentration of H^+ is lowered to maintain the same k_a. The k_a for acetic acid is stated as

$$k_a = \frac{[H^+][C_2H_3O^-]}{[HC_2H_3O]} = 1.8 \times 10^{-5}.$$

As previously shown for acetic acid equilibria in a solution of 0.10 molar acid, the concentration of acid after ionization is

$$[HC_2H_3O] = 0.10 - [H^+].$$

Because $[H^+]$ is very small compared to 0.10, $0.10 - [H^+] \approx 0.10$ and

$$[HC_2H_3O] = 0.10.$$

In this problem, we are told that the concentration of acetate is held constant at 1.0 molar by addition of sodium acetate. Because one now knows the concentrations of the acetate and the acid, the concentration of H^+ can be found.

$$\frac{[H^+][C_2H_3O^-]}{[HC_2H_3O]} = 1.8 \times 10^{-5}$$

$$\frac{[H^+][1.0]}{[0.10]} = 1.8 \times 10^{-5}$$

$$[H^+] = 1.8 \times 10^{-6}.$$

PROBLEM

Find the hydronium ion concentration of .1 M HOAC (acetic acid) solution. Assume $k_a = 1.75 \times 10^{-5}$ for acetic acid.

SOLUTION

You want to represent the equilibrium constant expression for the reaction, which necessitates a balanced equation. After writing the expression, you want to express the concentrations in terms of the same variables and solve for it. Begin by writing the balanced equation for the reaction of acetic acid in water. The acid will donate a proton (H^+) to the only available base, H_2O. Thus, $HOAC + H_2O \rightarrow H_3O + OAC^-$. $[H_3O^+]$, the hydronium concentration, is the quantity you are looking for. The equilibrium constant expression measures the ratio of the concentrations of the products to the reactants, each raised to the power of their respective coefficients in the chemical equation. Thus, the constant,

$$k_a = \frac{[OAC^-][H_3O^+]}{[HOAC]}.$$

Note: H_2O is omitted, since it is considered a constant. $k_a = 1.75 \times 10^{-5}$. Equating,

$$\frac{[OAC^-][H_3O^+]}{[HOAC]} = 1.75 \times 10^{-5}.$$

Let x = concentration of H_3O^+. According to the reaction, $[H_3O^+] = [OAC^-]$, thus, x = concentration of $[OAC^-]$, also. If the initial concentration of HOAC is .1 and x moles/liter of $[H_3O^+]$ are formed, then you have $(.1 - x)$ moles/liter of HOAC left. Substituting these variables into the equilibrium constant expression, you have

$$\frac{x^2}{.1 - x} = 1.75 \times 10^{-5}.$$

Solving, $x = [H_3O^+] = 0.0013$ M.

Drill 5: The Ionization Constant

1. $NH_3(aq) + H_2O(1) \leftrightarrow NH_4^+(aq) + OH^-(aq)$

    For the above equation, calculate the ionization constant if the $[NH_4^+]$ is 10^{-4} M, $[NH_3]$ is 1.0 M, and $[OH^-]$ is 0.18 M, respectively.

    (A) 1.8×10^{-5} (D) 1.8

    (B) 1.8×10^{-4} (E) 1.8×10^{-3}

    (C) 10^{-5}

2. The ionization constant for NH_4OH is 1.8×10^{-5}. Calculate the concentration of OH^- ions in a 1.0 molar solution of NH_4OH.

    (A) 4.2×10^{-3} (D) 6.2×10^{-1}

    (B) 3.9×10^{-2} (E) 2.9×10^{-5}

    (C) 5.4×10^{-4}

3. Find the degree of ionization of 0.05 M NH_3 in a solution of pH 11.0. $k_b = 1.76 \times 10^{-5}$.

    (A) 2.12×10^{-2} (D) 1.73×10^{-2}

    (B) 4.16×10^{-1} (E) 3.95×10^{-3}

    (C) 6.23×10^{-6}

4. Given k_i for acetic acid is 1.8×10^{-5}, calculate the percentage of ionization of 0.5 M acetic acid. The dissociation reaction is

    $$HC_2H_3O_2 \leftrightarrow H^+ + C_2H_3O_2^-.$$

    (A) 0.20%   (D) 0.54%

    (B) 0.60%   (E) 0.73%

    (C) 0.16%

---

6. The Dissociation Constant

Dissociation of Weak Electrolytes

For the equation

$$A^- + H_2O \leftrightarrow HA + OH^-,$$

$$k_b = \frac{[HA][OH^-]}{[A^-]}$$

where k_b is the base ionization constant.

$$k_a = \frac{[H^+][A^-]}{[HA]}$$

where k_a is the acid ionization constant.

$$k_b = \frac{k_w}{k_a}$$

for any conjugate acid/base pair, and therefore,

$k_w = [H^+] [OH^-]$.

Dissociation of Polyprotic Acids

For $H_2S \leftrightarrow H^+ + HS^-$,

$$k_{a_1} = \frac{[H^+][HS^-]}{[H_2S]}.$$

For $HS^- \leftrightarrow H^+ + S^{2-}$,

$$k_{a_2} = \frac{[H^+][S^{2-}]}{[HS^-]}$$

k_{a_1} is much greater than k_{a_2}. Also,

$$k_a = k_{a_1} \times k_{a_2} = \frac{[H^+][HS^-]}{[H_2S]} \times \frac{[H^+][S^{2-}]}{[HS^-]}$$

$$= \frac{[H^+]^2 [S^{2-}]}{[H_2S]}$$

This last equation is useful only in situations where two of the three concentrations are given and you wish to calculate the third.

PROBLEM

If 1 mole of HCl and 1 mole of $NaC_2H_3O_2$ are mixed in enough water to make one liter of solution, what will be the concentrations of the species in the final equilibrium? $k_{diss} = 1.8 \times 10^{-5}$ for $NaC_2H_3O_2$.

SOLUTION

To answer this question, you must consider what is happening at equilibrium. This necessitates defining $k_{dissociation}$, which is an equilibrium constant.

HCl and $NaC_2H_3O_2$ are strong electrolytes, which means that, in solution, they are completely dissociated. You have, therefore, H^+, Cl^-, Na^+, and $C_2H_3O_2^-$ ions present in the solution. The Na^+ and Cl^- do not associate and need not be considered. Thus, you must only consider the formation of $HC_2H_3O_2$ from H^+ and $C_2H_3O_2^-$. The equation for this reaction can be written

$$H^+ + C_2H_3O_2^- \leftrightarrow HC_2H_3O_2.$$

This reaction can proceed in both directions, an equilibrium exists, as the double arrow indicates. The equilibrium constant (k_{eq}) for this reaction is equal to

$$\frac{[HC_2H_3O_2]}{[H^+][C_2H_3O_2^-]}.$$

$k_{dissociation}$ measures the equilibrium quantitatively. The dissociation reaction for $HC_2H_3O_2$ can be written

$$HC_2H_3O_2 \leftrightarrow H^+ + C_2H_3O_2^-.$$

The dissociation constant,

$$k_{diss} = \frac{[H^+][C_2H_3O_2^-]}{[HC_2H_3O_2]} = 1.8 \times 10^{-5}.$$

By examination, you can see that k_{eq} for the association reaction is equal to $1/k_{diss}$. Thus,

$$k_{eq} = \frac{1}{k_{diss}} = \frac{1}{1.8 \times 10^{-5}} = \frac{[HC_2H_3O_2]}{[H^+][C_2H_3O_2^-]}.$$

To rewrite into a more convenient form for solving, take the reciprocal of each side.

$$1.8 \times 10^{-5} = \frac{[H^+][C_2H_3O_2^-]}{[HC_2H_3O_2]}$$

The final concentrations of the species, the unknowns, will be those at the equilibrium. Let y be the concentration of $HC_2H_3O_2$ at equilibrium. The concentrations of both H^+ and $C_2H_3O_2^-$ can be represented by $1 - y$. Initially, you started with 1 mole/liter of each, therefore, each y mole/liter that associates to form $HC_2H_3O_2$ must be subtracted from the initial concentration. You can now substitute these variables into the expression for k_{diss} to obtain

$$\frac{(1-y)(1-y)}{y} = 1.8 \times 10^{-5}.$$

Solving for y, using the quadratic formula, you obtain $y = .996$. Therefore, the concentrations of the species are

$$[H^+] = 1 - y = .004 \text{ M}$$

$$[C_2H_3O_2^-] = .004 \text{ M}$$

$$[HC_2H_3O_2] = .996 \text{ M}$$

PROBLEM

There exists a 0.5 M HF solution for which $k_{diss} = 6.71 \times 10^{-4}$. Determine how much dilution is necessary to double the percent dissociation of HF.

SOLUTION

Percent dissociation means the ratio of $[H^+]$ to original $[HF]$ concentration times 100. To find the amount of dilution necessary to double the percent dissociation of an acid, first, establish what the percent dissociation is before dilution. This can be determined from k_{diss}, the equilibrium dissociation constants, which measures the ratio of products to reactants (i.e., their concentrations), each raised to the power of their coefficients in the balanced chemical equation. The general reaction for the dissociation of an acid, e.g., HA, is

$$HA + H_2O \leftrightarrow H_3O^+ + A^-.$$

Thus, for HF,

$$HF + H_2O \leftrightarrow H_3O^+ + F^-$$

for the dissociation reaction. This means that

$$k_{diss} = 6.71 \times 10^{-4} = \frac{[H_3O^+][F^-]}{[HF]}.$$

It is given that the initial concentration of HF is 0.5. Thus, to find percent dissociation, one needs to know $[H_3O^+]$. To find this, perform the following operations:

Let $x = [H_3O^+]$. Since H_3O^+ and F^- are formed in equimolar amounts, as can be seen in the chemical reaction, $[H_3O^+] = [F^-] = x$. If the initial $[HF] = 0.5$, and x moles/liter dissociate to give $[H_3O^+]$ and $[F^-]$, then, at equilibrium, $[HF] = 0.5 - x$. Substituting these values:

$$\frac{x \times x}{0.5 - x} = 6.71 \times 10^{-4}.$$

Solving for x, using the quadratic formula, $x = 0.018 = [H_3O^+]$. Thus, percentage dissociation becomes

$$\frac{0.018}{0.5} \times 100 = 3.6\% \text{ HF dissociated.}$$

Then, to determine the final answer, 7.2% HF dissociation is needed when one dilutes. Recall that $[H^+] = [F^-]$. At 7.2% dissociation,

$$\frac{[F^-]}{[HF]} = \frac{0.072}{1 - .072} = \frac{0.072}{0.928}.$$

From the k_{diss} expression, one has

$$k_{diss} = 6.71 \times 10^{-4} = \frac{[H^+][F^-]}{[HF]}$$

or

$$\frac{6.71 \times 10^{-4}}{[H^+]} = \frac{[F^-]}{[HF]}.$$

$$\frac{[F^-]}{[HF]} = \frac{0.072}{0.928},$$

when percent dissociation is doubled to 7.2%. After substitution,

$$\frac{6.71 \times 10^{-4}}{[H^+]} = \frac{0.072}{0.928}.$$

Solving for $[H^+]$, $[H^+] = 8.65 \times 10^{-3} = [F^-]$. Substituting these actual molar concentrations into the equilibrium constant expression, one obtains

$$\frac{6.71 \times 10^{-4}}{8.65 \times 10^{-3}} = \frac{8.65 \times 10^{-3}}{[HF]}.$$

Solving: $[HF] = 0.112$ at equilibrium

$[H^+] = 8.65 \times 10^{-3}$.

Before dissociation $[HF]$ = amount at equilibrium plus amount dissociated = $8.65 \times 10^{-3} + 0.112 = 0.121$ M. Thus, when the percent dissociation is doubled, the initial amount equals 0.121 M for $[HF]$. The $[HF]$, when the percent dissociation was unchanged, was 0.5 M (given). Therefore, dilute by factor $0.500/0.121 = 4.13$. Remember that concentration or molarity is a parameter of volume, actually M = moles/liter. One is going from 0.5 M to 0.121 M, which means volume must be increased by a certain factor. When one dilutes a solution, one is adding volume to it and dividing the solution. This accounts for using division to obtain the factor of dilution, once initial concentrations are known.

Drill 6: The Dissociation Constant

1. In a 0.1 M solution of acetic acid in water, 1% of the acid is dissociated. Determine the pH of the solution.

    (A) 11   (D) 7

    (B) 3   (E) 8

    (C) 5

2. A chemist wants the percent dissociation of $HC_2H_3O_2$ to be 1%. If the k_{diss} of $HC_2H_3O_2$ is 1.8×10^{-5}, what concentration of $HC_2H_3O_2$ is required?

    (A) 2.13×10^{-2} M

    (B) 4.51×10^{-3} M

    (C) 1.76×10^{-1} M

    (D) 1.21×10^{-2} M

    (E) 3.92×10^{-1} M

3. Given a solution of 1.00 M $HC_2H_3O_2$, what is the concentration of all solute species? What is the percentage of acid that dissociated? Assume $k_{diss} = 1.8 \times 10^{-5}$.

    (A) 0.42%

    (B) 0.56%

    (C) 0.39%

    (D) 0.23%

    (E) 0.50%

7. The Hydrolysis Constant

Hydrolysis refers to the action of salts of weak acids or bases with water to form acidic or basic solutions.

Salts of Weak Acids and Strong Bases: Anion Hydrolysis

For

$$C_2H_3O_2^- + H_2O \leftrightarrow HC_2H_3O_2 + OH^-,$$

$$k_h = \frac{[HC_2H_3O_2][OH^-]}{[C_2H_3O_2^-]}$$

where k_h is the hydrolysis constant for the acetate ion, which is just k_b for acetate. Also,

$$k_a = \frac{[H^+][C_2H_3O_2^-]}{[HC_2H_3O_2]}$$

and $\quad k_h = \dfrac{k_w}{k_a}$

Salts of Strong Acids and Weak Bases: Cation Hydrolysis

For

$$NH_4^+ + H_2O \leftrightarrow H_3O^+ + NH_3,$$

$$k_h = \frac{[H_3O^+][NH_3]}{[NH_4^+]}$$

$$(k_h = k_a \text{ for } NH_4^+)$$

Also, $\quad k_h = \dfrac{k_w}{k_b}$

and $\quad k_b = \dfrac{[NH_4^+][OH^-]}{[NH_3]}$

Hydrolysis of Salts of Polyprotic Acids

For

$$S^{2-} + H_2O \leftrightarrow HS^- + OH^-,$$

$$k_{h_1} = \frac{k_w}{k_{a_2}} = \frac{[HS^-][OH^-]}{[S^{2-}]}$$

where k_{a_2} is the acid dissociation constant for the weak acid HS^-.

For

$$HS^- + H_2O \leftrightarrow H_2S + OH^-,$$

$$k_{h_2} = \frac{k_w}{k_{a_1}} = \frac{[H_2S][OH^-]}{[HS^-]}$$

where k_{a_1} is the dissociation constant for the weak acid H_2S.

Tables giving the values of k_A's and k_B's for various acids and bases at different temperatures are readily available. These can be used to calculate the degree of ionization. For example, formic acid has a k_A of 1.8×10^{-4}. What is the pH of a 0.0400 M solution? We know that formic acid dissociates to give us a hydrogen in:

$$HCOOH \leftrightarrow HCOO^- + H^+.$$

First, we set up the equilibrium equation:

$$\frac{[H^+][HCOO^-]}{[HCOOH]} = k_A$$

We know that the equilibrium concentration of HCOOH is 0.0400 M minus the amount that dissociates to $HCOO^-$ and H^+. If we term this amount "X", then $[HCOOH] = 0.0400 - X$. We also know from stoichiometry that $[H^+] = [HCOO^-]$

$= X$. Substituting this into the equation, we get:

$$\frac{(X)(X)}{(0.0400 - X)} = 1.8 \times 10^{-4}$$

$$\frac{X^2}{0.0400 - X} = 1.8 \times 10^{-4},$$

$$X^2 = (0.0400)(1.8 \times 10^{-4}) - X(1.8 \times 10^{-4})$$

$$X^2 + (1.8 \times 10^{-4})X - 7.2 \times 10^{-6} = 0$$

which can be solved as a quadratic. However, an even simpler solution is possible. Since we know from the K_A very little of the acid will dissociate, we can assume X is very small in terms of the 0.0400 molar concentration we subtract it from, and hence, can be ignored. This then gives us:

$$\frac{X^2}{0.0400} = 1.8 \times 10^{-4},$$

$$X^2 = 7.2 \times 10^{-6}$$

$$X = 2.7 \times 10^{-3} = [H^+] = [HCOO^-]$$

If the $[H^+] = 2.7 \times 10^{-3}$, then the pH ($- \log [H^+]$) = 2.57. Solving the original equation by a quadratic, we get:

$$[H^+] = 2.6 \times 10^{-3} \text{ and pH} = 2.58,$$

which shows that our initial assumption is correct in that X in the term $(0.0400 - X)$ was small enough to ignore.

PROBLEM

If the hydrolysis constant of Al^{3+} is 1.4×10^{-5}, what is the concentration of H_3O^+ in 0.1 M $AlCl_3$?

SOLUTION

Hydrolysis refers to the action of the salts of weak acids and bases with water to form acidic or basic solutions. Consequently, to answer this question, write out the reaction, which illustrates this hydrolysis, and write out an equilibrium constant expression. From this, the concentration of H_3O^+ can be defined. The net hydrolysis reaction is

$$AlCl_3 \rightarrow Al^{3+} + 3\ Cl^-$$

$$Al^{3+} + 2H_2O \leftrightarrow AlOH^{2+} + H_3O^+$$

$$k_{hyd} = 1.4 \times 10^{-5} = \frac{[H_3O^+][AlOH^{3+}]}{[Al^{2+}]}$$

Water is excluded in this expression since it is considered as a constant. Let x = the moles/liter of $[H_3O^+]$. Since H_3O^+ and $AlOH^{2+}$ are formed in equal mole amounts, the concentration of $[AlOH^{2+}]$ can also be represented by x. If one starts with 0.1 M of Al^{3+}, and x moles/liter of it forms H_3O^+ (and $AlOH^{2+}$), one is left with $0.1 - x$ at equilibrium. Substituting these representations into the k_{hyd} expression,

$$\frac{x \times x}{0.1 - x} = 1.4 \times 10^{-5}.$$

If one solves for x, the answer is $x = 1.2 \times 10^{-3}$ M, which equals $[H_3O^+]$.

PROBLEM

Calculate the hydrolysis constants of the ammonium and cyanide ions, assuming $k_w = 1 \times 10^{-14}$ and $k_a = 4.93 \times 10^{-10}$ for HCN and $K_b = 1.77 \times 10^{-5}$ for NH_3. For each, determine the percent hydrolysis in a .1 M solution.

SOLUTION

To find the hydrolysis constant, you must know what it defines. Hydrolysis is the process whereby an acid or base is regenerated from its salt by the action of water. The hydrolysis constant measures the extent of this process. Quantitatively, it is defined as being equal to

$$\frac{k_w}{k_a \text{ or } k_b},$$

where k_w = the equilibrium constant for the autodissociation of water, k_a = dissociation of acid, and k_b = dissociation of base. You are given k_w, k_a, and k_b. Thus, the hydrolysis constants can be easily found by substitution. Let k_h = hydrolysis constant.

For cyanide ion:

$$k_h = \frac{k_w}{k_a} = \frac{1 \times 10^{-14}}{4.93 \times 10^{-10}} = 2.02 \times 10^{-5}.$$

For ammonium ion:

$$k_h = \frac{k_w}{k_b} = \frac{1 \times 10^{-14}}{1.77 \times 10^{-5}} = 5.64 \times 10^{-10}.$$

To find the percent hydrolysis in a .1 M solution, write the hydrolysis reaction and express the hydrolysis constant just calculated in those terms. After this, represent the concentrations of the hydrolysis products in terms of variables and solve. For cyanide ion, the hydrolysis reaction is

$$CN^- + H_2O \leftrightarrow HCN + OH^-.$$

Therefore,

$$k_h = \frac{[HCN][OH^-]}{[CN^-]}.$$

But you calculated that $k_h = 2.02 \times 10^{-5}$.

Equating,

$$2.02 \times 10^{-5} = \frac{[HCN][OH^-]}{[CN^-]}.$$

You start with a .1 M solution of CN^-. Let $x = [HCN]$ formed. Thus, $x = [OH^-]$ also, since they are formed in equimolar amounts. If x moles/liter of substance are formed from CN^-, then at equilibrium you have $.1 - x$ moles/liter left. Substituting these values,

$$2.02 \times 10^{-5} = \frac{x \times x}{.1 - x}.$$

Solving, $x = 1.4 \times 10^{-3}$ M. The percent is just 100 times

$$\frac{x}{.10 \text{ M}}$$

since the initial concentration is .10 M, so that you have 1.4% hydrolysis in a .1 M solution.

For ammonium ion, the hydrolysis reaction is $NH_4^+ + H_2O \leftrightarrow NH_3 + H_3O^+$. k_h for this reaction equals

$$\frac{[NH_3][H_3O^+]}{[NH_4^+]}.$$

The calculated $k_h = 5.6 \times 10^{-10}$. Equating,

$$5.6 \times 10^{-10} = \frac{[NH_3][H_3O^+]}{[NH_4^+]}.$$

From this point, you follow the same reasoning as was used with the cyanide. Solving:

Let $x = [NH_3]$

$$5.6 \times 10^{-10} = \frac{(x)(x)}{(.1-x)}$$

$$x^2 = (5.6 \times 10^{-11}) - (5.6 \times 10^{-10})\, x$$

$$x^2 + (5.6 \times 10^{-10})\, x - 5.6 \times 10^{-11} = 0$$

Using the quadratic formula, one can solve for x, where $ax^2 + bx + c = 0$.

$$x = \frac{-b \pm \sqrt{b^2 - 4ac}}{2a}$$

$$x = \frac{-5.6 \times 10^{-10} \pm \sqrt{(5.6 \times 10^{-10})^2 - 4(1)(-5.6 \times 10^{-11})}}{2\,(1)}$$

$$x = \frac{-5.6 \times 10^{-10} \pm \sqrt{2.24 \times 10^{-10}}}{2}$$

$$x = \frac{-5.6 \times 10^{-10} \pm 1.50 \times 10^{-5}}{2}$$

$$x = \frac{-1.50 \times 10^{-5}}{2} \quad \text{or} \quad x = \frac{1.50 \times 10^{-5}}{2}$$

x cannot be negative, because concentration cannot be negative. Thus, $x = 7.5 \times 10^{-6}$.

Solving for the percent:

$$\frac{7.5 \times 10^{-6}}{.1} \times 100\% = 7.5\%.$$

Thus, you find that the percent hydrolysis is $7.5 \times 10^{-3}\%$.

Drill 7: The Hydrolysis Constant

1. Hydrolysis of sodium carbonate yields

    (A) a strong acid and a strong base.

    (B) a weak acid and a strong base.

    (C) a weak acid and a weak base.

    (D) a strong acid and a weak base.

    (E) none of the above.

2. If the hydrolysis constant of Al^{3+} is 1.4×10^{-5}, what is the concentration of H_3O^+ in 0.1 M $AlCl_3$?

(A) $\sqrt{1.4 \times 10^{-5}}$ M

(D) 1.4×10^{-4} M

(B) $\sqrt{1.4 \times 10^{-4}}$ M

(E) None of the above

(C) $\sqrt{1.4 \times 10^{-6}}$ M

3. Find an expression for the hydrolysis constant of the bicarbonate ion HCO_3^-, k_h (HCO_3^-), using only the first dissociation constant for H_2CO_3, k_{a_1} (H_2CO_3), and the water constant, k_w.

(A) k_{a_1}/k_w (H_2CO_3)

(D) $k_w\, k_{a_1}$ (H_2CO_3)

(B) k_w/k_{a_1} (H_2CO_3)

(E) k_w (H_2CO_3)$/k_{a_1}$

(C) k_{a_1} (H_2CO_3)$/k_w$

8. Neutralization

Acid-base neutralization occurs in aqueous solution when the H^+ ions of the acid react with the OH^- ions of the base to form water. This reaction occurs so fast that it is usually considered instantaneous, and the H^+ and OH^- concentrations always obey the equilibrium relation $k_w = [H^+][OH^-]$. The other product of the neutralization of an acid and base—from the cation from the base and the anion from the acid—is called a salt. Normal table salt (NaCl) is the salt formed from the neutralization of hydrochloric acid (HCl) and sodium hydroxide (NaOH). It is, however, only one of numerous salts that form from the neutralization of acids and bases. The key to working the neutralization problems lies in remembering that one mole of H^+ ions reacts with one mole of OH^- ions to form a neutral solution. The other ions will form a salt.

Diprotic or triprotic acids possess two or three hydrogens, respectively, that can ionize to form H^+ ions. In all such cases, the first H^+ ion is more readily formed than the second or third. A very common example is sulfuric acid which can form H^+ ions by both of the following reactions.

$$H_2SO_4 \leftrightarrow H^+ + HSO_4^- \tag{1}$$

$$HSO_4^- \leftrightarrow H^+ + SO_4^{-2} \tag{2}$$

The equilibrium constant for the first H^+ ion, equation 1, is very large and all the H_2SO_4 reacts. However, the equilibrium constant for the second is .012, so little H^+ is formed from this reaction. Let us calculate, for example, the SO_4^{-2} ion concentration in a 1 M aqueous solution of sulfuric acid. Follow the procedure outlined above.

1) Write the equation containing SO_4^{-2}.

$$HSO_4^- \leftrightarrow H^+ + SO_4^{-2} \tag{3}$$

2) Write the equation relating the equilibrium constant to compositions.

$$0.012 = (H^+)(SO_4^{-2})/(HSO_4^-) \tag{4}$$

3) Write the composition in terms of a single variable from the stoichiometry.

We can assume, since the equilibrium constant is small, that very little HSO_4^- is consumed and that, since equation 1 goes essentially to completion, HSO_4^- and H^+ concentrations are equal and equal to 1 M.

$$.012 = (1)\ (SO_4^{-2}0/(1) \tag{5}$$

4) Solve the equation for the single variable and calculate the equilibrium
compositions.

$$(SO_4^{-2}) = .012\ M \tag{6}$$

The assumption that very little HSO_4^- ion is consumed is reasonable. In this case, .012 moles or 1.2% of the HSO_4^- is consumed.

PROBLEM

Assuming complete neutralization, calculate the number of milliliters of 0.025 M H_3PO_4 required to neutralize 25 ml of 0.030 M $Ca(OH)_2$.

SOLUTION

This problem can be solved by two methods: mole method or equivalent method.

Mole Method

This method requires one to write out the balanced equation that illustrates the neutralization reaction. The balanced equation is

$$3Ca(OH)_2 + 2H_3PO_4 \rightarrow Ca_3\ (PO_4)_2 + 6H_2O.$$

From this equation, one can see that 2 moles of H_3PO_4 react for every 3 moles of $Ca(OH)_2$. This means that one must first calculate how many moles of $Ca(OH)_2$ are involved. The molarity of the $Ca(OH)_2$ is 0.030. (Molarity = no. of moles/ liters.)

As given, the $Ca(OH)_2$ solution is 25 ml or 0.025 liters. Therefore, the number of moles of $Ca(OH)_2$ is (0.030) (0.025) = 0.00075 moles. As the bal-anced equation indicates, the number of moles of H_3PO_4 is $2/3$ the moles of $Ca(OH)_2$ or $2/3$ (0.00075) = 0.00050 moles of H_3PO_4.

From the definition of molarity for H_3PO_4 one has

$$0.025\ M = \frac{0.00050\ \text{moles}}{\text{liters}}.$$

The molarity, 0.025, is given. Solving for liters, one obtains 0.020 liters or 20 ml. The key to solving this problem with the mole method is to write a balanced equation, which will indicate the relative amounts of moles required for completion neutralization.

Equivalent Method

This method requires that one consider normality and the definition of an equivalent. An equivalent is defined as the molecular weight or mass of an acid or base that furnished one mole of protons (H^+) or hydroxyl (OH^-) ions. For example, the number of equivalents contained in a mole of H_2SO_4 is $^{98}/_2$ or 49. Since each mole of H_2SO_4 produces two protons, divide the molecular weight by two.

The number of equivalents of an acid must equal that of the base in a neutralization reaction. Normality is defined as equivalents of solute per liter. In this problem, it is given that there are 25 ml of 0.03 $Ca(OH)_2$.

To solve the problem, determine how many equivalents are present. The number of moles of $Ca(OH)_2$ is (0.025) (0.03) or 0.00075 from the definition of molarity. The molecular weight of $Ca(OH)_2$ is 74.08. Therefore, there are (0.00075) (74.08) or 0.06 grams of $Ca(OH)_2$.

The number of equivalents per gram is 74.08/2 since two OH^- can be produced. The number of equivalents is

$$\frac{(0.00075)(74.08)\text{ g}}{\left(\dfrac{74.08}{2}\text{ g/equiv}\right)} = 0.0015 \text{ equiv of } Ca(OH)_2.$$

This indicates that 0.0015 equivalents of H_3PO_4 are required. The molarity of H_3PO_4 is 0.025 M, which means its normality is 0.075 N, because there are three ionizable protons per mole. Recalling the definition of normality, there are

$$0.075 \text{ N} = 0.0015 \text{ equiv/liters}$$

The reason one knows that there is 0.0015 equiv in the H_3PO_4 present is because one knows that for the neutralization to occur the number of equivalents of acid must equal the number of equivalents of base. In this problem, one has already calculated that there are 0.0015 equiv of base present. Thus,

$$\text{volume} = 0.20 \text{ } l \text{ or } 20 \text{ ml.}$$

PROBLEM

A 50 ml solution of sulfuric acid was found to contain 0.490 g of H_2SO_4. This solution was titrated against a sodium hydroxide solution of unknown concentration. 12.5 ml of the acid solution was required to neutralize 20.0 ml of the base. What is the concentration of the sodium hydroxide solution?

SOLUTION

At the neutralization point, the number of equivalents of acid in the 12.5 ml volume is equal to the number of equivalents of base in the 20.0 ml volume. Since the normality is defined as the number of equivalents per liter of solution, the number of equivalents is equal to the normality times the volume. At the neutralization point we have

$$N_a V_a = N_b V_b$$

where N_a = normality of acid, V_a = volume of acid, N_b = normality of base, and V_b = volume of base.

The normality of the 50.0 ml (0.050 *l*) sulfuric acid solution is

$$N_a = \frac{\text{number of equivalents in } 0.050 \, l}{0.05 \, l}$$

$$= \frac{\text{mass of acid/gram equivalent weight}}{0.05 \, l}$$

The gram equivalent weight of sulfuric acid is 49.0 g/equivalent, because there are two equiv per molecule. The MW of H_2SO_4 is 98 g/mole.

$$N_a = \frac{\text{mass of acid/gram equivalent weight}}{0.05 \, l}$$

$$= \frac{0.490 \, g/49.0 \, g/\text{equivalent}}{0.05 \, l}$$

$$= 0.200 \, \text{equivalent}/l$$

$$= 0.200 \, N$$

The normality of the base is then found as follows:

$$N_a V_a = N_b V_b$$

$$N_b = \frac{N_a V_a}{V_b} = \frac{0.200 \, N \times 12.5 \, ml}{20.0 \, ml} = 0.125 \, N.$$

Therefore, the sodium hydroxide solution is 0.125 N, which because there is one ionizable OH^- in NaOH, is equal to 0.125 M.

Drill 8: Neutralization

1. In a titration of Na_2CO_3 with HCl, which of the relations shown below would yield the molarity of HCl to the pH 4 end point?

    $$2HCl + Na_2CO_3 \leftrightarrow H_2CO_3 + 2NaCl$$

    (A) $\dfrac{(g\ Na_2CO_2)\,(2)}{(MW\ Na_2CO_3)\,(ml\ HCl\,/\,1,000)}$

    (B) $\dfrac{(g\ Na_2CO_3)\,(\frac{1}{2})}{(mol\ HCl\,/\,1,000)}$

    (C) $\dfrac{(moles\ Na_2CO_3)\,(\frac{1}{2})}{(ml\ HCl\,/\,1,000)}$

    (D) $\dfrac{moles\ Na_2CO_3}{(ml\ HCl\,/\,1,000)}$

    (E) $\dfrac{(g\ HCl)\,(2)}{(MW\ Na_2CO_3)\,(ml\ HCl\,/\,1,000)}$

2. What is the normality of a sulfuric acid solution if 50 milliliters completely neutralizes 1.00 liter of a 0.1 M potassium hydroxide solution?

    (A) 1.0 N   (D) 0.2 N

    (B) 0.1 N   (E) 10 N

    (C) 2.0 N

3. What volume of .5 M H_2SO_4 will neutralize 100 ml of .2 M NaOH?

    (A) 400 ml   (D) 20 ml

    (B) 200 ml   (E) 2 ml

    (C) 100 ml

4. 25.0 ml of 0.100 M HCl is titrated with 0.15 M NaOH. The pH of the acid solution after 10 ml of base added is

    (A) 2.39.   (D) 1.54.

    (B) 2.48.   (E) 1.4.

    (C) 1.9.

5. The volume in milliliters of a .3 M solution of NaOH needed to neutralize 3 liters of a .01 M HCl solution is

    (A) .1.   (D) 100.

    (B) 1.   (E) 1,000.

    (C) 10.

---

9. Buffers

Buffer solutions are equilibrium systems that resist changes in acidity and maintain constant pH when acids or bases are added to them.

Buffers are formed in two common ways:

1) By dissolving, in water, a weak acid and a soluble ionic salt containing the same anion as the weak acid

2) By dissolving, in water, a weak base and a soluble ionic salt containing the same cation as the weak acid

When the anion of a weak acid is maintained at a constant, relatively large concentration by dissolving a salt, it will control, through equilibrium, the H^+ ion concentration and, hence, the pH.

The most effective pH range for any buffer is at or near the pH where the acid and salt concentrations are equal (that is, pk_a).

The pH for a buffer is given by the Henderson-Hasselbach equation:

$$pH = pk_a + \log \frac{[A^-]}{[HA]} = pk_a + \log \frac{[base]}{[acid]}$$

which is obtained very simply from the equation for weak acid equilibrium,

$$k_a = \frac{[H^+][A^-]}{[HA]}.$$

PROBLEM

Explain the buffering action of a liter of 0.10 M acetic acid containing 0.1 mole of sodium acetate. In the explanation, use ionic equations.

SOLUTION

By buffering action, one means the ability of a substance to maintain relatively constant conditions of pH in the face of changes that might otherwise affect the acidity or basicity in solution. For example, a weak acid, such as acetic acid, will dissociate according to the following equation:

$$HOAc + H_2O \leftrightarrow H_3O^+ + OAc^-.$$

The sodium acetate particle will be completely ionized in solution:

$$NaOAc \leftrightarrow Na^+ + OAc^-.$$

These are the two processes that occur in this buffer. Suppose one increases the acetate concentration. By doing this, the equilibrium is shifted to the left, i.e., to acetic acid.

$$k_a = \frac{[H_3O^+][OAc^-]}{[HOAc]} \text{ or } [H_3O^+] = \frac{[HOAc]}{[OAc^-]} k_a$$

The key to the buffering action is this [HOAc] to [OAc$^-$] ratio. If [HOAc] changes, [OAc$^-$] will change accordingly so that the same value of the ratio is obtained. Thus, the ratio is a constant. This means, therefore, that [H$_3$O$^+$] is maintained as a constant. pH $= -\log$ [H$^+$], it is also constant.

PROBLEM

Design a buffer system that will function in the low, near neutral, and high pH levels, using combinations of the three sodium phosphates (Na$_3$PO$_4$, Na$_2$HPO$_4$, NaH$_2$PO$_4$) and phosphoric acid (H$_3$PO$_4$). The equilibrium constants for this polyprotic acid are $k_1 = 7.5 \times 10^{-3}$, $k_2 = 6.2 \times 10^{-8}$, and $k_3 = 4.8 \times 10^{-13}$. With this information, calculate the pH at both extremes of the buffer, assume 10 : 1 and 1 : 10 ratios, and at the mid-range of the buffer, assume a 1 : 1 ratio. Assume, also, that the acid-salt ratio in making up the buffer is equal to HA/A$^-$ in solution.

SOLUTION

Solutions that contain appreciable amounts of a weak acid, such as phosphoric acid, and its salt or salts, such as sodium phosphates, are called buffers. Their utility is in maintaining a relatively constant pH. This problem requires one to find various pH values for this buffer system. To do this, note that the pH of a buffer is given by

$$pH = pk_a + \log \frac{[A^-]}{[HA]} = pk_a + \log \frac{[salt]}{[acid]}.$$

One knows that to find the pH, one must determine [H$_3$O$^+$], since pH $= -\log$ [H$_3$O$^+$]. To do this, consider the ionization of the polyprotic acid, phosphoric acid, which can dissociate to produce three protons. k_1, the first ionization constant, corresponds to H$_3$PO$_4$ + H$_2$O $\leftrightarrow$ H$_3$O$^+$ + H$_2$PO$_4^-$ so that

$$k_1 = \frac{[H_3O^+][H_2PO_4^-]}{[H_3PO_4]}.$$

k_2, the second ionization constant, corresponds to H$_2$PO$_4^-$ + H$_2$O $\leftrightarrow$ H$_3$O$^+$ + HPO$_4^{2-}$ so that

$$k_2 = \frac{[H_3O^+][HPO_4^{2-}]}{[H_2PO_4^-]}.$$

k_3, the third ionization constant, corresponds to $HPO_4^{2-} + H_2O \leftrightarrow H_3O^+ + PO_4^{3-}$ and

$$k_3 = \frac{[H_3O^+][PO_4^{3-}]}{[HPO_4^{2-}]}.$$

In general, the k_a of any acid is given by the expression

$$k_a = \frac{[H_3O^+][A^-]}{[HA]} \quad \text{or} \quad [H_3O^+] = k_a \frac{[HA]}{[A^-]}.$$

It is from this equation that the $[H_3O^+]$ must be calculated and thus yield the pH value wanted. The three systems are as follows:

Buffer system with k_1

Consider the ratios of 10 : 1, 1 : 1, and 1 : 10 for the acid: salt ratios, given that the ratio equals $[HA]/[A^-]$ for the low, neutral, and high pH values, respectively. This means that $[H_3O^+] = 10\ k_a$, k_a, and $0.1\ k_a$ for low, neutral, and high, respectively. Here, $k_a = k_1$ which is 7.5×10^{-3}. $pk_1 = 2.12$. Since $[H_3O^+] = 10\ k_a$, k_a, and $0.1\ k_a$, $pH = -\log [H_3O^+] = (pk_a - 1)$, pk_a, and $(pk_a + 1)$ for low, middle, and high pH values, respectively. Thus, with $pk_1 = 2.12$, $pH = 1.12$ (low), 2.12 (middle), and 3.12 (high).

Buffer system with k_2

One still has $pH = (pk_a - 1)$, pk_a, and $pk_a + 1$. But now $k_a = k_2 = 6.2 \times 10^{-8}$ $pk_2 = 7.21$. Thus, $pH = 6.21$ (low), 7.21 (middle), and 8.21 (high).

Buffer system with k_3

One still has $pH = (pk_a - 1)$, pk_a, and $pk_a + 1$. Now, $k_a = k_3 = 4.8 \times 10^{-13}$ $pk_3 = 12.32$. Thus, $pH = 11.32$ (low), 12.32 (middle), and 13.32 (high).

Drill 9: Buffers

1. A buffer solution containing H_2CO_3 and $NaHCO_3$ is to be produced to maintain a pH of 7. What must the ratio $[NaHCO_3]/[H_2CO_3]$ be in order to realize such a pH if the ka of carbonic acid is 4.3×10^{-7}?

    (A) 43

    (B) 4.3

    (C) .43

    (D) 86

    (E) 1.29

2. Calculate the pOH of a solution made by mixing 70 ml of 0.10 M NH_3 and 60 ml of 0.05 M HCl. The reaction is $NH_3 + H_3O^+ \rightarrow NH_4^+ + H_2O$ with $k_{bNH3} = 1.8 \times 10^{-3}$.

    (A) $-\log (1.8 \times 10^{-3}) - \log \left(\dfrac{4.0}{3.0} \right)$

    (B) $-\log (1.8 \times 10^{-3}) + \log \left(\dfrac{4.0}{3.0} \right)$

    (C) $-\log (1.8 \times 10^{-3}) + \log \left(\dfrac{0.04}{0.03} \right)$

    (D) $\log (1.8 \times 10^{-3}) - \log \left(\dfrac{0.03}{0.04} \right)$

    (E) $-\log (1.8 \times 10^{-3}) - \log \left(\dfrac{0.04}{0.03} \right)$

3. The following bases, and their conjugate acids (as the chlorides), are available in the lab: ammonia, NH_3; pyridine, C_5H_5N; ethylamine, $CH_3CH_2NH_2$. A buffer solution of pH 9 is to be prepared, and the total concentration of buffering reagents is to be 0.5 mole/liter. Choose the best acid-base pair.

    (A) C_5H_5N and conjugate acid

    (B) NH_3 and conjugate acid

    (C) $CH_3CH_2NH_2$ and conjugate acid

    (D) (A) and (B) are equally good.

    (E) None of the above.

ACID AND BASE EQUILIBRIA DRILLS

ANSWER KEY

Drill 1 — Acids and Bases
1. (C)
2. (B)
3. (B)
4. (A)
5. (E)

Drill 2 — The Autoionization of Water
1. (A)
2. (A)
3. (B)

Drill 3 — Autoprotolysis
1. (E)
2. (B)
3. (A)

Drill 4 — pH
1. (C)
2. (C)
3. (C)
4. (C)
5. (E)

Drill 5 — The Ionization Constant
1. (A)
2. (A)

3. (D)
4. (B)

Drill 6 — The Dissociation Constant
1. (B)
2. (C)
3. (A)

Drill 7 — The Hydrolysis Constant
1. (B)
2. (B)
3. (B)

Drill 8 — Neutralization
1. (A)
2. (C)
3. (D)
4. (D)
5. (D)

Drill 9 — Buffers
1. (B)
2. (E)
3. (B)

GLOSSARY:
ACID AND BASE EQUILIBRIA

Acid

Any substance that gives a pH less than 7 when dissolved in water.

Amphoteric

A substance that is able to act as both an acid and a base.

Arrehenius Theory

States that acids are substances that ionize in water to form H^+ ions, and bases are substances that produce OH^- ions in water.

Autoionization

Water interacting with itself to form H_3O^+ and OH^-.

Autoprotolysis

The interaction of a substance with itself to create positive and negative ions.

Base

Any substance that gives a pH greater than 7 when dissolved in water.

Brönsted-Lowry Theory

States that acids are proton donors and bases are proton acceptors.

Buffer

Equilibrium systems that resist changes in acidity and maintain constant pH when acids or bases are added to them.

Common Ion Effect

The effects on equilibrium and other properties of a solution due to the addition of ions that are already present in the solution.

Conjugate Acid and Base

Pairs of acids and bases that are related by the addition or removal of an H^+, OH^-, or electron pair.

Electrolyte

Any substance that dissociates into ions when placed in solution.

Equilibrium

A situation in which the rate of a forward reaction is equal to that of the backward reaction. There is then no net change in the quantity of either products or reactants.

Equivalent Weight

The molecular weight of a substance divided by its valence.

Hydrolysis

The splitting of a water molecule into H^+ and OH^- through interaction with another species in solution.

Ionization Constant

Serves to measure the extent to which a weak acid or base dissociates in solution.

k_w

The dissociation constant for water, $k_w = [OH^-] [H^+] = 1 \times 10^{-14}$.

Lewis Theory

States that an acid is an electron-pair acceptor and a base is an electron-pair donor.

Neutralization

The reaction of an acid with a base that produces a neutral solution.

Normality

A concentration unit that is defined as the number of gram-equivalent weights of a solute per liter of solution.

pH

A method to express $[H^+]$. $pH = - \log [H^+]$.

Polyprotic Acid

An acid which is capable of donating more than one proton.

Strong Acids and Bases

Acids or bases that dissociate completely in solution.

Titration

The addition of one solution to another with the purpose of measuring the concentration of one of the solutions.

Weak Acids and Bases

Acids or bases which do not dissociate completely in solution.

CHAPTER 11

Thermodynamics I

➤ Diagnostic Test
➤ Thermodynamics I
Review & Drills
➤ Glossary

THERMODYNAMICS I
DIAGNOSTIC TEST

1. Ⓐ Ⓑ Ⓒ Ⓓ Ⓔ
2. Ⓐ Ⓑ Ⓒ Ⓓ Ⓔ
3. Ⓐ Ⓑ Ⓒ Ⓓ Ⓔ
4. Ⓐ Ⓑ Ⓒ Ⓓ Ⓔ
5. Ⓐ Ⓑ Ⓒ Ⓓ Ⓔ
6. Ⓐ Ⓑ Ⓒ Ⓓ Ⓔ
7. Ⓐ Ⓑ Ⓒ Ⓓ Ⓔ
8. Ⓐ Ⓑ Ⓒ Ⓓ Ⓔ
9. Ⓐ Ⓑ Ⓒ Ⓓ Ⓔ
10. Ⓐ Ⓑ Ⓒ Ⓓ Ⓔ
11. Ⓐ Ⓑ Ⓒ Ⓓ Ⓔ
12. Ⓐ Ⓑ Ⓒ Ⓓ Ⓔ
13. Ⓐ Ⓑ Ⓒ Ⓓ Ⓔ
14. Ⓐ Ⓑ Ⓒ Ⓓ Ⓔ
15. Ⓐ Ⓑ Ⓒ Ⓓ Ⓔ
16. Ⓐ Ⓑ Ⓒ Ⓓ Ⓔ
17. Ⓐ Ⓑ Ⓒ Ⓓ Ⓔ
18. Ⓐ Ⓑ Ⓒ Ⓓ Ⓔ
19. Ⓐ Ⓑ Ⓒ Ⓓ Ⓔ
20. Ⓐ Ⓑ Ⓒ Ⓓ Ⓔ

21. Ⓐ Ⓑ Ⓒ Ⓓ Ⓔ
22. Ⓐ Ⓑ Ⓒ Ⓓ Ⓔ
23. Ⓐ Ⓑ Ⓒ Ⓓ Ⓔ
24. Ⓐ Ⓑ Ⓒ Ⓓ Ⓔ
25. Ⓐ Ⓑ Ⓒ Ⓓ Ⓔ
26. Ⓐ Ⓑ Ⓒ Ⓓ Ⓔ
27. Ⓐ Ⓑ Ⓒ Ⓓ Ⓔ
28. Ⓐ Ⓑ Ⓒ Ⓓ Ⓔ
29. Ⓐ Ⓑ Ⓒ Ⓓ Ⓔ
30. Ⓐ Ⓑ Ⓒ Ⓓ Ⓔ
31. Ⓐ Ⓑ Ⓒ Ⓓ Ⓔ
32. Ⓐ Ⓑ Ⓒ Ⓓ Ⓔ
33. Ⓐ Ⓑ Ⓒ Ⓓ Ⓔ
34. Ⓐ Ⓑ Ⓒ Ⓓ Ⓔ
35. Ⓐ Ⓑ Ⓒ Ⓓ Ⓔ
36. Ⓐ Ⓑ Ⓒ Ⓓ Ⓔ
37. Ⓐ Ⓑ Ⓒ Ⓓ Ⓔ
38. Ⓐ Ⓑ Ⓒ Ⓓ Ⓔ
39. Ⓐ Ⓑ Ⓒ Ⓓ Ⓔ
40. Ⓐ Ⓑ Ⓒ Ⓓ Ⓔ

THERMODYNAMICS I
DIAGNOSTIC TEST

This diagnostic test is designed to help you determine your strengths and your weaknesses in thermodynamics. Follow the directions and check your answers.

Study this chapter for the following tests:
AP Chemistry, ASVAB, CLEP General Chemistry,
GED, MCAT, MSAT, PRAXIS II Subject Assessment:
Chemistry, SAT II: Chemistry, GRE Chemistry

---

40 Questions

DIRECTIONS: Choose the correct answer for each of the following problems. Fill in each answer on the answer sheet.

1. An equation for the electrolysis of water is

$$H_2O(I) + 68.3 \text{ kcal} \rightarrow H_2(g) + \frac{1}{2}O_2(g)$$

How many liters of gaseous product are produced by the addition of 273.2 kcal of electrical energy to the reaction above?

(A) 22.4 liters   (D) 96.8 liters

(B) 44.8 liters   (E) 119.2 liters

(C) 134.4 liters

2. $C_2H_6(g) + \frac{7}{2}O_2(g) \rightarrow 2CO_2(g) + 3H_2O(l)$

$$\Delta H_f° \text{ (kJ/mol)}$$

$C_2H_{6(g)}$	-85
$CO_{2(g)}$	-394
$H_2O_{(1)}$	-286

Given the $\Delta H_f°$ values above, $\Delta H°$ for the reaction above is

(A) $\Delta H° = -85 + 394 + 286.$

(B) $\Delta H^o = -394 - 286 + 85.$

(C) $\Delta H^o = -85 - 2(-394) - 3(-286).$

(D) $\Delta H^o = 2(-394) + 3(-286) - (-85).$

(E) ΔH^o cannot be calculated with the given information.

3. How much heat is released when 50 g of water at 0°C is changed to ice at 0° C?

(A) 50 calories

(B) 400 calories

(C) 500 calories

(D) 4,000 calories

(E) 5,000 calories

4. What is the ΔH_f for propane, if the heat of combustion of propane at 25°C is −530.6 kcal/mol?

$$C_3H_8 (g) + 5O_2 \rightarrow 3CO_2(g) + 4H_2O(l) \; \Delta H = -530.6$$

$$\Delta H_f (CO_2) = -94.1 \text{ kcal/mol } \Delta H_f (H_2O) = -68.3 \text{ kcal/mol}$$

(A) −24.9 kcal/mol

(B) −17.1 kcal/mol

(C) 0 kcal/mol

(D) 17.1 kcal/mol

(E) 24.9 kcal/mol

5. Calculate ΔH° for the reaction

$$2Ag_2S(s) + 2H_2O(1) \rightarrow 4Ag(s) + 2H_2S(g) + O_2(g)$$

if $\Delta H^\circ_{H_2S}(g) = -20.6 \text{ kJ/mol}, \; \Delta H^\circ_{Ag_2S}(s) = -32.6 \text{ kJ/mol},$

$$\Delta H^\circ_{H_2O}(1) = -285.8 \text{ kJ/mol}$$

(A) 595.6 kJ

(B) 495.6 kJ

(C) 585.6 kJ

(D) 485.6 kJ

(E) 600 kJ

6. From the data below, calculate the ΔH^o, expressed in kJ, for the reaction:

$$2Na(s) + 2H_2O(1) \rightarrow 2NaOH(s) + H_2(g)$$

Substance	ΔH^o kJ/mol
$H_2O(1)$	−285.8
$NaOH(s)$	−426.7
$H_2(g)$	−241.8

(A) -281.8 (D) $+712.5$

(B) $+140.9$ (E) zero

(C) -712.5

7. The specific heat of water is the heat required to

(A) raise the temperature of 1g of water 1°C.

(B) raise the temperature of 1 ml of water 1°C.

(C) raise the temperature of 1g of water to 100°C.

(D) freeze 1g of water at 0°C.

(E) vaporize 1g of water at 100°C.

8. Ammonia gas is made from nitrogen gas and hydrogen gas at 525°C according to the reaction:

$$\frac{1}{2} N_2(g) + \frac{3}{2} H_2(g) \rightarrow NH_3(g)$$

ΔH_f° for $NH_3(g)$ at 298°K is –46 kJ/mol. Using the following heat capacity values, what is ΔH for the reaction at 525°C?

	C_p (J/K × mol)
$NH_3(g)$	26
$N_2(g)$	27
$H_2(g)$	29

(A) –46 kJ/mol   (D) –30 kJ/mol

(B) –88 kJ/mol   (E) –102 kJ/mol

(C) –62 kJ/mol

9. A reaction that occurs only when heat is added is best described as

(A) exothermic.   (D) spontaneous.

(B) endothermic.   (E) non-spontaneous.

(C) an equilibrium process.

10. According to Le Châtelier's principle of the addition of heat to the following reaction

$$CO_2(g) + 2H_2O(g) \rightarrow CH_4(g) + 2O_2(g)$$

will cause it to shift to the right. This reaction can therefore be described as

(A) spontaneous.   (D) exothermic.

(B) endothermic.   (E) adiabatic.

(C) unimolecular.

11. $\Delta H_f\, CO_2(g) = -94.1$ kcal/mol

ΔH_f of ethene (gas) = 12.5 kcal/mol

ΔH_f of H_2O = -68.3 kcal/mol

What would be the ΔH_C for the combustion reaction

$C_2H_4 + 3O_2 \leftrightarrow 2CO_2 + 2H_2O$

(A) -487.4 kcal/mol   (D) -111.1 kcal/mol

(B) -337.3 kcal/mol   (E) -93.6 kcal/mol

(C) -183.6 kcal/mol

12. How much heat is absorbed by 100 g of water when its temperature decreases from 25°C to 5°C? (Take the specific heat of water to be 4.2 J/gK.)

(A) 84,000 J   (D) $\dfrac{-2{,}000}{4.2}$ J

(B) $-84{,}000$ J   (E) $\dfrac{-84}{100}$ J

(C) $\dfrac{2{,}000}{4.2}$ J

13. $\Delta H_{\text{formation}}$ of NOCl(g) from the gaseous elements:

$$\frac{1}{2}N_2(g) + \frac{1}{2}O_2(g) + \frac{1}{2}Cl_2(g) = NOCl(g)$$

is equal to 12.6 kcal/mol at 25°C. Assuming the gases are ideal, ΔE would equal which one of the following? (1 kcal = 4.184 kJ)

(A) 12.6 kJ

(B) 46,300 J

(C) 53.8 kJ

(D) 74,220 J

(E) Insufficient information given to find ΔE.

14. The complete combustion of 1 mole of propane (C_3H_8) results in the liberation of 488.7 kcal. What is the heat of formation of propane?

    The reaction is: $C_3H_8(g) + 5O_2(g) \rightarrow 3CO_2 + 4H_2O$

    ΔH_f° (kcal/mol): CO_2 is -94.1 and H_2O is -57.8

    (A) $+6.9$ kcal/mol

    (B) -19 kcal/mol

    (C) -24.8 kcal/mol

    (D) -63.6 kcal/mol

    (E) -143.2 kcal/mol

15. The following reaction coordinates cannot be associated with

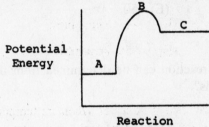

    (A) an endothermic reaction from A to C.

    (B) an exothermic reaction from A to C.

    (C) the activation energy for the reaction.

    (D) the energy for the intermediate.

    (E) the energy absorbed in the reaction from A to C.

16. A compound produced by adding heat to the reactants

    (A) will decompose upon addition of heat.

    (B) will liberate a net amount of heat when decomposed.

    (C) will decompose upon cooling.

    (D) Both (A) and (B)

    (E) (A), (B), and (C)

17. How much heat is liberated when 32 g of methane undergoes complete combustion?

    $$CH_4 + 2O_2 \rightarrow CO_2 + 2H_2O + 212.8 \text{ kcal}$$

    (A) -425.6 kcal

    (B) -212.8 kcal

    (C) $+106.4$ kcal

    (D) $+212.8$ kcal

    (E) $+425.6$ kcal

18. What amount of energy is required to convert 5 g of ice at 0°C to steam at 100°C?

(A) 450 cal

(D) 2,700 cal

(B) 900 cal

(E) 3,600 cal

(C) 1,800 cal

19. If the specific heat of aluminum is .89 J/°C g, how many joules are required to heat 23.2 g of aluminum from 30°C to 80°C?

(A) 45

(D) 1,032

(B) 1,160

(E) 50

(C) 20

20. The activation energy of a reaction can be determined from the slope of which of the following graphs?

(A) ln k vs T

(D) $\dfrac{T}{\ln k}$ vs $\dfrac{1}{T}$

(B) $\dfrac{\ln k}{T}$ vs T

(E) $\dfrac{\ln k}{T}$ vs $\dfrac{1}{T}$

(C) ln k vs $\dfrac{1}{T}$

21. The diagram below is used to describe a reaction path. What are y, x, and z respectively?

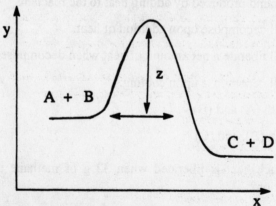

(A) Temperature, volume, pressure

(B) Activation energy, potential energy, temperature

(C) Potential energy, reaction coordinate, activation energy

(D) Distance, time, concentration

(E) Time, distance, concentration

22. In a Born-Haber cycle, the total energy involved in the preceding hypothetical preparation of NaCl is equal to the experimentally determined heat of formation (Q) of the compound from its elements. Which of the following thermochemical values is not used to calculate the total energy of formation of NaCl?

(1) Heat of fusion and vaporization (S)

(2) Dissociation energy of molecular chlorine (D)

(3) Ionization energy of sodium atom (I)

(4) Electron affinity of chlorine atom (E)

(5) Lattice energy of NaCl (U)

(A) (1) and (2)　　　　　　　(D) (1), (2), (3), and (5)

(B) (2), (3), and (5)　　　　　(E) All of the above

(C) (1), (2), (3), and (4)

23. Which of the following states of a compound has the greatest kinetic energy?

(A) A solid

(B) A liquid

(C) A gas

(D) A solid changing to a liquid at the melting point

(E) A liquid changing to a gas at the boiling point

24. Methane, CH_4, has a heat of formation of −18 in kcal per mole. Select the correct statement for a reaction producing 48 grams of methane.

(A) 9 kcal are absorbed.

(B) 18 kcal are absorbed.

(C) 18 kcal are released.

(D) 36 kcal are released.

(E) 54 kcal are released.

25. Determine ΔH (in kcal) for the reaction $2C(s) + O_2(g)\ 2CO(g)$. Additional information:

    (i) $CO_2(g) \rightarrow O_2(g) + C(s)$ $\Delta H = 94.1$ kcal

    (ii) $CO_2(g) \rightarrow \dfrac{1}{2}O_2(g) + CO(g)$ $\Delta H = 67.7$ kcal

    (A) -52.8

    (B) -26.4

    (C) $+26.4$

    (D) -161.8

    (E) Cannot be determined from the given information.

26. For nitrous dioxide, $\Delta H_f = +8$ (kcal/mol). Select the correct statement for a reaction producing 23 grams of this compound.

    (A) 4 kcal are absorbed.   (D) 16 kcal are absorbed.

    (B) 8 kcal are absorbed.   (E) 16 kcal are released.

    (C) 8 kcal are released.

27. For the reaction below

    $$2HN_3 + 2NO \rightarrow H_2O_2 + 4N_2$$

    with the following molar enthalpies (at 25°C):

    $HN_3 - H_m{}^\theta = +264.0$ kJ/mol

    $N_2 - H_m{}^\theta = 0$ kJ/mol

    $H_2O_2 - H_m{}^\theta = -187.8$ kJ/mol

    $NO - H_m{}^\theta = +90.25$ kJ/mol

    The change in the standard enthalpy for the reaction is

    (A) -896.3 kJ.   (D) $+742.6$ kJ.

    (B) $+937.4$ kJ.   (E) None of the above

    (C) -309.5 kJ

28. How many grams of CH_4 are required to produce 425.6 kcal of heat by the reaction

    $$CH_4 + 2O_2 \rightarrow CO_2 + H_2O + 212.8 \text{ kcal}$$

    (A) 8 g   (D) 32 g

    (B) 16 g   (E) 64 g

    (C) 24 g

29. Solid zinc oxide, ZnO, has a heat of formation of about −84 kcal per mole. Select the correct statement for a reaction producing 162 grams of zinc oxide.

    (A) 42 kcal are absorbed.   (D) 168 kcal are absorbed.

    (B) 81 kcal are absorbed.   (E) 168 kcal are released.

    (C) 81 kcal are released.

30. The temperature of a substance is an indication of its

    (A) kinetic energy.   (D) (A) and (B)

    (B) potential energy.   (E) (A), (B), and (C)

    (C) reaction energy.

31. The addition of 3.2 kcal of heat to 10 g of water at 50°C results in

    (A) 10 g of water at 80°C.   (D) 10 g of steam at 110°C.

    (B) 10 g of water at 100°C.   (E) 5 g of steam at 100°C.

    (C) 10 g of steam at 100°C.

32. From the heats of reaction, $\Delta H°$, for

$$C_{(graphite)} + O_2(g) \rightarrow CO_2(g) \qquad\qquad \Delta H° = -394 \text{ kJ}$$

$$CO(g) + \frac{1}{2}O_2 \rightarrow CO_2(g) \qquad\qquad \Delta H° = -284 \text{ kJ}$$

The calculated $\Delta H°$ for the reaction:

$$C_{(graphite)} + \frac{1}{2}O_2(g) \rightarrow CO(g)$$

is

    (A) 0.   (D) −678 kJ.

    (B) +18 kJ.   (E) −110 kJ.

    (C) −18 kJ.

33. Which of the following depicts an endothermic reaction (abscissa = reaction time; ordinate = energy)?

    (A)   (B)

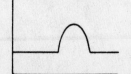 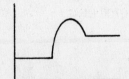

(C) 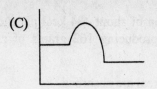   (D)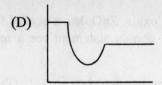

(E)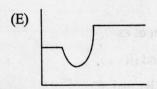

34. The least bond energy in kcals per mole is found with

(A) C–C.

(D) H–F.

(B) H–Br.

(E) O–H.

(C) H–Cl.

35. B_5H_9 ignites spontaneously in air according to the following reaction:

$$2B_5H_9(g) + 12O_2(g) \rightarrow 5B_2O_3(s) + 9H_2O(l)$$

What is the heat of reaction under standard conditions? The tabulated standard heats of formation of reactants and products are

Substance	ΔH°_{298}
$B_5H_9(g)$	$+60$ kJ mol^{-1}
$O_2(g)$	0.0 kJ mol^{-1}
$B_2O_3(s)$	$-1,260$ kJ mol^{-1}
$H_2O(1)$	-280 kJ mol^{-1}

(A) $-8,940$ kJ

(D) $+8,700$ kJ

(B) $+8,940$ kJ

(E) $+8,820$ kJ

(C) $-8,700$ kJ

36. What is the heat of combustion of octane at 25°C if ΔH_f for octane is -49.4 kcal/mole?

$$C_8H_{18} + \frac{25}{2}O_2 \rightarrow 8CO_2 + 9H_2O$$

$\Delta H_f(H_2O_{(1)}) = -68.3$ kcal/mol

$\Delta H_f(CO_{2(g)}) = -94.1$ kcal/mol

(A) $-1{,}318.1$ kcal/mol   (D) -543.6 kcal/mol

(B) -963.6 kcal/mol   (E) -232.7 kcal/mol

(C) -795.9 kcal/mol

37. How does the energy of the carbon-carbon bond of ethene pare to that of ethane?.

(A) It is less.   (D) They cannot be compared.

(B) It is greater.   (E) None of the above.

(C) They are identical.

Use the following information to answer questions 38, 39, and 40.

$$\frac{1}{2}N_2 + \frac{1}{2}O_2 \rightarrow NO \qquad \Delta H = 21.6 \text{ kcal/mol}$$

$$\frac{1}{2}N_2 + O_2 \rightarrow NO_2 \qquad \Delta H = 8.1 \text{ kcal/mol}$$

$$N_2 + 2O_2 \rightarrow N_2O_4 \qquad \Delta H = 4.3 \text{ kcal/mol}$$

38. The reaction $NO + \frac{25}{2}O_2 \rightarrow NO_2$ has a ΔH of

(A) -29.7 kcal/mol.   (D) 13.5 kcal/mol.

(B) -13.5 kcal/mol.   (E) 29.7 kcal/mol.

(C) 8.1 kcal/mol.

39. The reaction $N_2O_4 \rightarrow 2NO + O_2$ has a ΔH of

(A) -23.7 kcal/mol.   (D) 38.9 kcal/mol.

(B) -9.3 kcal/mol.   (E) 49.3 kcal/mol.

(C) 6.7 kcal/mol.

40. The reaction $2NO_2 \rightarrow N_2O_4$ has a ΔH of

    (A) −27.4 kcal/mol.   (D) 10.6 kcal/mol.

    (B) −11.9 kcal/mol.   (E) 63.4 kcal/mol.

    (C) 0.0 kcal/mol.

THERMODYNAMICS I
DIAGNOSTIC TEST

ANSWER KEY

1.	(C)	11.	(B)	21.	(C)	31.	(E)
2.	(D)	12.	(B)	22.	(E)	32.	(E)
3.	(D)	13.	(C)	23.	(C)	33.	(B)
4.	(A)	14.	(C)	24.	(E)	34.	(A)
5.	(A)	15.	(B)	25.	(A)	35.	(A)
6.	(A)	16.	(D)	26.	(A)	36.	(A)
7.	(A)	17.	(E)	27.	(A)	37.	(B)
8.	(C)	18.	(E)	28.	(D)	38.	(B)
9.	(B)	19.	(D)	29.	(E)	39.	(D)
10.	(B)	20.	(C)	30.	(A)	40.	(B)

DETAILED EXPLANATIONS
OF ANSWERS

1.   **(C)**   1.5 moles of gaseous product (one mole of H_2 and 0.5 mole of O_2) are produced for every 68.3 kcal of energy added to water. Since 4×68.3 kcal = 273.2 kcal, 4×1.5 or 6 moles of gaseous product are formed. Since molar volume is given as 22.4 liters at STP, 6×22.4 liters = 134.4 liters of gaseous product are evolved.

2.   **(D)**

$$\Delta H^\circ = \Delta H_f^\circ{}_{(products)} - \Delta H_f^\circ{}_{(reactants)}$$

ΔH_f° values must be multiplied by the molar amount of the reactant or product involved in the reaction:

$$\Delta H_f^\circ = 2(\Delta H_f^\circ{}_{CO_2(g)}) + 3(\Delta H_f^\circ{}_{H_2O(l)}) - (\Delta H_f^\circ{}_{C_2H_6(g)})$$

ΔH_f° for substances in their elemental state, e.g. $O_{2(g)}$, is by definition zero.

3.   **(D)**   The heat (q) transferred during the freezing of water is given by the product of ΔH_{fusion} (80 cal/g for water) and the mass (50 g):

$$q_{fusion} = \Delta H_{fusion} \times mass$$

$$q_{fusion} = 80 \text{ cal/g} \times 50 \text{ g}$$

$$q_{fusion} = 4{,}000 \text{ cal}$$

4.   **(A)**   We have

$$-530.6 = 3\Delta H_{f\,CO_2(g)} + 4\Delta H_{f\,H_2O(l)} - \Delta H_{f\,C_3H_8(g)}$$

since $\Delta H = \Sigma \Delta H_f$ products $- \Sigma \Delta H_f$ reactants and ΔH of the elements in their natural state is zero

$$(\Delta H_{f\,O_2(g)} = 0).$$

Solving for

$$\Delta H_{f\,C_3H_8(g)}$$

$$\Delta H_{f\,C_3H_8(g)} = 3\Delta H_{f\,CO_2(g)} + 4\Delta H_{f\,H_2O(l)} + 530.6$$

Substituting values, we have

$$\Delta H_{f\,C_3H_8\,(g)} = 3(-94.1) + 4(-68.3) + 530.6$$

$$\Delta H_{f\,C_3H_8\,(g)} = -24.9 \text{ kcal/mol}$$

5. **(A)**

$$2Ag_2S(s) + 2H_2O(1) \rightarrow 4Ag(s) + 2H_2S(g) + O_2(g)$$

$$\Delta H^o = 4\Delta H^o_{f\,Ag\,(s)} + 2\Delta H^o_{f\,H_2S\,(g)} + \Delta H^o_{f\,O_2\,(g)}$$
$$- 2\Delta H^o_{f\,Ag_2S\,(s)} - 2\Delta H^o_{f\,H_2O\,(1)}$$

$$= 4 \text{ mol Ag(s)} \times 0 + 2 \text{ mol H}_2S \times (-20.6 \text{ kJ/mol}) +$$

$$1 \text{ mol O}_2 \times 0 - 2 \text{ mol Ag}_2S \times (-32.6 \text{ kJ/mol}) -$$

$$2 \text{ mol H}_2O \times (-285.8 \text{ kJ/mol})$$

$$= [-41.2 - (-65.2) - (-571.6)]kJ = 595.6 \text{ kJ}$$

6. **(A)** The standard heat of reaction, ΔH^o is equal to the sum of the heats of the products minus the sum of the heats of the reactants:

$$\Delta H^o = \Sigma \text{ heats of products} - \Sigma \text{ heats of reactants}$$

$$\Delta H^o = [2(\Delta H^o_{NAOH}) + \Delta H^o_{H_2}] - [2(\Delta H^o_{H_2O} + 2(\Delta H^o_{Na})]$$

By convention, the standard heats of formation of an element in its most stable form is zero. Thus, both $H_2(g)$ and Na(a) equal zero, then:

$$\Delta H^o = 2\,(-426.7) - 2\,(-285.8)\,{}^*$$

$$= -853.4 - (-571.6)$$

$$= -853.4 + 571.6$$

$$= -281.8 \text{ kJ}$$

*Note: The enthalpies given in the table above are in units of kJ/mol; therefore, it is necessary to multiply by the number of moles of each substance as represented in the balanced equation to calculate the ΔH^o for the overall reaction.

7. **(A)** The specific heat of a substance is defined as the heat required to raise the temperature of 1 g of that substance by 1°C.

8. **(C)**

Reaction path I and reaction path II should have the same ΔH. Thus, to find ΔH for our reaction at 798K (525°C), we can sum the changes in enthalpy involved in cooling the reactants to 298K, forming the products, and heating the products to 798°K.

$$-\left[\left(\frac{1}{2}C_{pN_2} + \frac{3}{2}C_{pH_2}\right)\Delta T\right] + \Delta H_f^\circ + [C_{pNH_3} \times \Delta T] = \Delta H_{798}$$

$$-[(13.5 + 43.5)\,(798 - 298)] + (-46 \text{ kJ/mol}) + [26(798 - 298)]$$

$$= \Delta H_{798}$$

$$-[28{,}500 \text{ J/mol}] + (-46 \text{ kJ/mol}) + [13{,}000 \text{ J/mol}] = \Delta H_{798}$$

$$-28.5 \text{ kJ/mol} - 46 \text{ kJ/mol} + 13 \text{ kJ/mol} = \Delta H_{798}$$

$$-62 \text{ kJ/mol} = \Delta H_{798}$$

9. **(B)** An endothermic reaction is one in which heat may be considered one of the "reactants." An exothermic reaction releases heat upon formation of the products. An equilibrium reaction may be either exothermic or endothermic. The same holds true for spontaneity; spontaneity can only be determined if one also knows the entropy change (ΔS) for the reaction.

10. **(B)** Le Châtelier's principle states: If a system at equilibrium is disturbed by a change in temperature, pressure, or the concentration of one of the components, the system will shift its equilibrium position so as to counteract the effect of the disturbance. If the addition of heat caused the reaction to shift to the right (i.e., away from the reactants), this implies that heat is required for the reaction to "go." This is indicative of an endothermic reaction.

11. **(B)** The heat of combustion is defined as the heat of reaction of a compound when it reacts completely with oxygen to produce CO_2 and H_2O. Therefore, the heat of the reaction

$$C_2H_4 + 3O_2 \rightarrow 2CO_2 + 2H_2O$$

is

$$\Delta H_c = (2\Delta H_{f(CO_2)} + 2\Delta H_{f(H_2O)}) - (\Delta H_{f(C_2H_4)} + 3\Delta H_{f(O_2)})$$

$$\Delta H_c = [2(-94.1) + 2(-68.3)] - [12.5 + 3(0)]$$

$$\Delta H_c = -337.3 \text{ kcal/mole}$$

12. **(B)** Q (heat absorbed) $= m \times c \times \Delta T$

where m = mass,

$$\Delta T = T_{final} - T_{initial}$$

hence: $Q = 100(4.2)\,(5 - 25) = -84,000\ J$

Note that it's not even necessary to convert °C to °K, the difference between temperatures in any scale would be the same.

13.   **(C)** $\Delta E = \Delta H - \Delta(PV)$

However, for gases $PV = nRT$ so

$$\Delta PV = RT\Delta n_{gas}$$

$$\Delta n_{gas} = n_{NOCl} - \frac{1}{2}n_{N_2} - \frac{1}{2}n_{O_2} - \frac{1}{2}n_{Cl_2}$$

$$-1 - \frac{1}{2}(1) - \frac{1}{2}(1) - \frac{1}{2}(1) = -\frac{1}{2}\ \text{mole}$$

Hence, $\Delta E = \Delta H - \Delta\,(PV)$

$$= [(12.57\ \text{kcal})(4.184\ \text{kJ/kcal})] - [(8.314 \times 10^{-3}\ \text{kJ/mol K})$$

$$(298K)(-\frac{1}{2}\ \text{mole})]$$

14.   **(C)**   Recalling that the change in enthalpy for a reaction is given by the sum of the heats of formation of the reactants subtracted from the sum of the heats of formation of the products, we have

$$\Delta H^\circ = \Sigma\ \Delta H_{f\,products} - \Sigma\Delta H_{f\,reactants}$$

or $-488.7 = 3(-94.1) + 4(-57.8) - (x)$

$$x = 3(-94.1) + 4(-57.8) + 488.7$$

$$x = -24.8\ \text{kcal/mol}$$

15.   **(B)**   Heat was absorbed by the system during the reaction as indicated by the products having a greater potential energy than the reactants. A reaction in which heat is absorbed in order to produce the products is said to be endothermic.

16.   **(D)**   A compound produced by addition of heat to the reactants will easily be decomposed upon addition of heat equal to the activation energy and will liberate a quantity of heat greater than that added.

17.   **(E)**   The heat of combustion of methane is given as -212.8 kcal/mol. This indicates that 212.8 kcal of heat are liberated when one mole of methane undergoes complete reaction with oxygen. Converting to moles

$$32 \text{ g} \times \frac{1 \text{ mole}}{16 \text{ g of CH}_4} = 2 \text{ moles of CH}_4$$

we find that

$$2 \text{ moles} \times \frac{212.8 \text{ kcal}}{1 \text{ mole}}$$

or 425.6 kcal of heat is liberated.

18.   **(E)**   The heat required to melt one gram of a solid is the heat of fusion (80 calories for water). Thus, the heat required to melt 5 g of ice at 0°C is

$$5 \text{ g} \times 80 \text{ cal/g} = 400 \text{ cal.}$$

The heat required to raise the temperature of one gram of a substance 1C° is the specific heat (1 calorie for water). Therefore, the heat required to increase the temperature of 5 g of water from 0°C to 100°C is

$$5 \text{ g} \times (100\text{C}° - 0\text{C}°) \times 1 \text{ cal/gC}° = 500 \text{ cal.}$$

The heat required to vaporize one gram of a liquid is the heat of vaporization (540 calories for water). Therefore, the heat required to vaporize 5 g of water at 100°C is

$$5 \text{ g} \times 540 \text{ cal/g} = 2,700 \text{ cal.}$$

The heat required to convert 5 g of ice at 0°C to steam at 100°C is then

$$400 \text{ cal} + 500 \text{ cal} + 2,700 \text{ cal} = 3,600 \text{ cal}$$

or $3,600 \text{ cal} \times \dfrac{1 \text{ kcal}}{1,000 \text{ cal}} = 3.6 \text{ kcal}$

19.   **(D)**   The specific heat of a substance is the amount of heat required to produce a change of 1°C in 1 g of the substance. For aluminum, it requires .89 joules to raise 1 g of Al 1°C; therefore, the heat required to raise 23.2 g from 30°C to 80°C, or a change of 50°, will be equal to

$$q = \text{mass} \times \text{specific heat} \times \text{temperature change}$$

$$q = 23.2 \text{ g} \times .89 \text{ J/°C g} \times 50°$$

$$q = 1,032 \text{ joules}$$

20.   **(C)**   The Arrhenius law states that

$$k = A \exp (-E_a/RT).$$

This can be rearranged to:

$$\ln k = \ln A - E_a/RT.$$

If we plot $\ln k$ vs $1/T$, the slope of the graph is $-E_a/R$.

21. **(C)**

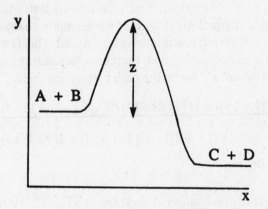

The above diagram is a plot of the potential energy (E) possessed by molecules as a function of the reaction coordinate. As A and B approach each other, the potential energy of the system increases to a maximum, which corresponds to an activated complex. The difference in energy between $A + B$ and the activated complex is called the activation energy. In the diagram above, $y = E = $ potential energy, $x = $ reaction coordinate, and $z = $ activation energy.

22. **(E)** All of the choices, since

$$Q = S + \frac{1}{2}D + I - E - U.$$

23. **(C)** Temperature is a direct measure of the kinetic energy of a species. Thus, the state of the species at the highest temperature has the greatest kinetic energy. A substance has its highest temperature when it is in the gas phase.

24. **(E)** The negative sign to the heat per mole in kilocalories means that the heat is liberated. One mole of methane is 16 grams: $12 + 4(1)$. Therefore, three moles liberate 54 kcal or 3×18.

25. **(A)** The reaction equation is obtained as $-2(i) + 2 (ii)$:

$$2C + 2O_2 \rightarrow 2CO_2 - 188.2 \text{ kcal}$$

$$2CO_2 \rightarrow O_2 + 2CO + 135.4 \text{ kcal}$$

$$\overline{2C + O_2 \rightarrow 2CO - 52.8 \text{ kcal}}$$

26. **(A)** The free energy amount is positive, meaning that energy is absorbed

rather than released (negative sign). One mole of NO, nitrogen monoxide, is 46 grams $[(16 \times 2) + 14]$. Twenty-three grams is one-half of a mole and, therefore, absorbs one-half the molar energy, which is 4.

27.   **(A)**   Hess's Law (of constant heat summation) states that the enthalpy change of any reaction is equal to the sum of the enthalpy changes of a series of reactions into which the overall reaction may be divided. This law is based on the fact that enthalpy is a state function and, therefore, dependent only on the initial and final states themselves and not on the path connecting them.

$$\Delta H^\theta = [H^\theta_{m\,H_2O_2} + 4(H^\theta_{m\,N_2}0] - [2(H^\theta_{m\,NH_3}) + 2(H^\theta_{m\,NO})]$$

$$= [(-187.8\ kJ) + 4(0)] - [2(264.0\ kJ) + 2(90.25\ kJ)]$$

$$= -896.3\ kJ$$

28.   **(D)**   The heat of combustion of methane is -212.8 kcal/mol according to the reaction equation. To determine how many grams of CH_4 are required to produce 425.6 kcal, we have

$$-425.6\ kcal \times \frac{1\ mole\ of\ CH_4}{-212.8\ kcal} \times \frac{16\ g\ of\ CH_4}{1\ mole\ of\ CH_4} = 32\ g\ of\ CH_4$$

Note the negative values used for ΔH_c to indicate that heat is evolved (exothermic) when methane undergoes combustion.

29.   **(E)**   Zinc oxide's formula weight is 81 (65 + 16). Therefore, 162 grams is twice this weight or two moles. If one mole liberates 84 calories, two liberate twice that amount of energy. The minus sign indicates energy liberation.

30.   **(A)**   The temperature of a substance is directly related to its kinetic energy.

31.   **(E)**   The heat (q) absorbed or released during the process is related to the change in temperature (ΔT) by the equation $q = mc\Delta T$, where m is mass, c is specific heat (1 cal/g°C for water), and ΔT is the change in temperature. The heat required for water to raise from 50°C to 100°C (just before actually boiling) is obtained from

$$q = mc\Delta T = (10\ g)\,(1\ cal/g°C)\,(50°C) = 500\ calories$$

Since we have 3,200 cal (3.2 kcal), then all the water would reach 100°C, and the extra energy (3,200 − 500 = 2,700 cal) would be used to vaporize water molecules. The heat required for the vaporization of water is obtained by the product of ΔH_{vapor} and the mass:

$$q_{vapor} = \Delta H_{vapor} \times m = (540\ cal/g)\,(m) = 5,400\ cal$$

Since we only have 2,700 cal, only one-half of the sample (5 g) would evaporate. The net result is 5 g of water at 100°C and 5 g of steam at 100°C.

32. **(E)**   According to Hess's law, the enthalpy change for a reaction is the same whether the reaction takes place in one step or a series of steps. We can utilize this law to determine the enthalpy change ΔH^o for the reaction

$$C_{(graphite)} + \frac{1}{2}O_2 \rightarrow CO(g).$$

First by reversing the reaction

$$CO_2(g) + \frac{1}{2}O_2 \rightarrow CO_2(g) \ \Delta H^o = -284 \text{ kJ}$$

we have

$$CO_2(g) \ CO(g) + \frac{1}{2}O_2 \qquad\qquad \Delta H^o = +284 \text{ kJ}$$

and adding it to the reaction

$$C_{(graphite)} + \frac{1}{2}O_2(g) \rightarrow CO_2(g) \qquad\qquad \Delta H^o = -394 \text{ kJ}$$

$$CO_2(g) \rightarrow CO(g) + \frac{1}{2}O_2 \qquad\qquad \Delta H^o = +284 \text{ kJ}$$

we get $C_{(graphite)} + \frac{1}{2}O_2(g) \rightarrow CO(g) \qquad\qquad \Delta H^o = -110 \text{ kJ}$

Note: When reversing the reaction we must also change the sign of ΔH^o.

33.   **(B)**   An endothermic reaction results in reaction products that have a greater energy than the reactants. The activation energy for a reaction may be small but it is always positive. Thus, we have

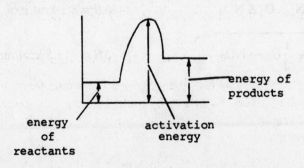

34. **(A)** Carbon is neither a strong metal nor a strong nonmetal and has less electron-attracting power than the strong nonmetals bromine, chlorine, fluorine, and oxygen, in their covalent bond formation.

35. **(A)** The standard heat of reaction (ΔH^o_{298}) is (the summation of the standard heats of formation of products) – (the summation of the standard heats of formation of reactants).

Therefore,

$$\Delta H^o_{298} = 5(-1260) + (-280) - 2(60)$$

$$= -8{,}940 \text{ kJ}$$

36. **(A)** The heat of combustion of octane is given by

$$\Delta H_c = (8\Delta H_{f\,(CO_2)} + 9\Delta H_{f\,(H_2O)}) - (\Delta H_{f\,(C_8H_{18})} - \frac{25}{2}\,\Delta H_{f\,(O_2)})$$

Substituting values given ($H_2O(l)$ is used because $T = 25°C$:

$$\Delta H_c = [8(-94.1) + 9(-68.3)] - [-49.4 + \frac{25}{2}(O)]$$

$$\Delta H_c = -1{,}318.1 \text{ kcal/mol}$$

Note that ΔH_f of an element is zero.

37. **(B)** The energy of a C–C double bond is greater than that of a single bond. The energy of the triple bond is greater still.

38. **(B)** The heat of a particular reaction can be determined by the summation of two or more reactions which added together give the ΔH of the particular reaction. Thus, we have

$$NO \rightarrow \frac{1}{2}N_2 + \frac{1}{2}O_2 \qquad\qquad \Delta H = -21.6 \text{ kcal/mol}$$

$$+ \frac{1}{2}N_2 + O_2 \rightarrow NO_2 \qquad\qquad \Delta H = 8.1 \text{ kcal/mol}$$

$$\overline{NO + \frac{1}{2}O_2 \rightarrow NO_2} \qquad\qquad \Delta H = -13.5 \text{ kcal/mol}$$

39. **(D)** In a similar manner

$$N_2O_4 \rightarrow N_2 + 2O_2 \qquad\qquad \Delta H = -4.3 \text{ kcal/mol}$$

$$+ 2(\frac{1}{2} N_2 + \frac{1}{2} O_2 \rightarrow NO) \qquad \Delta H = 2(21.6) \text{ kcal/mol}$$

$$N_2O_4 \rightarrow 2NO + O_2 \qquad\qquad \Delta H = 38.9 \text{ kcal/mol}$$

40. **(B)**

$$2(NO_2 \rightarrow \frac{1}{2}N_2 + O_2) \qquad\qquad \Delta H = 2(-8.1) \text{ kcal/mol}$$

$$+ N_2 + 2O_2 \rightarrow N_2O_4 \qquad\qquad \Delta H = 4.3 \text{ kcal/mol}$$

$$2NO_2 \rightarrow N_2O_4 \qquad\qquad \Delta H = -11.9 \text{ kcal/mol}$$

THERMODYNAMICS I
REVIEW

Some Commonly Used Terms in Thermodynamics

A **system** is that particular portion of the universe on which we wish to focus our attention.

Everything else is called the **surroundings.**

An **adiabatic process** occurs when the system is thermally isolated so that no heat enters or leaves.

An **isothermal process** occurs when the system is maintained at the same temperature throughout an experiment ($t_{final} = t_{initial}$).

An **isopiestic (isobaric) process** occurs when the system is maintained at constant pressure (i.e., $P_{final} = P_{initial}$).

The **state of the system** is some particular set of conditions of pressure, temperature, number of moles of each component, and their physical form (for example, gas, liquid, solid, or crystalline form).

State functions depend only on the present state of the substance and not on the path by which the present state was attained. Enthalpy, energy, Gibbs free energy, and entropy are examples of state functions.

Heat capacity is the amount of heat energy required to raise the temperature of a given quantity of a substance by $1.0°C$.

Specific heat is the amount of heat energy required to raise the temperature of 1g of a substance by $1°C$.

Molar heat capacity is the heat necessary to raise the temperature of 1 mole of a substance by $1°C$.

1. Bond Energies

For diatomic molecules, the bond dissociation energy, $\Delta H°_{diss}$, is the amount of energy per mole required to break the bond and produce two atoms, with reactants and products being ideal gases in their standard states at $25°C$.

The heat of formation of an atom is defined as the amount of energy required to form one mole of gaseous atoms from its element in its common physical state at $25°C$ and 1 atm pressure.

In the case of diatomic gaseous molecules of elements, the ΔH_f of an atom is equal to one-half the value of the dissociation energy, that is,

$$ -\Delta H_f = \frac{1}{2} \Delta H°_{diss} $$

where the minus sign is needed since one process liberates heat, and the other requires heat.

For the reaction $HO - OH(g) \rightarrow 2OH(g)$,

$$\Delta H^{\circ}{}_{\text{diss}} = 2(\Delta H^{\circ}{}_{f}\text{OH}) - (\Delta H^{\circ}{}_{f}\text{OH} - \text{OH}).$$

The energy needed to reduce a gaseous molecule to neutral gaseous atoms, called the atomization energy, is the sum of all the bond energies in the molecule.

For polyatomic molecules, the average bond energy, $\Delta H^{\circ}{}_{\text{diss}}$ avg., is the average energy per bond required to dissociate one mole of molecules into their constituent atoms.

PROBLEM

> Given for hydrogen peroxide, H_2O_2:
>
> $$\text{HO} - \text{OH(g)} \rightarrow 2\text{OH(g)} + \Delta H^{\circ}{}_{\text{diss}} = 51 \text{ kcal/mol}$$
>
> From this value and the following data calculate: (a) ΔH_f° of OH(g). (b) C–O bond energy; $\Delta H^{\circ}{}_{\text{diss}}$ in CH_3OH(g). ΔH_f° of H_2O_2(g) = –32.58 kcal/mol. ΔH_f° of CH_3(g) = 34.0 kcal/mol. ΔH_f° of CH_3OH(g) = –47.96 kcal/mol.

SOLUTION

(a) $\Delta H^{\circ}{}_{\text{diss}}$ for this reaction is equal to the ΔH_f° of the reactants subtracted from the ΔH_f° of the products, where ΔH° equals enthalpy or heat content.

$$\Delta H^{\circ}{}_{\text{diss}} = 2 \times \Delta H_f^{\circ}{}_{\text{OH}} - \Delta H_f^{\circ}{}_{\text{HO–OH}} = 51 \text{ kcal/mol}$$

$$2 \times \Delta H_f^{\circ}{}_{\text{OH}} = 51 \text{ kcal/mol} + \Delta H_f^{\circ}{}_{\text{HO–OH}}$$

$$\Delta H_f^{\circ}{}_{\text{OH}} = \frac{5 \text{ kcal/mol} - 32.58 \text{ kcal/mol}}{2}$$

$$\Delta H_f^{\circ}{}_{\text{OH}} = \frac{18.42 \text{ kcal/mol}}{2} = 9.21 \text{ kcal/mol}$$

(b) CH_3OH dissociates by breaking the C–O bond:

$$CH_3OH(g) \rightarrow CH_3^{+}(g) + OH^{-}(g)$$

Therefore, the bond energy of the C–O bond equals $\Delta H^{\circ}{}_{\text{diss}}$. $\Delta H^{\circ}{}_{\text{diss}}$ equals the sum of the ΔH's of the products minus the ΔH's of the reactants. Thus,

$$\text{bond energy of C–O} = \Delta H_f^{\circ} \text{ of } CH_3^{+}(g) + \Delta H_f^{\circ} \text{ of } OH^{-}(g)$$

$$- \Delta H_f^{\circ} \text{ of } CH_3OH(g)$$

$$\text{bond energy of C–O} = 34.0 \text{ kcal/mol} + 9.21 \text{ kcal/mol}$$

$$+ 47.96 \text{ kcal/mol}$$

$$= 91.17 \text{ kcal/mol}$$

PROBLEM

Using the following table of bond energies, calculate the energy change in the following:

a) $2H_2(g) + O_2(g) \rightarrow 2H_2O(g)$

b) $CH_4(g) + 2O_2(g) \rightarrow CO_2(g) + 2H_2O(g)$

c) $CH_4(g) + Cl_2(g) \rightarrow CH_3Cl(g) + HCl(g)$

d) $C_2H_6(g) + Cl_2(g) \rightarrow C_2H_5Cl(g) + HCl(g)$

Bond Energies (in kcal per mole)

H–H	104	C–O	83
H–F	135	C=O	178
H–Cl	103	C–Cl	79
H–Br	88	C–F	105
H–I	71	Si–O	106
Li–H	58	Si–F	136
Cl–Cl	58	C–C	83
C–H	87	C=C	146
O–H	111	C≡C	199
O=O	118	N≡N	225
P–Cl	78		

SOLUTION

When using the bond energies to calculate the net energy change in a reaction, the net energy change is equal to the total bond energy of the bonds formed subtracted from the bond energy of bonds broken. This method is illustrated in the following examples.

a) $2H_2(g) + O_2(g) \rightarrow 2H_2O(g)$

There are two H_2 bonds and one O_2 bond broken and four O–H bonds formed here.

$$2H - H(g) + O = O(g) \rightarrow 2H - O - H(g)$$

net bond energy = (2 × bond energy of H–H) +

(1 × bond energy of O=O) – (4 × bond energy of O–H)

The following bond energies can be found in the table.

Bond	Bond energy (kcal/mol)
H–H	104
O=O	118
O–H	111

The net energy or heat evolved in the reaction can be found using these values.

$$\text{net bond energy} = [(2 \times 104) + (1 \times 118)] - (4 \times 111)$$

$$\Delta H = (208 + 118) - 444 = -118 \text{ kcal}$$

Thus, when this reaction occurs 118 kcal are released. The following examples will be solved in a similar manner.

Drill 1: Bond Energies

Questions 1–3 use the following table of bond energies.

Bond Energies (in kcal per mole)

H–H	104	C–O	83
H–F	135	C=O	178
H–Cl	103	C–Cl	79
H–Br	88	C–F	105
H–I	71	Si–O	106
Li–H	58	Si–F	136
Cl–Cl	58	C–C	83
C–H	87	C=C	146
O–H	111	C≡C	199
O=O	118	N≡N	225
P–Cl	78		

1. $CH_4(g) + 2O_2(g) \rightarrow CO_2(g) + 2H_2O(g)$

   (A) –111 kcal

   (B) –178 kcal

   (C) +200 kcal

   (D) –216 kcal

   (E) –145 kcal

2. $CH_4(g) + Cl_2(g) \rightarrow CH_3Cl(g) + HCl(g)$

   (A) –37 kcal

   (B) –87 kcal

   (C) –79 kcal

   (D) –105 kcal

   (E) –64 kcal

3. $C_2H_6(g) + Cl_2(g) \rightarrow C_2H_5Cl(g) + HCl(g)$

   (A) –87 kcal

   (B) –79 kcal

   (C) –105 kcal

   (D) –37 kcal

   (E) –64 kcal

2. Heat Capacity

For any phase transition, there is an associated energy change, e.g.,

$$S \rightarrow liq, \Delta H°_T \text{ (fusion); } liq \rightarrow S, \Delta H°_T \text{ (crystallization);}$$

$$S \rightarrow gas, \Delta H°_T \text{ (sublimation); } liq \rightarrow gas, \Delta H°_T \text{ (vaporization);}$$

$$\Delta H°_T \text{ (fusion)} = -\Delta H°_T \text{ (crystallization)}$$

Approximation of the heats of transition

For substances that are not highly joined in the liquid state, the following relationships have been noticed:

$$\Delta H°_T \text{ (vaporization)} \cong (88 \text{ Jk}^{-1} \text{ mol}^{-1})T_{bp}$$

where T_{bp} is the normal boiling point of the liquid. For elements,

$$\Delta H°_T \text{ (fusion)} \cong 9.2 \text{ Jk}^{-1} \text{ mol}^{-1})T_{mp}$$

where T_{mp} is the melting point of the solid.

Thus, the ΔH for heating is

$$\Delta H° = \overset{\text{phases}}{\underset{i}{\sum}} \int Cp_i° dT + \overset{\text{transitions}}{\underset{j}{\sum}} \Delta H°_T \text{(transition } j).$$

The following example shows how the above equation is applied when heating ice from $-15°C$ to steam at $110°C$.

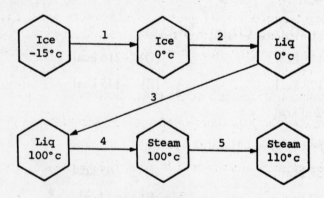

Figure 2.1

$$\Delta H° = \Delta H°_{(1)} + \Delta H°_{(2)} + \Delta H°_{(3)} + \Delta H°_{(4)} + \Delta H°_{(5)}$$

$$= \int_{258k}^{273k} Cp° dT + \Delta H°_{273} \text{ (fusion)} + \int_{273k}^{373k} Cp° dT$$

$$+ \Delta H°_{373} \text{ (vaporization)} + \int_{373k}^{383k} Cp° dT$$

PROBLEM

At constant volume, the heat capacity of gas differs from the heat capacity at constant pressure. Why?

SOLUTION

Heat capacity is defined as the quantity of heat that will result in a temperature rise of 1°C per mole of material. When heat is added at constant volume, all the heat goes into increasing the kinetic energy of the molecules. However, with constant pressure, the volume can expand. This means the added heat must do two things. It must be able to increase the kinetic energy of the molecules AND do work against the pressure to expand the volume. The constant pressure case would require more heat. In fact, the difference between C_p and C_v, where C_p = heat capacity at constant pressure, and C_v = heat capacity at constant volume, is the universal gas constant R. In summary, then, the difference stems from the fact that C_p must increase kinetic energy and do work to expand the volume, while C_v only has to increase the kinetic energy.

PROBLEM

What is the heat capacity of uranium? Its atomic heat is 26 J/mol and its atomic weight is 238 g/mol.

SOLUTION

The atomic heat of a solid element at room temperature is defined as the amount of heat required to raise the temperature of one mole of an element by 1°C. The heat capacity, or specific heat, is the amount of heat required to raise the temperature of one gram of a substance by 1°C. These two quantities can be related by the equation,

atomic heat = atomic weight × heat capacity.

Thus, to find the heat capacity of uranium, substitute the given values of atomic heat and weight.

heat capacity of uranium = 26 J/mol °C/238 g/mol

$$= 0.11 \text{ J/g°C}$$

Drill 2: Heat Capacity

1. How many calories of heat must be added to 3.0 liters of water to raise the temperature of the water from 20°C to 80°C?

    (A) 160 kcal   (B) 180 kcal

(C)   300 kcal   (D)   45 kcal

(E)   220 kcal

2.   Calculate the quantity of heat required to raise the temperature of one gallon of water (3.78 liters) from 10°C to 80°C.

(A)   265 kcal   (D)   290 kcal

(B)   230 kcal   (E)   320 kcal

(C)   340 kcal

3.   A piece of iron weighing 20.0 g at a temperature of 95.0°C was placed in 100.0 g of water at 25.0°C. Assuming that no heat is lost to the surround-ings, what is the resulting temperature of the iron and water? Specific heats: iron = .108 cal/g–°C; water = 1.0 cal/g–°C.

(A)   32.5°C   (D)   29.23°C

(B)   26.48°C   (E)   16.45°C

(C)   19.60°C

4.   A container has the dimensions 2.5 m × 50 cm × 60 mm. If it is filled with water at 32°F, how many kilocalories will be necessary to heat the water to the boiling point (212°F)?

(A) 9.2×10^3 kcal   (D) 6.5×10^2 kcal

(B) 2.3×10^6 kcal   (E) 7.5×10^3 kcal

(C) 4.5×10^4 kcal

5.   The following reaction using hydrogen and oxygen is carried out in a bomb calorimeter: $2H_2(g) + O_2(g) \rightarrow 2H_2O(l)$. The following data are recorded: weight of water in calorimeter = 2.650 kg, initial temperature of water = 24.442°C, final temperature of water after reaction = 25.635°C, specific heat of reaction vessel is 0.200 kcal/°C–kg, the weight of the calorimeter is 1.060 kg, and the specific heat of water is 1.00 kcal/°C–kg. Calculate the heat of reaction.

(A)   3.414 kcal   (D)   5.452 kcal

(B)   6.213 kcal   (E)   1.912 kcal

(C)   2.215 kcal

3. Enthalpy

In chemical reactions, there are two energies that are very common and enter into almost all energy calculations associated with reactions. The first is the internal energy, E. This is the bond energy associated with chemical bonds plus the kinetic energy of molecular species. The second is the PV energy. Since so many reactions take place at conditions that cause the system to either expand or contract against the atmospheric surroundings, the effect of PV work on the atmosphere becomes important. The enthalpy is the thermodynamic quantity defined to include both the internal energy and the PV work against the atmosphere.

$$H = E + PV$$

where H is the enthalpy, E is the internal energy, and PV is the pressure and volume terms. For condensed phases (i.e., liquids and solids), the volume change is often negligible and $H \approx E,$ but for gases the difference is significant. Only changes in enthalpy and internal energy have physical significance; there is no absolute zero point for these quantities. Therefore, tabulation of enthalpy is the difference between the state for which it is tabulated and some reference state either implied or specified.

The heat content of a substance is called enthalpy, H. A heat change in a chemical reaction is termed a difference in enthalpy, or ΔH. The term "change in enthalpy" refers to the heat change during a process carried out at a constant pressure:

$$\Delta H = q_p \; ; q_p \text{ means "heat at constant pressure."}$$

The change in enthalpy, $\Delta H,$ is defined

$$\Delta H = \Sigma H_{products} - \Sigma H_{reactants}$$

When more than one mole of a compound is reacted or formed, the molar enthalpy of the compound is multiplied by the number of moles reacted (or formed).

Enthalpy is a state function. Changes in enthalpy for exothermic and endothermic reactions are shown below:

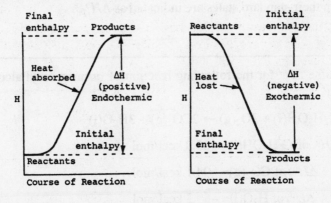

Figure 3.1

The ΔH of an endothermic reaction is positive, while that for an exothermic reaction is negative.

Heats of Reaction

ΔE is equal to the heat absorbed or evolved by the system under conditions of constant volume:

$$\Delta E = q_v, \quad q_v \text{ means "heat at constant volume."}$$

Since $\quad H = E + PV$

at constant pressure $\Delta H = \Delta E + P \Delta V$. Note that the term $P \Delta V$ is just the pressure-volume work $((\Delta n)RT)$ for an ideal gas at constant temperature, where Δn is the number of moles of gaseous products minus the number of moles of gaseous reactants. Therefore,

$$\Delta H = \Delta E + \Delta nRT$$

for a reaction which involves gases. If only solid and liquid phases are present, ΔV is very small, so that $\Delta H \approx \Delta E$.

Hess's Law of Heat Summation

Hess's law of heat summation states that when a reaction can be expressed as the algebraic sum of two or more reactions, the heat of the reaction, ΔH_r, is the algebraic sum of the heats of the constituent reactions.

The enthalpy changes associated with the reactions that correspond to the formation of a substance from its free elements are called heats of formation, ΔH_r.

$$\Delta H^\circ_r = \Sigma \, \Delta H^\circ_f \, (\text{products}) - \Sigma \, \Delta H^\circ_f \, (\text{reactants})$$

Standard States

The standard state corresponds to 25°C and 1 atm. Heats of formation of substances in their standard states are indicated as ΔH°_f.

PROBLEM

Determine ΔH° for the following reaction of burning ethyl alcohol in oxygen:

$$C_2H_5OH(l) + 3O_2(g) \rightarrow 2CO_2(g) + 3H_2O(l)$$

ΔH°_f of $C_2H_2OH(l) = -65.9$ kcal/mol

ΔH°_f of $CO_2(g) = -94.1$ kcal/mol

ΔH°_f of $H_2O(l) = -68.3$ kcal/mol

SOLUTION

The heat reaction ($\Delta H°$) may be found from the heats of formation ($\Delta H°_f$) by subtracting the sum of the heats of formation of all reactants from the sum of the heats of formation of all products. The heat of formation of all pure elements is zero. When more than one mole of a compound is either reacted or formed, the heat of formation is multiplied by the stoichiometric coefficient for the specific compound in the equation.

heat of reaction

$= [(2 \text{ moles} \times \Delta H°_f \text{ of } CO_2) + (3 \text{ moles} \times \Delta H°_f \text{ of } H_2O)]$

$\quad - [(1 \text{ mole } \Delta H°_f \text{ of } C_2H_2OH) + (3 \text{ moles} \times \Delta H°_f \text{ of } O_2)]$

$= [(2 \text{ moles} \times -94.1 \text{ kcal/mol}) + (3 \text{ moles} \times -68.3 \text{ kcal/mol})]$

$\quad - [(1 \text{ mole} \times -68.3 \text{ kcal/mol}) + 3 \text{ moles} \times 0 \text{ kcal/mol})]$

$= -327.2 \text{ kcal/mol}$

PROBLEM

Using the information in the following table, determine $\Delta H°$ for the reactions:

a) $3Fe_2O_3(s) + CO(g) \rightarrow 2Fe_2O_4(s) + CO_2(g)$

b) $Fe_3O_4(s) + CO(g) \rightarrow 3FeO(s) + CO_2(g)$

c) $FeO(s) + CO(g) \rightarrow Fe(s) + CO_2(g)$

Heats of Formation

Compound	$\Delta H°$ (kcal/mol)
$CO(g)$	−26.4
$CO_2(g)$	−94.1
$Fe_2O_3(s)$	−197
$Fe_3O_4(s)$	−267
$FeO(s)$	−63.7

SOLUTION

The energy change involved in the formation of one mole of a compound from its elements in their normal state is called the heat of formation. Thus, $\Delta H°$ for a reaction can be found by subtracting the sum of the $\Delta H°$'s for the reactants from the sum of the $\Delta H°$'s for the products.

$\Delta H° = \Delta H°$ of the products $- \Delta H°$ of the reactants

a) $\quad 3Fe_2O_3(s) + CO(g) \rightarrow 2Fe_3O_4(s) + CO_2(g)$

When more than one mole of a compound is either reacted or formed in a reaction, the ΔH^o for that compound is multiplied by the number of moles present in solving for the ΔH^o of the reaction.

For this reaction:

$$\Delta H^o = (2 \times \Delta H^o \text{ of } Fe_3O_4 + \Delta H^o \text{ of } CO_2) - (3 \times \Delta H^o \text{ of } Fe_2O_3 + \Delta H^o \text{ of } CO)$$

$$\Delta H^o = [2 \times (-267) + (-94.1)] - [3 \times (-197) + -26.4)]$$

$$= -10.7 \text{ kcal}$$

b) $Fe_3O_4(s) + CO(g) \rightarrow 3FeO(s) + CO_2(g)$

$$\Delta H^o = (3 \times \Delta H^o \text{ of } FeO + \Delta H^o \text{ of } CO_2) - (\Delta H^o \text{ of } Fe_3O_4 + \Delta H^o \text{ of } CO)$$

$$\Delta H^o = (3 \times (-63.7) + (-94.1)) - ((-267) + (-26.4))$$

$$= 8.2 \text{ kcal}$$

c) $FeO(s) + CO(g) \rightarrow Fe(s) + CO_2(g)$

$$\Delta H^o = (\Delta H^o \text{ of } Fe + \Delta H^o \text{ of } CO_2) - (\Delta H^o \text{ of } FeO + \Delta H^o \text{ of } CO)$$

The ΔH^o of any element is 0. Thus, the ΔH^o of Fe is 0.

$$\Delta H^o = [0 + (-94.1)] - [(-63.7) + (-26.4)]$$

$$= -4.0 \text{ kcal}$$

Drill 3: Enthalpy

1. Given the following reactions:

$$S(s) + O_2(g) \rightarrow SO_2(g) \qquad\qquad \Delta H = -71.0 \text{ kcal}$$

$$SO_2(g) + \frac{1}{2}O_2(g) \rightarrow SO_3(g) \qquad\qquad \Delta H = -23.5 \text{ kcal}$$

calculate ΔH for the reaction:

$$S(s) + 1\frac{1}{2}O_2(g) \rightarrow SO_3(g)$$

(A) −25.6 kcal   (D) −57.5 kcal

(B) −94.5 kcal   (E) +23.5 kcal

(C) −72.6 kcal

2. Given that $\Delta H^\circ{}_{CO_2\,(g)} = -94.0$, $\Delta H^\circ{}_{CO\,(g)} = -26.4$, $\Delta H^\circ{}_{H_2O\,(l)} = -68.4$ and $\Delta H^\circ{}_{H_2O\,(g)} = -57.8$ in kcal/mol, determine the heats of reaction of $CO(g) + \frac{1}{2}O_2(g) \rightarrow CO_2(g)$

(A) –30.2 kcal/mol   (D) 22.6 kcal/mol

(B) –67.6 kcal/mol   (E) 27.9 kcal/mol

(C) –42.2 kcal/mol

3. Using the information in the previous question, determine the heat of reaction for

$$H_2(g) + \frac{1}{2}O_2(g) \rightarrow H_2O(l)$$

(A) –61.7 kcal/mol   (D) –45.6 kcal/mol

(B) –28.3 kcal/mol   (E) –29.7 kcal/mol

(C) –68.4 kcal/mol

4. Using the information from question 2, what is the heat of reaction for $H_2O(l) \rightarrow H_2O(g)$ in kcal/mol?

(A) 4.23   (D) 12.4

(B) 6.10   (E) 10.6

(C) 4.11

5. You are given the following reactions at 25°C: $2NaHCO_3(s) \rightarrow Na_2CO_3(s) + CO_2(g) + H_2O(l)$, $\Delta H = 30.92$ kcal/mol. $\Delta H°_{Na_2CO_3\,(s)} = -270.3$, $\Delta H°_{CO_2\,(g)} = -94.0$ and $\Delta H°_{H_2O\,(l)} = -68.4$ kcal/mol, what is the standard enthalpy of formation for $NaHCO_3(s)$ in kcal/mol?

(A) –231.8   (D) 80.62

(B) –110.0   (E) 50.70

(C) 440.0

4. Enthalpy Calculations and the First Law of Thermodynamics

The First law of thermodynamics states that the change in internal energy is equal to the difference between the energy supplied to the system as heat and the energy removed from the system as work performed on the surroundings:

$$\Delta E = q - w$$

where E represents the internal energy of the system (the total of all the energy possessed by the system). ΔE is the energy difference between the final and initial states of the system:

$$\Delta E = E_{\text{final}} - E_{\text{initial}}$$

The quantity q represents the amount of heat that is added to the system as it passes from the initial to the final state, and w denotes the work done by the system on the surroundings.

Heat added to a system and work done by a system are considered positive quantities (by convention).

For an ideal gas at constant temperature, $\Delta E = 0$ and $q - w = 0$ ($q = w$).

Considering only work due to expansion of a system, against constant external pressure:

$$w = P_{\text{external}} \times \Delta V$$

$$\Delta V = V_{\text{final}} - V_{\text{initial}}$$

PROBLEM

> You have 1 liter of an ideal gas at 0°C and 10 atm pressure. You allow the gas to expand against a constant external pressure of 1 atm, while the temperature remains constant. Assuming, 24.217 cal/liter-atm, find q, w, ΔE, and ΔH in calories, (a) in these values, if the expansion took place in a vacuum, and (b) if the gas were expanded to 1 atm pressure.

SOLUTION

The solution to the parts of this problem require a combination of thermodynamics and ideal gas law theory. The gas expands because its pressure is greater than the external pressure. When the pressure of the gas falls to 1 atm, which is the pressure being applied externally, gas expansion will terminate. This allows for determination of the volume change using Boyle's law. You want ΔV (volume change), since $w = \text{work} = P\Delta V$. Boyle's law states that $PV = \text{constant}$ for ideal gas. Thus, $P_1V_1 = P_2V_2$. Originally, $P_1 = 10$ atm with $V_1 = 1$ liter. You end up with $P_2 = 1$ atm. Thus,

$$V_2 = \frac{P_1V_1}{P_2} = \frac{(10 \text{ atm})(1 \text{ liter})}{(1 \text{ atm})} = 10 \text{ liters.}$$

$\Delta V = 10$ liters (final) $- 1$ liter (original) $= 9$ liters. Therefore, $w = P\Delta V = (1 \text{ atm}) (9 \text{ liters}) = 9$ liters-atm. But, there are 24.217 cal per liter-atm, so that, in calories,

$$w = 9 \text{ liter-atm} \times 24.217 \text{ cal/liter-atm} = 218 \text{ cal.}$$

q = the heat absorbed by the system. To find its value, you employ the First law of thermodynamics, which says $\Delta E = q - w$, where ΔE = change in energy of system and w = work performed. Since the gas is ideal, E = energy is only a function of temperature. As such, $\Delta E = 0$, since the temperature is constant. You have, therefore, $0 = q - w$ or $q = w$. But, you just found w, which means $q = 218$ cal.

To find ΔH, the enthalpy change, remember that $\Delta H = \Delta E + \Delta(PV)$. You know that $\Delta E = 0$. To find $\Delta(PV)$, recall that $\Delta(PV) = P_2V_2 - P_1V_1$, as derived from Boyle's law. However, $P_1V_1 = P_2V_2$, thus $P_2V_2 - P_1V_1 = 0$. This means $\Delta H = 0$.

In a vacuum, there is no external pressure. This means that no work can be done, so that $w = 0$. This is derived from the fact that work is defined as performing an action against a surrounding environment. If there is no environment, there cannot be any work. Since ΔE remains 0 and $q = \Delta E + w$, then $q = 0$.

Since there is zero pressure, $PV = 0$. Since $\Delta(PV) = 0$, $\Delta E = 0$ and $\Delta H = \Delta E + \Delta(PV)$, then $\Delta H = 0$.

You now find the work (w), required in going from zero pressure to 1 atm pressure; use the equation for the reversible expansion of a gas in terms of work.

$$w = 2.303 \, n \, RT \log \frac{V_2}{V_1}, \text{ where } V_1 = \text{initial}$$

volume, V_2 = final volume, R = universal gas constant, T = temperature in Kelvin (Celsius plus 273°), and n = number of moles. V_2 was calculated from Boyle's law. The rest is given. Thus, you substitute to obtain

$$w = 2.303 \, (1 \text{ mole}) \, (1.987) \, (273) \log \left(\frac{10}{1}\right) = 1{,}249 \text{ cal.}$$

ΔE^o and ΔH^o are still zero, but since $q = \Delta E + w$, $q = 0 + 1{,}249 = 1{,}249$ cal.

PROBLEM

Calculate ΔE^o and ΔH^o for the following reactions at 25°C:

(a) $4NH_3(g) + 5O_2(g) \rightarrow 4NO(g) + 6H_2O(l)$

(b) $H_2S(g) + 1^1/_2O_2(g) \rightarrow SO_2(g) + H_2O(l)$

Use the following values for ΔH^o of the compounds:

Compound	H^o (kcal/mol)
$H_2O(l)$	$- 68.3$
$SO_2(g)$	-71.0
$H_2S(g)$	-5.3
$NO(g)$	21.6
$NH_3(g)$	-11.0

SOLUTION

ΔH^o may be defined as the heat released or absorbed as a reaction proceeds. ΔH^o for a particular reaction is found by subtracting the sum of the ΔH^o's of the compounds reacting from the sum of the ΔH^o's of the products being formed, i.e., $\Delta H^o = (\Delta H^o \text{ of products}) - (\Delta H^o \text{ of reactants})$.

(1) Determination of the ΔH^o's for the reactions:

(a) $4NH_3(g) + 5O_2(g) \rightarrow 4NO(g) + 6H_2O(l)$. When more than one mole of a compound is reacted or formed, the ΔH^o of the compound is multiplied by the number of moles present.

$$\Delta H^o = (4 \times \Delta H^o \text{ of NO} + 6 \times \Delta H^o \text{ of } H_2O)$$

$$- (4 \times \Delta H^o \text{ of } NH_3 + 5 \times \Delta H^o \text{ of } O_2)$$

The ΔH^o of any element is always 0. Thus, the ΔH^o of O_2 is zero. Substituting,

$$\Delta H^o = [(4 \times 21.6) + 6 \times (-68.3)] - [(4 \times (-11.0)) + (5 \times 0)]$$

$$= -279.40 \text{ kcal.}$$

(b) $H_2S(g) + 1\frac{1}{2}O_2(g) \rightarrow SO_2(g) + H_2O(l)$

$$\Delta H^o = (\Delta H \text{ of } SO_2 + \Delta H^o \text{ of } H_2O) - (\Delta H^o \text{ of } H_2S + 1\frac{1}{2} \Delta H^o \text{ of } O_2)$$

$$\Delta H^o = [(-71.0) + (-68.3)] - [(-5.3) + (1\frac{1}{2} \times O)]$$

$$= -134.0 \text{ kcal}$$

(2) Determination of ΔE: ΔE^o is the change in energy of a given reaction. It is defined as the difference of the ΔH^o's of the reaction and the pressure times the change in volume occurring during the reaction, i.e.,

$$\Delta E^o = \Delta H^o - P\Delta V,$$

where P is the pressure and ΔV is the change in volume. One is not given the values for pressure or volume here, thus, one uses the Ideal Gas Law to substitute ΔnRT for $P\Delta V$ where Δn is the change in the number of moles present, R is the gas constant (1.99 cal/mol–K), and T is the absolute temperature. The Ideal gas law states that

$$P\Delta V = \Delta nRT$$

Here, one is told that the reactions both occur at 25°C. This temperature can be converted to the absolute scale by adding 273 to 25°C.

$$T = 25 + 273 = 298 \text{ K}$$

Δn is found by subtracting the number of moles of compounds reacting from the number of moles formed. Caution: Only moles of gases are taken into account. Δn = number of moles formed – number of moles reacted.

One can solve for ΔE^o in these two reactions.

(a) $4NH_3(g) + 5O_2(g) \rightarrow 4NO(g) + 6H_2O(l)$

ΔH^o for this reaction was found to be -279.40 kcal. ΔH^o must be converted to calories when R is used. Kcal are converted to cal by multiplying the number of kcal by 1,000 cal/1 kcal.

$$\Delta n = (4 \text{ moles NO}) - (4 \text{ moles NH}_3 + 5 \text{ moles O}_2)$$

$$= (4) - (4 + 5) = -5 \text{ mole}$$

$$\Delta E^o = \Delta H^o - \Delta nRT$$

$$\Delta H^o = -279.40 \text{ kcal}$$

$$\Delta n = 5 \text{ moles}$$

$$R = 1.99 \text{ cal/mole–}^oK$$

$$T = 298^oK$$

$$\Delta E^o = -279.4 \text{ kcal} \times \left(\frac{1,000 \text{ cal}}{1 \text{ kcal}}\right) - (-5 \text{ mole} \times 1.99 \text{ cal/mole } ^oK \times 298^oK)$$

$$= -279,400 \text{ cal} + 2,965 \text{ cal}$$

$$= -276,435 \text{ cal} = -276.4 \text{ kcal}$$

(b) $H_2S(g) + 1\frac{1}{2}O_2(g) \rightarrow SO_2(g) + H_2O(l)$

$$\Delta H^o = -134.0 \text{ kcal}$$

ΔH^o should be expressed in calories:

$$\Delta H^o \text{ in cal} = -134.0 \text{ kcal} \times 1,000 \text{ cal/1 kcal}$$

$$= -134,000 \text{ cal}$$

$$\Delta n = (\text{no. of moles of SO}_2) - (\text{no. of moles of H}_2S$$

$$+ \text{ no. of moles of O}_2)$$

$$\Delta n = (1 \text{ mole of SO}_2) = (1 \text{ mole of H}_2S + 1\frac{1}{2} \text{ moles of O}_2)$$

$$= (1) - (1 + 1\frac{1}{2}) = -1.5 \text{ moles}$$

$$R = 1.99 \text{ cal/mole K}$$

$$t = 298^oK$$

$$\Delta E^o = \Delta H^o - \Delta nRT$$

$$= -134,000 \text{ cal} - (-1.5 \text{ moles} \times 1.99 \text{ cal/mole}^oK \times 298^oK)$$

$$= -134,000 \text{ cal} + 890 \text{ cal}$$

$$= -133,110 \text{ cal} = -133.1 \text{ kcal}$$

Drill 4: Enthalpy Calculations and the First Law of Thermodynamics

1. The equation for the burning of naphthalene is $C_{10}H_8(s) + 12O_2(g) \rightarrow 10CO_2 + 4H_2O(l)$. For every mole of $C_{10}H_8$ burned, $-1,226.7$ kcal is evolved at 25°C in a fixed volume combustion chamber. ΔH° for $H_2O(l) = -64.4$ kcal/mol and $H_2O(g) = -57.8$ kcal/mol. Calculate the heat of reaction in Kcal/mole at constant temperature.

    (A) $-1,459.2$ (D) $2,145.6$

    (B) $-1,227.9$ (E) $3,017.8$

    (C) $-1,268.3$

2. In the reaction, $CaCO_3(s) \rightarrow CaCO(s) + CO_2(g)$ at 950°C and CO_2 pressure of 1 atm, the ΔH is found to be 176 kJ/mol. Assuming that the volume of the solid phase changes little by comparison with the volume of gas generated, calculate the ΔE in kJ for this reaction. Hint: $\Delta E = \Delta H - P\Delta V$.

    (A) 166   (D) 609

    (B) 245   (E) 413

    (C) 102

3. You have the reaction $H_2O(l) \rightarrow H_2O(g)$. For both states, 1 mole of water is at 100°C and 1 atm pressure. The volume of 1 mole of water = 18 ml, $\Delta H = 9,710$ cal/mol and there are 24.2 cal/liter-atm. Calculate the value of ΔE in cal for this conversion.

    (A) 2,016.29   (D) 8,969.44

    (B) 4,162.61   (E) 5,912.63

    (C) 9,841.63

---

5. Heats of Fusion and Vaporization

The heat of vaporization of a substance is the number of calories required to convert 1 g of liquid to 1 g of vapor without a change in temperature.

The reverse process, changing 1 g of gas into a liquid without change in temperature, requires the removal of the same amount of heat energy (the heat of condensation).

The heat needed to vaporize 1 mole of a substance is called the molar heat of vaporization, or the molar enthalpy of vaporization, ΔH_{vap}, which is also represented as

$$\Delta H_{vaporization} = H_{vapor} - H_{liquid}$$

The magnitude of ΔH_{vap} provides a good measure of the strengths of the attractive forces operative in a liquid.

The number of calories needed to change 1 g of a solid substance (at the melting point) to 1 g of liquid (at the melting point) is called the heat of fusion.

The total amount of heat that must be removed in order to freeze 1 mole of a liquid is called its molar heat of crystallization. The molar heat of fusion, ΔH_{fus}, is equal in magnitude but opposite in sign to the molar heat of crystallization and is defined as the amount of heat that must be supplied to melt 1 mole of a solid:

$$\Delta H_{fus} = H_{liquid} - H_{solid}$$

PROBLEM

It is known that the heat of vaporization of water is 5 times as great as the heat of fusion. Explain this fact.

SOLUTION

The heat of vaporization is the quantity of heat necessary to vaporize 1^4 g of a liquid substance at its boiling point at constant temperature. The heat of fusion is the quantity of heat necessary to liquify 1^4 g of a solid substance at constant temperature at its melting point. With this in mind, let us consider what goes on in each of these phase changes. The volume-increase of the gas that is going from a liquid state to a vapor state is much greater than the volume-increase accompanying a solid to liquid transformation. When you increase volume, work is required to overcome the existing external pressure that hinders the volume expansion. Work is defined as the product of the pressure and the change in volume. Since the volume increase is larger in converting from liquid to vapor, more energy is necessary because more work is done.

More importantly, however, is that the molecules of a substance in the gaseous state are so much further apart than those in the liquid state. To bring about this separation of particles requires tremendous energy. This is the main reason that the heat of vaporization exceeds the heat of fusion.

PROBLEM

What weight of ice could be melted at 0°C by the heat liberated by condensing 100 g of steam at 100°C to liquid> Heat of vaporization = 540 cal/g, heat of fusion = 80 cal/g.

SOLUTION

The quantity of heat necessary to convert 1 g of a liquid into a vapor is termed the heat of vaporization. For water, 540 calories are necessary to change liquid water at 100°C into vapor at 100°C. In this problem, vapor is condensed, thus 540 cal of heat are *evolved* for each gram of liquid condensed.

no. of cal evolved in condensation = 540 cal/g × weight of vapor

Here 100 g of vapor is condensed. Thus,

no. of cal evolved = 540 cal/g × 100 g = 54,000 cal

When ice melts, heat is absorbed. About 80 calories of heat are required to melt 1 g of ice, the heat of fusion. Here, 54,000 cal are evolved in the condensation. Therefore, this is the amount of heat available to melt the ice. Because 80 cal are needed to melt 1 g of ice, one can find the number of grams of ice that can be melted by 54,000 cal, by dividing 54,000 cal by 80 cal/g.

$$\text{no. of grams of ice melted} = \frac{54,000 \text{ cal}}{80 \text{ cal/g}} = 675 \text{ g}$$

Drill 5: Heats of Fusion and Vaporization

1. The following are physical properties of methyl alcohol, CH_3OH: freezing point – 98°C; boiling point 65°C; specific heat of liquid 0.570 cal/g – degree; heat of fusion 22.0 cal/g; and heat of vaporization 263 cal/g. Calculate the number of kilocalories required to convert one mole of methyl alcohol solid at – 98°C to vapor at 65°C.

    (A) 12.09   (D) 30.16

    (B) 14.12   (E) 9.45

    (C) 17.95

2. 40 g of ice at 0°C is mixed with 100 g of water at 60°C. What is the final temperature after equilibrium has been established? Heat of fusion of H_2O = 80 cal/g, specific heat = 1 cal/g –°C.

    (A) 40°C   (D) 80°C

    (B) 20°C   (E) 100°C

    (C) 60°C

3. Determine the quantity of heat (in calories) required to convert 10 g of ice at 0°C to vapor at 100°C. For water, heat of fusion – 80 cal/g, heat of vaporization = 540 cal/g, and specific heat = 1 cal/g –°C.

    (A) 6,530   (D) 4,000

    (B) 8,000   (E) 7,200

    (C) 10,000

6. Activation Energy

The activation energy is the energy necessary to cause a reaction to occur. It is equal to the difference in energy between the transition state (or "activated complex") and the reactants:

$$\Delta E = \Sigma E_{products} = - \Sigma E_{reactants}$$

In an exothermic process, energy is released and ΔE of reaction is negative; in an endothermic process, energy is absorbed and ΔE is positive.

For a reversible reaction, the energy liberated in the exothermic reaction equals the energy absorbed in the endothermic reaction. (The energy of the reaction, ΔE, is equal also to the difference between the activation energies of the opposing reactions, $\Delta E = Ea - Ea'$.)

A catalyst affects a chemical reaction by lowering the activation energy for both the forward and the reverse reactions, equally.

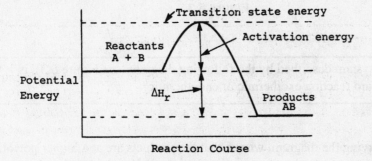

$$\Delta E = \Sigma E \text{ products} - \Sigma E \text{ reactants}$$

Figure 6.1

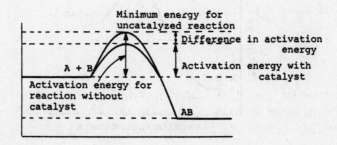

Figure 6.2

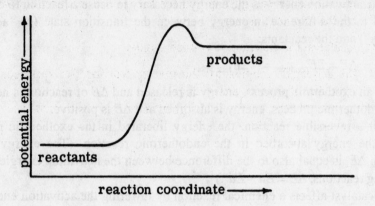

Figure 6.3

PROBLEM

For the system described by the following diagram (see Figure 6.3), is the forward reaction exothermic or endothermic?

SOLUTION

By observing the diagram, we see that the products are at a higher potential energy than the reactants, thus energy must have been absorbed during the course of the reaction. The only mechanism by which this can occur is absorption of heat and subsequent conversion of heat into potential energy. Hence, the reaction absorbs heat and is, therefore, endothermic.

Drill 6: Activation Energy

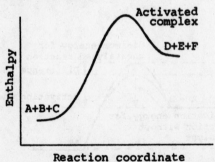

1. The enthalpies associated with the reaction $A + B + C \rightarrow D + E + F$ is shown above. The fact that the enthalpy of $D + E + F$ is higher than that of $A + B + C$ tells us that

    (A) the reaction is exothermic.

(B) the reaction is endothermic.

(C) the activation energy required for the reverse reaction is higher than for the forward reaction.

(D) a catalyst for the reaction is unnecessary.

(E) the activated complex for the reverse reaction is a different species from that of the forward reaction.

THERMODYNAMICS I
DRILLS

ANSWER KEY

Drill 1 — Bond Energies
1. (D)
2. (A)
3. (D)

Drill 2 — Heat Capacity
1. (B)
2. (A)
3. (B)
4. (E)
5. (A)

Drill 3 — Enthalpy
1. (B)
2. (B)
3. (C)
4. (E)
5. (A)

Drill 4 — Enthalpy Calculations and the First Law of Thermodynamics
1. (B)
2. (A)
3. (D)

Drill 5 — Heats of Fusion and Vaporization
1. (A)
2. (B)
3. (E)

Drill 6 — Activation Energy
1. (B)

GLOSSARY:
THERMODYNAMICS I

Activation Energy

The energy necessary to cause a reaction to occur.

Adiabatic Process

A process that occurs without the entrance or exit of heat.

Bond Energy

Amount of energy per mole required to break a bond between two atoms.

Calorie

A unit of energy.

Endothermic

A reaction that proceeds by utilizing heat is termed endothermic.

Enthalpy

The heat content of a substance.

Exothermic

A reaction that proceeds by evolving heat is termed exothermic.

First Law of Thermodynamics

States that the change in internal energy is equal to the difference between the energy supplied to the system as heat and the energy removed from the system as work performed on the surroundings.

Heat Capacity

The amount of heat energy required to raise the temperature of a substance 1°C.

Heat of Formation

The amount of energy required to form one mole of gaseous atoms from its element in its common physical state at 25°C and one atmosphere pressure.

Heat of Fusion

The number of calories needed to change 1 g of a solid substance (at the melting point) to 1 g of liquid (at the melting point).

Heat of Vaporization

The number of calories required to convert 1 g of liquid to 1 g of vapor without a change in temperature.

Hess's Law

States that when a reaction can be expressed as the algebraic sum of two or more reactions, the heat of the reaction is the algebraic sum of the heats of the constituent reactions.

Isobaric Process

A process that occurs when the system is maintained at constant pressure. Also termed isopiestic.

Isopiestic

See isobaric process.

Isothermal Process

A process that occurs when the system is maintained at the same temperature throughout an experiment.

Joule

A unit of energy.

Molar Heat Capacity

The amount of heat necessary to raise the temperature of one mole of a substance 1°C.

Specific Heat

The amount of heat energy required to raise the temperature of a given quantity of a substance 1°C.

Standard State

25°C and 1 atm pressure.

State Function

A parameter of a system that depends only on the present state of the substance and not on the path by which the present state was attained.

Surroundings

Everything that is not included in the system that is being studied.

System

The particular portion of the universe on which we wish to focus our attention.

CHAPTER 12

Thermodynamics II

➤ Diagnostic Test
➤ Thermodynamics II
Review & Drills
➤ Glossary

THERMODYNAMICS II
DIAGNOSTIC TEST

1. Ⓐ Ⓑ Ⓒ Ⓓ Ⓔ   21. Ⓐ Ⓑ Ⓒ Ⓓ Ⓔ
2. Ⓐ Ⓑ Ⓒ Ⓓ Ⓔ   22. Ⓐ Ⓑ Ⓒ Ⓓ Ⓔ
3. Ⓐ Ⓑ Ⓒ Ⓓ Ⓔ   23. Ⓐ Ⓑ Ⓒ Ⓓ Ⓔ
4. Ⓐ Ⓑ Ⓒ Ⓓ Ⓔ   24. Ⓐ Ⓑ Ⓒ Ⓓ Ⓔ
5. Ⓐ Ⓑ Ⓒ Ⓓ Ⓔ   25. Ⓐ Ⓑ Ⓒ Ⓓ Ⓔ
6. Ⓐ Ⓑ Ⓒ Ⓓ Ⓔ   26. Ⓐ Ⓑ Ⓒ Ⓓ Ⓔ
7. Ⓐ Ⓑ Ⓒ Ⓓ Ⓔ   27. Ⓐ Ⓑ Ⓒ Ⓓ Ⓔ
8. Ⓐ Ⓑ Ⓒ Ⓓ Ⓔ   28. Ⓐ Ⓑ Ⓒ Ⓓ Ⓔ
9. Ⓐ Ⓑ Ⓒ Ⓓ Ⓔ   29. Ⓐ Ⓑ Ⓒ Ⓓ Ⓔ
10. Ⓐ Ⓑ Ⓒ Ⓓ Ⓔ   30. Ⓐ Ⓑ Ⓒ Ⓓ Ⓔ
11. Ⓐ Ⓑ Ⓒ Ⓓ Ⓔ   31. Ⓐ Ⓑ Ⓒ Ⓓ Ⓔ
12. Ⓐ Ⓑ Ⓒ Ⓓ Ⓔ   32. Ⓐ Ⓑ Ⓒ Ⓓ Ⓔ
13. Ⓐ Ⓑ Ⓒ Ⓓ Ⓔ   33. Ⓐ Ⓑ Ⓒ Ⓓ Ⓔ
14. Ⓐ Ⓑ Ⓒ Ⓓ Ⓔ   34. Ⓐ Ⓑ Ⓒ Ⓓ Ⓔ
15. Ⓐ Ⓑ Ⓒ Ⓓ Ⓔ   35. Ⓐ Ⓑ Ⓒ Ⓓ Ⓔ
16. Ⓐ Ⓑ Ⓒ Ⓓ Ⓔ   36. Ⓐ Ⓑ Ⓒ Ⓓ Ⓔ
17. Ⓐ Ⓑ Ⓒ Ⓓ Ⓔ   37. Ⓐ Ⓑ Ⓒ Ⓓ Ⓔ
18. Ⓐ Ⓑ Ⓒ Ⓓ Ⓔ   38. Ⓐ Ⓑ Ⓒ Ⓓ Ⓔ
19. Ⓐ Ⓑ Ⓒ Ⓓ Ⓔ   39. Ⓐ Ⓑ Ⓒ Ⓓ Ⓔ
20. Ⓐ Ⓑ Ⓒ Ⓓ Ⓔ   40. Ⓐ Ⓑ Ⓒ Ⓓ Ⓔ

THERMODYNAMICS II
DIAGNOSTIC TEST

This diagnostic test is designed to help you determine your strengths and your weaknesses in advanced thermodynamics. Follow the directions and check your answers.

Study this chapter for the following tests:
AP Chemistry, CLEP General Chemistry,
GED, MCAT, MSAT, PRAXIS II Subject Assessment:
Chemistry, SAT II: Chemistry, GRE Chemistry

---

40 Questions

DIRECTIONS: Choose the correct answer for each of the following problems. Fill in each answer on the answer sheet.

1. Consider the two reactions below:

    (a) $H_2O(s) \rightarrow H_2O(l)$; $\Delta H = 6.0$ kJ

    (b) $CaCO_3(s) \rightarrow CaO(s) + CO_2(g)$; $\Delta H = 178$ kJ

    Which of the following statements concerning these reactions is(are) correct?

    I.   The reactions are spontaneous and exothermic.

    II.   Both reactions occur with an increase in the system's disorder.

    III.   The entropy change (ΔS^o) for equation (a) is most likely positive.

    (A) (I) only

    (B) (I) and (II) only

    (C) (II) and (III) only

    (D) None of the statements

    (E) All of the statements

2. For the reaction

    $N_2(g) + O_2(g) \rightarrow 2NO(g)$

    ΔG^o for the reaction is + 174 kJ. This reaction would:

    (A) be at equilibrium.

    (B) proceed spontaneously.

    (C) have a negative ΔH^o.

    (D) not proceed spontaneously.

    (E) be exothermic.

3. A reaction is said to be at equilibrium when

   (A) $\Delta H = 0$. (D) $\Delta H < 0$.

   (B) $\Delta G = 0$. (E) $\Delta H > 0$.

   (C) $\Delta S = 0$.

4. Determine ΔG° (free energy at standard conditions) for the reaction below, considering $k_P = 8$ at 25°C.

   $$2A(g) \leftrightarrow B(g)$$

   (A) +50.82 kJ (D) −4,830 J

   (B) 5,082J (E) −5.153 kJ

   (C) −1,240 J

5. A perfect crystal at °K has

   (A) zero enthalpy.

   (B) zero entropy.

   (C) maximum entropy.

   (D) reached a state of equal entropy and enthalpy; pressure however must be constant.

   (E) reached a state where the entropy is positive, fixed, and unchangeable.

6. Which of the following statements about chemical equilibrium is accurate?

   I. Equilibrium is reached when ΔG (free energy change) equals zero.

   II. ΔG° (standard free energy charge) = $-RT \ln k_{eq}$ (where T = temperature, R = gas constant, k_{eq} = equilibrium constant).

   III. ΔH° is independent of pressure.

   (A) I only (D) II only

   (B) II and III only (E) All of the statements

   (C) I and II only

7. When ΔG for an isolated system is equal to zero, the system is said to be

   (A) ideal. (D) in a phase transition.

   (B) in equilibrium. (E) inert.

   (C) adiabatic.

8. The free energy of a gas depends on its partial pressure according to the expression $G = G^o + RT \ln (P/p^o)$. Which of the following plots below shows agreement with this relationship?

(A)

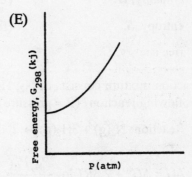

(D)

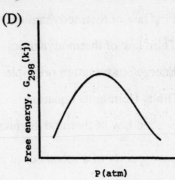

(B)

(E)

(C)

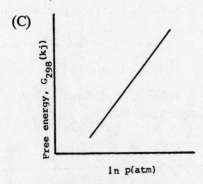

9. A student calculated that for a given reaction $\Delta G = 22$ J, $\Delta S = 22.0$ J/K, and $\Delta H = 6,028$ J. She neglected to report the temperature for which these values were valid. What was it?

(A) 173°C

(D) 273°K

(B) 1.0°C

(E) 137°K

(C) 275°K

10. A perpetual motion machine capable of generating increasing amounts of energy without interacting with its surroundings cannot exist. This is best explained by the

    (A) First law of thermodynamics.

    (B) Third law of thermodynamics.

    (C) Energy conservation principle.

    (D) Gibbs-Helmholtz equation.

    (E) Second law of thermodynamics.

11. The heat capacity of a substance at constant volume is directly related to the

    (A) enthalpy, H.   (D) internal energy, U or E.

    (B) entropy, S.   (E) Helmholz free energy, A.

    (C) free energy, G.

12. A reaction mixture consists of N_2, H_2, and NH_3. At 298°K, what is ΔG for the following reaction? (P = pressure)

    Reaction: $N_2(g) + 3H_2(g) \rightarrow 2NH_3(g)$

    (A) $\Delta G = \Delta G° + 2.3RT \log \dfrac{P_{N_2} P_{H_2}^3}{P_{NH_3}^2}$

    (B) $\Delta G = 2.3RT \log \dfrac{P_{NH_3}^2}{P_{N_2} P_{H_2}^3}$

    (C) $\Delta G = \Delta G° + 2.3RT \log \dfrac{P_{N_2} P_{H_2}^3}{P_{NH_3}^2}$

    (D) $\Delta G = \dfrac{\Delta G° + 2.3RT}{\dfrac{P_{NH_3}^2}{P_{H_2}^3 P_{N_2}}}$

    (E) $\Delta G = \Delta G° + 2.3RT \left(\dfrac{P_{NH_3}^2}{P_{N_2} P_{H_2}^2} \right)$

13. The standard free energy, ΔG°, expressed in kJ for the reaction below at 25°C is

    $Cd + Pb^{2+} \rightarrow Cd^{2+} + Pb$ $\qquad E^\circ = +.28V$

    (A) −27.

    (D) −108.

    (B) −54.

    (E) 0.

    (C) +27.

14. An ice cube is placed in an open glass of water at room temperature. Describe the resultant effect on its energy content and entropy.

	Energy	Entropy
(A)	decrease	decrease
(B)	increase	increase
(C)	increase	decrease
(D)	increase	remains constant
(E)	remains constant	increase

15. Which of the following is not a state function of a system?

    (A) H, the enthalpy

    (D) P, the pressure

    (B) S, the entropy

    (E) A, the Helmholtz energy.

    (C) w, the work

16. How would you express the absolute temperature at which a reaction occurs in terms of the free energy change of the reaction ΔG, the enthalpy change ΔH, and the entropy change ΔS?

    (A) $\dfrac{\Delta G - \Delta H}{\Delta S}$

    (D) $\Delta H - \Delta G + \Delta S$

    (B) $\dfrac{\Delta H - \Delta S}{\Delta G}$

    (E) None of the above

    (C) $\dfrac{\Delta H - \Delta G}{\Delta S}$

17. Which of the following reactions would be expected to involve the largest increase in entropy?

    (A) $H_2O(l) \rightarrow H_2O(g)$ (D) $H_2(s) \rightarrow H_2(l)$

    (B) $H_2O(s) \rightarrow H_2O(l)$ (E) $He(l) \rightarrow He(g)$

    (C) $H_2(l) \rightarrow H_2(s)$

18. For the following reaction the entropies of $Al(s)$, $Cl_2(g)$, and $AlCl_3(s)$ are 28.3 J/kmol, 222.96 J/kmol, and 110.7 J/kmol, respectively. Calculate the standard entropy change for the reactive system.

    $$2Al(s) + 3Cl_2(g) \rightarrow 2AlCl_3 (s)$$

    (A) -221.4 J/kmol   (D) -504.08 J/kmol

    (B) 725.48 J/kmol   (E) -56.6 J/kmol

    (C) -668.88 J/kmol

19. A reaction is more thermodynamically favored or spontaneous if

    (A) ΔH is positive.

    (B) ΔS is positive.

    (C) ΔH is negative and ΔS is positive.

    (D) ΔH is positive and ΔS is negative.

    (E) $\Delta H = 0$ and $\Delta S = 0$.

20. Consider the following reaction at 25°C:

    $$N_2(g) + O_2(g) \rightarrow 2NO(g)$$

    What is the value of $\Delta S°$ if $\Delta G° = + 88.4$ Kj and $\Delta H° = 92.0$ Kj?

    (A) 0   (D) .144

    (B) .012   (E) 180.4

    (C) .065

21. The equation $\Delta G = \Delta H - T\Delta S$ tells us that an exothermic reaction will be associated with which of the following?

    I.   Negative ΔH

    II.   Positive ΔH

    III. More disordered positive ΔS

    IV.   A spontaneous reaction

    V.   More ordered negative ΔS

(A) I and V

(D) I, III, and IV

(B) II and V

(E) None of the above

(C) II and III

22. At 25°C the entropy of diamond is 2.44 JK^{-1} mol^{-1} and the entropy of diamond-structure tin is 44.8 JK^{-1} mol^{-1}. As the temperature is reduced to 0°K, the difference in entropy between these substances will

(A) increase to infinity.

(B) increase to a finite number.

(C) remain constant.

(D) decrease to a non-zero number.

(E) decrease to zero.

23. The reaction $2A(g) + 2B(g) \rightarrow 2C(l)$ is a spontaneous and exothermic reaction. What are the signs of $\Delta G, \Delta H, \Delta S,$ and ΔE?

	ΔG	ΔH	ΔS	ΔE
(A)	+	+	+	−
(B)	+	+	−	+
(C)	−	−	+	−
(D)	−	−	−	+
(E)	−	−	−	−

24.

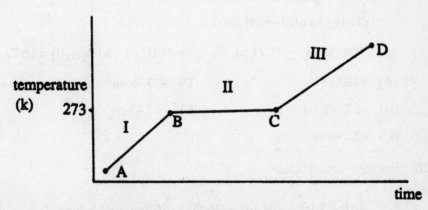

The above sketch represents a warming curve for water at 1 atm pressure. Select the one *incorrect* statement. (Note: C_I = heat capacity of water in region I, ΔT = change in temperature between points specified, ΔH_f = heat of fusion.)

(A) Water is a solid in region I of the curve.

(B) Water exists in a solid/liquid equilibrium in region II.

(C) The energy absorbed by a sample of mass N in the temperature range A to C is $(C_I)(\Delta T_{A'B})(N) + (\Delta H_f)(\Delta T_{B'C})(N)$.

(D) The energy absorbed by a sample in the range B to C is $\Delta H_f N$.

(E) In region II water doesn't gain any kinetic energy but does gain in entropy.

25. The Law of entropy states that

(A) energy is neither created nor destroyed, but changed from one form to another.

(B) gas pressures are determined independently in a mixture.

(C) heat flows to a more concentrated medium.

(D) matter is neither created nor destroyed.

(E) systems tend toward increasing disorder.

26. Based on the Third law of thermodynamics, we know that all perfect crystals at absolute zero have:

(A) the same enthalpy.   (D) the same crystal lattices.

(B) differing ΔA values.   (E) Both (A) and (C).

(C) the same entropy.

27. Calculate $\Delta S°_{298}$ for the reaction below:

$$2H_2(g) + O_2(g) \rightarrow 2H_2O(1)$$

given that $S°_{H_2}(g) = 31.21$ eu, $S°_{O_2}(g) = 49.00$ eu, and $S°_{H_2O}(l) = 16.72$ eu.

(A) +122.86 eu   (D) −77.98 eu

(B) −122.86 eu   (E) −63.49 eu

(C) −33.44 eu

28. For the reaction below:

$$\frac{1}{2}Hg_2Cl_2(s) + \frac{1}{2}H_2(g) \rightarrow Hg(1) + HCl(aq) \text{ (not balanced).}$$

$\Delta S_m^\theta = -31.0$ J/kmol

$\Delta G_m^\theta = -25.82$ kJ/mol

$$T = 298°K$$

$$\Delta E_m^\theta = -27.61 \text{ kJ/mol}$$

Using the appropriate information, calculate the enthalpy of the reaction.

(A) −38.69 kJ/mol   (D) −21.76 kJ/mol

(B) −35.06 kJ/mol   (E) 0 kJ/mol

(C) −16.58 kJ/mol

29. Which of the following statements is false?

    (A) The energy of the universe is constant.

    (B) $\Delta E = w$ for adiabatic processes.

    (C) The entropy of the universe tends toward a maximum.

    (D) Entropy is a measure of increasing disorder.

    (E) The entropy of a perfect crystal of a pure substance at 0°C is zero.

30. At 298°K, $\Delta G° = 90.3$ kJ for the reaction

    $$HgO(s) \leftrightarrow Hg(g) + \frac{1}{2}O_2(g).$$

    Calculate Kp at 298°K

    (A) 1.48×10^{-16} atm $^{3/2}$ (D) -5.42×10^{-12} atm $^{3/2}$

    (B) 2.51×10^{-14} atm $^{3/2}$ (E) 6.31×10^5 atm $^{3/2}$

    (C) -2.51×10^{-14} atm $^{3/2}$

31. An increase in pressure will change the equilibrium constant by

    (A) shifting to the side where a smaller volume results.

    (B) shifting to the side where a larger volume results.

    (C) favoring the exothermic reaction.

    (D) favoring the endothermic reaction.

    (E) None of the above.

32. $\Delta G°$ for the reaction below is 15 kJ/mol. At standard conditions, the reaction

    $$A \xrightarrow{\text{STP}} B$$

(A) will be endothermic.   (D) will proceed quickly.

(B) will be exothermic.   (E) will not go forward.

(C) will proceed slowly.

33. A certain reaction has an equilibrium constant of 10 at 300°K and 100 at 400°K. The ratio of ΔG at 300°K to ΔG at 400°K is

(A) $\dfrac{1}{10}$ (D) $\dfrac{3}{4}$

(B) $\dfrac{3}{8}$ (E) 2

(C) $\dfrac{1}{2}$

34. The expression for w, in the First law of thermodynamics, if negative, implies all of the following, except

(A) work has been done by the system.

(B) the total internal energy has decreased.

(C) a negative amount of work has been done on the system.

(D) the system has lost heat.

(E) work has been done on the outside world.

35. In an adiabatic system, if work is done, the temperature must

(A) increase.   (D) increase then decrease.

(B) decrease.   (E) decrease then increase.

(C) remain the same.

36. All of the following statements about ΔS are true except

(A) it is a measure of the energy dispersal.

(B) the natural tendency is for it to increase.

(C) it is not a state function under all conditions.

(D) it can be defined both thermodynamically and statistically.

(E) its calculation is only possible for isothermal processes.

37. $\Delta H^{\circ}_{298} = -46.19$ kJ $\times$ mol^{-1} for the reaction below. Which statement is true about the equilibrium constant k_{eq} for this reaction?

    $$N_2(g) + 3H_2(g) \leftrightarrow 2NH_3(g)$$

    (A) K_{eq} increases with increasing temperature.

    (B) K_{eq} decreases with increasing temperature.

    (C) K_{eq} increases with increasing pressure.

    (D) K_{eq} decreases with increasing pressure.

    (E) K_{eq} is independent of temperature and pressure.

38. ΔG for a nonspontaneous reaction is

    (A) negative.   (D) small.

    (B) zero.   (E) large.

    (C) positive.

39. The change in ΔS when ice is melted to water is

    (A) zero.

    (B) positive.

    (C) negative.

    (D) Cannot tell from the information given.

    (E) always 0.

40. ΔG_f° for $NH_3(g)$ is -16.6 kJ/mol. If we assume that this reaction occurs with ideal gases at standard conditions ($T = 298°$K), what is the equilibrium constant K_{eq}?

    $$\frac{1}{2}N_2\ (g) + \frac{3}{2}H_2\ (g) \leftrightarrow NH_3(g)$$

    (A) $K_{eq} = \exp \dfrac{16.6}{R \times 298}$

    (B) $K_{eq} = \exp -\dfrac{16.6}{R \times 298}$

    (C) $K_{eq} = \exp \dfrac{R \times 298}{-16.6}$

    (D) $K_{eq} = \ln -\dfrac{16.6}{R \times 298}$

    (E) Cannot be determined with given information.

THERMODYNAMICS II
DIAGNOSTIC TEST

ANSWER KEY

1. (C)	11. (D)	21. (D)	31. (E)
2. (D)	12. (C)	22. (E)	32. (E)
3. (B)	13. (B)	23. (D)	33. (B)
4. (E)	14. (B)	24. (C)	34. (D)
5. (B)	15. (C)	25. (E)	35. (B)
6. (E)	16. (C)	26. (C)	36. (C)
7. (B)	17. (A)	27. (D)	37. (B)
8. (D)	18. (D)	28. (B)	38. (C)
9. (D)	19. (C)	29. (E)	39. (B)
10. (A)	20. (B)	30. (A)	40. (A)

DETAILED EXPLANATIONS
OF ANSWERS

1.   **(C)**   A positive ΔH indicates an endothermic reaction. Hence, both reactions are endothermic. The criteria for spontaneity depends on ΔH^o values, negative ΔH values indicate spontaneity. Water molecules that make up an ice crystal are held in fixed positions in the crystal, when ice melts, the molecules move through the liquid freely (more random distribution of molecules). Thus, the amount of order is higher in the solid than in the liquid. By similar reasoning, we conclude that for reaction (B) also, there's an increase in disorderliness. The amount of the disorder is called the system's entropy; and the greater the randomness (or disorder) the greater the entropy. Now, we define ΔS^o as

$$S^o_{H_2O}(l) - S^o_{H_2O}(s),$$

we would expect ΔS^o to be positive here, since

$$S^o_{H_2O}(l) > S^o_{H_2O}(s).$$

This same reasoning may be applied to reaction (B).

2.   **(D)** ΔG^o is the Gibbs free energy and may be calculated from the values of ΔH^o (enthalpy) and ΔS^o (entropy) by the relationship.

$$\Delta G^o = \Delta H^o - T\Delta S^o$$

The value of ΔG^o will predict if the reaction is spontaneous, non-spontaneous, or at equilibrium.

If ΔG^o is positive the reaction is non-spontaneous

$\Delta G^o = 0$ the system is at equilibrium

ΔG^o is negative the reaction is spontaneous

Since the values of ΔG^o for the reaction given is $+174$ kJ the reaction will not proceed as written (non-spontaneous) but the reverse reaction

$$2NO(g) \rightarrow N_2(g) + O_2(g)$$

$$\Delta G^o = -174 \text{ kJ}$$

would be spontaneous.

3.   **(B)**   A reaction is said to be at equilibrium when the Gibbs free energy change, ΔG, is equal to zero. A reaction is spontaneous when $\Delta G < 0$ and not spontaneous when $\Delta G > 0$. An enthalpy change, ΔH, of zero indicates that heat is neither released nor absorbed by the system. $\Delta H < 0$ indicates that the reaction

evolves heat while $\Delta H > 0$ indicates absorption of heat. The change in entropy, ΔS, is zero when the randomness of the system remains unchanged by the reaction, while $\Delta S > 0$ indicates an increase in randomness and $\Delta S < 0$ a decrease.

4.   **(E)** ΔG^o is given by the expression below:

$$\Delta G^o = -2.303RT \log k_p$$

$$\Delta G^o = -2.303 \times (8.314) \times 298 \times \log 8$$

$$= -2.303 \times 8.314 \times 298 \times 0.9031$$

$$= -5,152.9 \text{ J} = -5.153 \text{ kJ}$$

5.   **(B)**   This is actually the Third law of thermodynamics.

6.   **(E)**   All three statements are accurate, note that ΔG, not ΔG^o, is dependent on pressure.

7.   **(B)**   Since $\Delta G = 0$, there is no free energy present in the system, it is not a spontaneous reaction, and therefore the system is in equilibrium.

8.   **(D)**   Note G vs. ln p should be in the form of a straight line. Since $G = G^o + RT\ln(P/p^o)$ is a linear function of the form $y = mx + b$. Note that (C) is incorrect. Since the p axis is expressed in the form of natural logarithms.

9.   **(D)**   We can rearrange $\Delta G = \Delta H - T\Delta S$ to:

$$T = \frac{\Delta H - \Delta G}{\Delta S}$$

Substituting given values:

$$T = \frac{608 - 22}{22} = 273°\text{K}$$

10.   **(A)**   According to the First law of thermodynamics the energy of an isolated system is constant. Only by work or heat passing through the walls of such a system can its energy be changed. Thus, the energy of a closed system cannot increase without interacting with its surroundings and, therefore, a perpetual motion machine as described cannot exist.

11.   **(D)**   The heat capacity at constant volume, C_v, is directly related to the internal energy, U, as

$$C_v = \left(\frac{\partial U}{\partial T}\right)_v \quad \text{or} \quad \Delta U = \int C_v dT$$

12. **(C)**   The formula for ΔG is:

$$\Delta G = \Delta G^o + 2.3RT \log Q,$$

but Q for $N_2(g) + 3H_2(g) \rightarrow 2NH_3(g)$

is $\dfrac{P^2_{NH_3}}{P_{N_2} P^3_{H_2}}$

Hence, $\Delta G = \Delta G^o + 2.3RT \log \dfrac{P^2_{NH_3}}{P_{N_2} P^2_{H_2}}$

13.   **(B)**   The standard free energy for an electrochemical cell is related to the standard electrode potential E^o by:

$$\Delta G^o = -nFE^o$$

where n is the number of electrons involved in the oxidation-reduction reaction.

E^o is the standard electrode potential.

F is Faraday's constant, 96.487 kJ/V.

The value of ΔG^o is then equal to

$$\Delta G^o = -2 \times 96.487 \text{ kJ/V} \times .28\text{V}$$

$$\Delta G^o = -54 \text{ kJ}$$

14.   **(B)**   The ice cube will melt by gaining heat from the water. The temperature of the water drops below room temperature; therefore, heat flows from the surroundings into the water until room temperature is attained. The resultant effect is an increase in energy for the ice and water system. When the ice melts, it changes from an ordered to a disordered system. Entropy is a measure of disorder (randomness). The higher the disorder, the higher the entropy. In this case, energy and entropy both increase.

15.   **(C)**   The work obtained from a system is not solely a function of the state of the system. It is also dependent on the "path" travelled by the system. For example, more useful work is done in the reversible expansion from V_1 to V_2 than in the irreversible expansion. For a system expanding against a vacuum, no work is done at all.

16.   **(C)**   Using the fact that the change in Gibbs free energy is related to ΔH and ΔS by the equation

$$\Delta G = \Delta H - T\Delta S$$

$$\therefore \qquad T = \frac{\Delta H - \Delta G}{\Delta S}$$

17. **(A)** The largest increase in entropy corresponds with the largest loss of order. As expected, this occurs when water makes the transition from a strongly hydrogen-bonded liquid to a free gas.

18. **(D)**

$$2Al(s) + 3Cl_2(g)\ 2AlCl_3(s)$$

$$\Delta S^o = 2(S^o_{[AlCl_3\,(s)]}) - 2(S^o_{[Al\,(s)]}) - 3(S^o_{[Cl_2\,(s)]})$$

$$= 2 \times 110.7 - 2 \times 28.3 - 3 \times 222.96$$

$$= 221.4 - 56.6 - 668.88 = -504.08$$

19. **(C)** A process is thermodynamically favored if ΔG is negative. Since $\Delta G = \Delta H - T\Delta S$, then the more negative ΔH (enthalpy) and the more positive ΔS (entropy) the more favorable the process.

20. **(B)** ΔS^o is the entropy change the reaction undergoes when the products and reactants are in their standard state. Entropy is a measure of the randomness or the amount of disorder in the system. A positive value of ΔS indicates an increase in the randomness of the system whereas a negative entropy indicates the system is becoming more ordered. To solve this problem we recognize that the entropy ΔS^o is related to the ΔH^o and ΔG^o of the reaction by the expression

$$\Delta G^o = H^o - T\Delta S^o$$

Substituting the values for ΔH^o and ΔG^o and solving for ΔS^o:

$$-\Delta S^o = \frac{\Delta G^o - \Delta H^o}{T}$$

$$-\Delta S^o = \frac{88.4\,kJ - 92.0\,kJ}{298°\,K}$$

$$-\Delta S^o = \frac{-3.6\,kJ}{298°\,K}$$

$$\Delta S^o = .012\,kJ/°K$$

21. **(D)** An exothermic reaction is a spontaneous change in which the Gibbs free energy is negative. A negative ΔG can only be possible if ΔH is negative and ΔS is positive.

22. **(E)** The third law states that all perfect crystals have the same entropy at 0°K.

23. **(D)** A spontaneous, exothermic reaction must have, by definition, a negative ΔH and negative ΔG. If ΔG is negative then, also by definition, ΔE is positive. Since 3 mol of gas forms 2 mol of liquid, the entropy has decreased, thus, ΔS is negative.

24. **(C)** The second part of the equation is incorrect. The energy gained in region II equals $\Delta H_f N$. There is no temperature change during the melting process. All the energy gained is involved in the change from solid to liquid structure—an entropy gain.

25. **(E)** This is a strict statement of the law, predicting that the universe is gradually approaching randomness. Chemical reactions are a part of this.

26. **(C)** The phrase that is key to answering this question correctly is "based on the Third law of thermodynamics." The third law states: All perfect crystals have the same entropy at absolute zero.

27. **(D)**

$$\Delta S^{\circ} = 2S^{\circ}_{H_2O(l)} - (2S^{\circ}_{H_2(g)} + 2S^{\circ}_{O_2(g)})$$

$$= 2(16.72) - [2(31.21) + 49.00]$$

$$= 33.44 - 111.42 = -77.98 \text{ eu}$$

28. **(B)** In order to solve this problem a non-calorimetric means of determining the enthalpy of a reaction is required.

$$\Delta G^{\theta}_m = \Delta H^{\theta}_m - T\Delta S^{\theta}_m$$

is the equation that meets these requirements. Solving for ΔH^{θ}_m, we obtain:

$$\Delta H^{\theta}_m = \Delta G^{\theta}_m + T\Delta S^{\theta}_m$$

After changing the ΔS^{θ}_m from J/kmol to kJ/kmol, we can simply substitute and solve to find the enthalpy.

$$\Delta H^{\theta}_m = -25.82 \text{ kJ/mol} + (298°K)(-0.032 \text{ kJ/kmol})$$

$$\Delta H^{\theta}_m = -35.06 \text{ kJ/mol}$$

29. **(E)** Choice (E) would correctly state the Third law of thermodynamics if the temperature was listed as 0°K.

30. **(A)** Use $\Delta G = -RT \ln k$ where $\Delta G = 90{,}300$ J mol^{-1}, $T = 298°$K, and $R = 8.314$ J mol^{-1} K^{-1}.

31. **(E)** The equilibrium constant is independent of pressure and volume but dependent on temperature.

32. **(E)** "Endothermic" and "exothermic" are terms which describe the direction of ΔH. Positive ΔH corresponds to an endothermic reaction and negative ΔH indicates an exothermic reaction.

The rate of a reaction is not related to ΔG values. ΔG values can tell us whether a reaction at constant temperature and pressure will tend to go forward or not. Negative ΔG values indicate reactions that are favored in the forward direction.

33. **(B)** k_{eq} and ΔG are related:

$$\Delta G = -RT \ln k_{eq}$$

$$\Delta G = -2.303RT \log k_{eq}$$

We are looking for the ratio:

$$\frac{\Delta G_{300}}{\Delta G_{400}} = \frac{-2.303R(300)\log 10}{-2.303R(400)\log 100} = \frac{(300)(1)}{(400)(2)} = \frac{3}{8}$$

34. **(D)** The First law of thermodynamics is

$$U = Q + W.$$

If W is negative, work has been done by the system on the outside world and this has caused a decrease in the internal energy of the system. Nothing can be said about the heat of the system unless more information is given about U and Q.

35. **(B)** An adiabatic system is one which has zero heat flow (in or out). If work is done, the temperature must decrease since internal energy is transformed into work.

36. **(C)** A state function is one that is a property of the present state of the system; its value is completely independent of the way in which the state was attained. The entropy (ΔS) is a state function.

37. **(B)** Since ΔH is negative the reaction is exothermic. We might consider the reaction in this way:

$$N_2(g) + 3H_2(g) \leftrightarrow 2NH_3(g) + \text{heat}$$

Increasing the temperature drives the reaction to the left, and thus k_{eq} decreases with increasing T. k_{eq} is independent of pressure; it varies only with temperature.

38. **(C)** The change in free energy, ΔG, for a non-spontaneous reaction is positive. A spontaneous reaction has $\Delta G < 0$ and a system in equilibrium is characterized by $\Delta G = 0$.

39. **(B)** The increasing disorder of water when it is converted from the solid to the liquid state indicates $\Delta S > 0$.

40. **(A)** ΔG° for the reaction is:

$$\Delta G^\circ = \Delta G^\circ_{f\,NH_3} - \frac{1}{2}\Delta G^\circ_{f\,N_2} - \frac{3}{2}\Delta G^\circ_{f\,H_2}$$

$$\Delta G^\circ = -16.6 - 0 - 0$$

k_{eq} is related to ΔG° by:

$$-\Delta G^\circ = RT \ln k_{eq}$$

therefore,

$$k_{eq} = \exp(-\Delta G^\circ/RT)$$

$$k_{eq} = \exp(16.6/R \times 298)$$

THERMODYNAMICS II
REVIEW

1. Entropy

The degree of randomness of a system is represented by a thermodynamic quantity called the entropy, S. The greater the randomness, the greater the entropy.

A change in entropy or disorder associated with a given system is

$$\Delta S = S_2 - S_1$$

The entropy of the universe increases for any spontaneous process:

$$\Delta S_{universe} = (\Delta S_{system} + \Delta S_{surroundings}) \geq 0$$

When a process occurs reversibly at constant temperature, the change in entropy, ΔS, is equal to the heat absorbed divided by the absolute temperature at which the change occurs:

$$\Delta S = \frac{q_{reversible}}{T}$$

The Second Law of Thermodynamics

The Second law of thermodynamics states that in any spontaneous process there is an increase in the entropy of the universe ($\Delta S_{total} > 0$).

$$\Delta S_{universe} = \Delta S_{total} = \Delta S_{system} + \Delta S_{surroundings}$$

$$\Delta S_{surroundings} = \frac{-\Delta H_{system}}{T} \text{ at constant } P \text{ and } T$$

$$T\Delta S_{total} = - (\Delta H_{system} - T\Delta S_{system})$$

The entropy of a substance, compared to its entropy in a perfect crystalline form at absolute zero, is called its absolute entropy, S^{o}.

The Third law of thermodynamics states that the entropy of any pure, perfect crystal at absolute zero is equal to zero.

PROBLEM

For the following reaction at 25°C

$$CuO(s) + H_2(g) \rightarrow Cu(s) + H_2O(g)$$

Values of S^o, the absolute entropies for the substances, are:

$CuO(s) = 10.4$ cal/mol

$H_2(g) = 31.2$ cal/mol

$Cu(s) = 8.0$ cal/g-atm

$H_2O(g) = 45.1$ cal/mol

Assuming standard conditions, find out if the reaction will proceed spontaneously.

SOLUTION

Entropy change is often used to predict the spontaneity of a reaction. A process will occur spontaneously if there is an increase in entropy, i.e., ΔS^o is positive.

$$\Delta S^o = S^o{}_{(products)} - S^o{}_{(reactants)}$$

For the above reaction

$$\Delta S^o = S^o_{Cu(s)} + S^o_{H_2O(g)} - S^o_{CuO(s)} - S^o_{H_2(g)}$$

$$= (8 + 45.1 - 10.4 - 31.2) \text{ cal/deg-mol}$$

$$= +11.5 \text{ cal/deg-mol}$$

ΔS^o is positive thus this is a spontaneous reaction.

Drill 1: Entropy

1. Which of the following would have the lowest entropy at 25°C?

    (A) NaCl(s)   (D) He(g)

    (B) $H_2(g)$ (E) Both H_2 and He

    (C) $H_2O(l)$

2. Determine the entropy change in cal/deg-mol that takes place when one mole of ammonia as a gas at –33°C comes to room temperature, 25°C. Assume heat capacity is constant at 8.9 cal/deg-mol for this range.

    (A) 2.36   (B) 1.91

(C) 4.50   (D) 6.92

(E) 0.142

3. Determine the entropy change in cal/deg-mol that takes place when one mole of ammonia passes from the liquid state to the gaseous state at its boiling point, $-33°C$; $\Delta H_{vap} - 5,570$ cal/mol.

(A) 36.5   (D) 20.2

(B) 19.4   (E) 14.5

(C) 23.2

---

2. Free Energy

The standard free energy of formation, $\Delta G°_f$, of a substance is defined as the change in free energy for the reaction in which one mole of a compound is formed from its elements under standard conditions:

$$\Delta G°_f = \Delta H°_f - T\Delta S°_f$$

$$\Delta S°_f = \Sigma S°_{f\,products} - \Sigma S°_{f\,reactants}$$

and $\Delta G°_r = \Sigma \Delta G°_{f\,products} - \Sigma \Delta G°_{f\,reactants}$

The maximum amount of useful work that can be done by any process at constant temperature and pressure is called the change in Gibbs free energy, ΔG:

$$\Delta G = \Delta H - T\Delta S$$

Another way in which the second law is stated is that in any spontaneous change, the amount of free energy available decreases.

Thus, if $\Delta G = 0$, then the system is at equilibrium.

PROBLEM

Determine $\Delta G°$ for the reaction

$$4\,NH_3(g) + 5O_2(g) \rightarrow 4NO(g) + 6H_2O(l)$$

$\Delta G°_f$ of $NH_3(g) = -4.0$ kcal/mol

$\Delta G°_f$ of $NO(g) = 20.7$ kcal/mol

$\Delta G°_f$ of $H_2O(l) = -56.7$ kcal/mol

SOLUTION

The change in free energy ($\Delta G°$) may be found by subtracting the sum of free energies ($\Delta G°_f$) of the reactants from the free energies of the products. The free energy of formation of pure elements is always 0. When more than one mole of a compound is either reacted or formed, the ΔG°_f of that compound must be multiplied by the stoichiometric coefficient for the specific compound.

$$\Delta G^\circ = (4 \text{ moles} \times \Delta G^\circ_f \text{ of } NO(g) + 6 \text{ moles} \times \Delta G^\circ_f \text{ of } H_2O(l))$$

$$- (4 \text{ moles} \times \Delta G^\circ_f \text{ of } NH_3(g) + 5 \text{ moles} \times \Delta G^\circ_f \text{ of } O_2)$$

$$\Delta G^\circ = (4 \text{ moles} \times 20.7 \text{ kcal/mol} + 6 \text{ moles} \times (-56.7 \text{ kcal/mol}))$$

$$- (4 \text{ moles} \times (-4.0 \text{ kcal/mol}) + 5 \text{ moles} \times 0 \text{ kcal/mol})$$

$$= -241.4 \text{ kcal}$$

Drill 2: Free Energy

1. For sublimation of iodine crystals,

    $$I_2(s) \leftrightarrow I_2(g)$$

    at 25°C and atmospheric pressure, it is found that the change in enthalpy, $\Delta H = 9.41$ kcal/mol and the change in entropy, $\Delta S = 20.6$ cal/deg-mol. At what temperature will solid iodine be in equilibrium with gaseous iodine?

    (A) 184°C   (D) 60°C

    (B) 190°C   (E) 75°C

    (C) 85°C

2. Two moles of hydrogen chloride are to be made from one mole each of hydrogen (H_2) and chlorine (Cl_2) at 25° and one atm. Calculate the ΔG° for this chemical reaction. The ΔG°_f of HCl is -95.3 kJ/mol.

    (A) −10,612 cal   (D) −79,421 cal

    (B) −45,532 cal   (E) 16,143 cal

    (C) −63,132 cal

3. If it takes 30.3 kJ/mole of heat to melt NaCl, calculate the melting point of NaCl, assuming the entropy increase is 28.2 Jmol^{-1} deg^{-1}.

    (A) 6,120°K   (D) 1,070°K

    (B) 120°K   (E) 378°K

    (C) 295°K

3. Equilibrium Calculations

Solvation is the interaction of solvent molecules with solute molecules or ions to form aggregates, the particles of which are loosely bonded together.

When water is used as the solvent, the process is also called aquation or hydration.

When one substance is soluble in all proportions with another substance, then the two substances are completely miscible. Ethanol and water are a familiar pair of completely miscible substances.

A saturated solution is one in which solid solute is in equilibrium with dissolved solute.

The solubility of a solute is the concentration of dissolved solute in a saturated solution of that solute.

Unsaturated solutions contain less solute than required for saturation.

Supersaturated solutions contain more solute than required for saturation. Supersaturation is a metastable state; the system will revert spontaneously to a saturated solution (stable state).

Solubility and Temperature

The solubility of most solids in liquids usually increases with increasing temperature.

For gases in liquids, the solubility usually decreases with increasing temperature.

A positive ΔH^o indicates that solubility increases with increasing temperature.

$$\log \frac{k_2}{k_1} = \frac{-\Delta H^o}{2.303R} \left[\frac{1}{T_2} - \frac{1}{T_1} \right]$$

where k_2 = solubility constant at T_2, k_1 = solubility constant at T_1, and ΔH^o = enthalpy change at standard conditions. For most substances, when a hot concentrated solution is cooled, the excess solid crystallizes. The overall process of dissolving the solute and crystallizing it again is known as recrystallization, and is useful in purification of the solute.

Another related equation is the following:

$$\Delta G = \Delta G^o + 2.303RT \log Q$$

The symbol Q represents the mass action expression for the reaction. For gases, Q is written with partial pressures. ΔG is the free energy.

At equilibrium $Q = k_{eq}$, and the products and reactants have the same total free energy, such that $\Delta G = 0$.

$$\Delta G^o = -2.303RT \log k_{eq} = -RT \ln k_{eq}$$

For the equation $2NO_2(g) \leftrightarrow N_2O_4(g)$,

$$\Delta G^\circ = -2.303RT \log\left(\frac{P_{N_2O_4}}{(P_{NO_2})^2}\right) \text{ eq, } k_c = \frac{[N_2O_4]}{[NO_2]^2}$$

Effects of Pressures on Solubility

Pressure has very little effect on the solubility of liquids or solids in liquid solvents.

The solubility of gases in liquid (or solid) solvents always increases with increasing pressure.

PROBLEM

Assuming ΔH° remains constant, calculate the equilibrium constant, K, at 373°K, if it equals 1.6×10^{12} at 298°K for the reaction $2NO(g) + O_2(g) \leftrightarrow 2NO_2(g)$. The standard enthalpy change for this reaction is -113 kJ/mole.

SOLUTION

This problem can be solved by employing the Van't Hoff equation. This equation allows a determination of an equilibrium constant, if the ΔH° is known (and constant) and the value of the equilibrium constant at another temperature is known. It states

$$\log\frac{k_2}{k_1} = \frac{\Delta H^\circ}{19.15R}\left(\frac{1}{T_1} - \frac{1}{T_2}\right)$$

where k_2 and k_1 are equilibrium constants, ΔH° = the standard enthalpy formation, and T_1 and T_2 are temperatures in Kelvin. Let $k_1 = 1.6 \times 10^{12}$ at $T_1 = 298°K$; $T_2 = 373°K$; and $\Delta H = -113$ J/mole. Substitute these values into the equation and solve for k_2, which will be the solution to this problem. Thus,

$$\log\frac{k_2}{1.6 \times 10^{12}} = \frac{-113,000}{19.15}\left(\frac{1}{298} - \frac{1}{373}\right)$$

Solving for k_2, one obtains $k_2 = 1.7 \times 10^8$.

Drill 3: Equilibrium Calculations

1. Given that k = 8.85 at 298°K and k = 0.0792 at 373°K calculate the ΔH° in J/mole for the reaction of the dimerization of NO_2 to N_2O_4. Namely, $2NO_2(g) \leftrightarrow N_2O_4(g)$.

   (A) −62,000   (D) −75,000

   (B) 43,000   (E) −58,200

   (C) 16,000

2. Calculate the enthalpy change, ΔH in cal/mol, for the reaction

   $$N_2(g) + O_2(g) = 2NO(g)$$

   given the equilibrium constants 4.08×10^{-4} for a temperature of 2,500°K.

   (A) 62,900   (D) 43,240

   (B) −45,610   (E) 94,750

   (C) 17,110

3. For the reaction, $PbSO_4(s) \rightarrow Pb^{2+} + SO_4^{2-}$

   ΔH = +2,990 cal/mol, and

   $k_{sp} = 1.8 \times 10^{-8}$ at 25°C.

   What is the reaction's k_{sp} at 55°C?

   (A) 1.8×10^{-8} (D) 10×10^{-2}

   (B) 3.07×10^{-4} (E) 4.6×10^{-6}

   (C) 2.8×10^{-8}

THERMODYNAMICS II
DRILLS

ANSWER KEY

Drill 1 — Entropy

1. (A)
2. (B)
3. (C)

Drill 2 — Free Energy

1. (A)
2. (B)
3. (D)

Drill 3 — Equilibrium Calculations

1. (E)
2. (D)
3. (C)

GLOSSARY:
THERMODYNAMICS II

Entropy

The degree of randomness in a system.

Gibbs Free Energy

The maximum amount of useful work that can be done by any process at constant temperature and pressure.

Second Law of Thermodynamics

States that in any spontaneous process there is an increase in the entropy of the universe.

Spontaneous

A reaction that is able to proceed without the addition of work.

Standard Free Energy of Formation

The change in free energy for the reaction in which one mole of a compound is formed from its elements under standard conditions.

CHAPTER 13

Electrochemistry

➤ Diagnostic Test
➤ Electrochemistry
Review & Drills
➤ Glossary

ELECTROCHEMISTRY DIAGNOSTIC TEST

1. Ⓐ Ⓑ Ⓒ Ⓓ Ⓔ
2. Ⓐ Ⓑ Ⓒ Ⓓ Ⓔ
3. Ⓐ Ⓑ Ⓒ Ⓓ Ⓔ
4. Ⓐ Ⓑ Ⓒ Ⓓ Ⓔ
5. Ⓐ Ⓑ Ⓒ Ⓓ Ⓔ
6. Ⓐ Ⓑ Ⓒ Ⓓ Ⓔ
7. Ⓐ Ⓑ Ⓒ Ⓓ Ⓔ
8. Ⓐ Ⓑ Ⓒ Ⓓ Ⓔ
9. Ⓐ Ⓑ Ⓒ Ⓓ Ⓔ
10. Ⓐ Ⓑ Ⓒ Ⓓ Ⓔ
11. Ⓐ Ⓑ Ⓒ Ⓓ Ⓔ
12. Ⓐ Ⓑ Ⓒ Ⓓ Ⓔ
13. Ⓐ Ⓑ Ⓒ Ⓓ Ⓔ
14. Ⓐ Ⓑ Ⓒ Ⓓ Ⓔ
15. Ⓐ Ⓑ Ⓒ Ⓓ Ⓔ
16. Ⓐ Ⓑ Ⓒ Ⓓ Ⓔ
17. Ⓐ Ⓑ Ⓒ Ⓓ Ⓔ
18. Ⓐ Ⓑ Ⓒ Ⓓ Ⓔ
19. Ⓐ Ⓑ Ⓒ Ⓓ Ⓔ
20. Ⓐ Ⓑ Ⓒ Ⓓ Ⓔ

21. Ⓐ Ⓑ Ⓒ Ⓓ Ⓔ
22. Ⓐ Ⓑ Ⓒ Ⓓ Ⓔ
23. Ⓐ Ⓑ Ⓒ Ⓓ Ⓔ
24. Ⓐ Ⓑ Ⓒ Ⓓ Ⓔ
25. Ⓐ Ⓑ Ⓒ Ⓓ Ⓔ
26. Ⓐ Ⓑ Ⓒ Ⓓ Ⓔ
27. Ⓐ Ⓑ Ⓒ Ⓓ Ⓔ
28. Ⓐ Ⓑ Ⓒ Ⓓ Ⓔ
29. Ⓐ Ⓑ Ⓒ Ⓓ Ⓔ
30. Ⓐ Ⓑ Ⓒ Ⓓ Ⓔ
31. Ⓐ Ⓑ Ⓒ Ⓓ Ⓔ
32. Ⓐ Ⓑ Ⓒ Ⓓ Ⓔ
33. Ⓐ Ⓑ Ⓒ Ⓓ Ⓔ
34. Ⓐ Ⓑ Ⓒ Ⓓ Ⓔ
35. Ⓐ Ⓑ Ⓒ Ⓓ Ⓔ
36. Ⓐ Ⓑ Ⓒ Ⓓ Ⓔ
37. Ⓐ Ⓑ Ⓒ Ⓓ Ⓔ
38. Ⓐ Ⓑ Ⓒ Ⓓ Ⓔ
39. Ⓐ Ⓑ Ⓒ Ⓓ Ⓔ
40. Ⓐ Ⓑ Ⓒ Ⓓ Ⓔ

ELECTROCHEMISTRY
DIAGNOSTIC TEST

This diagnostic test is designed to help you determine your strengths and your weaknesses in electrochemistry. Follow the directions and check your answers.

Study this chapter for the following tests:
AP Chemistry, CLEP General Chemistry,
GED, MCAT, MSAT, PRAXIS II Subject Assessment:
Chemistry, SAT II: Chemistry, GRE Chemistry

---

40 Questions

DIRECTIONS: Choose the correct answer for each of the following problems. Fill in each answer on the answer sheet.

1. Consider the following balanced equation:

    $$Zn^+ + H_2SO_4 \rightarrow ZnSO_4 + H_2$$

    Zinc's oxidation number changes in this reaction from

    (A) 0 to +2. 　　　　　　(D) +2 to –2.

    (B) 0 to +4. 　　　　　　(E) +2 to –4.

    (C) +2 to +4.

2. A strip of zinc is dipped in a solution of copper sulfate. Select the correct occurring half-reaction.

    (A) $Co^{++} + 2$ electrons $\rightarrow Co$, reduction

    (B) $Cu + 2$ electrons $\rightarrow Cu^{++}$, reduction

    (C) $Cu \rightarrow Cu^{++} + 2$ electrons, reduction

    (D) $Zn \rightarrow Zn^{++} + 2$ electrons, oxidation

    (E) $Zn + 2$ electrons $\rightarrow Zn^{++}$, oxidation

3. In the reaction

    $$Zn(s) + 2HCl \rightarrow ZnCl_2(aq) + H_2(g)$$

    (A) zinc is oxidized.

(B)   the oxidation number of chlorine remains unchanged.

(C)   the oxidation number of hydrogen changes from +1 to 0.

(D)   Both (A) and (B)

(E)   (A), (B), and (C)

Questions 4–5 refer to the following equation.

$$HClO + Sn^{2+} + H^+ \rightarrow Cl^- + Sn^{4+} + H_2O$$

4.   In the above unbalanced reaction, which reactant acts as an oxidizing agent?

(A) H^+ (D) $HClO$

(B) Sn^{4+} (E) Cl^-

(C) Sn^{2+}

5.   What is the oxidation state of O in ClO?

(A) -2 (D) 1

(B) -1 (E) 2

(C) 0

Questions 6 and 7 refer to the list of elements in the activity series below.

Activity Series

K
Ba
Ca
Na
Mg
Al
Mn
Zn
Cr
Fe
Co
Ni
Sn
Pb
H
Sb
Bi
Cu
Hg
Ag
Pt
Au

More likely
to lose electrons
in this direction.

6. Among manganese, lead, iron, chromium, and copper, the least easily oxidized metal is

   (A) chromium.　　　　　　　(D) lead.

   (B) copper.　　　　　　　　(E) manganese.

   (C) iron.

7. Select the *incorrect* statement among the following:

   (A) Gold is the least active metal.

   (B) Iron will replace manganese in a compound.

   (C) Potassium is most easily oxidized.

   (D) Silver is relatively unreactive.

   (E) Sodium will replace nickel in a compound.

8. One faraday of electricity is passed through an HCl electrolyte solution. Select the correct electrode result.

   (A) 1 gram of chloride ions is deposited at the anode.

   (B) 1 gram of hydrogen ions is deposited at the cathode.

   (C) 5 grams of hydrogen ions are deposited at the anode.

   (D) 35 grams of chloride ions are deposited at the anode.

   (E) 36.5 grams of chloride ions are deposited at the cathode.

9. The oxidation number of sulfur in $NaHSO_4$ is

   (A) 0.　　　　　　　　　　(D) +4.

   (B) +2.　　　　　　　　　　(E) +6.

   (C) –2.

Question 10 refers to the chart below.

Half Cell Reaction	Standard Electrode Potentials (volts)
$K \leftrightarrow K^+ + e^-$	+ 2.92
$Ca \leftrightarrow Ca^{++} + 2e^-$	+ 2.87
$Mg \leftrightarrow Mg^{++} + 2e^-$	+ 2.34
$Zn \leftrightarrow Zn^{++} + 2e^-$	.762
$Cu \leftrightarrow Cu^{++} + 2e^-$	– .344
$Ag \leftrightarrow Ag^+ + e^-$	– .7995
$Pt \leftrightarrow Pt^{++} + 2e^-$	– 1.2

10. A potential difference of 1.1068 is produced in an electrolysis set-up with the two-half cell elements of

    (A) calcium and silver.   (D) potassium and silver.

    (B) copper and zinc.   (E) zinc and silver.

    (C) magnesium and platinum.

11. Select the *incorrect* statement about the chemical activity at electrodes during electrolysis.

    (A) Anions give up electrons.

    (B) Cations take up electrons.

    (C) Oxidation occurs at the anode.

    (D) Proton transfer occurs in the reactions.

    (E) Reduction occurs at the cathode.

12. Which of the following equations can be used to calculate the emf of a voltaic cell at various concentrations?

    (A) $E = E^o \dfrac{-0.05915}{n} \log Q$

    (B) $E = q - w$

    (C) $E = E^o$ products $- E^o$ reactants

    (D) $E = E^o \dfrac{-0.05915}{n} \ln Q$

    (E) None of the above

13. Based on the following information, which will be the most effective oxidizing agent?

$$Na^+ + e^- \rightarrow Na \qquad\qquad E^o = -2.71$$
$$O_2 + 4e^- + 2H_2O \rightarrow 4OH^- \qquad E^o = +0.40$$
$$Cl_2 + 2e^- \rightarrow 2Cl^- \qquad\qquad E^o = +1.36$$

    (A) Na   (D) Cl^{-1}

    (B) Na^+ (E) Cl_2

    (C) O_2

14. Given

$$Zn^{+2} + 2e^- \rightarrow Zn \qquad E^o = -0.76$$

$$Mn^{+2} + 2e^- \rightarrow Mn \qquad E^o = -1.03$$

$$Al^{+3} + 3e^- \rightarrow Al \qquad E^o = -1.66$$

$$Cr^{+3} + 3e^- \rightarrow Cr \qquad E^o = -0.74$$

$$Ni^{+2} + 2e^- \rightarrow Ni \qquad E^o = -0.14$$

Which of the following is most easily oxidized?

(A) Zn

(B) Mn

(C) Al

(D) Cr

(E) Ni

15. Which one of the following compounds contains chlorine in a positive oxidation state?

(A) HCl

(B) KCl

(C) $HClO_3$

(D) PCl_3

(E) NH_4Cl

16. Copper metal will replace silver ions in solution, resulting in the production of silver metal and copper ions. This indicates that

(A) silver has a higher oxidation potential than copper.

(B) a combustion reaction is occurring.

(C) copper has a higher oxidation potential than silver.

(D) silver is much less soluble than copper.

(E) copper metal is readily reduced.

17. Predict the theoretical voltage of an electrochemical cell consisting of a zinc anode in 1 M $ZnSO_4$ and a copper cathode in 1 M $CuSO_4$ if

$$Zn \rightarrow Zn^{+2} + 2e^-, \qquad E^o = +0.76V$$

$$Cu \rightarrow Cu^{+2} + 2e^-, \qquad E^o = -0.34V$$

(A) −0.76V

(B) −0.42V

(C) +0.34V

(D) +0.42V

(E) +1.10V

18. Calculate the emf of a cell with the given standard electrodes:

    Cu; Cu^{2+} || Ag^+; Ag [use the table below]

Half -Reaction	Standard Reduction Potential, v
$e^- + Ag^+ \rightarrow Ag(s)$	+0.80
$2e^- + Cu^{2+} \rightarrow Cu(s)$	−0.34

    (A) −4.6V

    (B) −0.46V

    (C) +0.46V

    (D) +1.14V

    (E) −1.14V

19. How many moles of electrons must be removed from 0.5 mole of Fe^{2+} to produce Fe^{3+}?

    (A) 0.5

    (B) 1.0

    (C) 1.5

    (D) 2.0

    (E) 2.5

20. The salt bridge in the electrochemical cell serves to

    (A) increase the rate at which equilibrium is attained.

    (B) increase the voltage of the cell.

    (C) maintain electrical neutrality.

    (D) increase the oxidation/reduction rate.

    (E) supply a pathway for electrons to travel along.

21. Which of the following shows a metal being oxidized?

    (A) $2Na + 2H_2O \rightarrow 2NaOH + H_2$

    (B) $Cu \rightarrow Cu^{2+} + 2e^-$

    (C) $Cu^{2+} + 2e^- \rightarrow Cu$

    (D) Both (A) and (B)

    (E) (A), (B), and (C)

22. Which of the following correctly describes a change in oxidation state for the reaction $H_2 + S \rightarrow H_2S$?

    (A) $H_2 + 2e^- \rightarrow 2H^-$

    (B) $H_2 \rightarrow 2H$

    (C) $S + 2e^- \rightarrow S^{2-}$

    (D) $S^{2-} \rightarrow S + 2e^-$

    (E) None of the above.

23. Balancing the oxidation-reduction reaction

    $$KMnO_4 + KCl + H_2SO_4 \rightarrow MnSO_4 + K_2SO_4 + H_2O + Cl_2$$

    gives the coefficients

    (A) 4, 12, 10, 4, 10, 8, 6.   (D) 2, 10, 8, 2, 6, 8, 5.

    (B) 2, 6, 10, 4, 8, 10, 6.   (E) 2, 6, 10, 4, 6, 5, 8.

    (C) 2, 10, 8, 4, 6, 5, 8.

24. The reaction $Cu^+ + e^- \rightarrow Cu$ ($E^o = +0.52$) would not spontaneously occur with

    (A) $Li^+ + e^- \rightarrow Li$ ($E^o = -3.00$).

    (B) $Na \rightarrow Na^+ e^-$ ($E^o = +2.71$).

    (C) $Ag^+ + e^- \rightarrow Ag$ ($E^o = +0.80$).

    (D) Both (A) and (B)

    (E) (A), (B), and (C)

25. In a galvanic cell the following reaction takes place:

    $$2H_2O \leftrightarrow O_2(g) + 4H^+ + 4e^-$$

    It occurs at the

    (A) cathode.   (D) external conductor.

    (B) anode.   (E) None of the above

    (C) cathode and anode.

26. The standard electrode in electrochemistry is composed of

    (A) gold.   (D) magnesium.

    (B) platinum.   (E) hydrogen.

    (C) copper.

27. The oxidation state of manganese in $KMnO_4$ is

    (A) +1.   (D) +4.

    (B) +2.   (E) +7.

    (C) +3.

28. A galvanic cell can be represented as $Pt(s) \mid Sn^{2+}$ (aq, 1 M), Sn^{4+} (aq, 1 M) $\parallel$ Fe^{2+} (aq, 1 M) Fe^{3+} (Aq, 1 M) $\mid Pt(s)$. What reaction is occurring at the anode?

(A) $Pt \rightarrow Pt^{2+} + 2e^-$

(B) $Sn^{2+} \rightarrow Sn^{4+} + 2e^-$

(C) $Pt \rightarrow Sn^{2+} + 2e^-$

(D) $Fe^{2+} \rightarrow Fe^{3+} + 1e^-$

(E) $Fe^{3+} + 1e^- \rightarrow Fe^{2+}$

29. What is the cell voltage of $Zn \mid Zn^{2+}$ (0.1 M) $\parallel Ag^+$ (0.1 M) $\mid Ag$?

The standard reduction potential for $Zn^{2+} + 2e^- \rightarrow Zn$ is –0.76 V and for $Ag^+ + e^- \rightarrow Ag$ is +0.80 V.

(A) –0.76 V

(B) +0.80 V

(C) +1.53 V

(D) +1.59 V

(E) –1.59 V

30. Which statement below is not true for the reaction?

$$Fe^{3+} + e^- \rightarrow Fe^{2+}$$

(A) Fe^{3+} is being reduced.

(B) The oxidation state of Fe has changed.

(C) Fe^{3+} could be referred to as an oxidizing agent in this reaction.

(D) The above reaction is similar (in type) to that of an oxide losing oxygen.

(E) Both Fe^{3+} and Fe^{2+} are called anions.

31. The site of oxidation in an electrochemical cell is

(A) the anode.

(B) the cathode.

(C) the electrode.

(D) the salt bridge.

(E) none of the above.

32.
$$Zn \rightarrow Zn^{2+} + 2e^- \qquad \varepsilon^o = +0.76 \text{ V}$$

$$Cr^{3+} + 3e^- \rightarrow Cr \qquad \varepsilon^o = -0.74 \text{ V}$$

The anode in this cell is

(A) Zn.

(B) Cr.

(C) Zn^{2+}.

(D) Cr^{3+}.

(E) none of the above.

33. During the electrolysis of a $CuCl_2$ solution, which of the following reactions is possible at the anode?

    (A) $2H_2O(l) = O_2(g) + 4H^+(aq) + 4e^-$

    (B) $Cu^{2+}(aq) + 2e^- = Cu(s)$

    (C) $2H^+(aq) + 2e^- = H_2(g)$

    (D) $Cu(s) = Cu^{2+}(aq) + 2e^-$

    (E) None of the above

34. Which of the following atomic numbers describes an element with a probable oxidation number of +2?

    (A) 4   (D) 19

    (B) 6   (E) 21

    (C) 14

35. How many grams of copper will be deposited from a solution of $CuSO_4$ by a current of 3 amperes in 2 hours?

    (A) 5 g   (D) 11 g

    (B) 7 g   (E) 15 g

    (C) 8 g

36. What is the potential of a half cell consisting of a platinum wire dipped into a solution 0.01M in Sn^{2+} and 0.001M in Sn^{4+} at 25°C?

    (A) $E^\circ_{oxid.} + 0.059$ (D) $E^\circ_{oxid.} - 0.059$

    (B) $E^\circ_{red} - \dfrac{0.059}{2}$ (E) E°_{red}

    (C) $E^\circ_{red} + \dfrac{0.059}{2}$

37. How many faradays of charge are required to electroplate 127 g of copper from a 2 M cuprous chloride solution?

    (A) 1   (D) 6

    (B) 2   (E) 8

    (C) 4

Questions 38–40 refer to the following diagram.

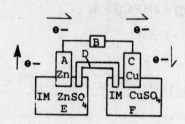

38. Which region is the anode?

    (A) *A*   (D) *D*

    (B) *C*   (E) *B*

    (C) *E*

39. Which region is responsible for maintaining electrical neutrality?

    (A) *A*   (D) *D*

    (B) *C*   (E) *B*

    (C) *E*

40. Which region(s) is(are) eventually depleted by the cell reaction?

    (A) *A*   (D) *C* and *E*

    (B) *C*   (E) *E* and *F*

    (C) *A* and *F*

ELECTROCHEMISTRY DIAGNOSTIC TEST

ANSWER KEY

1.	(A)	11.	(D)	21.	(D)	31.	(A)
2.	(D)	12.	(A)	22.	(C)	32.	(A)
3.	(E)	13.	(E)	23.	(D)	33.	(A)
4.	(D)	14.	(C)	24.	(C)	34.	(A)
5.	(A)	15.	(C)	25.	(B)	35.	(B)
6.	(B)	16.	(C)	26.	(E)	36.	(B)
7.	(B)	17.	(E)	27.	(E)	37.	(B)
8.	(B)	18.	(C)	28.	(B)	38.	(A)
9.	(E)	19.	(A)	29.	(C)	39.	(D)
10.	(B)	20.	(C)	30.	(E)	40.	(C)

DETAILED EXPLANATIONS
OF ANSWERS

1.   **(A)**   Zinc lacks an oxidation number initially in an uncombined state. It then loses two electrons and becomes +2 to couple with the −2 sulfate radical in $ZnSO_4$.

2.   **(D)**   Zinc, a more active metal than copper, will replace it in the sulfate salt. It will alter its state from a neutral atom, Zn, to a cation Zn^{++} as it combines with the minus two sulfate radical. To become positive two, it loses electrons which is oxidation.

3.   **(E)**   Writing oxidation numbers for each of the species

$$\overset{0}{Zn} + 2\overset{+1-1}{HCl} \rightarrow \overset{+2-1}{ZnCl_2} + \overset{0}{H_2}$$

we find that Zn is oxidized to Zn^{2+}, the oxidation number of chlorine remains −1, and H^+ is reduced to H_2.

4.   **(D)**   By definition, an oxidizing agent is a species that oxidizes another species. Oxidizing agents are reduced. HClO is the oxidizing agent. Sn^{2+} is the reducing agent as it is oxidized.

5.   **(A)**   The normal oxidation state for oxygen is −2 when it is combined with other atoms. However, oxygen does have a −1 state in peroxide compounds such as hydrogen peroxide (H_2O_2).

6.   **(B)**   Oxidation is the loss of electrons. Copper, Cu, is lowest on the series, among the list of choices, with a tendency to do this.

7.   **(B)**   Gold is at the bottom of the series list, and silver is near the bottom. Their tendency to oxidize and react is low. Potassium, at the top, is most likely to react. A metal must be above another to replace it in a compound. Sodium (Na) is above nickel (Ni). Iron (Fe) is not above manganese (Mn).

8.   **(B)**   One faraday of current deposits one gram equivalent of a substance at its attracting electrode. Hydrogen has a mass of one. As a positive ion, it will travel to the cathode (negative electrode) during electrolysis.

9.   **(E)**   The oxidation state of sulfur in sodium bisulfate may be determined by recalling that the oxidation states of sodium, hydrogen, and oxygen are usually +1, +1, and −2, respectively. Since the sum of the oxidation states for the

atoms of a neutral compound are zero, we have:

oxidation state of S + 1 + 1 + 4(–2) = 0

∴ oxidation state of S = +6

So, the oxidation state of sulfur in $NaHSO_4$ is +6.

10.   **(B)**   The potential difference is calculated by determining the more easily oxidized metal's potential and subtracting the standard electrode potential
from it. Only a subtraction of .762 (Zn) – (–.344) (Cu) yields a difference of
1.1068. Zinc loses electrons more easily than copper.

11.   **(D)**   During electrolysis, anions, or negative ions, move to an anode and
release electrons (oxidation). The *electrons* move through a metallic conductor to
the cathode where the electrons are accepted (reduction). Thus, the electrons
move from one electrode to another.

12.   **(A)**   In order to calculate the emf value of a voltaic cell at various concentrations, we use the Nernst equation

$$E = E° - \frac{-0.05915}{n} \log Q$$

where E = the emf for the reaction at the new concentration

$E°$ = the standard electrode potential

n = the number of moles of electrons involved in the half-reactions

Q = the reaction quotient

13.   **(E)**   The strongest oxidizing agent is the one most easily reduced. Based
on the reactions given, Cl_2 should be the most easily reduced. Notice: Based on
the reactions given O_2 requires the presence of H_2O for reduction.

14.   **(C)**   Aluminum is the least readily reduced metal of the given choices.
Therefore, it is the most easily oxidized.

15.   **(C)**   Group VIIA elements assume the –1 oxidation state in all their
binary compounds with metals and with the ammonium ion, NH_4^+. However, Cl,
Br, and I can exist in positive oxidation states of +1, +3, +5, and +7 in covalently
bonded species that contain more electronegative elements. An example of such
species are:

	Oxidation State of Chlorine
ClO^-	+1
ClO_2^-	+3
ClO_3^-	+5
ClO_4^-	+7

The compound $HClO_3$ is the only compound listed that contains chlorine in a positive oxidation state.

16. **(C)** The unbalanced reaction occurring is $Cu + Ag^+ \rightarrow Cu^{2+} + Ag$. For the reaction to proceed, copper must be easier to oxidize than silver. Thus, copper must have a higher oxidation potential than silver. Choice (B) is a redox reaction. (D) is incorrect because both silver and copper metal are insoluble. (E) is incorrect because metals are not readily reduced.

17. **(E)** The half-cell reactions and their corresponding electrode potentials are

$$Zn \rightarrow Zn^{2+} + 2e^- \qquad E^\circ = +0.76 \text{ V}$$
$$Cu^{2+} + 2e^- \rightarrow Cu \qquad E^\circ = +0.34 \text{ V}$$

The half-cells were written in these directions because Zn is oxidized (definition of the anode) and Cu is reduced. The cell reaction is the sum of the half cell reactions

$$Zn + Cu^{2+} \rightarrow Zn^{2+} + Cu \qquad E^\circ = +1.10V$$

18. **(C)** From the given cell

$$Cu; Cu^{2+} \| Ag^{2+}; Ag$$
$$Cu \rightarrow Cu^{2+} + 2e^- \qquad \text{oxidation}$$
$$2Ag^+ + 2e^- \rightarrow 2Ag \qquad \text{reduction}$$
$$Cu + Ag^{2+} \rightarrow Cu^{2+} + Ag$$
$$E^\circ_{cel} = -(E^\circ_{red\ Cu}) + (E^\circ_{red\ Ag^{2+}})$$
$$= -0.34 + 0.80 = 0.46 \text{ V}$$

19. **(A)** Iron loses one mole of electrons when one mole of Fe^{2+} reacts to produce Fe^{3+}. The removal of 0.5 mole of electrons is required to oxidize iron from the +2 to the +3 state.

20. **(C)** The salt bridge prevents mixing of the electrolyte solutions but allows ions to flow through in order to maintain electrical neutrality.

21. **(D)** One may determine if a metal is being oxidized by observing the charge on the metal before and after the reaction. Oxidation is indicated by an increase in positive charge on an atom corresponding to a loss of electrons. Sodium is oxidized to sodium hydroxide by water as may be seen by the charge on sodium going from zero to +1. The first reaction shown for copper ($Cu \rightarrow Cu^{2+}$) is an oxidation while the second ($Cu^{2+} \rightarrow Cu$) is a reduction (gain of electrons).

22. **(C)** Recalling that the oxidation states of the elements are zero, we have

$$H_2 \rightarrow 2H^+ + 2e^-$$

$$\underline{S + 2e^- \rightarrow S^{2-}}$$

$$H_2 + S \rightarrow H_2S$$

23. **(D)** The oxidation state of the manganese atom changes from +7 to +2 and the oxidation state of the chlorine atom changes from −1 to 0. Thus,

$$Mn^{+7} + 5e^- \rightarrow Mn^{2+}$$

$$Cl^{-1} \rightarrow Cl^\circ + e^-$$

Multiplying the chlorine half-cell by five balances the electrons gained with the electrons lost and gives

$$Mn^{+7} + 5Cl^{-1} = 5Cl^\circ + Mn^{2+}$$

Using this in the original equation:

$$KMnO_4 + 5KCl + H_2SO_4 \rightarrow MnSO_4 + K_2SO_4 + H_2O + \frac{5}{2}Cl_2$$

Note that 5Cl is identical to $^5/_2Cl_2$ stoichiometrically. The remaining coefficients are easily obtained. Since there are 6K on the left side, we place a 3 in front of K_2SO_4. This gives us 4S on the right side so we place a 4 in front of H_2SO_4. We now have 20 O on the left side so we place a 4 in front of H_2O. This gives us the balanced equation:

$$KMnO_4 + 5KCl + 4H_2SO_4 \rightarrow MnSO_4 + 3K_2SO_4 + 4H_2O + \frac{5}{2}Cl_2$$

Multiplying by two to remove the fraction before Cl_2 gives us the coefficients 2, 10, 8, 2, 6, 8, 5.

24. **(C)** Balancing the equations and adding emf's gives

$Cu^+ + Li \rightarrow Cu + Li^+$	$+0.52 + 3.00 = +3.52$
$Cu^+ + Na \rightarrow Cu + Na^+$	$+0.52 + 2.71 = +3.23$
$Cu^+ + Ag \rightarrow Cu + Ag^+$	$+0.52 - 0.80 = -0.28$

A negative emf indicates that the reaction does not occur spontaneously.

25. **(B)** By definition, the electrode where reduction takes place is the cathode and the oxidation takes place at the anode. The reaction shown here is an oxidation so (B) is the right answer.

26. **(E)** The hydrogen electrode has been chosen as the standard electrode with an assigned value of $E^o = 0.00$ V.

27. **(E)** The oxidation states of the atoms of a neutral compound must add up to equal zero. For $KMnO_4$, the oxidation state of K must be +1 since it is in Group IA and the oxidation state of O must be −2 since it is in Group VIA. Thus, we have:

$$1 + Mn + 4(-2) = 0 \text{ and } Mn = +7$$

28. **(B)** In cell notation, the anodic (oxidation) reaction is indicated to the left of the double vertical bars. Sn^{2+} is oxidized to Sn^{4+}. Note that the Pt is inert, and it serves as the electrode. Choice (E) represents the cathodic (reduction) reaction.

29. **(C)** The standard cell potential is calculated using the half-cell potentials for the reactions

$$Zn \rightarrow Zn^{2+} + 2e^- \qquad E^o = +0.76 \text{ V}$$

$$\underline{2(Ag^+ + e^- \rightarrow Ag) \qquad E^o = +0.80 \text{ V}}$$

$$Zn + 2Ag^+ \rightarrow Zn^{2+} + 2Ag \qquad E^o = +1.56 \text{ V}$$

Using the Nernst equation

$$E = E^o - \frac{0.059}{n} \log \frac{[C]^c[D]^d}{[A]^a[B]^b}$$

where n = the number of electrons. For the general reaction $aA + bB \leftrightarrow cC + dD$ and substituting given values

$$E = 1.56 \text{ V} - \frac{0.059}{2} \log \frac{[Zn^{2+}]}{[Ag^+]^2} \text{ V}$$

$$E = 1.56 \text{ V} - 0.0295 \log \frac{(0.1)}{(0.1)^2} \text{ V}$$

$$E = 1.56 \text{ V} - (0.0295)(\log 10) \text{ V}$$

$$E = 1.56 \text{ V} - (0.0295)(1) \text{V}$$

$$E = 1.56 \text{ V} - 0.0295 \text{ V}$$

$$E = +1.53 \text{ V}$$

30. **(E)** Fe^{3+} and Fe^{2+} are actually cations (positively charged ions) not anions. Note that (D) is true since the loss of oxygen is also referred to as reduction.

31. **(A)**   The anode of an electrochemical cell is defined as the oxidation site while the cathode is the reduction site.

32. **(A)**   The anode of any electrochemical cell is defined to be the site of oxidation. Thus, since Zn is being oxidized to Zn^{2+} in this cell, it is determined to be the anode. The cathode, the site of reduction, is Cr in this cell. The solutions of the metal ions are not the anode nor the cathode but rather the electrolytic medium.

33. **(A)**   During electrolysis anions migrate toward the anode, and cations migrate to the cathode. Hence, in the electrolysis of $CuCl_2$, the following oxidation processes take place at the anode:

(1) $Cl^-(aq) = \dfrac{1}{2}Cl_2(g) + 1e^-$ and

(2) $2H_2O(l) = O_2(g) + 4H^+(aq) + 4e^-$

making (A) the only correct choice.

34. **(A)**   The most probable oxidation number of an element is indicated by the group number, G, or by $G - 8$. Thus, an element with oxidation number +2 is most likely to be found in Group II. Thus, we have atomic number 4 (beryllium) as our solution. Using the same principles for the other choices, we obtain oxidation states of ±4, ±4, +1, and +3, respectively. Note that oxidation states of a magnitude greater than 4 are less probable than those less than 4.

35. **(B)**   Copper is being reduced from Cu^{2+} to the metal according to

$$Cu^{2+} + 2e^- \rightarrow Cu$$

The amount of electricity that allows one mole of electrons to undergo reaction is the faraday (F) which is equal to 96,500 coulombs. Thus, two faradays of charge are required to reduce one mole of Cu^{2+} to the metal. Now we must calculate the number of coulombs provided by the applied current

$$3.0 \text{ amps} \times 2 \text{ hours} \times \frac{3,600 \text{ sec}}{1 \text{ hour}} \times \frac{1 \text{ coulomb}}{1 \text{ amp sec}} = 21,600 \text{ coulombs}$$

Calculating the number of faradays donated to the copper, we obtain

$$21{,}600 \text{ coulombs} \times \frac{1\,F}{96{,}500 \text{ coulombs}} = 0.22\,F$$

Now we may compute the amount of copper deposited by this amount of charge since we know that 2 F of charge reduces one mole of Cu^{2+} to Cu^{o}. Thus, we have

$$0.22 F \times \frac{1 \text{ mole Cu}}{2F} \times \frac{63.5 \text{ g Cu}}{1 \text{ mole Cu}} =$$

7.0 g of copper deposited.

36. **(B)**   Using the Nernst equation to calculate the potential

$$E_{red} = E^{o}_{red} - \frac{0.059}{n} \log\left[\frac{(Products)}{(Reactants)}\right]$$

$$E_{red} = E^{o}_{red} - \frac{0.059}{2} \log\left(\frac{0.01}{0.001}\right)$$

$$E_{red} = E^{o}_{red} - \frac{0.059}{2} \log 10; \text{ but } \log 10 = 1$$

$$\therefore \qquad E_{red} = E^{o}_{red} - \frac{0.059}{2}$$

37. **(B)**   Using the following conversions and recalling that cuprous chloride is CuCl and cupric chloride is $CuCl_2$, we have

$$127 \text{ g of Cu} \times \frac{1 \text{ mole of Cu}}{63.5 \text{ g of Cu}} = 2 \text{ moles of Cu}$$

Since one faraday of charge corresponds to one mole of electrons, we require

$$2 \text{ moles of } Cu^{+} \times \frac{1 \text{ mole of electrons}}{1 \text{ mole of } Cu^{+} \rightarrow Cu}$$

$$= 2 \text{ moles of electrons} = 2 \text{ faradays}$$

38. **(A)**   The anode is the site of oxidation. An electrochemical cell consisting of Zn and Cu will result in Zn being oxidized to Zn^{2+} and Cu being reduced from Cu^{2+}. Thus, zinc is the anode.

39. **(D)**   The salt bridge contains a concentrated soluble electrolyte such as KCl. As electrons are removed from the anode and are transported to the cathode, Cl^{-} moves towards the anode and K^{+} towards the cathode to maintain neutrality.

40. **(C)**   The half cell reactions are

$$Zn \rightarrow Zn^{2+} + 2e^-$$ oxidation

$$Cu^{2+} + 2e^- \rightarrow Cu$$ reduction

Thus, the Zn electrode and the Cu^{2+} electrolyte will eventually be depleted.

ELECTROCHEMISTRY
REVIEW

1. Redox Reactions

The electrons found in the outermost shell of an atom are called valence electrons. When these electrons are lost or partially lost (through sharing), the oxidation state is assigned a positive value for the element. If valence electrons are gained or partially gained by an atom, its oxidation number is taken to be negative.

Oxidation is defined as a reaction in which atoms or ions undergo an increase in oxidation state. The agent that caused oxidation to occur is called the oxidizing agent and is itself reduced in the process.

Reduction is defined as a reaction in which atoms or ions undergo decrease in oxidation state. The agent that caused reduction to occur is called the reducing agent and is itself oxidized in the process.

An oxidation number can be defined as the charge that an atom would have if both of the electrons in each bond were assigned to the more electronegative element. The term "oxidation state" is used interchangeably with the term "oxidation number."

The following are the basic rules for assigning oxidation numbers:

1. The oxidation number of any element in its elemental form is zero.

2. The oxidation number of any simple ion (one atom) is equal to the charge on the ion.

3. The sum of all of the oxidation numbers of all of the atoms in a neutral compound is zero.

(More generally, the sum of the oxidation numbers of all of the atoms in a given species is equal to the net charge on that species.)

Balancing Oxidation-Reduction Reactions Using the Oxidation Number Method

The Oxidation-Number-Change Method:

1. Assign oxidation numbers to each atom in the equation.

2. Note which atoms change oxidation number, and calculate the number of electrons transferred, per atom, during oxidation and reduction.

3. When more than one atom of an element that changes oxidation number is present in a formula, calculate the number of electrons transferred per formula unit.

4. Make the number of electrons gained equal to the number lost.

5. Once the coefficients from step 4 have been obtained, the remainder of the equation is balanced by inspection, adding H^+ (in acid solution), OH^- (in basic solution), and H_2O, as required.

PROBLEM

Balance the following reaction in basic aqueous solution:

$$SO_3^{2-} + CrO_4^{2-} \rightarrow SO_4^{2-} + Cr(OH)_3$$

SOLUTION

Three rules can be used to balance oxidation-reduction reactions: (1) Balance charge by adding H^+ (in acid) or OH^- (in base). (2) Balance oxygen by adding water. (3) Balance atoms (of hydrogen) by adding hydrogen to the appropriate side. These three rules will balance the redox equation. You proceed as follows:

Reduction: $CrO_4^{2-} \rightarrow Cr(OH)_3$.

Add $2OH^-$ to the right side so that the charge is balanced. You obtain

$$CrO_4^{2-} \rightarrow Cr(OH)_3 + 2OH^-.$$

Balance oxygens by adding one water molecule to the left side. Thus,

$$H_2O + CrO_4^{2-} \rightarrow Cr(OH)_3 + 2OH^-.$$

Balance H's by adding three H's to the right side. You have

$$\frac{3}{2}H_2 + H_2O + CrO_4^{2-} \rightarrow Cr(OH)_3 + 2OH^-$$

Oxidation: $SO_3^{2-} \rightarrow SO_4^{2-}$

Charges are already balanced. To balance oxygen, add water to the left side. As such,

$$H_2O + SO_3^{2-} \rightarrow SO_4^{2-}.$$

Now balance hydrogens to obtain

$$H_2O + SO_3^{2-} \rightarrow SO_4^{2-} + H_2.$$

In summary, the balanced half-reactions are

oxid: $H_2O + SO_3^{2-} \rightarrow SO_4^{2-} + H_2$

red: $\frac{3}{2}H_2 + H_2O + CrO_4^{2-} \rightarrow Cr(OH)_3 + 2OH^-$

So that no free H's appear in the overall reaction, multiply the oxidation reaction by 3 and reduction by 2. You obtain

oxid: $3H_2O + 3SO_3^{2-} \rightarrow 3SO_4^{2-} + 3H_2$

red: $3H_2 + 2H_2O + 2CrO_4^{2-} \rightarrow 2Cr(OH)_3 + 4OH^-$

overall (oxid + red):

$$5H_2O + 3SO_3^{2-} + 2CrO_4^{2-} \rightarrow 3SO_4^{2-} + 2Cr(OH)_3 + 4OH^-.$$

Notice: The H_2's dropped out.

PROBLEM

> Balance the following reaction in acidic aqueous solution:
>
> $ClO_3^- + Fe^{2+} \rightarrow Cl^- + Fe^{3+}$

SOLUTION

Reactions in which electrons are transferred from one atom to another are known as oxidation-reduction reactions or as redox reactions. To balance this type of reaction, you want to conserve charge and matter, i.e., one side of the equation must not have an excess of charge or matter. To perform this balancing, you need to (1) Balance charge by adding electrons. (2) Balance oxygen by adding water. (3) Balance atoms (of hydrogen) by adding H^+ (in acid) or OH^- (in base). These three rules will balance the redox equation. These rules apply to balancing only the half-reactions. The overall reaction, the sum of these, will be balanced by their addition. Proceed as follows: Fe^{2+} goes to Fe^{3+}. It lost an electron; thus, it's the oxidation half-reaction. To balance charge, add e_. Thus,

$Fe^{2+} \rightarrow Fe^{3+} + e^-$.

The reduction must be $ClO_3^- \rightarrow Cl^-$. Chlorine changes oxidation state from +5 to −1 so $6e^-$ must be added. To balance the 3 oxygen atoms on the left side, add 3 water molecules on the right side. You obtain

$ClO_3^- + 6e^- \rightarrow Cl^- + 3H_2O$.

Since we know the reaction occurs in acidic media, add $6H^+$ to the reactants as the source of hydrogen in the water produced.

$ClO_3^- + 6e^- + 6H^+ \rightarrow Cl^- + 3H_2O$

In summary, you have

oxid: $Fe^{2+} \rightarrow Fe^{3+} + e^-$

red: $ClO_3^- + 6e^- + 6H^+ \rightarrow Cl^- + 3H_2O$

To balance the number of electrons appearing in the equations, multiply the oxidation reaction by six. You obtain

$$6Fe^{2+} \rightarrow 6Fe^{3+} + 6e^-$$

Thus,

oxid: $\quad 6Fe^{2+} \rightarrow 6Fe^{3+} + 6e^-$

red: $\quad ClO_3^- + 6e^- + 6H^+ \rightarrow Cl^- + 3H_2O$

overall: $\quad 6Fe^{2+} + ClO_3^- + 6H^+ \rightarrow 6Fe^{3+} + Cl^- + 3H_2O$

Notice: The electrons dropped out.

Drill 1: Redox Reactions

1. Which of the following statements is correct concerning the reaction:

    $$Fe^{2+} + 2H^+ + NO_3^- \rightarrow Fe^{3+} + NO_2 + H_2O$$

    (A) Fe^{3+} is oxidized and H^+ is reduced.

    (B) Fe^{2+} is oxidized and nitrogen is reduced.

    (C) Fe^{2+} and H^+ are oxidized.

    (D) Oxygen is oxidized.

    (E) H^+ and oxygen are reduced.

2. The reaction $2H^+ + 2e^- \rightarrow H_2(g)$ is an example of

    (A) an oxidation.   (D) the reaction at the hydrogen anode.

    (B) a reduction.   (E) an addition reaction.

    (C) an oxidation-reduction.

3. The oxidation state of nitrogen in nitric acid (HNO_3) is

    (A) +1.   (D) +4.

    (B) +2.   (E) +5.

    (C) +3.

4. Consider the following balanced equation:

    $$2K + 2HCl \rightarrow 2KCl + H_2$$

    The respective oxidation numbers for K, H, and Cl before and after reaction

    (A) go from 0, −1, +1 to −1, +1, 0.

    (B) go from 0, +1, −1 to +1, −1, 0.

(C) go from 1, –1, 0 to –1, +1, –1.

(D) go from 1, –1, 0 to –1, –1, 0.

(E) go from 0, 0, 1 to 1, 1, –1.

5. Consider the reaction:

$$2Al + 3S \rightarrow Al_2S_3$$

The oxidation numbers of aluminum and sulfur in the product are, respectively,

(A) 1, 1.   (D) 3, 2.

(B) –2, 3.   (E) 3, –2.

(C) 2, –3.

---

2. Electrode Potential

An important concept in electrochemistry is that of the half-cell standard potential or emf. The table below lists a sample of half-cell reactions and their standard potentials. This is the potential produced by the half-cell reaction when balanced against a standard hydrogen electrode with all species present at standard (molar) concentrations.

Balancing Redox Equations: The Ion-Electron Method

The Ion-Electron Method:

1. Determine which of the substances present are involved in the oxidation-reduction.

2. Break the overall reaction into two half-reactions, one for the oxidation step and one for the reduction step.

3. Balance for mass (i.e., make sure there is the same number of each kind of atom on each side of the equation) for all species except H and O.

4. Add H^+ and H_2O as required (in acidic solutions), or OH^- and H_2O as required (in basic solutions) to balance O first, then H.

5. Balance these reactions electrically by adding electrons to either side so that the total electric charge is the same on the left and right sides.

6. Multiply the two balanced half-reactions by the appropriate factors so that the same number of electrons is transferred in each.

7. Add these half-reactions to obtain the balanced overall reaction. (The electrons should cancel from both sides of the final equation.)

In the table of standard electrode potentials found on page 430 all concentrations are 1 molar, all gases are at 1 atmosphere, and the temperature is 25°C. The first reaction on the table is the reduction of a lithium ion to a lithium:

$$Li^+ + e^- \rightarrow Li$$

with a potential listed of –3.045 volts. Were we to reverse reaction (e.g., oxidize a lithium atom to an ion):

$$Li \rightarrow Li^+ + e^-$$

the potential would be +3.045 volts. The potentials (voltages) for each of these half-reactions are summed up. If the sum of the potentials are positive, the reaction is spontaneous and will run on its own. If it is negative, the energy has to be supplied to make the reaction go.

For example, if iron metal is placed in a copper (II) sulfate solution, will the copper displace the iron spontaneously, or will no reaction occur?

We know that copper (II) sulfate, C_4SO_4, when in water exists as ions (e.g., $C_4^{++} + SO_4^-$). To reduce copper (II) ions to copper metal would require the copper ions to each gain two electrons, and would be written as:

$$Cu^{++} + 2e^- \rightarrow Cu°.$$

Looking at our table, we can see the potential for this half-reaction is +0.337 volts. If the copper is replacing the iron, then the iron is being oxidized to ferrous (iron II) ions, e.g.,

$$Fe° \rightarrow Fe^{++} + 2e^-.$$

The potential for this half-reaction is +0.440 volts (remember to change the sign when reversing a reaction!). Summing the potentials for the two half-reactions we get:

$$+0.337 + 0.440 = +0.777 \text{ volts.}$$

Since the voltage is positive, it means the reaction will run by itself and give off energy. If we were to simply place an iron bar in a copper (II) sulfate solution, this energy would be given up as heat; but with a proper arrangement of the material, we would be able to use this energy as electric current (with a potential of 0.777 volts under standard conditions). Such a device is called a voltaic cell, or battery. When the reaction reaches equilibrium (in this case we will have copper metal and iron (II) sulfate), we say the battery is dead. The two half-reactions above can be summed to give us the total reaction:

$$Cu^{++} + 2e^- \rightarrow Cu°$$
$$+ Fe° \rightarrow Fe^{++} + 2e$$
$$\overline{Cu^{++} + Fe° + 2e^- \rightarrow 2e^- + Fe^{++} + Cu°}$$

The two electrons on each side of the arrow can be cancelled out to give us:

$$Cu^{++} + Fe° \rightarrow Cu° + Fe^{++}$$

Notice that in a redox equation, the number of charges on each side of the arrow are equal.

What about the reaction of aluminum chloride in water ($AlCl_3$) with the formation of chlorine gas (Cl_2) to give us metallic aluminum? Will that reaction go spontaneously? Again, we know that $AlCl_3$ in water exists as $Al^{+++} + Cl^-$. To reduce the aluminum ions to aluminum metal would require the reaction

$$Al^{+++} + 3e^- \rightarrow Al^\circ,$$

and from the table it can be seen that the potential for this reaction is –1.66 volts. The only donor for electrons would be the chloride ions, by the reaction:

$$2Cl^- \rightarrow 2e^- + Cl_2.$$

The potential for this reaction is –1.3595 volts. Summing these potentials we get –3.02 volts. Because this value is negative, we know that the reaction will not go on its own (in other words, aluminum chloride will *not* spontaneously break down into aluminum metal and chlorine gas). However, if we were to dissolve $AlCl_3$ in water and place it in a special container called an electrolytic cell, the reaction will occur whenever a potential greater than or equal to 3.02 volts is applied. Again, we can sum the two half-reactions to obtain the full reactions as follows:

$$Al^{+++} + 3e^- \rightarrow Al^\circ$$

$$2Cl^- \rightarrow Cl_2 + 2e^-$$

We first need to get the total number of electrons equal. Aluminum requires two, but each chlorine molecule formed only gives up two electrons. Were we to multiply the Al reaction by the number of electrons chlorine liberates and vise versa, we obtain:

$$2 \times (Al^{+++} + 3e^- \rightarrow Al^\circ)$$

and $\quad 3 \times (2Cl^- \rightarrow Cl_2 + 2e^-)$

Thus,

$$2Al^{+++} + 6e^- \rightarrow 2Al^\circ$$

and $\quad 6Cl^- \rightarrow 3Cl_2 + 6e^-$

which, when added yield:

$$2Al^{+++} + 6Cl^- + 6e^- \rightarrow 6e^- + 2Al^\circ + 3Cl_2$$

The six electrons cancel, giving:

$$2Al^{+++} + 6Cl^- \rightarrow 2Al^\circ + 3Cl_2.$$

Table — Standard Electrode Potentials in Aqueous Solutions at 25°C

Electrode	Electrode Reaction	E°(V)
Acid Solutions		
Li | Li$^+$	Li$^+$ + e$^-$ $\leftrightarrow$ Li	−3.045
K | K$^+$	K$^+$ + e$^-$ $\leftrightarrow$ K	−2.925
Ba | Ba^{2+}	Ba^{2+} + 2e$^-$ $\leftrightarrow$ Ba	−2.906
Ca | Ca^{2+}	Ca^{2+} + 2e$^-$ $\leftrightarrow$ Ca	−2.87
Na | Na$^+$	Na$^+$ + e$^-$ $\leftrightarrow$ Na	−2.714
La | La^{3+}	La^{3+} + 3e$^-$ $\leftrightarrow$ La	−2.52
Mg | Mg^{2+}	Mg^{2+} + 2e$^-$ $\leftrightarrow$ Mg	−2.363
Th | Th^{4+}	Th^{4+} + 4e$^-$ $\leftrightarrow$ Th	−1.90
U | U^{3+}	U^{3+} + 3e$^-$ $\leftrightarrow$ U	−1.80
Al | Al^{3+}	Al^{3+} + 3e$^-$ $\leftrightarrow$ Al	−1.66
Mn | Mn^{2+}	Mn^{2+} + 2e$^-$ $\leftrightarrow$ Mn	−1.180
V | V^{2+}	V^{2+} + 2e$^-$ $\leftrightarrow$ V	−1.18
Zn | Zn^{2+}	Zn^{2+} + 2e$^-$ $\leftrightarrow$ Zn	−0.763
Tl | Tl I | I$^-$	TlI(s) + e$^-$ $\leftrightarrow$ Tl + I$^-$	−0.753
Cr | Cr^{3+}	Cr^{3+} + 3e$^-$ $\leftrightarrow$ Cr	−0.744
Tl | TlBr | Br$^-$	TlBr(s) + e$^-$ $\leftrightarrow$ Tl + Br$^-$	−0.658
Pt | U^{3+}, U^{4+}	U^{4+} + e$^-$ $\leftrightarrow$ U^{3+}	−0.61
Fe | Fe^{2+}	Fe^{2+} + 2e$^-$ $\leftrightarrow$ Fe	−0.440
Cd | Cd^{2+}	Cd^2 + 2e$^-$ $\leftrightarrow$ Cd	−0.403
Pb | PbSO$_4$ | SO$_4$$^{2-}$	PbSO$_4$ + 2e$^-$ $\leftrightarrow$ Pb + SO$_4$$^{2-}$	−0.359
Tl | Tl$^+$	Tl$^+$ + e$^-$ $\leftrightarrow$ Tl	−0.3363
Ag | AgI | I$^-$	AgI + e$^-$ $\leftrightarrow$ Ag + I$^-$	0.152
Pb | Pb^{2+}	Pb^{2+} + 2e$^-$ $\leftrightarrow$ Pb	−0.126
Pt | D$_2$ | D$^+$	2D$^+$ + 2e$^-$ $\leftrightarrow$ D$_2$	−0.0034
Pt | H$_2$ | H$^+$	2H$^+$ + 2e$^-$ $\leftrightarrow$ H$_2$	−0.0000
Ag | AgBr | Br$^-$	AgBr + e$^-$ $\leftrightarrow$ Ag + Br$^-$	+0.071
Ag | AgCl | Cl$^-$	AgCl + e$^-$ $\leftrightarrow$ Ag + Cl$^-$	+0.2225
Pt | Hg | Hg$_2$Cl$_2$ | Cl$^-$	Hg$_2$Cl$_2$ + 2e$^-$ $\leftrightarrow$ 2Cl$^-$ + 2Hg(l)	+0.2676
Cu | Cu^{2+}	Cu^{2+} + 2e$^-$ $\leftrightarrow$ Cu	+0.337
Pt | I$_2$ | I$^-$	I$_2^-$ + 2e$^-$ $\leftrightarrow$ 3I$^-$	+0.536
Pt | O$_2$ | H$_2$O$_2$	O$_2$ + 2H$^+$ + 2e$^-$ $\leftrightarrow$ H$_2$O$_2$	+0.682
Pt | Fe^{2+}, Fe^{3+}	Fe^{3+} + e$^-$ $\leftrightarrow$ Fe^{2+}	+0.771
Ag | Ag$^+$	Ag$^+$ + e$^-$ $\leftrightarrow$ Ag	+0.7991
Au | AuCl$_4^-$, Cl$^-$	AuCl$_4^-$ + 3e$^-$ $\leftrightarrow$ Au + 4Cl$^-$	+1.00
Pt | Br$_2$ | Br$^-$	Br$_2$ + 2e$^-$ $\leftrightarrow$ 2Br$^-$	+1.065

Electrode	Electrode Reaction	E°(V)
Pt | Ti$^+$, Ti^{3+}	Ti^{3+} + 2e$^-$ ↔ Ti$^+$	+1.25
Pt | H$^+$, Cr$_2$O$_7^{2-}$, Cr^{3+}	Cr$_2$O$_7^{2-}$ + 14H$^+$ + 6e$^-$ ↔ 2Cr^{3+} + 7H$_2$O	+1.33
Pt | Cl$_2$ | Cl$^-$	Cl$_2$ + 2e$^-$ ↔ 2Cl$^-$	+1.3595
Pt | Ce^{4+}, Ce^{3+}	Ce^{4+} + e$^-$ ↔ Ce^{3+}	+1.45
Au | Au^{3+}	Au^{3+} + 3e$^-$ ↔ Au	+1.50
Pt | Mn^{2+}, MnO$_4^-$	MnO$_4^-$ + 8H$^+$ + 5e$^-$ ↔ Mn^{2+} + 4H$_2$O	+1.51
Au | Au$^+$	Au$^+$ + e$^-$ ↔ Au	+1.68
PbSO$_4$ | PbO$_2$ | H$_2$SO$_4$	PbO$_2$ + SO$_4$ + 4H$^+$ + 2e$^-$ ↔ PbSO$_4$ + 2H$_2$O	+1.685
Pt | F$_2$ | F$^-$	F$_2$(g) + 2e$^-$ ↔ 2F$^-$	+2.87

Basic Solutions

Electrode	Electrode Reaction	E°(V)
Pt | SO$_3^{2-}$, SO$_4^{2-}$	SO$_4^{2-}$ + H$_2$O + 2e$^-$ ↔ SO$_3^{2-}$ + 2OH$^-$	−0.93
Pt | H$_2$ | OH$^-$	2H$_2$O + 2e$^-$ ↔ H$_2$ + 2OH$^-$	−0.828
Ag | Ag(NH$_3$)$_2^+$, NH$_3$(aq)	Ag(NH$_3$)$_2^+$ + e$^-$ ↔ Ag + 2NH$_3$(aq)	+0.373
Pt | O$_2$ | OH$^-$	O$_2$ + 2H$_2$O + 4e$^-$ ↔ 4OH$^-$	+0.401
Pt | MnO$_2$ | MnO$_4^-$	MnO$_4^-$ + 2H$_2$O + 3e$^-$ ↔ MnO$_2$ + 4OH$^-$	+0.588

PROBLEM

Using the tables of standard electrode potentials, arrange the following substances in decreasing order of ability as reducing agents: Al, Co, Ni, Ag, H$_2$, Na.

SOLUTION

The tables of standard electrode potentials list substances according to their ability as oxidizing agents. The greater the standard electrode potential, $E°$, of a substance, the more effective it is as an oxidizing agent and the less effective it is as a reducing agent. From the table of standard electrode potentials,

$$Al^{3+} + 3e^- ↔ Al(s) \qquad E° = -1.66 \text{ V}$$

$$Co^{2+} + 2e^- ↔ Co(s) \qquad E° = -0.28 \text{ V}$$

$$Ni^{2+} + 2e^- ↔ Ni(s) \qquad E° = -0.25 \text{ V}$$

$$Ag^+ + e^- ↔ Ag(s) \qquad E° = +0.80 \text{ V}$$

$$2H^+ + 2e^- ↔ H_2(g) \qquad E° = 0 \text{ V}$$

$$Na^+ + e^- ↔ Na(s) \qquad E° = -2.71 \text{ V}$$

Thus, in increasing ability as oxidizing agents,

$$Na^+ < Al^{3+} < Co^{2+} < Ni^{2+} < H^+ < Ag^+.$$

But if Na^+ has a greater tendency to oxidize (gain electrons) than Al^{3+}, then, from looking at the reverse reactions, the "conjugate oxidant" Na must have a greater tendency to reduce (lose electrons than the "conjugate oxidant" Al. Thus, Na is a better reducing agent than Al, and so on. The substances, in order of decreasing ability as reducing agents, are therefore

$$Na > Al > Co > Ni > H_2 > Ag.$$

PROBLEM

Using the tables of standard electrode potentials list the following ions in order of decreasing ability as oxidizing agents: Fe^{3+}, F_2, Pb^{2+}, I_2, Sn^{4+}, O_2.

Half-reaction	E°, V
$Li^+ + e^- \leftrightarrow Li$	−3.05
$K^+ + e^- \leftrightarrow K$	−2.93
$Na^+ + e^- \leftrightarrow Na$	−2.71
$Mg^{2+} + 2e^- \leftrightarrow Mg$	−2.37
$Al^{3+} + 3e^- \leftrightarrow Al$	−1.66
$Mn^{2+} + 2e^- \leftrightarrow Mn$	−1.18
$Zn^{2+} + 2e^- \leftrightarrow Zn$	−0.76
$Cr^{3+} + 3e^- \leftrightarrow Cr$	−0.74
$Fe^{2+} + 2e^- \leftrightarrow Fe$	−0.44
$Cd^{2+} + 2e^- \leftrightarrow Cd$	−0.40
$Co^{2+} + 2e^- \leftrightarrow Co$	−0.28
$Ni^{2+} + 2e^- \leftrightarrow Ni$	−0.250
$Sn^{2+} + 2e^- \leftrightarrow Sn$	−0.14
$Pb^{2+} + 2e^- \leftrightarrow Pb$	−0.13
$Fe^{3+} + 3e^- \leftrightarrow Fe$	−0.04
$2H^+ + 2e^- \leftrightarrow H_2$	0 (definition)
$Sn^{4+} + 2e^- \leftrightarrow Sn^{2+}$	0.15
$Cu^{2+} + 2e^- \leftrightarrow Cu$	0.34
$Fe(CN)_6^{3-} + e^- \leftrightarrow Fe(CN)_6^{4-}$	0.46
$I_2 + 2e^- \leftrightarrow 2I^-$	0.54
$O_2 + 2H^+ + 2e^- \leftrightarrow H_2O_2$	0.68
$Fe^{3+} + e^- \leftrightarrow Fe^{2+}$	0.77
$Hg_2^{2+} + 2e^- \leftrightarrow 2Hg$	0.79
$Ag^+ + e^- \leftrightarrow Ag$	0.80
$2Hg^{2+} + 2e^- \leftrightarrow Hg_2^{2+}$	0.92

(Table continued)

$Br_2 + 2e^- \leftrightarrow 2Br^-$	1.09
$O_2(g) + 4H^+ + 4e^- \leftrightarrow 2H_2O$	1.23
$Cr_2O_7^{2-} + 14H^+ + 6e^- \leftrightarrow 2Cr^{2+} + 7H_2O$	1.33
$Cl_2 + 2e^- \leftrightarrow 2Cl^-$	1.36
$MnO_4^- + 8H^+ + 5e^- Mn^{2+} + 4H_2O$	1.51
$Ce^{4+} + e^- \leftrightarrow Ce^{2+}$	1.61
$MnO_4^- + 4H^+ + 3e^- \leftrightarrow MnO_2(s) + 2H_2O$	1.68
$H_2O_2 + 2H^+ + 2e^- \leftrightarrow 2H_2O$	1.77
$O_2 + 2H^+ + 2e^- \leftrightarrow O_2 + H_2O$	2.07
$F_3 + 2e^- \leftrightarrow 2F^-$	2.87

SOLUTION

The best oxidizing agent will be the one with the greatest ability to gain electrons (be reduced) and therefore will have the most positive standard electrode potential, E^o. From the tables,

$$Fe^{3+} + e^- \leftrightarrow Fe^{2+} \qquad E^o = +0.77 \text{ V}$$

$$F_2(g) + 2e^- \leftrightarrow 2F^- \qquad E^o = +2.87 \text{ V}$$

$$Pb^{2+} + 2e^- \leftrightarrow Pb(s) \qquad E^o = -0.13 \text{ V}$$

$$I_2(s) + 2e^- \leftrightarrow 2I^- \qquad E^o = +0.54 \text{ V}$$

$$Sn^{4+} + 2e^- \leftrightarrow Sn^{2+} \qquad E^o = +0.15 \text{ V}$$

$$O_2(g) + 2H^+ + 2e^- \leftrightarrow H_2O_2(l) \qquad E^o = +068 \text{ V}$$

Thus, the substances, in order of decreasing ability as oxidizing agents, are

$$F_2 > Fe^{3+} > O_2 > I_2 > Sn^{4+} > Pb^{2+}.$$

Drill 2: Electrode Potential

1. The standard electrode potential, E°, for the half-reaction

   $$Sn^{4+} + 2e \rightarrow Sn^{2+}$$

   is +.150 V. What would be the electrode potential of this cell if the Sn^{2+} concentration is five times the Sn^{4+} concentration?

   (A) + .129

   (D) + .191

   (B) + .171

   (E) + .091

   (C) − .171

2.   Based on these standard reduction potentials

E^o (Fe) –0.44

E^o (Cu) +0.34

The reaction $Fe^{2+} + Cu \rightarrow Cu^{2+} + Fe$ will occur

(1)   spontaneously.

(2)   if the concentration of Fe^{2+} is decreased.

(3)   if an electric current is applied to the cell.

(A)   If 1, 2, and 3 are correct.

(B)   Only 1 and 2 are correct

(C)   If only 2 and 3 are correct

(D)   If only 1 is correct

(E)   If only 3 is correct

3.   Using the data below, which of the given reactions are spontaneous?

$Co^{+2} + 2e^- \rightarrow Co$ $E^o = -0.28$ V

$Zn^{+2} + 2e^- \rightarrow Zn$ $E^o = -0.76$ V

$Mg^{+2} + 2e^- \rightarrow Mg$ $E^o = -2.37$ V

$Cn^{+2} + 2e^- \rightarrow Mn$ $E^o = -1.18$ V

$2(Ag^{+1} + e^- \rightarrow Ag)$ $E^o = +0.80$ V

$Sn^{+2} + 2e^- \rightarrow Sn$ $E^o = -0.14$ V

$Fe^{+2} + 2e^- \rightarrow Fe$ $E^o = -0.44$ V

$F_2 + 2e^- \rightarrow 2F^-$ $E^o = +2.87$ V

$2 (Li^{+1} + e^- \rightarrow Li)$ $E^o = -3.00$ V

(A) $Co^{2+} + Zn \rightarrow Zn^{2+} + Co$ (D) $Sn^{2+} + Fe \rightarrow Fe^{2+} + Sn$

(B) $Mg^{2+} + Mn \rightarrow Mn^{2+} + Mg$ (E) $F_2 + 2Li \rightarrow 2Li^+ + 2F^-$

(C) $2Ag^+ + H_2 \rightarrow 2H^+ + 2Ag$

4.   Based on the following potentials, which metal would best react with hydrochloric acid using this setup in order to produce hydrogen gas?

$Fe^{+2} + 2e^- \rightarrow Fe$ $E^o = -0.41$

$Pb^{+2} + 2e^- \rightarrow Pb$ $E^o = -0.13$

$Cu^{+2} + 2e^- \rightarrow Cu$ $E^o = +0.34$

$$Hg_2^{+2} + 2e^- \rightarrow Hg \qquad\qquad E^\circ = +0.80$$

$$Ag^{+1} + e^- \rightarrow Ag \qquad\qquad E^\circ = +0.80$$

(A) Fe   (D) Hg

(B) Pb   (E) Ag

(C) Cu

---

3. Faraday's Law

The amount of electricity that produces a specific amount of reduction (or oxidation) is related by $q = nF$ (Faraday's Law, the quantity of electricity in coulombs, n = number of equivalents oxidized or reduced, and F = faradays. The number of equivalents equals the weight of material oxidized or reduced (m) divided by the gram-equivalent weight of the material (M_{eq}) i.e.,

$$N = \frac{M}{M_{eq}}.$$

A faraday = 96,490 coulombs or one mole of electrons.

One coulomb is the amount of charge that moves past any given point in a circuit when a current of one ampere (amp) is supplied for one second. (Alternatively, one ampere is equivalent to one coulomb/second.)

PROBLEM

How much electricity will be required to decompose 454 g of water by electrolysis? The overall reaction is

$$2H_2O \rightarrow 2H_2 + O_2.$$

SOLUTION

Whenever a problem deals with weights and electricity, the solution involves an application of Faraday's law: The passage of one faraday of electricity (96,500 coulombs) causes one equivalent weight of matter to be oxidized (the loss of one electron) at one electrode and the reduction (the gain of one electron) of one equivalent weight at the other electrode. Equivalent weight may be defined as molecular weight divided by a number of moles of hydrogen transferred. To solve this problem, therefore, calculate the number of equivalents present in 454 g of water. Water has a molecular weight of 18 g/mol, but since 2 H's are transferred, water has an equivalent weight of 9 g. Therefore, the number of equivalents is

$$\frac{\text{total weight}}{\text{equivalent weight}} = \frac{454}{9} = 50.4 \text{ equiv.}$$

Recalling that one faraday of electricity is used per equivalent, 50.4 equivalents times 1 faraday/equivalent = 50.4 faradays of electricity required to decompose 454 g of water by electrolysis.

PROBLEM

> The flashlight battery is a zinc and manganese dioxide (MnO_2) dry cell. The anode consists of a zinc can and the cathode of a carbon rod surrounded by a moist mixture of MnO_2, carbon, ammonium chloride (NH_4Cl_2), and zinc chloride ($ZnCl_2$). The cathode reaction is written
>
> $$MnO_2(s) + Zn^{2+} + 2e^- \rightarrow ZnMn_2O_4(s).$$
>
> If the cathode in a typical zinc and manganese dioxide dry cell contains 4.35 g of MnO_2, how long can it deliver a steady current of 2.0 milli-amperes (mA) before all its chemicals are exhausted?

SOLUTION

The problem is solved by calculating the amount of charge required to exhaust the supply of MnO_2 and, from this, determining the lifetime of the battery using the relationship charge (coulombs) = current (A) × time (sec),

or, $$\text{time (sec)} = \frac{\text{charge (coulombs)}}{\text{current (A)}}.$$

The cathode reaction indicates that two moles of MnO_2, 2 F MnO_2, are consumed for every two moles of electrons present (2 F e⁻). The number of moles of MnO_2 present is

$$\frac{\text{mass, } MnO_2}{\text{molecular weight, } MnO_2} = \frac{4.35 \text{ g}}{87 \text{ g/mol}} = 0.05 \text{ mole } MnO_2.$$

Hence, it requires 0.05 mole of electrons (or 0.05 F) to consume the 0.05 mole of MnO_2 in the cathode. Converting faradays to coulombs (there are 96,500 coulombs in 1 F), 0.05 F is equivalent to

$$0.05 \text{ F} \times 96,500 \text{ coulombs/F} = 4.8 \times 10^3 \text{ coulombs.}$$

The battery is supposed to deliver 2.0×10^{-3} amp. Therefore, the lifetime of the battery is

$$\text{time} = \frac{\text{charge}}{\text{current}} = \frac{4.8 \times 10^3 \text{ coulombs}}{2.0 \times 10^{-3} \text{ amp}} = 2.4 \times 10^6 \text{ sec.}$$

Therefore, the battery lasts 2.4×10^6 sec (about 30 days).

Drill 3: Faraday's Law

1. How many grams of Cu could be produced from $CuSO_4$ by 0.5 faradays of charge?

    (A) 15.9

    (B) 63.5

    (C) 127.0

    (D) 31.75

    (E) 252.0

2. How many grams of copper would be produced by the reduction of Cu^{2+} if 3.0 amperes of current are passed through a copper (II) nitrate solution for one hour?

    (A) 18.20

    (B) 3.56

    (C) 31.80

    (D) 7.12

    (E) 63.50

3. How many grams of Ni can be electroplated from a solution of nickel chloride by four faradays of electricity?

    (A) 29.3

    (B) 58.7

    (C) 117.4

    (D) 176.1

    (E) 234.8

4. Electrochemical Cell Reactions

One of the most common voltaic cells is the ordinary "dry cell" used in flashlights. It is shown in the drawing below, along with the reactions occurring during the cell's discharge:

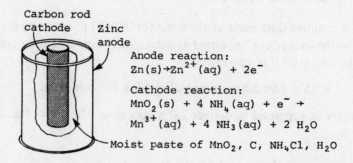

```
Carbon rod
cathode     Zinc
            anode
                    Anode reaction:
                    Zn(s)→Zn²⁺(aq) + 2e⁻

                    Cathode reaction:
                    MnO₂(s) + 4 NH₄(aq) + e⁻ →

                    Mn³⁺(aq) + 4 NH₃(aq) + 2 H₂O

                  Moist paste of MnO₂, C, NH₄Cl, H₂O
```

Figure 4.1

In galvanic or voltaic cells, the chemical energy is converted into electrical energy.

In galvanic cells, the anode is negative and the cathode is positive (the opposite is true in electrolytic cells).

The force with which the electrons flow from the negative electrode to the positive electrode through an external wire is called the electromotive force, or emf, and is measured in volts (V):

$$1\,V = \frac{1\,J}{coul.}$$

The greater the tendency or potential of the two half-reactions to occur spontaneously, the greater will be the emf of the cell. The emf of the cell is also called the cell potential, E_{cell}. The cell potential for the Zn/Cu cell can be written

$$E^{\circ}_{cell} = E^{\circ}_{Cu} - E^{\circ}_{Zn}$$

where the E°s are standard reduction potentials.

The overall standard cell potential is obtained by subtracting the smaller reduction potential from the larger one. A positive emf corresponds to a negative ΔG and therefore to a spontaneous process.

PROBLEM

For the following voltaic cell, write the half-reactions, designate which is oxidation and which is reduction. Write the cell reaction and calculate the voltage of the cell made from standard electrodes. The cell is Co; $Co^{+2} \parallel Ni^{+2}$; Ni.

SOLUTION

The cell reaction is the algebraic sum of the reactions that take place at the electrodes. Every cell has two electrodes an anode and a cathode. Oxidation, which is the loss of electrons, occurs at the anode. Reduction, which is the gain of electrons, takes a place at the cathode.

The cell is always written as solid; ion in solution $\parallel$ ion in solution; solid (anode).   (cathode)

Oxidation and reduction are the half-reactions that take place in the cell. For this cell, they are

$$Co \rightarrow Co^{+2} + 2e^{-} \text{ (oxidation at anode)}$$

$$Ni^{+2} + 2e^{-} \rightarrow Ni \text{ (reduction at cathode)}$$

Sum: $Ni^{+2} + Co \rightarrow Ni + Co^{+2}$ (cell reaction).

Since Co is losing electrons, it provides the oxidation reaction and Ni^{+2}, gaining these electrons, takes part in the reduction reaction.

The voltage of a cell is the sum of the oxidation and reduction potentials in units of volts and is designated by E^o (under standard conditions).

$$E^o_{cell} = E^o_{oxidation} + E^o_{reduction}.$$

The voltages of half-cell reactions are usually given as the reduction potentials. The oxidation potential is opposite in sign to the reduction potential. The potentials can be obtained from a table of standard reduction potentials.

$$Ni^{+2} + 2e^- \rightarrow Ni, \text{ the potential is } -.25 \text{ V}, E^o_{red} = -.25 \text{ V}.$$

For

$$Co^{+2} + 2e^- \rightarrow Co, \text{ the potential is } -.277 \text{ V}.$$

Since $Co \rightarrow Co^{+2} + 2e^-$ is the oxidation reaction, E^o_{ox} equals the negative of $-.277$ V, or $E^o_{ox} = .277$ V.

Substituting these values into the question $E^o = E^o_{oxid} + E^o_{red}$, one obtains

$$E^o = +.277 + (-.25) = .027 \text{ V}.$$

Since E^o is positive, the reaction proceeds spontaneously and can be used to supply current.

PROBLEM

For the following voltaic cell, write the half-reactions, designating which is oxidation and which reduction. Write the cell reaction and calculate the voltage (E^o) of the cell from the given electrodes. The cell is

$$Cu; Cu^{+2} \| Ag^{+2}; Ag.$$

SOLUTION

In a voltaic cell, the flow of electrons creates a current. Their flow is regulated by two types of reactions occurring concurrently, oxidation and reduction. Oxidation is a process where electrons are lost and reduction where electrons are gained. The equation for these are the half-reactions. From the cell diagram, the direction of the reaction is always left to right.

$$Cu \rightarrow Cu^{+2} + 2e^- \qquad \text{oxidation}$$

$$Ag^{+2} + 2e^- \rightarrow Ag \qquad \text{reduction}$$

Therefore, the combined cell reaction is

$$Cu + Ag^{+2} \rightarrow Cu^{+2} + Ag.$$

To calculate the total E^o, look up the value for the E^o of both half-reactions as reductions. To obtain E^o for oxidation, reverse the sign of the reduction E^o. Then, substitute into $E^o_{cell} = E^o_{red} + E^o_{ox}$. If you do this, you find

$$E^o_{cell} = -(E^o_{red} \text{ Cu}) + E^o_{red} \text{ Ag}^{+2}$$

$$= -.34 + .80$$

$$= .46 \text{ volt}$$

Electrolytic Cells

Reactions that do not occur spontaneously can be forced to take place by supplying energy with an external current. These reactions are called electrolytic reactions. In electrolytic cells, electrical energy is converted into chemical energy.

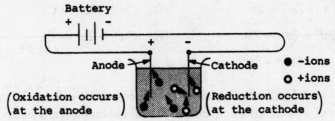

ELECTROCHEMICAL REACTIONS
In electrolytic cells, electrical energy is
converted into chemical energy.

Drill 4: Electrochemical Cell Reactions

Given two solutions, $ZnSO_4$ (1 M) and $CuSO_4$ (1 M), answer the following based on this diagram.

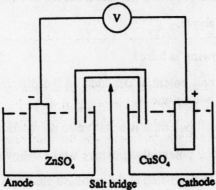

1.   What reaction, if any, takes place at the cathode?

(A) $Cu^{2+} + 2e^- \rightarrow Cu$ (D) $Zn \rightarrow Zn^{2+} + 2e^-$

(B) $Zn^{2+} + 2e^- \rightarrow Zn$ (E) None of the above

(C) $Cu \rightarrow Cu^{2+} + 2e^-$

2. Which of the following is produced at the anode of an electrolytic cell containing a solution of HCl?

    (A) H_2O (D) H_2

    (B) O_2 (E) N_2

    (C) Cl_2

3. The cathode reaction for $Cu^{2+} + Zn \leftrightarrow Cu + Zn^{2+}$ is

    (A) $Zn \rightarrow Zn^{2+} + 2e^-$.   (D) $Cu^{2+} + 2e^- \rightarrow Cu$.

    (B) $Zn^{2+} + 2e^- \rightarrow Zn$.   (E) $Cu^{2+} + 2e^- \rightarrow Zn$.

    (C) $Cu \rightarrow Cu^{2+} + 2e^-$.

Questions 4 and 5 refer to the following statement: A voltaic cell consists of a combination of a standard silver electrode and another silver electrode in which the concentration of silver ions is 10^{-3} M, $E^o_{Ag+/Ag} = 0.80V$.

4. Which of the following is true?

    (A) The standard electrode is the cathode.

    (B) The standard electrode is the anode.

    (C) There will be no electron transfer.

    (D) This cell will work like a perpetual source of energy.

    (E) None of the above.

5. Which of the following is false?

    (A) The overall cell potential depends on the difference in concentration between the two solutions.

    (B) The overall cell potential is negative so no reaction takes place.

    (C) The overall cell potential is positive so the cell works.

    (D) Copper will be deposited at the standard electrode.

    (E) The electrode immersed in the 10^{-3} M solution will lose weight.

6. The cell standard potential E^o_{cell} is

    (A) 0.00 V   (D) +1.6 V

    (B) +0.80 V   (E) −1.6 V

    (C) −0.80 V

5. Nernst Equation

For a cell at concentrations and conditions other than standard, a potential can be calculated using the following Nernst equation:

$$E_{cell} = E°_{cell} - \frac{.059}{n} \log Q$$

where $E°_{cell}$ is the standard-state cell voltage, n is the number of electrons exchanged in the equations for the reaction, and Q is the mass action quotient (which is similar in form to an equilibrium constant).

For the cell reaction

$$Zn + Cu^{2+} \rightarrow Cu + Zn^{2+}, \text{ the term } Q = \frac{.059}{n} \log \frac{[Zn^{2+}]}{[Cu^{2+}]}$$

The Nernst equation takes the form:

$$E = E° = \frac{.059}{n} \log \frac{[Zn^{2+}]}{[Cu^{2+}]}$$

The Gibbs free energy change ($\Delta G°$) is related to the standard electrode potential, $E°$ by the following equation:

$$\Delta G° = -nFE°$$

Using $\Delta G° = \Delta G° + RT \ln Q$

the equation

$$E_{cell} = E°_{cell} - \frac{RT}{nF} \ln Q$$

where n is the number of electrons involved in the half-reaction and F is the Faraday constant, which has a value of 23,061 calories/volt, or 96,487 coulombs.

For example, calculate the Gibbs free energy change under standard conditions at 25° for the following reaction:

$$Br_2 + Pb \rightarrow PbBr_2$$

Consulting our table of half-reactions we get:

$$2e^- + Br_2 \rightarrow 2Br^- = + 1.065 \text{ volts}$$

$$Pb° \rightarrow Pb^{++} + 2e^- = + 0.126 \text{ volts}$$

This gives a total of +1.191 volts, and a total of two electrons changing atoms. Putting this data into the equation yields:

$$\Delta G° = -(2)(23,061)(1.191) = -54,930 \text{ calories.}$$

Notice that although a positive potential (+ 1.191 volts) designates a spontaneous reaction, a negative ΔG^o (– 54,930 calories) designates the same phenomena.

ΔG^o can also be related to the equilibrium constant, by the formula $\Delta G^o = - RT \ln k$, where R is the ideal gas law constant (1.987 calories/mole Kelvin),* T is the temperature in Kelvin, and k is the equilibrium constant (k_A, k_{sp}, etc.). What is the k of the above reaction?

$$\Delta G^o = -54,930, T = 25 + 273 = 278°K,$$

so $- 54,930 = -(1.987)(278) \ln K, 99.4 = \ln k,$

$$e^{99.4} = e^{\ln k}, 2 \times 10^{+43} = k.$$

*The numerical value for R differs here because the units are different. R can be expressed as 0.08205 l atm/mol Kelvin, 1.987 cal/mol Kelvin, etc. The actual value of R is still the same.

PROBLEM

A cell possesses two electrodes. Both half-cells are .01 M MnO_4^- 10. One cell is .01 M H_3O^+ ion, while the other has a H_3O^+ concentration of .10 M. The electrode reaction for the reduction half-cell may be written:

$$MnO_4^- + 4H^+ + 3e^- \rightarrow MnO_2 + 2H_2O.$$

The oxidation half-cell is the reverse of this reaction. 1) Write the net equation for the spontaneous cell process taking place; 2) Find ΔE for the reaction; 3) Find the value of the equilibrium constant.

SOLUTION

1) The net reaction, in such a situation, is the sum of the balanced half-reactions, i.e., oxidation reaction plus reduction reaction. Since, you are given both reactions, add the equations together to find the overall reaction. You have, with concentrations included,

oxid: $MnO_2 + 2H_2O \rightarrow MnO_4^-$ (.01 M) + $4H^+$ (.01 M) + $3e^-$

red: MnO_4^- (.01 M) + $4H^+$ (.10 M) + $3e^- \rightarrow MnO_2 + 2H_2O$

Net Reaction: $4H^+$ (.10 M) $\rightarrow 4H^+$ (.01 M)

Notice that all species cancelled out, except H^+ (actually H_3O^+). This is the net equation for the spontaneous reaction taking place.

2) To find ΔE for the reaction, use the Nernst equation, which states

$$\Delta E = \Delta E^o - \frac{.059}{n} \log K$$

for a temperature at 25°C, where ΔE = potential for cells under other than standard conditions, ΔE^o = standard cell potential, n = number of electrons transferred, and k = equilibrium concentration expression. $n = 3$, since from either the oxidation or reduction reaction, three electrons are being transferred. $\Delta E^o = 0$, since $\Delta E^o = E^o_{prod} = E^o_{reactants}$, and both the product and reactant are the same species. K is defined as the ratio of products to reactants, each raised to the power of their coefficients in the net equation. Substituting these values into the Nernst equation, you have

$$\Delta E = 0 - \frac{.059}{3} \log \frac{(.01)^4}{(.10)^4} = 0 - .0197 \,(\log 10^{-4})$$

$$= (-.0197)\,(-4) = 0.079 \text{ volt}$$

3) To find the value of the equilibrium constant, note that there exists a relationship between ΔE^o and the constant at 25°C. Namely,

$$\Delta E^o = \frac{.059}{n} \log k \,.$$

From part 2, you found $\Delta E^o = 0$. Thus, k = unity (one), since this is the only value that permits log k = 0, which then allows $\Delta E^o = 0$, as it must.

PROBLEM

You have the following cell process:

$Fe(s) + Co^{2+}\,(.5 \text{ M}) \rightarrow Fe^{2+}\,(1.0 \text{ M}) + Co(s)$.

$Fe^{2+} + 2e^- \leftrightarrow Fe(s)$ with $E^o = -.44e$ and

$Co^{2+} + 2e^- \leftrightarrow Co(s)$ with $g^+ = -.28$,

find the standard cell potential ΔE^o, the cell potential ΔE, and the concentration ratio at which the potential generated by the cell is exactly zero.

SOLUTION

Assume that the reaction proceeds spontaneously. This means, therefore, that the reaction must have a positive value for ΔE^o. With this in mind, you proceed as follows: You are given two half-reactions:

$Fe^{2+} + 2e^- \leftrightarrow Fe(s)$ $\qquad E^o = -.44 \text{ eV}$

$Co^{2+} + 2e^- \leftrightarrow Co(s)$ $\qquad E^o = -.28 \text{ eV}$

Both reactions represent reduction (gain of electrons). But, the overall reaction in a cell is a combination of both a reduction and an oxidation reaction. Thus, you must reverse one, keeping in mind that the ΔE^o must be a positive value. Recall, also, that $\Delta E^o = E^o_{oxid} + E^o_{red}$. You can write

$$Fe(s) \leftrightarrow 2e^- + Fe^{2+} \qquad\qquad E^\circ = -(-.44) = .44 \text{ eV}$$

$$\underline{Co^{2+} + 2e^- \leftrightarrow Co(s) \qquad\qquad E^\circ = -.28 \text{ eV}}$$

$$Fe(s) + Co^{2+} \leftrightarrow Fe^{2+} + Co(s) \qquad \text{(overall reaction)}$$

with $\Delta E^\circ = .44 - .28 = .16$ eV. Notice that by reversing the Fe reaction and combing it with the other, you obtained the overall reaction with a $\Delta E^\circ = .16$ eV, a positive value, which indicates that the reaction proceeds spontaneously. To find ΔE, use the Nernst equation, which states

$$\Delta E = \Delta E^\circ - \frac{.059}{n} \log k,$$

where n = number of electrons transferred and k = equilibrium constant of reaction. In this problem, the number of electrons transferred is two, so that $n = 2$.

$$k = \frac{[Fe^{2+}]}{[Co^{2+}]},$$

i.e., the ratio of the concentrations of products to reactants, each raised to the power of its respective coefficient in the chemical equation. Note that Co(s) and Fe(s) are omitted, because they are solids and, thus, considered constants themselves. You are given $[Fe^{2+}]$ and $[Co^{2+}]$ and you have calculated ΔE°. Therefore,

$$\Delta E = \Delta E^\circ - \frac{.059}{n} \log \frac{[Fe^{2+}]}{[Co^{2+}]}$$

$$= .16 - \frac{.059}{2} \log = .16 - .0295 \log 2$$

$$= .16 - .01 = .15 \text{ eV}$$

To find the concentration ratio, k, when $\Delta E = 0$, use the Nernst equation:

$$\Delta E = \Delta E^\circ - \frac{.059}{n} \log k$$

$\Delta E = 0$, $\Delta E^\circ = .16$, $n = 2$, and

$$k = \frac{[Fe^{2+}]}{[Co^{2+}]},$$

so that

$$0 = 0.16 - \frac{0.059}{2} \log \frac{[Fe^{2+}]}{[Co^{2+}]}$$

or $\qquad 0 = 0.16 - 0.0295 \log \frac{[Fe^{2+}]}{[Co^{2+}]}.$

$$\frac{[Fe^{2+}]}{[Co^{2+}]} = \text{antilog}\left(\frac{0.16}{0.0295}\right) = \text{antilog } 5.4 = 2.5 \times 10^5$$

Thus, $\qquad \dfrac{[Fe^{2+}]}{[Co^{2+}]} = 2.5 \times 10^5,$

when $\qquad \Delta E = 0$

Drill 5: The Nernst Equation

1. Calculate the voltage of the cell Fe; $Fe^{+2} \parallel H^+$; H_2 if the iron half-cell is at standard conditions but the H^+ ion concentrations is .001 M.

   (A) 0.85 volt

   (B) 1.23 volt

   (C) 0.69 volt

   (D) 2.16 volt

   (E) 0.00 volt

2. Calculate the voltage (E) of a cell with $E^o = 1.1$ volts, if the copper half-cell is at standard conditions but the zinc ion concentration is only .001 molar. Temperature is 25°C. The overall reaction is

   $$Zn + Cu^{+2} \rightarrow Cu + Zn^{+2}.$$

   (A) 0.39 volt

   (B) 1.43 volt

   (C) 6.19 volt

   (D) 1.19 volt

   (E) 10.01 volt

3. Given $Zn \rightarrow Zn^{+2} + 2e^-$ with $E^o = +.763$, calculate E for a Zn electrode in which $Zn^{+2} = .025$ M.

   (A) 1.00 V

   (B) 0.621 V

   (C) 0.810 V

   (D) 0.124 V

   (E) 0.513 V

ELECTROCHEMISTRY DRILLS

ANSWER KEY

Drill 1 — Redox Reactions

1. (B)
2. (B)
3. (E)
4. (B)
5. (E)

Drill 2 — Electrode Potential

1. (A)
2. (E)
3. (B)
4. (A)

Drill 3 — Faraday's Law

1. (A)
2. (B)
3. (C)

Drill 4 — Electrochemical Cell Reactions

1. (A)
2. (C)
3. (D)
4. (A)
5. (B)
6. (A)

Drill 5 — Nernst Equation

1. (A)
2. (D)
3. (C)

GLOSSARY: ELECTROCHEMISTRY

Anode

Electrode at which oxidation occurs.

Cathode

Electrode at which reduction occurs.

Electrolytic Cells

A device in which reaction that does not occur spontaneously can be forced to take place by supplying energy with an external current.

Faraday

One mole of electrons or 96,494 coulombs of electric charge.

Faraday's Law

States that during electrolysis, the passage of one faraday through the circuit brings about the oxidation of one equivalent weight of a substance at one electrode and the reduction of one equivalent weight at the other electrode.

Galvanic Cell

See Voltaic Cell.

Nernst Equation

An equation used to relate cell voltage to its standard potential and the concentrations of reactants and products.

Oxidation

A reaction in which atoms or ions undergo an increase in oxidation state.

Oxidation Number

The charge that an atom would have if both of the electrons in each bond were assigned to the more electronegative element. Also called oxidation state.

Oxidation State

See Oxidation Number.

Reduction

A reaction in which atoms or ions undergo a decrease in oxidation state.

Salt Bridge

A device used in electrochemical cells to allow the passage of ions from one part of the cell to another.

Standard Electrode Potential

The voltage that is associated with an oxidation-reduction reaction at an electrode.

Voltaic Cell

A device in which chemical energy is converted to electric energy.

CHAPTER 14

Atomic Theory

➤ Diagnostic Test
➤ Atomic Theory
Review & Drills
➤ Glossary

ATOMIC THEORY DIAGNOSTIC TEST

1. Ⓐ Ⓑ Ⓒ Ⓓ Ⓔ	21. Ⓐ Ⓑ Ⓒ Ⓓ Ⓔ
2. Ⓐ Ⓑ Ⓒ Ⓓ Ⓔ	22. Ⓐ Ⓑ Ⓒ Ⓓ Ⓔ
3. Ⓐ Ⓑ Ⓒ Ⓓ Ⓔ	23. Ⓐ Ⓑ Ⓒ Ⓓ Ⓔ
4. Ⓐ Ⓑ Ⓒ Ⓓ Ⓔ	24. Ⓐ Ⓑ Ⓒ Ⓓ Ⓔ
5. Ⓐ Ⓑ Ⓒ Ⓓ Ⓔ	25. Ⓐ Ⓑ Ⓒ Ⓓ Ⓔ
6. Ⓐ Ⓑ Ⓒ Ⓓ Ⓔ	26. Ⓐ Ⓑ Ⓒ Ⓓ Ⓔ
7. Ⓐ Ⓑ Ⓒ Ⓓ Ⓔ	27. Ⓐ Ⓑ Ⓒ Ⓓ Ⓔ
8. Ⓐ Ⓑ Ⓒ Ⓓ Ⓔ	28. Ⓐ Ⓑ Ⓒ Ⓓ Ⓔ
9. Ⓐ Ⓑ Ⓒ Ⓓ Ⓔ	29. Ⓐ Ⓑ Ⓒ Ⓓ Ⓔ
10. Ⓐ Ⓑ Ⓒ Ⓓ Ⓔ	30. Ⓐ Ⓑ Ⓒ Ⓓ Ⓔ
11. Ⓐ Ⓑ Ⓒ Ⓓ Ⓔ	31. Ⓐ Ⓑ Ⓒ Ⓓ Ⓔ
12. Ⓐ Ⓑ Ⓒ Ⓓ Ⓔ	32. Ⓐ Ⓑ Ⓒ Ⓓ Ⓔ
13. Ⓐ Ⓑ Ⓒ Ⓓ Ⓔ	33. Ⓐ Ⓑ Ⓒ Ⓓ Ⓔ
14. Ⓐ Ⓑ Ⓒ Ⓓ Ⓔ	34. Ⓐ Ⓑ Ⓒ Ⓓ Ⓔ
15. Ⓐ Ⓑ Ⓒ Ⓓ Ⓔ	35. Ⓐ Ⓑ Ⓒ Ⓓ Ⓔ
16. Ⓐ Ⓑ Ⓒ Ⓓ Ⓔ	36. Ⓐ Ⓑ Ⓒ Ⓓ Ⓔ
17. Ⓐ Ⓑ Ⓒ Ⓓ Ⓔ	37. Ⓐ Ⓑ Ⓒ Ⓓ Ⓔ
18. Ⓐ Ⓑ Ⓒ Ⓓ Ⓔ	38. Ⓐ Ⓑ Ⓒ Ⓓ Ⓔ
19. Ⓐ Ⓑ Ⓒ Ⓓ Ⓔ	39. Ⓐ Ⓑ Ⓒ Ⓓ Ⓔ
20. Ⓐ Ⓑ Ⓒ Ⓓ Ⓔ	40. Ⓐ Ⓑ Ⓒ Ⓓ Ⓔ

ATOMIC THEORY
DIAGNOSTIC TEST

This diagnostic test is designed to help you determine your strengths and your weaknesses in atomic theory. Follow the directions and check your answers.

Study this chapter for the following tests:
AP Chemistry, ASVAB, CLEP General Chemistry,
GED, MCAT, MSAT, PRAXIS II Subject Assessment:
Chemistry, SAT II: Chemistry, GRE Chemistry

---

40 Questions

DIRECTIONS: Choose the correct answer for each of the following problems. Fill in each answer on the answer sheet.

1. The ionization energy of an element is

    (A) a measure of its mass.

    (B) the energy required to remove an electron from the element in its gaseous state.

    (C) the energy released by the element in forming an ionic bond.

    (D) the energy released by the element upon receiving an additional electron.

    (E) None of the above.

2. Which of the following elements has the highest ionization energy?

    (A) Ne   (D) Na

    (B) Cl   (E) Li

    (C) Si

3. The high boiling point of water considering its low molecular weight could best be explained in terms of

    (A) polar covalent bonding.   (D) Van der Waals' forces.

    (B) hydrogen bonding.   (E) London forces.

    (C) dipole attraction.

4. Which of the following molecules exhibits more nonpolar bond character?

   (A) NH_3

   (B) CH_4

   (C) CF_4

   (D) CCl_4

   (E) H_2O

5. The contribution of the electron to the atomic weight is

   (A) zero.

   (B) $1/1,837$ that of a proton or a neutron.

   (C) equal to that of a proton.

   (D) equal to that of a neutron.

   (E) None of the above.

6. Which of the following has the smallest mass?

   (A) A hydrogen nucleus

   (B) An alpha particle

   (C) A neutron

   (D) A helium nucleus

   (E) A beta particle

Questions 7 and 8 refer to the figure below.

7. The relationship between these three atoms is that they are

   (A) isobars.

   (B) isomers.

   (C) isometric.

   (D) isotonic.

   (E) isotopes.

8. All three atoms represent the element

   (A) C.

   (B) H.

   (C) He.

   (D) N.

   (E) O.

9. Which of the following are true statements with regard to the periodic table?

    (A) Electronegativity increases from left to right.

    (B) Ionization energy decreases from left to right.

    (C) Electronegativity increases from top to bottom.

    (D) Both (A) and (B)

    (E) (A), (B), and (C)

Questions 10-12 refer to the valence electron dot formulas in the figure below. The letters merely identify the different atoms. They do not stand for actual known elements.

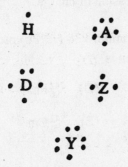

10. The most active nonmetal is

    (A) A.   (D) Y.

    (B) D.   (E) Z.

    (C) H.

11. A likely bonding association is

    (A) HA_2.   (D) ZH_4.

    (B) HD_4.   (E) YZ.

    (C) DH_5.

12. Element D has a valence of

    (A) 1.   (D) 5.

    (B) 3.   (E) 7.

    (C) 4.

13. Water is immiscible with most organic compounds due to differences in

    (A) molecular weight.

    (B) atomic composition.

    (C) density.

    (D) polarity.

    (E) None of the above.

14. Select the *incorrect* statement about radiation.

    (A) Alpha rays exhibit low penetrating power.

    (B) Alpha ray particles consist of two neutrons and two protons.

    (C) Beta ray particles can move close to the speed of light.

    (D) Beta ray particles possess a positive charge.

    (E) Gamma rays lack a possession of charge.

15. The radioactive decay of plutonium-238 ($^{238}_{94}$Pu) produces an alpha particle and a new atom. That new atom is

    (A) $^{234}_{92}$Pu.

    (B) $^{234}_{92}$U.

    (C) $^{234}_{92}$Cm.

    (D) $^{242}_{96}$Pu.

    (E) $^{242}_{96}$Cm.

16. Neutral atoms of F (fluorine) have the same number of electrons as

    (A) B^{3-}.

    (B) N^+.

    (C) Ne^-.

    (D) Na^-.

    (E) Mg^{3+}.

17. Which of the following are soluble in water?

    (A) ethyl ether

    (B) hexane

    (C) ethyl alcohol

    (D) Both (A) and (B)

    (E) (A), (B), and (C)

18. Which of the following cations will be solvated to the greatest extent in water solution?

    (A) H^+

    (B) Li^+

    (C) Na^+

    (D) K^+

    (E) Rb^+

19. The extremely high melting point of diamond (carbon) may be explained by large numbers of

    (A) covalent bonds.

    (B) ionic bonds.

    (C) hydrogen bonds.

    (D) Van der Waals' forces.

    (E) None of the above.

20. Which of the following exhibit hydrogen bonding?

    (A) NH_3

    (B) CH_4

    (C) BH_3

    (D) Both (A) and (B)

    (E) (A), (B), and (C)

21. The normal electronic configuration of chlorine gas is

    (A) Cl : Cl

    (B) : Cl : Cl :

    (C) Cl : : Cl

    (D) : Cl : Cl :

    (E) : Cl : : : Cl :

22. Which of the following contains a coordinate covalent bond?

    (A) HCl

    (B) H_2O

    (C) H_2

    (D) H_3O^+

    (E) NaCl

23. Boron is bombarded by alpha particles. Complete the products formed in the following equation by transmutation.

    $$_5B^{11} + _2He^4$$

    (A) $_6C^{12} + _0n^1$

    (B) $_3L^6 + _0n^1$

    (C) $_7N^{14} + _0n^1$

    (D) $_{15}P^{31} + _0n^1$

    (E) $_{14}Si^{28} + _1H^1$

24. The next disintegration product in the given radioactive decay of uranium is

    $$_{92}U^{238} \rightarrow _{90}Th^{234} \rightarrow _{91}Pa^{234} \rightarrow$$

    (A) $_{88}Ra^{226}$.

    (B) $_{86}Pa^{226}$.

    (C) $_{90}Th^{230}$.

    (D) $_{92}U^{234}$.

    (E) $_{90}U^{232}$.

25. An example of a dipole molecule is

   (A) CH_4.

   (B) H_2.

   (C) H_2O.

   (D) NaCl.

   (E) O_2.

26. The electronic configuration of N_2 is best represented as

   (A) $: N : N :$

   (B) $: N : : N$

   (C) $\cdot N : : N \cdot$

   (D) $: N : : : N :$

   (E) $N : : : N$

27. Which of the following elements is the *least* electronegative?

   (A) Al

   (B) Br

   (C) F

   (D) Na

   (E) Li

28. When a beta particle is emitted from an atomic nucleus,

   (A) the atomic number increases by one.

   (B) the atomic number decreases by one.

   (C) the atomic mass increases by one.

   (D) the atomic mass decreases by one.

   (E) None of the above

29. A deuterium nucleus contains how many protons and neutrons?

	protons	neutrons
(A)	1	0
(B)	2	0
(C)	1	1
(D)	1	2
(E)	2	2

30. An atom, $_A^B$X, undergoes nuclear radioactive decay by emitting two beta particles and an alpha particle. The atom's new identity is given by

    (A) $_{A+2}^{Y-2}$X.

    (B) $_{A+2}^{Y-2}$Y.

    (C) $_{A-4}^{B-2}$X.

    (D) $_A^{B-4}$Y.

    (E) $_A^{B-4}$X.

31. An atom has an atomic mass of 45 and an atomic number of 21. Select the correct statement about its atomic structure.

    (A) The number of electrons is 24.

    (B) The number of neutrons is 21.

    (C) The number of protons is 24.

    (D) The number of electrons and neutrons is equal.

    (E) The number of protons and neutrons is unequal.

32. An ionic bond is best described as

    (A) an equal sharing of electrons.

    (B) an unequal sharing of electrons.

    (C) the gain of one or more electrons on one atom with the loss of one or more electrons on the other atom.

    (D) the attraction of one atom's nucleus to the electrons of another atom.

    (E) the mutual repulsion of a pair of electrons by two nuclei.

33. The type of bonding in carbon tetrachloride (CCl_4) is

    (A) ionic.

    (B) covalent.

    (C) polar covalent.

    (D) coordinate covalent.

    (E) hydrogen.

34. When the electrons of a bond are shared unequally by two atoms, the bond is said to be

    (A) covalent.

    (B) polar covalent.

    (C) coordinate covalent.

    (D) ionic.

    (E) metallic.

35. Which of the following describes an alpha particle?

    (A) Helium nucleus

    (B) +2 charge

    (C) Atomic mass of 4

    (D) Both (A) and (B)

    (E) (A), (B), and (C)

36. Which of the following describes the bond between potassium and bromine?

    (A) Covalent

    (B) Polar covalent

    (C) Ionic

    (D) Hydrogen

    (E) Metallic

37. How many protons would be found in a nucleus of atomic weight 80 if it contains 43 neutrons?

    (A) 37

    (B) 43

    (C) 60

    (D) 80

    (E) 123

38. What type of bonding would you expect to find in compounds containing Group IVA elements?

    (A) Covalent

    (B) Ionic

    (C) Hydrogen

    (D) Coordinate covalent

    (E) Metallic

39. An element's atom most commonly has the following listing of subatomic particles:

    6 protons, 6 neutrons, 6 electrons.

    Select the following listing of particles that reveal an isotope to this given atom.

    (A) 6 protons, 6 neutrons, 4 electrons

(B) 6 protons, 6 neutrons, 8 electrons

(C) 6 protons, 8 neutrons, 6 electrons

(D) 8 protons, 6 neutrons, 6 electrons

(E) 8 protons, 8 neutrons, 6 electrons

40. In which period is the least electronegative element found?

(A) 1   (D) 4

(B) 2   (E) 5

(C) 3

ATOMIC THEORY
DIAGNOSTIC TEST

ANSWER KEY

1. (B)	11. (D)	21. (B)	31. (E)
2. (A)	12. (B)	22. (D)	32. (C)
3. (B)	13. (D)	23. (C)	33. (C)
4. (B)	14. (D)	24. (D)	34. (B)
5. (B)	15. (B)	25. (C)	35. (E)
6. (E)	16. (E)	26. (D)	36. (C)
7. (E)	17. (C)	27. (D)	37. (A)
8. (B)	18. (B)	28. (A)	38. (A)
9. (A)	19. (A)	29. (C)	39. (C)
10. (D)	20. (A)	30. (E)	40. (E)

DETAILED EXPLANATIONS
OF ANSWERS

1. **(B)** The ionization energy is defined as the energy required to remove the most loosely bound electron from an element in the gaseous state. The energy released by an element in forming an ionic solid with another element is the lattice energy of that ionic compound. The electronegativity of an element gives the relative strength with which the atoms of that element attract valence electrons in a chemical bond.

2. **(A)** Ionization energy (I.E.) is the energy required to remove an electron from an elemental atom. The lowest I.E. would occur for the alkali metals (Li, Na, etc.), since by removing an electron these atoms will have a highly stable noble gas configuration. The highest I.E. would occur for noble gases (e.g., Ne), since the noble gas configuration would be destroyed by removing an electron.

3. **(B)** The abnormally high boiling point of water is due to hydrogen bonding. The mutual attraction between the hydrogens of one water molecule and the oxygen of another represent a significant force which must be overcome if water is to be converted to steam.

4. **(B)** The type of bonding in a molecule is determined by the difference in electronegativities of the atoms comprising the bond. In summary:

Bond type	Electronegativity difference
nonpolar covalent	< 0.5
polar covalent	0.5 – 1.7
ionic	> 1.7

Using this and the table of electronegativity values we find that CH_4 is nonpolar covalent $(2.5 - 2.1 = 0.4 < 0.5)$ and that NH_3, CF_4, CCl_4, and H_2O are polar covalent.

5. **(B)** The mass of an electron is $1/1,837$ that of a proton or that of a neutron.

6. **(E)** A beta particle is a fast electron of mass 9.11×10^{-28} g while a proton and a neutron both have a mass of 1.67×10^{-24} g. A hydrogen nucleus is a proton, and an alpha particle is a helium nucleus (two protons and two neutrons). Thus, the electron (beta particle) has the smallest mass of the choices given.

7.   **(E)**

8.   **(B)**   All of these atoms vary in neutron number, thus changing the atomic mass. This is a definition of an element's isotopes, hydrogen's in this case.

9.   **(A)**   Electronegativity increases from left to right across a period and decreases from top to bottom along a group. There is a general trend for ionization energy to follow the same pattern as electronegativity. However, we encounter several discrepancies for ionization energy due to the larger amount of energy required to remove an electron from the relatively stable completed s orbital or the half-filled p and d orbitals.

10.   **(D)**   Nonmetals tend to accept electrons to obey the octet rule. They have five or more electrons in their valence shell. Atom Y has seven, being very active and close to fulfillment.

11.   **(D)**   Z has four electrons. Four hydrogens, each with one electron, can share and fulfill Z with its remaining four. Hydrogen is also satisfied, gaining a second electron for fulfillment of its only energy shell. Four covalent bonds are formed.

12.   **(B)**   With five valence electrons, D is in need of three more for octet fulfillment.

13.   **(D)**   Water and most organic solvents are immiscible due to differences in polarity; water molecules are polar while most organic solvents are nonpolar (recall the principle of "like dissolves like").

14.   **(D)**   All statements are true except this one. Beta ray particles, known as high-speed electrons, have a negative charge.

15.   **(B)**   Plutonium-238 has a mass of 238 and an atomic number of 94. The atomic mass tells us the number of protons and neutrons in the nucleus while the atomic number tells us the number of protons. An alpha particle ($^4_2\alpha$) is a helium nucleus composed of two neutrons and two protons (atomic mass of 4). Hence, upon emitting an alpha particle, the atomic number decreases by two and the atomic mass decreases by 4. This gives us $^{234}_{92}X$. Examining the periodic table we find that element 92 is uranium. Thus, our new atom is $^{234}_{92}U$. $^{234}_{92}Pu$ and $^{234}_{92}Cm$ are impossible since the atomic number of plutonium is 94 and that of curium is 96. $^{242}_{96}Pu$ and $^{242}_{96}Cm$ are impossible since these nuclei could only be produced by fusion of $^{238}_{94}Pu$ with an alpha particle. In addition, $^{242}_{96}Pu$ is incorrectly named.

∴   The reaction (decay) is $_{94}Pu^{238} \rightarrow {}_{92}U^{234} + {}_2\alpha^4$.

16. **(E)**   Neutral fluorine atoms have 9 electrons as determined by their atomic number. Magnesium atoms have 12 electrons so Mg^{3+} has 9 electrons. Boron has 5 electrons so B^{3-} has 8 electrons (the same as oxygen). Nitrogen has 7 electrons so N^+ has 6 electrons (the same as carbon). Neon has 10 electrons so Ne^- has 11 electrons (the same as sodium). Sodium has 11 electrons so Na^- has 12 electrons (the same as magnesium).

17. **(C)**   Ethyl alcohol is soluble in water due to its polar ^-OH group while ethyl ether and hexane are insoluble since they are nonpolar.

18. **(B)**   The lithium ion would be solvated to the greatest extent due to its large charge to size ratio.

19. **(A)**   Diamond, composed solely of carbon, cannot have ionic bonds or hydrogen bonds. Van der Waals attraction between the nucleus of one atom and the electrons of an adjacent atom are relatively weak compared to the covalent bonding network (sp^3 hybrid) between the carbon atoms in diamond. On the other hand, graphite (another allotropic form of carbon) is sp^2 hybrid and not strongly bonded as compared to diamond.

20. **(A)**   Only compounds containing nitrogen, oxygen, or fluorine covalently bonded to hydrogen exhibit hydrogen bonding.

21. **(B)**   The most stable electronic configuration of a molecule is that in which each atom has a complete octet of electrons surrounding it. Chlorine, being in Group VIIA has seven electrons in its valence shell. Therefore, Cl_2 has 14 electrons. This leads to the structure

$$: Cl : Cl :$$

22. **(D)**   Coordinate covalent bonds result when one of the atoms supplies both bonding electrons.

Electronegativity differences between the bonded atoms describes the type of bonding. Differences greater than 1.7 result in ionic bonds, those less than 0.5 result in nonpolar covalent bonds and those between result in polar covalent bonds. Thus, HCl and H_2O are polar, H_2 is nonpolar, and NaCl is ionic.

23. **(C)**   An alpha particle consists of two protons and two neutrons as in a helium nucleus, $_2He^4$. In the bombardment, boron incorporates the two protons

for an atomic number increase from 5 to 7. These two protons plus one captured neutron yield an atomic mass increase from 11 to 14, thus $_7N^{14}$ as in nitrogen. The second neutron remains free.

24.   **(D)**   The first step results from emission of an alpha particle, loss of two protons and two neutrons. Thus, atomic number decreases by two (two protons) and atomic mass drops by four. The second step involves emission of a beta particle (electron), which increases atomic number by one, but does not affect atomic mass. Note that the same alteration occurs from protactinium, Pa, to $_{92}U^{234}$ in this well-known sequence from uranium's isotope of 238.

25.   **(C)**   A dipole is an electrically asymmetrical molecule due to the unequal sharing of electron pairs between the spheres of bonding atoms. The two shared electron pairs of water spend more time in the command of oxygen's sphere than hydrogen's with its lower attracting power. Sodium chloride is not molecular but ionic. Methane (CH_4), hydrogen gas and oxygen gas, share electron pairs equally and are thus nonpolar molecules.

26.   **(D)**   The prime consideration in representing the bonding of a polyatomic element or compound is that each atom bonded should have a complete valence shell (eight electrons except hydrogen and helium which have two). Since nitrogen is in Group VA, it has five valence electrons illustrated as

     : N•

Diatomic nitrogen must have the structure

     : N : : : N : (or :N ≡ N :)

to completely fill the valance shells of both atoms.

27.   **(D)**   Electronegativity is a measure of the attraction of an atom for electrons. It increases as we move up and to the right on the periodic table (noble gases are not assigned electronegativity values). Sodium is the farthest down-left on the table of the choices given.

28.   **(A)**   A beta particle is a high-energy electron. In beta decay, a neutron in the nucleus emits an electron and becomes a proton. Thus, the mass number of the nucleus remains the same since there is no change in the number of nucleons (neutrons or protons). The nuclear charge, however, increases by one with the addition of a proton. The atomic number thus increases by one.

29.   **(C)**   Deuterium is an isotope of hydrogen with one extra neutron. Hydrogen usually has only one proton and no neutrons in its nucleus hence, deuterium has one proton and one neutron. Tritium is an isotope with one proton and two neutrons.

30. **(E)** An alpha particle is a helium nucleus, $_2^4He$, and a beta particle is an electron. Beta particles are produced in the nucleus by the decomposition of a neutron into a proton and an electron. Thus, if a beta particle is emitted, the atomic number increases by one while the mass number remains unchanged. For this radioactive decay, we have:

$$_A^B X \xrightarrow[_{-1}^{0}\beta]{} {}_{A+1}^{B} Q \xrightarrow[_{-1}^{0}\beta]{} {}_{A+2}^{B} R \xrightarrow[_{2}^{4}\alpha]{} {}_{A}^{B-4} X$$

Different letters were chosen to represent the nuclei since the number of protons in the nucleus determines the atomic identity. The initial and final nuclei are represented by the same letter, since they are isotopes of the same element.

31. **(E)** With an atomic number of 21, the electron and proton numbers are each 21. For a mass of 45, 24 neutrons must exist with the 21 protons.

32. **(C)** Ionic bonds are characterized by one of the bonded atoms gaining an electron while the other atom loses an electron. Covalent bonding occurs when the bonded atoms share an electron pair equally while polar covalent bonding occurs with an unequal sharing. The attraction of one atom's nucleus to the electrons of another atom is known as van der Waal's forces.

33. **(C)** The type of bonding in an element or a compound may be determined by the difference in electronegativities of the atoms engaged in the bond. The bonding is covalent if the electronegativity difference is less than 0.5, polar covalent if from 0.5 to 1.7, and ionic if greater than 1.7. Coordinate covalent bonding is characterized by one atom supplying two electrons to the bond while the other supplies none. Hydrogen bonding occurs between protons of one molecule and highly electronegative (especially oxygen) atoms of another molecule. The electronegativities of carbon and chlorine are 2.5 and 3.0, respectively. The electronegativity difference is 0.5 making the bonding polar covalent.

34. **(B)** The unequal sharing of electrons between two atoms is a polar covalent bond. Covalent (or nonpolar covalent) bonds are manifested by the equal sharing of bonding electrons. Coordinate covalent bonds are the result of one atom supplying both bonding electrons. Ionic bonds occur when one atom involved in the bond has control of both electrons (the atoms are bonded together by the attraction of one atom's positive charge to the other atom's negative charge). Metallic bonds are characterized by free electrons which travel from nucleus to nucleus.

35.   **(E)**   An alpha particle is identical to a helium nucleus and is composed of two protons and two neutrons, giving it an atomic mass of four.

36.   **(C)**   Because of the large electronegativity difference they form an ionic bond.

37.   **(A)**   The atomic weight gives the number of protons plus the number of neutrons. Thus, the number of protons (atomic number) is $80 - 43 = 37$.

38.   **(A)**   The elements of Group IVA would be expected to engage in covalent bonding (both polar and nonpolar) due to their intermediate electronegativities.

39.   **(C)**   Isotopes of an atom have the same atomic number but a different atomic mass, due to a varying number of neutrons. Choice (C) fits this requirement.

40.   **(E)**   Electronegativity decreases as one moves from top to bottom along a group. Therefore, the least electronegative element would be found in period 5.

ATOMIC THEORY REVIEW

1. Atomic Weight and Components of Atomic Structure

The nucleus is made up of very small positively-charged particles called protons and neutral particles called neutrons. The proton mass is approximately equal to the mass of the neutron and is 1,837 times the mass of the electron.

The number of protons and neutrons in the nucleus is called the mass number, which corresponds to the isotopic atomic weight. The atomic number is the number of protons found in the nucleus.

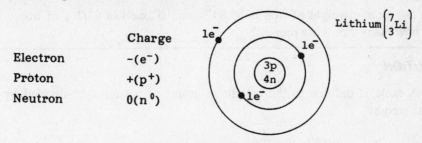

	Charge
Electron	$-(e^-)$
Proton	$+(p^+)$
Neutron	$0(n^0)$

Lithium $\left(\begin{smallmatrix} 7 \\ 3 \end{smallmatrix} Li\right)$

Figure 1.1

The atomic weight is defined as the weight of an atom compared to ^{12}C being exactly 12 atomic mass units (amu). This translates into a mole of ^{12}C having a mass of exactly 12 grams. Tabulated values of atomic weights are the weighted averages of the naturally occurring isotopes. (Isotopes have the same number of protons but different numbers of neutrons.) For example, the atomic weight of chlorine is tabulated as 35.45, natural chlorine is 75.5% ^{35}Cl (atomic weight 34.97), and 24.5% ^{37}Cl (atomic weight 36.97).

$$\text{Avg. AW} = .755(34.97) + .245(36.97) = 35.45$$

PROBLEM

> If the atomic weight of carbon 12 is exactly 12 amu, find the mass of a single carbon-12 atom.

SOLUTION

To solve this problem, one must first define the mole concept. A mole is defined as the weight in grams divided by the atomic weight (or molecular weight) of the atom or compound.

$$\text{mole} = \frac{\text{weight in grams}}{\text{atomic or molecular weight}}$$

If one has a mole of carbon 12, then there are 12 grams of it present. One mole of any substance contains 6.022×10^{23} particles.

The mass of a single carbon atom is found by dividing 12 g/mol by 6.022×10^{23} atoms/mol.

$$\text{mass of 1 C atom} = \frac{12 \text{ g/mol}}{6.022 \times 10^{23} \text{ atoms/mol}}$$

$$= 2.0 \times 10^{-23} \text{ g/atom}$$

PROBLEM

The atomic weight of iron is 55.847 amu. If one has 6.02 g of iron, how many atoms are present?

SOLUTION

A mole is defined as the weight in grams of a substance divided by its atomic weight:

$$\text{mole} = \frac{\text{amount in grams}}{\text{atomic weight}}.$$

If one calculates the number of moles of iron, then the number of atoms present can be calculated. There are 6.02×10^{23} atoms per mole.

$$\text{no. of moles of Fe} = \frac{6.02 \text{ g}}{55.846 \text{ g/mol}}$$

$$= 1.08 \times 10^{-1} \text{ moles}$$

$$\text{no. of Fe atoms present} = (1.08 \times 10^{-1} \text{ moles}) \times (6.02 \times 10^{23} \text{ atoms/mol})$$

$$= 6.49 \times 10^{22} \text{ atoms}$$

Drill 1: Atomic Weight and Components of Atomic Structure

1.   Every atom consists of electrons, protons, and neutrons except a(an)

   (A) helium atom.   (D) boron atom.

   (B) sodium atom.   (E) calcium atom.

   (C) ordinary hydrogen atom.

2. Nitrogen reacts with hydrogen to form ammonia (NH_3). The weight-percent of nitrogen in ammonia is 82.25. The atomic weight of hydrogen is 1.008. Calculate the atomic weight of nitrogen.

(A) 6.12   (D) 4.02

(B) 14.01   (E) 5.17

(C) 29.3

3. John Dalton found water to be 12.5% hydrogen by weight. Calculate the atomic weight of oxygen if Dalton assumed water contained two hydrogen atoms for every three oxygen atoms. Assume 1 H atom weighs 1 amu.

(A) 12.0   (D) 4.67

(B) 14.6   (E) 2.31

(C) 1.51

2. Valence and Electron Dot Diagrams

In the first orbital of an atom there are two valence electrons. The neutral atom with one electron is hydrogen; the one with two is helium. No more electrons can enter the first orbital. For all other orbitals, up to eight electrons can exist. The tendency or driving force for reaction is to fill the outermost orbital by sharing or acquiring electrons from other atomic species. Therefore, atoms react in a way to get eight electrons in the outermost or valence orbit (except for hydrogen which reacts to get two).

For example:

H : H

or

: Cl : Cl :

The structures above are called Lewis structures. Lewis structures include only the valence electrons and enable the depiction of a covalent bond by a pair of electrons between the two atoms

Sharing two pairs of electrons produces a double bond. An example:

The sharing of three electron pairs results in a triple bond. An example:

Greater energy is required to break double bonds than single bonds, and triple bonds are harder to break than double bonds. Molecules which contain double and triple bonds have smaller interatomic distances and greater bond strength than molecules with only single bonds. Thus, in the series,

$$H_3C - CH_3, H_2C = CH_2, HC \equiv CH,$$

the carbon-carbon distance decreases, and the C-C bond energy increases because of increased bonding.

Steps in drawing Lewis structures are as follows:

1) Draw a skeleton.

2) Count the valence electrons.

3) Subtract two valence electrons for each single bond in the first step.

4) Distribute the remaining electrons.

If there are too few electrons, convert single bonds to double bonds.

In cases where the Lewis structure doesn't completely describe a situation, resonance structures may be needed.

The resonance structures for sulfur dioxide are as follows:

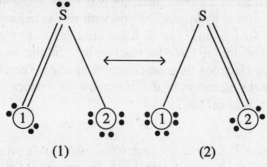

(1)   (2)

The actual electronic structure of SO_2 does not correspond to either A or B, but, instead, to an "average" structure somewhere in between. This true structure is known as a resonance hybrid of the contributing structures A and B.

PROBLEM

On the basis of valence, predict the formulas of the compounds formed between the following pairs of elements: (a) Sn and F, (b) P and H, (c) Si and O.

SOLUTION

Valence may be defined as a number which represents the combining capacity of an atom or radical, based on hydrogen as a standard.

For molecules containing two kinds of atoms, the product of the number of times one kind of atom appears in a molecule and the valence of that kind of atom must be equal to the product of the number of times the other kind of atom appears multiplied by the valence of this second kind of atom.

(a) The valence of Sn is 4 and that of F is 1. Hence, the compound is SnF_4 (1 atom Sn $\times$ 4 = 4 atoms F $\times$ 1).

(b) The valence of P is 3 and that of H is 1. Hence, the compound is PH_3 (1 atom P $\times$ 3 = 3 atoms H $\times$ 1).

(c) The valence of Si is 4 and that of O is 2. Hence, the compound is SiO_2 (1 atom Si $\times$ 4 = 2 atoms O $\times$ 2).

PROBLEM

The atomic weight of element X is 58.7. 7.34 g of X displaces 0.25 g of hydrogen from hydrochloric acid (HCl). What is the valence of element X?

SOLUTION

Chlorine has a valence of -1, thus to determine the valence of X, one must calculate the number of moles of Cl^- that will bind to each mole of X. The valence will be equal to this number. There are two moles of Cl^- present for every mole of H_2 formed. To find the number of moles of H_2 formed, one must divide 0.25 g by the molecular weight of H_2 (MW of H_2 = 2).

$$\text{moles of } H_2 = \frac{0.25 \text{ g}}{2 \text{ g/mol}} = .125 \text{ moles of } H_2$$

no. of moles of $Cl^- = 2 \times .125 = .250$ moles

One should now determine the number of moles of X present. This is done by dividing the number of grams by the molecular weight (MW of X = 58.7).

$$\text{no. of moles} = \frac{7.34 \text{ g}}{58.7 \text{ g/mol}} = .125 \text{ moles.}$$

From this, one sees that .125 moles of X combines with .250 moles of Cl^-. The number of moles of Cl^- that bind to each mole of X is equal to the number of moles of Cl^- present divided by the number of moles of X.

$$\text{no. of } Cl^- \text{ that combine with each X} = \frac{.250 \text{ moles of } Cl^-}{.125 \text{ moles of X}}$$

$$= 2 \text{ moles/}Cl^-\text{/mole X}$$

The formula for the resulting compound is XCl_2. Because Cl^- has a valence of -1 and 2 Cl^- combine with each X, X must have a valence of $+2$ for a neutral molecule to be formed.

Drill 2: Valence and Electron Dot Diagrams

1. Select the element with an atomic number of 19 and one electron in its valence shell.

    (A) Calcium   (D) Potassium

    (B) Chlorine   (E) Sodium

    (C) Hydrogen

2. Identify the incorrect Lewis structure(s).

    (A) CH_4

    $$H - \overset{\overset{\displaystyle H}{|}}{\underset{\underset{\displaystyle H}{|}}{C}} - H$$

    (B) NH_3O $H - \overset{\overset{\displaystyle H}{|}}{\underset{\underset{\displaystyle H}{|}}{N}} - \ddot{\underset{..}{O}}:$ $\leftrightarrow$ $H - \overset{\overset{\displaystyle H}{|}}{\underset{\underset{\displaystyle H}{|}}{N}} = \ddot{\underset{..}{O}}$

    (C) C_2H_3N $H - \overset{\overset{\displaystyle H}{|}}{\underset{\underset{\displaystyle H}{|}}{C}} - C \equiv N:$

    (D) NH_4^+ $\left[H - \overset{\overset{\displaystyle H}{|}}{\underset{\underset{\displaystyle H}{|}}{N}} - H \right]^+$

    (E) NO_2^- $[O = N - O:] \leftrightarrow [:O - N = O]^-$ $[\ddot{O} = N - \ddot{O}]^- \leftrightarrow [:\ddot{O} - \ddot{N} = \ddot{O}]^-$

3. H_2O_3, hydrogen trioxide, a close relative of hydrogen peroxide, has recently been synthesized. It is extremely unstable and can be isolated only in very small quantities. Write a Lewis electron dot structure for H_2O_3.

    (A) $H:O:O:O:H$ (D) $H:O:O:O:H$

    (B) $H:O:H:O:H:O:$ (E) None of the above

    (C) $H:H:O:O:O:$

3. Ionic and Covalent Bonding

An ionic bond occurs when one or more electrons are transferred from the valence shell of one atom to the valence shell of another.

The atom that loses electrons becomes a positive ion (cation), while the atom that acquires electrons becomes a negatively-charged ion (anion). The ionic bond results from the coulomb attraction between the oppositely-charged ions.

The octet rule states that atoms tend to gain or lose electrons until there are eight electrons in their valence shell.

A covalent bond results from the sharing of a pair of electrons between atoms.

Metallic Bonds

In a metal the atoms all share their outer electrons in a manner that might be thought of as an electron "atmosphere." The electrons hop freely from one atom to another, and it is this property that makes metals good conductors of electricity.

PROBLEM

Distinguish a metallic bond from an ionic bond and from a covalent bond.

SOLUTION

The best way to distinguish between these bonds is to define each and provide an illustrative example of each.

When an actual transfer of electrons results in the formation of a bond, it can be said that an ionic bond is present. For example,

$$2K \quad + \quad S: \quad \rightarrow \quad 2K^+ \quad + \quad :S:^{2-} \rightarrow \quad K_2S$$

potassium atoms	sulfur atom	potassium ions (unlike ions due to transfer of electrons from potassium to sulfur)	sulfur ion	ionic bonds due to the attraction of unlike ions

When a chemical bond is the result of the sharing of electrons, a covalent bond is present. For example:

$$: Br \bullet + \bullet F : \rightarrow : Br : F :$$

These electrons are shared with both atoms.

A pure crystal of elemental metal consists of millions of atoms held together by metallic bonds. Metals possess electrons that can easily ionize, i.e., they can be easily freed from the individual metal atoms. This free state of electrons in metals binds all the atoms together in a crystal. The free electrons extend over all the atoms in the crystal and the bonds formed between the electrons and positive nucleus are electrostatic in nature. The electrons can be pictured as a "cloud" that surrounds and engulfs the metal atoms.

PROBLEM

A chemist possesses KCl, PH_3, $GeCl_4$, H_2S, and CsF. Which of these compounds do not contain a covalent bond?

SOLUTION

A covalent bond is defined as one in which electrons are shared. The stability of covalent bonds in molecules depends on the difference in electronegativity values of the two atoms which make up the molecule. Electronegativity refers to the tendency of an atom to attract shared electrons in a chemical bond. If the electronegativity difference of two elements is greater than 1.7, an ionic bond is formed; if it is less than 1.7, a covalent bond is formed.

To solve this problem, consult a table of electronegativity values and compute the electronegativity difference of the atoms in each of the given compounds. Proceed as follows:

	Electronegativity Values		**Difference**
KCl	K = 0.8	Cl = 3.0	2.2
PH_3	P = 2.1	H = 2.1	0
$GeCl_4$	Ge = 1.8	Cl = 3.0	1.2
H_2S	H = 2.1	S = 2.5	0.4
CsF	Cs = 0.7	F = 4.0	3.3

Thus, only KCl and CsF exceed 1.7. They possess ionic bonds. The remainder of the molecules possess covalent bonds.

Drill 3: Ionic and Covalent Bonding

1. A covalent bond is unlikely to exist in the product of which of the following reactions?

(A) $H^+ + H^+ \rightarrow H_2$

(D) $Si + 2F_2 \rightarrow SiF_4$

(B) $Br^- + Br^- \rightarrow Br_2$

(E) $Ca + \frac{1}{2}O_2 \rightarrow CaO$

(C) $Se + H_2 \rightarrow SeH_2$

2. Consider a covalent bond between hydrogen and arsenic. It is known that the radii of hydrogen and arsenic atoms are respectively: 0.37 and 1.21 Angstroms. What is the approximate length of the hydrogen-arsenic bond?

(A) 2.13Å   (D) 0.24Å

(B) 1.58Å   (E) 1.79Å

(C) 0.12Å

3. Which of the following best describes a molecule that contains an ionic bond?

(A) F_2 (D) CH_4

(B) NaCl   (E) NH_3

(C) NO_2

4. Electronegativity and Ionization Energy

The electronegativity of an element is a number that measures the relative strength with which the atoms of the element attract valence electrons in a chemical bond. This electronegativity number is based on an arbitrary scale from 0 to 4. Metals have electronegativities less than 2. Electronegativity increases from left to right in a period and decreases as you go down a group.

Ionization energy is defined as the energy required to remove an electron from an isolated atom in its ground state. As we proceed down a group, a decrease in ionization energy occurs. Proceeding across a period from left to right, the ionization energy increases. As we proceed to the right, base-forming properties decrease and acid-forming properties increase.

PROBLEM

Assuming the ionization potential of sodium is 5.1 eV and the electron affinity of chlorine is 3.6 eV, calculate the amount of energy required to transfer one electron from an isolated sodium (Na) atom to an isolated chlorine (Cl) atom.

SOLUTION

Ionization potential is the amount of energy required to pull an electron off an isolated atom. Electron affinity is the amount of energy released when an electron is added to an isolated neutral atom. In this problem, one must add energy to remove an electron from Na and energy will be released upon the addition of an electron to Cl. 5. 1eV are needed to expel an electron from Na and 3.5 eV are released when Cl accepts an electron. Thus, the amount of energy required for the overall process to occur is the difference between the ionization potential of Na and the electron affinity of Cl.

The energy necessary to be added to the system for this reaction to occur is 5.1 eV – 3.6 eV or 1.5 eV.

PROBLEM

40.0 kJ of energy is added to 1.00 gram of magnesium atoms in the vapor state. What is the composition of the final mixture? The first ionization potential of Mg is 735 kJ/mol and the second ionization potential is 1,447 kJ/mol.

SOLUTION

Ionization potential may be defined as the energy required to pull an electron away from an isolated atom. The second ionization potential is the amount of energy required to pull off a second electron after the first has been removed.

The composition of the final mixture is determined by calculating the number of electrons that will be removed from the magnesium ions. To do this one must determine the number of moles of Mg present to 1 g. From this, one can determine the number of electrons that will be liberated by using the values for the first and second ionization potentials of Mg.

The atomic weight of Mg is 24.3. Since moles = grams/atomic weight, there are in 1 gram of Mg, 1/24.30 or 4.11×10^{-2} moles present. The first ionization potential of Mg is 735 kJ/mol. Therefore, 4.11×10^{-2} moles of Mg requires 4.11 ¥ 10^{-2} moles × 735 kJ/mole or 30.2 kJ to ionize all of the atoms once. 40 kJ was added to the system leaving 40 kJ – 30.2 kJ or 9.8 kJ to remove the second electron. If one has 9.8 kJ and 1,447 kJ/mol is required to remove the second electrons, then

$$\frac{9.8 \text{ kJ}}{1,447 \text{ kJ/mol}} = 6.77 \times 10^{-3}$$

moles of atoms can have their second electron removed. 4.11×10^{-2} moles of Mg are present.

$$\frac{6.77 \times 10^{-3}}{4.11 \times 10^{-2}} \times 100 = 16.5\%$$

This means that 16.5% of the atoms can have a second electron removed. Therefore, the composition of the mixture is: Mg^{++} 16.5%, Mg^{+} 100 – 16.5 or 83.5%.

Drill 4: Electronegativity and Ionization Energy

1. The most active metal of the alkali metals is

    (A) Li.

    (B) Mg.

    (C) K.

    (D) Sr.

    (E) Cs.

2. Which of the following pairs of elements does *not* have approximately the same electronegativity?

   (A) C, S   (D) U, Pu

   (B) Co, Ni   (E) Fe, Ni

   (C) B, Al

3. Which of the following best describes ionization energy?

   (A) Energy needed to remove the most loosely bound electron from its ground state.

   (B) It is represented by

   $$x + e^- \rightarrow x^- + energy.$$

   (C) It decreases from left to right across a period.

   (D) It increases down the periodic table.

   (E) None of the above

4. A correct ranking of elements in order of increasing electronegativity is

   (A) Ca-Li-Ba-K-Ca.   (D) F-Br-H-Al-Rb.

   (B) H-I-Na-K-Ca.   (E) Na-O-C-Ca-Li.

   (C) Ca-Al-P-S-F.

5. Which of the following best explains the large electronegativity value of fluorine?

   (A) The fluorine atom requires only one electron in order to attain a complete valence shell.

   (B) The fluorine atom must have a charge of −1 if it is to react to form a stable compound.

   (C) The diatomic fluorine gas molecule is unstable.

   (D) All of the above

   (E) None of the above

5. Polarity of Bonds

In a nonpolar covalent bond, the electrons are shared equally.

Nonpolar covalent bonds are characteristic of homonuclear diatomic molecules. For example, the fluorine molecule:

$$\bullet F : \bullet F : \rightarrow : F : F :$$

Fluorine atoms → Fluorine molecule

Where there is an unequal sharing of electrons between the atoms involved, the bond is called a polar covalent bond. An example:

H $\overset{\bullet}{\underset{\bullet\bullet}{\text{:}\overset{\times}{C}\text{l}}}$:   × hydrogen electron
                          • chlorine electrons

H $\overset{\bullet\bullet}{\underset{\bullet\bullet}{\text{:}\overset{}{O}\text{x}}}$ H   × hydrogen electrons
                          • oxygen electrons

Because of the unequal sharing, the bonds shown are said to be polar bonds (dipoles). The more electronegative element in the bond is the negative end of the bond dipole. In each of the molecules shown here, there is also a non-zero molecular dipole moment, given by the vector sum of the bond dipoles.

In the example of methane, CH_4, slight differences in electronegativity cause a polar bond to be formed between the carbon and each hydrogen. Each bond has its associated dipole moment, but as shown below, each dipole cancels and the molecule has no net dipole:

$$H \rightleftharpoons C \rightleftharpoons H$$

Hence, a molecule that contains polar bonds may not be polar. One must check to see if symmetry causes cancellation of dipoles.

A dipole consists of a positive and negative charge separated by a distance. A dipole is described by its dipole moment, which is equal to the charge times the distance between the positive and negative charges:

net dipole moment = charge × distance

In polar molecular substances, the positive pole of one molecule attracts the negative pole of another. The force of attraction between polar molecules is called a dipolar force.

When a hydrogen atom is bonded to a highly electronegative atom, it will become partially positively-charged, and will be attracted to neighboring electron pairs. This creates a hydrogen bond. The more polar the molecule, the more effective the hydrogen bond is in binding the molecules into a larger unit.

The relatively weak attractive forces between molecules are called Van der Waals' forces. These forces become apparent only when the molecules approach one another closely (usually at low temperatures and high pressure). They are due to the way the positive charges of one molecule attract the negative charges of another molecule. Compounds of the solid state that are bound mainly by this type of attraction have soft crystals, are easily deformed, and vaporize easily. Because of the low intermolecular forces, the melting points are low and evaporation takes place so easily that it may occur at room temperature. Examples of substances with this last characteristic are iodine crystals and naphthalene crystals.

If we neglect the row of inert gases on the periodic table, the further to the right and up an element is, the greater its electronegativity. (Electronegativity is the ability of an atom to pull electrons off another atom.) Looking at the periodic chart, one can easily see that fluorine is the most electronegative element, while cesium is the least. Elements with very great differences in electronegativity (i.e., from opposite sides of the periodic table) are more likely to combine by an ionic bond, e.g., NaF. Elements with *no* difference in electronegativity will combine by a nonpolar covalent bond, e.g., Cl_2. Elements with slight differences in electronegativity (usually from the same side of the periodic table) will combine by a polar covalent bond, e.g., HCL, BrCl, SO_2.

PROBLEM

Determine which of the atoms in each pair possess a partial positive charge and which a partial negative. (a) the O-F bond, (b) the O-N bond, (c) the O-S bond. Electronegativity values for these elements can be found from a table of electronegativities.

SOLUTION

Electronegativity is the tendency of an atom to attract shared electrons in a chemical covalent bond. Since electrons are negatively charged, to find the partial charge on each atom of the pair consult a table for the electronegativity values of the atoms in the molecules. The atom with the higher value will have a greater tendency to attract electrons, and will, thus, have a partial negative charge. Because the overall bond is neutral, the other atom must have a partial positive charge.

For part (a), F has a higher electronegativity than 0, which means that F will have a partial negative charge (δ^-) and O will have a partial positive charge ($^+\delta$). To show this the molecule can be written

$$\begin{pmatrix} \delta^+ & \delta^- \\ O & - & F \end{pmatrix}$$

Similar logic is used in working out parts (b) and (c).

(b) O-N bond: The electronegativity of O is 3.5 and of N is 3.0, thus O is more negative than N. The molecule is written

$$\begin{pmatrix} \overset{\delta^-}{O} - \overset{\delta^+}{N} \end{pmatrix}$$

(c) O-S bond: The electronegativity of O is 3.5 and of S is 2.5; therefore, O is the more negative of this pair

$$\begin{pmatrix} \overset{\delta^-}{O} - \overset{\delta^+}{S} \end{pmatrix}$$

PROBLEM

> Which molecule of each of the following pairs would exhibit a higher degree of polarity: HCl and HBr, H_2O and H_2S, BrCl and IF?

SOLUTION

Polarity indicates that there is an uneven sharing of electrons between two atoms. This creates a charge distribution in the molecule where one atom is partially positive and the other is partially negative. The degree of polarity is measured by finding the difference in the abilities of the two atoms to attract electrons. This tendency to accept electrons is called the electronegativity. The greater the electronegativity difference, the greater the degree of polarity.

From the table of electronegativity values, the following electronegativities can be obtained:

	Compounds	Electronegativity Difference
(1)	HCl	$3.0 - 2.1 = .9$
	HBr	$2.8 - 2.1 = .7$
(2)	H_2O	$3.5 - 2.1 = 1.4$
	H_2S	$2.5 - 2.1 = .4$
(3)	BrCl	$3.0 - 2.8 = .2$
	IF	$4.0 - 2.5 = 1.5$

In pair (1) (HCl and HBr), HCl has the larger electronegativity. Hence, HCl has a greater degree of polarity than HBr. For the same reason, H_2O in pair (2) and IF in pair (3) have the greater degrees of polarity.

Drill 5: Polarity of Bonds

1. Compounds that can form hydrogen bonds with water include

    (A) CH_4.

    (B) CH_4, HF.

    (C) CH_3OH, HF, CCl_4.

    (D) NH_3, HF, CH_3OH.

    (E) None of the above.

2. Which of the following substances has the least ionic character in its bond?

    (A) CCl_4

    (B) KCl

    (C) $MgCl_2$

    (D) NaCl

    (E) $BaCl_2$

3. Sodium chloride (NaCl) would be most soluble in

    (A) ether.

    (B) benzene.

    (C) water.

    (D) carbon tetrachloride.

    (E) gasoline.

4. Which of the following can best be described as a polar covalent molecule?

    (A) Cl_2

    (B) HCl

    (C) KCl

    (D) H_2

    (E) NaCl

5. Which of the following compounds have only nonpolar bonds?

    (A) KO_2

    (B) NaF

    (C) HF

    (D) KBr

    (E) I_2

6. The high boiling point of H_2O relative to the boiling points of H_2S, H_2Se, and H_2Te can be attributed to

    (A) the molecular weight of H_2O.

    (B) the covalent bonds between H and O.

    (C) the atomic number of oxygen.

    (D) the ability of water to absorb oxygen.

    (E) the ability of water to form hydrogen bonds.

7. The pair of atoms that are most likely to form a covalent compound are:

    (A) H and He.

    (D) Li and F.

    (B) Na and F.

    (E) Na and Cl.

    (C) H and Cl.

8. Consider the outline of a water molecule in the drawing below. The regions most likely to be attracted towards another water dipole's negative end are:

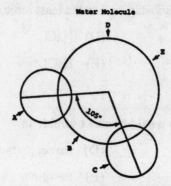

Water Molecule

    (A) A and B.

    (D) C and D.

    (B) A and C.

    (E) D and E.

    (C) B and D.

6. Nuclear Chemistry

There are three main types of particles that can escape from an atom's nucleus, named after the first three letters of the Greek alphabet, α, β, and γ (alpha, beta, and gamma). α particles consist of two neutrons and two protons bound together. This arrangement might remind one of a helium nucleus, and that is exactly what it is. After an α particle slows down, it can capture two electrons and can become a helium atom. Decay by α emission results in a decrease of atomic mass by four and atomic number by two.

PROBLEM

The radioactive gas radon-222 decays by alpha emissions. What does it decay to?

SOLUTION

We can write the equation as

$$^{222}_{86}\text{Rn} \rightarrow \alpha + ?$$

It is easier to write the equation using the symbol for helium, rather than α, this gives us:

$$^{222}_{86}\text{Rn} \rightarrow\ ^{4}_{2}\text{He} + ?$$

We can now balance our equation, as the mass number (superscript) must have the same total on each side of the equation. If we subtract 4 from 222, we get 218, this is the atomic mass (weight) of the new atom. Likewise, the atomic number (subscript) must have the same total on each side of the arrow. If we subtract 2 from 86, we get 84. This is our new atomic number, and by looking at the periodic table, we can see it belongs to polonium. We can now properly write the equation as

$$^{222}_{86}\text{Rn} \rightarrow\ ^{4}_{2}\text{He} +\ ^{218}_{84}\text{Po} \text{ or } ^{222}_{86}\text{Rn} \rightarrow \alpha +\ ^{218}_{84}\text{Po}.$$

Beta particles are electrons that are emitted from a nucleus when a neutron is converted to a proton and an electron. The electron is shot out of the nucleus sometimes at a speed very close to that of light. Once it slows down, outside of the nucleus, it becomes an ordinary electron and behaves accordingly. Beta (β^-) decay does not cause any change in atomic weight, but due to the conversion of a neutron into a proton, it increases the atomic number by 1.

PROBLEM

The Polonium-218 from the previous example decays by β^- emissions. What does it decay to?

SOLUTION

We would start our equation with what we know.

$$^{218}_{84}\text{Po} \rightarrow \beta^- + ?$$

It is useful to substitute the electron symbol for β^-, resulting in the following:

$$^{218}_{84}\text{Po} \rightarrow\ ^{0}_{-1}\text{e} + ?$$

Notice that the atomic weight (mass) of the electron is 0 (it isn't exactly, it's really $1/1832$, but for this purpose 0 is close enough), and the "atomic number" is −1. The atomic number is the measure of positive charges (protons); an electron has a negative charge; hence −1. Again, as in a decay, the atomic numbers and atomic weights (mass) must total the same on both sides of the arrow. If we subtract 0 from 218, we still have 218, our mass remains unchanged. If we subtract −1 from 84, we get $84 - (-1) = 84 + 1 = 85$. If we look at the periodic table, element number 85 is astatine. We can then write the correct equation as:

$$^{218}_{84}\text{Po} \rightarrow\ ^{0}_{-1}\text{e} +\ ^{218}_{85}\text{At}, \text{ or } ^{218}_{84}\text{Po} \rightarrow\ ^-\beta +\ ^{218}_{85}\text{At}.$$

Gamma (γ) rays are made up of massless particles (this time the mass really is 0) and resemble X-rays very closely. γ rays are usually emitted along with β⁻ and α radiation (but not always), and they are just not shown in the equations. Pure γ decay can occur from a nucleus in a "metastable" state, when it undergoes a transformation from a higher energy state to a lower one. The resultant atom may or may not be radioactive. The metastable state is designated by the lower-case letter "m" after the atomic mass (e.g., Tc-99m, as opposed to Tc-99). Since the γs emitted from the nucleus possess neither charge nor mass, they change neither the atomic number nor mass.

PROBLEM

How would one write the decay of indium-113m to stable (non-radio-active) indium?

SOLUTION

$$^{113m}_{49}In \rightarrow \gamma + ^{113}_{49}In$$

Penetrating Power—α particles penetrate the least, since they are relatively slow moving and large. A sheet of paper can stop αs, as does the outer layer of one's skin. β⁻ particles are smaller and more energetic, usually a sheet of lead or several inches of wood will stop β⁻s. Normal clothing will stop most β⁻ rays. γ rays are much harder to stop. They require several inches of lead or several feet of concrete. Since they can easily penetrate plant and animal tissues, they are useful in "tracer" studies.

Half-Life—A useful term in nuclear chemistry is "half-life," which is the amount of time it takes for half of the material to decay away. Half-lives can range from fractions of seconds to trillions of years. They are constant for each isotope, and cannot be altered by changes in temperature, oxidation state, or any other factors.

PROBLEM

Strontium-85 has a half-life of 65.2 days. If we start with 10 grams of Sr-85, how much will be left after 130.4 days?

SOLUTION

$130.4 \div 65.2 = 2$, or two half-lives. After one half-life, 5 grams would be left; after two half-lives, half of that would be left, or 2.5 grams. If we waited another half-life, or a total of 195.6 days, there would be 1.25 grams left.

PROBLEM

Given that the masses of a proton, neutron, and electron are 1.00728, 1.00867, and .000549 amu respectively, how much missing mass is there in $^{19}_{9}F$ (atomic weight = 18.9984)?

SOLUTION

The total number of particles in $^{19}_9\text{F}$ and their total weight can be calculated. The amount of missing mass in $^{19}_9\text{F}$ will be the difference of this calculated weight and the given atomic weight of $^{19}_9\text{F}$. The subscript number 9, in $^{19}_9\text{F}$, indicates the atomic number of fluorine (F). Because the atomic number equals the number of protons, there are nine protons in F. The superscript, 19, indicates the total number of particles in the nucleus. Since the nucleus is composed of protons and neutrons, and there are nine protons, there are 10 neutrons present. In a neutral atom, the number of electrons equals the number of protons. Thus, there are nine electrons. The total number of particles is, thus, 28. The mass and quantity of each particle is now known. Calculating the total weight contribution of each type of particle:

$$\text{Protons:} \quad 9 \times 1.00728 = 9.06552$$

$$\text{Neutrons:} \quad 10 \times 1.00867 = 10.0867$$

$$\text{Electrons:} \quad 9 \times .000549 = \underline{.004941}$$

$$\text{Total mass} = 19.1572$$

It is given that the mass of the fluorine atom is 18.9984. Therefore, the missing mass is $19.1572 - 18.9984 = 0.1588$ amu.

PROBLEM

Complete the following nuclear equations:

(a) $\quad _7\text{N}^{14} + _2\text{H}^4 \rightarrow _8\text{O}^{17} + \dots$

(b) $\quad _4\text{Be}^9 + _2\text{He}^4 \rightarrow _6\text{C}^{12} + \dots$

(c) $\quad _{15}\text{P}^{30} \rightarrow _{14}\text{Si}^{30} + \dots$

(d) $\quad _1\text{H}^3 \rightarrow _2\text{He}^3 + \dots$

SOLUTION

The rules for balancing nuclear equations are: (1) the superscript assigned to each particle is equal to its mass number and the subscript is equal to its atomic number or nuclear charge; (2) a free proton is the nucleus of a hydrogen atom, and is therefore written as $_1\text{H}^1$; (3) a free neutron has no charge and is therefore assigned zero atomic number. Its mass number is one and its notation is $_0\text{n}^1$; (4) an electron, β^-, has zero mass and its atomic number is -1, hence the notation $_{-1}\text{e}^0$; (5) a positron has zero mass and its atomic number is $+1$, hence the notation $_{+1}\text{e}^0$; (6) an alpha particle (α-particle) is a helium nucleus and is represented by $_2\text{He}^4$ or α; (7) gamma radiation (γ) is a form of light, and has no mass and no charge; (8) in a balanced equation, the sum of the subscripts must be the same on both sides of the equation; the sum of the superscripts must also be the same on both sides of the equation.

In equation (a), $_7N^{14} + _2He^4 \rightarrow _8O^{17} + ...$, the sum of the subscripts on the left is $(7 + 2) = 9$. The subscript of one of the products is 8, thus, the other product must have a subscript or net charge of 1. The sum of the superscripts on the left is $(14 + 4) = 18$. The superscript of one of the products is 17, thus the other product on the right must have a superscript or mass number of 1. The particle with a +1 nuclear charge and a mass number of 1 is the proton, $_1H^1$.

In equation (b), $_4Be^9 + _2He^4 \rightarrow _6C^{12} + ...$, the nuclear charge of the second product particle (that is, its subscript) is $(4 + 2) - 6 = 0$. The mass number of the particle (its superscript) is $(9 + 4) - 12 = 1$. Thus, the particle must be the neutron, $_0n^1$.

In equation (c), $_{15}P^{30} \rightarrow _{14}Si^{30} + ...$, the nuclear charge of the second particle is $15 - 14 = +1$. Its mass number is $30 - 30 = 0$. Thus, the particle must be the positron, $_{+1}e^0$.

In equation (d), $_1H^3 \rightarrow _2He^3 + ...$, the nuclear charge of the second product is $1 - 2 = -1$. Its mass number is $3 - 3 = 0$. Thus, the particle must be a β^{-1} or an electron, $_{-1}e^0$.

Drill 6: Nuclear Chemistry

1. How many protons are there in the nucleus of an uncharged atom containing 13 electrons and 14 neutrons?

   (A) 1
   (B) 12
   (C) 13
   (D) 14
   (E) 27

2. What is the value of "x" in the nuclear reaction $^{31}_{14}Si \rightarrow ^{x}_{15}P + ^{0}_{-1}\beta$?

   (A) 28
   (B) 30
   (C) 31
   (D) 32
   (E) 33

3. Identify the element which is converted to the phosphorous isotope and neutron when it collides with alpha particles.

   $$\underline{\qquad} + _2He^4 \rightarrow _{15}P^{30} + _0H^1$$

   (A) $_{13}Al^{27}$
   (B) $_7N^{14}$
   (C) $_{11}Na^{23}$
   (D) $_{25}Mn^{55}$
   (E) $_{15}P^{31}$

4. The radioactive decay of plutonium-241 $^{241}_{94}Pu$ to neptunium (Np) takes place in two steps; first a beta emission followed by an alpha emission. The symbol of Np after the radioactive emissions should be

(A) $^{236}_{93}Np.$

(D) $^{235}_{92}Np.$

(B) $^{237}_{93}Np.$

(E) $^{236}_{92}Np.$

(C) $^{241}_{95}Np.$

5. What type of radiation is emitted when uranium decays by the nuclear reaction

$$^{238}_{92}U \rightarrow ^{234}_{90}Th + ? \text{ radiation}$$

(A) No radiation is emitted.

(D) Cosmic

(B) Beta

(E) Gamma

(C) Alpha

ATOMIC THEORY
DRILLS

ANSWER KEY

Drill 1 — Atomic Weight and Components of Atomic Structure

1. (C)
2. (B)
3. (E)

Drill 2 — Valence and Electron Dot Diagrams

1. (D)
2. (B)
3. (A)

Drill 3 — Ionic and Covalent Bonding

1. (E)
2. (B)
3. (B)

Drill 4 — Electronegativity and Ionization Energy

1. (E)
2. (C)
3. (A)
4. (C)
5. (D)

Drill 5 — Polarity of Bonds

1. (D)
2. (A)
3. (C)
4. (B)
5. (E)
6. (E)
7. (C)
8. (B)

Drill 6 — Nuclear Chemistry

1. (C)
2. (C)
3. (A)
4. (B)
5. (C)

GLOSSARY: ATOMIC THEORY

Alpha Particle

A particle that consists of two neutrons and two protons bound together.

Atomic Number

The number of electrons in an atom of an element or the number of protons in the nucleus of that atom.

Atomic Weight

The average weight of an element based on a weighted average of its isotopes.

Beta Particle

Electrons that are emitted from a nucleus when a neutron is converted to a proton and an electron.

Covalent Bond

Bond resulting from the sharing of a pair of electrons between atoms.

Dipole

A positive charge and a negative charge separated by a distance.

Electron

An entity that carries one unit of negative charge. It is generally found in the outer layers of an atom.

Electronegativity

A number that measures the relative strength with which the atoms of the element attract valence electrons in a chemical bond.

Gamma Rays

Rays composed of massless particles. They resemble X-rays.

Hydrogen Bond

When a hydrogen is bonded to a highly electronegative atom, it will become partially positively-charged and will be attracted to neighboring electron pairs. This situation is termed a hydrogen bond.

Ionic Bond

Bond created by the transfer of one or more electrons from the valence shell of one atom to the valence shell of another.

Ionization Energy

The energy required to remove an electron from an isolated atom in its ground state.

Isotope

An atom that has the same number of protons as another atom, but a different number of neutrons.

Lewis Structure

A pictorial representation of the electronic structure of an atom or molecule, in which the octet rule is obeyed.

Metallic Bond

The sharing of all valence electrons by all the atoms of a metal.

Neutron

Particles having no charge that are contained within the nucleus of the atom.

Nucleon

A general term for a particle in a nucleus.

Nucleus

The dense center of an atom that contains protons and neutrons.

Octet Rule

States that atoms will obtain or donate electrons in order to have eight outer electrons.

Polarity

The unequal sharing of electrons between the atoms of a bond.

Proton

A positively-charged entity that is found in the nucleus of an atom.

Resonance Structures

An average of two or more Lewis structures which differ in the placement of electrons.

Valence Electrons

Electrons in the outer shell of an atom.

Van der Waals' Forces

Weak attractive forces between molecules.

CHAPTER 15

Quantum Chemistry

➤ Diagnostic Test
➤ Quantum Chemistry
Review & Drills
➤ Glossary

QUANTUM CHEMISTRY DIAGNOSTIC TEST

1. Ⓐ Ⓑ Ⓒ Ⓓ Ⓔ   21. Ⓐ Ⓑ Ⓒ Ⓓ Ⓔ
2. Ⓐ Ⓑ Ⓒ Ⓓ Ⓔ   22. Ⓐ Ⓑ Ⓒ Ⓓ Ⓔ
3. Ⓐ Ⓑ Ⓒ Ⓓ Ⓔ   23. Ⓐ Ⓑ Ⓒ Ⓓ Ⓔ
4. Ⓐ Ⓑ Ⓒ Ⓓ Ⓔ   24. Ⓐ Ⓑ Ⓒ Ⓓ Ⓔ
5. Ⓐ Ⓑ Ⓒ Ⓓ Ⓔ   25. Ⓐ Ⓑ Ⓒ Ⓓ Ⓔ
6. Ⓐ Ⓑ Ⓒ Ⓓ Ⓔ   26. Ⓐ Ⓑ Ⓒ Ⓓ Ⓔ
7. Ⓐ Ⓑ Ⓒ Ⓓ Ⓔ   27. Ⓐ Ⓑ Ⓒ Ⓓ Ⓔ
8. Ⓐ Ⓑ Ⓒ Ⓓ Ⓔ   28. Ⓐ Ⓑ Ⓒ Ⓓ Ⓔ
9. Ⓐ Ⓑ Ⓒ Ⓓ Ⓔ   29. Ⓐ Ⓑ Ⓒ Ⓓ Ⓔ
10. Ⓐ Ⓑ Ⓒ Ⓓ Ⓔ   30. Ⓐ Ⓑ Ⓒ Ⓓ Ⓔ
11. Ⓐ Ⓑ Ⓒ Ⓓ Ⓔ   31. Ⓐ Ⓑ Ⓒ Ⓓ Ⓔ
12. Ⓐ Ⓑ Ⓒ Ⓓ Ⓔ   32. Ⓐ Ⓑ Ⓒ Ⓓ Ⓔ
13. Ⓐ Ⓑ Ⓒ Ⓓ Ⓔ   33. Ⓐ Ⓑ Ⓒ Ⓓ Ⓔ
14. Ⓐ Ⓑ Ⓒ Ⓓ Ⓔ   34. Ⓐ Ⓑ Ⓒ Ⓓ Ⓔ
15. Ⓐ Ⓑ Ⓒ Ⓓ Ⓔ   35. Ⓐ Ⓑ Ⓒ Ⓓ Ⓔ
16. Ⓐ Ⓑ Ⓒ Ⓓ Ⓔ   36. Ⓐ Ⓑ Ⓒ Ⓓ Ⓔ
17. Ⓐ Ⓑ Ⓒ Ⓓ Ⓔ   37. Ⓐ Ⓑ Ⓒ Ⓓ Ⓔ
18. Ⓐ Ⓑ Ⓒ Ⓓ Ⓔ   38. Ⓐ Ⓑ Ⓒ Ⓓ Ⓔ
19. Ⓐ Ⓑ Ⓒ Ⓓ Ⓔ   39. Ⓐ Ⓑ Ⓒ Ⓓ Ⓔ
20. Ⓐ Ⓑ Ⓒ Ⓓ Ⓔ   40. Ⓐ Ⓑ Ⓒ Ⓓ Ⓔ

QUANTUM CHEMISTRY DIAGNOSTIC TEST

This diagnostic test is designed to help you determine your strengths and your weaknesses in quantum chemistry. Follow the directions and check your answers.

Study this chapter for the following tests:
AP Chemistry, CLEP General Chemistry,
GED, MCAT, MSAT, PRAXIS II Subject Assessment:
Chemistry, SAT II: Chemistry, GRE Chemistry

---

40 Questions

DIRECTIONS: Choose the correct answer for each of the following problems. Fill in each answer on the answer sheet.

1. An sp^3 configuration is represented by which orientation?

   (A) Tetrahedral

   (B) Trigonal planar

   (C) Linear

   (D) Trigonal bipyramidal

   (E) Octahedral

2. Carbon's valence shell electron configuration can be symbolized as

   (A) s^1, p^1.

   (B) s^1, p^2.

   (C) s^2, p^2.

   (D) s^2, p^4.

   (E) s^4, p^2.

3. An atom of calcium has the same number of electrons as all of the following except

   (A) K^-.

   (B) Sc^+.

   (C) Kr^{2-}.

   (D) Cl^{3-}.

   (E) V^{3+}.

4. Which of the following molecular geometries is typical of sp bonding?

   (A) Tetrahedral

   (B) Trigonal planar

(C) Linear   (D) Trigonal bipyramidal

(E) Octahedral

5. A tetrahedral geometry with one corner occupied by a lone electron pair best represents

(A) H_2O.   (D) $AlCl_3$.

(B) H_3O^+.   (E) NH_4^+.

(C) BF_3.

6. Which of the following is the usual bond hybridization exhibited by carbon in alkanes (i.e., C–C single bonds only)?

(A) sp (D) sp^3d

(B) sp^2 (E) sp^3d^2

(C) sp^3

7. The structure shown here best resembles

(A) methane.

(B) a hydronium ion.

(C) water.

(D) ammonia.

(E) carbon dioxide.

8. The valence of tellurium is

(A) 1.   (D) 4.

(B) 2.   (E) 5.

(C) 3.

9. Which of the following best describes the bonding usually found in Group IIIA elements?

(A) s (D) sp^3

(B) sp (E) sp^3d^2

(C) sp^2

10. Which hybridization best describes boron in compounds of the type BF_3 and BH_3?

    (A) s (D) sp^3

    (B) sp (E) sp^3d^2

    (C) sp^2

11. Which of the following best describes four equivalent bonds having $^1/_4s$ character and $^3/_4p$ character?

    (A) sp (D) sp^3d

    (B) sp^2 (E) sp^3d^2

    (C) sp^3

12. A dsp^2 configuration is represented by which orientation?

    (A) Tetrahedral   (D) Trigonal bipyramidal

    (B) Trigonal planar   (E) Square planar

    (C) Octahedral

13. Which of the following represents a resonance structure for CO_2?

    (A) $O = C - O :$

    (B) $O \equiv C - O :$

    (C) $O - C = O :$

    (D) $: O - C - O :$

    (E) CO_2 doesn't form resonance structures.

14. The transition metals are characterized by

    (A) completely filled d subshells.

    (B) completely filled f subshells.

    (C) partially filled d subshells.

    (D) partially filled f subshells.

    (E) Both (A) and (C).

15. Calcium has an atomic number of 20. Its electron configuration can be summarized as

    (A) $1s^2, 2s^2, 2p^6, 3s^2, 3p^6, 4s^2.$

(B) $1s^2, 2s^2, 2p^6, 3s^2, 3p^4, 4s^4$.

(C) $1s^2, 2s^4, 2p^4, 3s^2, 3p^6, 4s^2$.

(D) $1s^1, 2s^4, 2p^4, 3s^2, 3p^6, 4s^4$.

(E) $1s^1, 2s^3, 2p^5, 3s^1, 3p^6, 4s^2$.

16. Which subshell of the transition metals is incomplete in most cases?

(A) *s*   (D) *f*

(B) *p*   (E) They are all complete.

(C) *d*

17. Which of the following elements can form bonds with sp^3 hybridization?

(A) Sodium   (D) Oxygen

(B) Nitrogen   (E) Fluorine

(C) Carbon

18. The lanthanide and actinide series are characterized by incomplete

(A) *s* subshells.   (D) *f* subshells.

(B) *p* subshells.   (E) *d* and *f* subshells.

(C) *d* subshells.

19. The valence shell of all alkaline earth metals can be designated by

(A) ns^1.   (D) np^1.

(B) ns^2.   (E) nd^1.

(C) ns^2np^1.

20. Which of the following represents the final state of the process

$$\frac{1}{\text{⥮}} \rightarrow X + \text{electromagnetic radiation}$$

(A)   (B)

(C)   (D)

(E) $\overline{}$
$\underset{\text{\Large ⥮}}{}$

21. The bond angles between carbon and hydrogen in methane (CH_4) are best labeled as

    (A) covalent.   (D) tetrahedral.

    (B) ionic.   (E) trihybrid.

    (C) linear.

22. The metalloids have a valence shell of

    (A) $2p^1$ and $6p^5$.   (D) Both (A) and (B)

    (B) $3p^1$ and $5p^4$.   (E) (A), (B), and (C)

    (C) $3p^2$ and $5p^3$.

23. Which type of bond hybridization is identified by a tetrahedral structure?

    (A) sp (D) sp^3d^2

    (B) sp^2 (E) sp^2d^3

    (C) sp^3

24. The shell configuration $1s^2\ 2s^2\ 2p^6\ 3s^2\ 3p^4$ is that of

    (A) O.   (D) Cl^-.

    (B) S.   (E) Ar.

    (C) P^+.

25. Pi bonds are found in all of the following except

    (A) alkanes.   (D) Both (A) and (B)

    (B) alkynes.   (E) (A), (B), and (C)

    (C) aldehydes.

26. For the reaction:

$$FeCl_2 + KMnO_4 + HCl \rightarrow FeCl_3 + MnCl_2 + KCl + H_2O$$
$$\text{(aq)} \qquad \text{(aq)} \qquad \text{(aq)} \qquad \text{(aq)} \qquad \text{(aq)} \qquad \text{(aq)} \qquad \text{(l)}$$

    the net ionic equation is

    (A) $Fe^{2+} \rightarrow Fe^{3+}$.

(B) $Fe^{2+} + MnO_4^- + H^+ \rightarrow Fe^{3+} + Mn^{2+} + H_2O$.

(C) $MnO_4^- + H^+ \rightarrow Mn^{2+} + H_2O$.

(D) $FeCl_2 + MnO_4^- \rightarrow FeCl_3 + Mn^{2+}$.

(E)   None of the above

27. sp^2 hybridization will be found for carbon in

(A) CH_4.

(D) CH_3OH.

(B) C_2H_4.

(E) CH_3OCH_3.

(C) C_2H_2.

28.   What is the electronic configuration of sulfur?

(A) $1s^2\, 2s^2\, 2p^8\, 3s^2\, 3p^2$

(D) $1s^2\, 2s^2\, 2p^8\, 3s^2\, 3p^8$

(B) $1s^2\, 2s^2\, 2p^6\, 3s^2\, 3p^4$

(E) $1s^2\, 2s^2\, 3s^2\, 2p^8\, 3p^8$

(C) $1s^2\, 2s^2\, 3s^2\, 2p^8\, 3p^2$

29.   A triple bond may best be described as

(A)   two sigma bonds and one pi bond.

(B)   one sigma bond and two pi bonds.

(C)   two sigma bonds and two pi bonds.

(D)   three sigma bonds.

(E)   three pi bonds.

30.   Select the element with an atomic number of 15 and 5 electrons in its valence shell.

(A)   Chlorine

(D)   Phosphorus

(B)   Nitrogen

(E)   Sulfur

(C)   Oxygen

31.   How many orbitals can one find in a p subshell?

(A)   2

(D)   7

(B)   3

(E)   14

(C)   6

32. VSEPR theory predicts that the geometry of the ICl_4^- ion is

    (A) tetrahedal.

    (D) linear.

    (B) square planar.

    (E) trigonal pyramidal.

    (C) octahedral.

33. The number of electrons in a sulfur atom associated with the primary quantum number n is

    (A))2)8)6.

    (D))2)2)2)6)4.

    (B))2)10)4.

    (E))2)2)2)4)6.

    (C))2)2)6)2)4.

Questions 34–37 refer to energy levels and the maximum number of electrons they can hold.

    (A) 8

    (D) 12

    (B) 18

    (E) 2

    (C) 32

34. *K*

35. *L*

36. *M*

37. *N*

38. Oxygen's valence shell electron configuration can be symbolized as:

    (A) s^2p^2.

    (D) s^4p^4.

    (B) s^2p^4.

    (E) s^4p^6.

    (C) s^4p^2.

39. Which of the following molecular geometries is typical of sp^2 bonding?

    (A) Tetrahedral

    (D) Trigonal planar

    (B) Square planar

    (E) Trigonal bipyramidal

    (C) Linear

40. Aluminum has three electrons in its valence shell. Sulfur has six electrons at its similar site. A probable compound between them has the formula.

(A) AlS.

(D) Al_2S_3.

(B) AlS_2.

(E) Al_3S_2.

(C) AlS_3.

QUANTUM CHEMISTRY DIAGNOSTIC TEST

ANSWER KEY

1.	(A)	11.	(C)	21.	(D)	31.	(B)
2.	(C)	12.	(E)	22.	(E)	32.	(B)
3.	(C)	13.	(E)	23.	(C)	33.	(A)
4.	(C)	14.	(E)	24.	(B)	34.	(E)
5.	(B)	15.	(A)	25.	(A)	35.	(A)
6.	(C)	16.	(C)	26.	(B)	36.	(B)
7.	(D)	17.	(C)	27.	(B)	37.	(C)
8.	(B)	18.	(D)	28.	(B)	38.	(B)
9.	(C)	19.	(B)	29.	(B)	39.	(D)
10.	(C)	20.	(E)	30.	(D)	40.	(D)

DETAILED EXPLANATIONS
OF ANSWERS

1. **(A)** An sp^3 configuration is represented by the tetrahedral orientation because it looks as shown below.

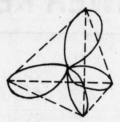

2. **(C)** Carbon has four valence electrons. Two occupy the smaller s orbital and the two remaining fill the larger p orbital.

3. **(C)** Calcium, atomic number 20, has 20 electrons. Potassium has 19 electrons, so K^- has 20. Scandium has 21 electrons, so Sc^+ has 20. Krypton has 36 electrons so Kr^{2-} has 38. Chlorine has 17 electrons so Cl^{3-} has 20. Vanadium has 23 electrons so V^{3+} has 20.

4. **(C)** In sp hybridization we are once again separating the orbitals as much as possible. The two sp orbitals point in opposite directions along a line. Note that hybridization does not always result in maximum separation of orbitals. dsp^2 hybridization results in a square, planar geometry:

5. **(B)** Water, the hydronium ion, and the ammonium ion have tetrahedral geometries; except lone electron pairs occupy two corners for a water molecule and there are no lone electrons in the ammonium ion.

   Both BF_3 and $AlCl_3$ assume trigonal planar geometries since there are no lone electrons on boron or aluminum.

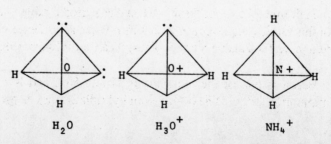

6.   **(C)**   Carbon generally exhibits sp^3 bond hybridization in alkanes.

7.   **(D)**   Ammonia occurs in a trigonal bipyramidal geometry so that the three electron clouds of the N–H bonds and the nonbonding electron pair on nitrogen are farthest apart from each other.

Methane assumes a regular tetrahedral geometry in order to maximize these distances.

Water assumes a bent geometry,

and the hydronium ion is similar to ammonia.

All of the previous examples may be seen as regular tetrahedral by considering the nonbonding electron pairs as "something" bonded to the central atom. In contrast, carbon dioxide forms a linear molecule since there are no nonbonded electrons associated with the central carbon.

$$\ddot{O} = C = \ddot{O}$$

8.   **(B)**   Tellurium, Te, is in family VI of the periodic table. As with other nonmetals in this family, e.g., oxygen and sulfur, it has six valence electrons and thus needs two more for a stable outer 8. Thus, it can form two covalent bonds.

9.   **(C)**   Boron is one of the elements in Group IIIA and has the valence electron configuration $2s^2 2p^1$. The boron atom hybridizes from its ground state as

follows:

$$2p \overset{1}{\underline{}} \,{-}\,{-}\; \text{energy} \qquad 2p \overset{1}{\underline{}}\overset{1}{\underline{}} \,{-} \longrightarrow 2p \,{-}$$

$$2x \underline{\underline{\downarrow}} \qquad\qquad\qquad 2s \underline{1} \qquad\qquad sp^2 \overset{1}{\underline{}}\overset{1}{\underline{}}\overset{1}{\underline{}}$$

The hybrid orbitals are composed of one original s orbital and two original p orbitals. Thus, the hybridization is sp^2.

10. **(C)** We can choose boron, one of the elements in Group IIIA ($2s^2p^1$), to explain the hybridization. The boron atom hybridizes from its ground state as follows:

$$2p \;\overset{1}{\underline{}}\,{-}\,{-} \xrightarrow{\;\text{energy}\;} 2p \;\overset{1}{\underline{}}\,\overset{1}{\underline{}}\,{-} \longrightarrow 2p \;{-}$$

$$2s \;\underline{\underline{1}} \qquad\qquad\qquad 2s \;\underline{1} \qquad\qquad sp^{2}\underline{1}\,\underline{1}\,\underline{1}$$

The hybrid orbitals are composed of one original s orbital and two original p orbitals. Thus, the hybridization is sp^2.

11. **(C)** Two sp hybrid bonds are produced by the combination of one s orbital and one p orbital. Three sp^2 hybrid orbitals result from the combination of one s orbital and two p orbitals. Four sp^3 hybrid orbitals are produced from one s orbital and three p orbitals. In a likewise manner, the combination of one s orbital, three p orbitals, and two d orbitals give six equivalent sp^3d^2 orbitals.

12. **(E)** A dsp^2 configuration is represented by the square planar orientation and it looks as shown here.

13. **(E)** Carbon dioxide does not have any resonance structures that provide eight electrons around each carbon atom. The structure is

$$\ddot{O} = C = \ddot{O}$$

14. **(E)** The transition metals may have either completely filled or partially filled, but not empty, d subshells ($3d$, $4d$, and $5d$). Lanthanides and actinides are

characterized by the electrons in the $4f$ and $5f$ subshells, respectively.

15.   **(A)**   Since oxygen has atomic number 20, the first shell can hold a maximum of two electrons. Eight plus eight fill out the next two shells with two left over for a fourth outer shell. s and p refer to the two different shaped orbital spheres within the three energy levels beyond the first.

16.   **(C)**   The transition metals are characterized by incomplete d subshells. The elements of the lanthanide and actinide series have incomplete f subshells.

17.   **(C)**   The element in question must be able to engage in four covalent bonds if it is to have sp^3 hybridization. The only element given which fulfills this criteria is carbon (for example, in CH_4). Nitrogen may also form four bonds, as in NH_4^+, but one of these bonds is coordinate covalent, with nitrogen donating both bonding electrons.

18.   **(D)**   The lanthanide and actinide series are characterized by incomplete f subshells. Transition metals have incomplete d subshells. Elements in Groups IIIA-VIIA are characterized by incomplete p subshells while elements in Group IA have incomplete s subshells.

19.   **(B)**   The alkaline earth metals are those in the second column of the periodic table. They have two s electrons in their valence sell, designated ns^1 and ns^2, in which n is the principal quantum number. ($n = 2$ for Be, $n = 3$ for Mg, etc.)

20.   **(E)**   An excited electron always falls back to the lowest unfilled energy level upon emission of light (electromagnetic radiation).

21.   **(D)**   Methane, CH_4, is a symmetrical molecule in terms of the direction of its four covalent bonds. Each C–H bond is at an approximate 109° bond angle, oriented toward the corner of an imaginary tetrahedron (a four-sided figure).

22.   **(E)**   The metalloids are indicated by the heavy-line staircase towards the right of the periodic table. All elements in contact with this line are metalloids. Referring to the periodic table, we find metalloids with valence shells of $2p^1$, $3p^1$, $3p^2$, $4p^2$, $4p^3$, $5p^3$, $5p^4$, $6p^4$, and $6p^5$.

23.   **(C)**   A linear geometry is obtained when the central atom is sp hybridized. sp^2 hybridization results in a trigonal planar geometry and a regular tetrahedron is obtained when sp^3 hybridization occurs. sp^3d^2 hybridization is characterized by an octahedral geometry.

24.   **(B)**   The given electronic configuration indicates an atom with 16 electrons. Sulfur, atomic number 16, has 16 electrons. Oxygen, atomic number 8, has

8 electrons, P^+, atomic number 15, has 14 electrons, Cl^-, atomic number 17, has 18 electrons, and Ar, atomic number 18, has 18 electrons.

25. **(A)** Pi bonds, or multiple bonds, are found in the carbon skeleton of alkynes and the carbonyl functional group of aldehydes.

26. **(B)** In the net ionic equation only ions in solution that appear unchanged on both sides of the equation are omitted. If an ion changes in any way during the reaction, it must be shown in the net ionic equation. In the reaction the chloride ions are not oxidized or reduced, and do not undergo any changes and are therefore omitted. On the other hand, the Fe in $FeCl_2$ is oxidized to Fe^{3+} in $FeCl_3$ and must be shown in the net ionic equation as Fe^{2+} going to Fe^{3+}. Also, the Mn in MnO_4^- changes from Mn^{7+} to Mn^{2+} and must also be shown in the net ionic equation. Since the reaction is taking place in aqueous acid solution, the H^+ ion in HCl is shown on the reactant side of the equation, and the water, is shown on the product side. Both the reactants and products are soluble in water, and they are written in their ionic form in the net ionic equation, e.g. $KMnO_4$ is written as MnO_4^-. In view of these changes in oxidation states, the only correct net ionic equation is choice (B).

27. **(B)** A simple method for determining hybridization in carbon compounds is by determining how many atoms are attached to the carbon atom. If two atoms are attached, the hybridization is sp, if three: sp^2, and if four: sp^3. Thus,

$$H - C \equiv C - H \qquad\qquad \text{has } sp \text{ hybridization}$$

has sp^2 hybridization

has sp^3 hybridization

28. **(B)** Consulting the periodic table, we find that sulfur (atomic number 16) has the electronic configuration

$$1s^2\, 2s^2\, 2p^6\, 3s^2\, 3p^4.$$

29. **(B)** A single bond consists of one sigma bond while double and triple bonds consist of one sigma bond and one and two pi bonds, respectively.

30. **(D)** The electron array is 2–8–5 among three energy levels. Checking phosphorus in the periodic table shows phosphorus in family V with the atomic number 15.

31. (B) The number of orbitals in a subshell is described by the quantum number m_l in the following manner:

subshell	l	$m_l = 2l + 1$
s	0	1
p	1	3
d	2	5
f	3	7

32. (B) Valence Shell Electron Pair Repulsion (VSEPR) is a convenient method of predicting molecular geometries based on the fact that electron pairs, either bonded pairs or lone pairs, tend to orient themselves as far from one another as possible to reduce electron pair – electron pair repulsion. Iodine has seven valence electrons and since the charge on the ion, ICl_4^- is negative, it is viewed as having eight electrons associated with it. Of the eight electrons, four are involved in bonding (four shared electron pairs) leaving four electrons or two pairs that are nonbonding):

$$\text{total electron pairs} = 2(\text{lone pairs}) + 4(\text{bonding pairs}) = 6$$

The possible structure for the arrangement of the six pairs of electrons (4 bonding pairs and two lone pairs) is then either (A) or (B).

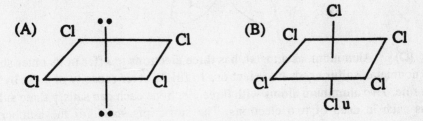

To choose the correct structure we evaluate the lone pair – lone pair interaction, and then the lone pair – bonding pair interaction.

Structure (A)	**Structure (B)**
0 LP – LP @ 90°	1 LP – LP @ 90°
8 LP – BP @ 90°	6 LP – BP @ 90°

Evaluating these interactions we see that structure (B) has a lone pair – lone pair interaction at 90°. This is a highly unfavorable situation and is not expected to occur. Hence, structure (A) is favored. The atoms in ICl_4^- are all in a plane, and the geometry of the molecule is best described as square planar.

Note: Although structure (B) has two less lone pair – bonding pair interactions, it is not sufficient to overcome the repulsion of the lone pair – lone pair interactions at 90°.

33. **(A)** The electronic configuration of sulfur is $1s^2\,2s^2\,2p^6\,3s^2\,3p^4$ as given by the periodic table. Thus, we have two electrons in the first energy level, eight in the second, and six in the third. This configuration is represented as)2)8)6.

34. **(E)**
35. **(A)**
36. **(B)**
37. **(C)** *K*, *L*, *M*, and *N* rank the energy levels from the inside out with a 2-8-18-32 capacity for electrons.

38. **(B)** Oxygen, in family VI of the periodic table, has six valence electrons. Two of them saturate the smaller *s* orbital with four remaining for the larger *p* orbital.

39. **(D)** In sp^2 hybridization, three bonding orbitals are positioned as far from each other as possible. This requirement yields trigonal-planar geometry:

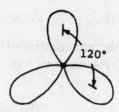

40. **(D)** Aluminum, as a metal, has three electrons to offer in its outer shell. As a nonmetal, sulfur needs two electrons to fill its outer capacity of eight by the octet rule. *Two* aluminum atoms with three electrons each can satisfy three sulfur atoms each in need of two electrons. The subscripts stand for the number of atoms of each element.

QUANTUM CHEMISTRY REVIEW

1. Pauli Exclusion Principle, Hund Rules, and Electronic Configuration

The name "quantum" comes from the concept that an electron (or any other particle) can only have discrete energies. Most values of the energy are forbidden so that if an electron changes energy it must do so in a quantum jump.

All electrons in the atoms can be described by four quantum numbers and no two electrons can have the same set of quantum numbers. The quantum numbers are, for historical reasons, called respectively n, l, m_l, and m_s. n is the principal quantum number, l is the orbital quantum number, m_l is the magnetic quantum number, and m_s is the spin. The allowed quantum numbers are as follows:

Quantum Number	Permitted Values
n	1, 2, 3, ... etc.
l	0, 1, 2, ... $n-1$
m_l	$+l$ to $-l$
m_s	$+\frac{1}{2}, -\frac{1}{2}$ only

The electronic configuration designations, for example $1s^2, 2s^2, 2p^2$ for carbon, are, in fact, a tabulation of the quantum numbers of each of the electrons in the atom. Tables 1 and 2 show the relationship between the electronic configuration designation and the quantum numbers associated with each electron. In Table 1 the following electronic configuration for neon is tabulated.

Table 1: Electronic Configuration of Neon

Element	Atomic Number	$1s$	$2s$	$2p$	$3s$	$3p$	Electronic Configuration
Ne	10	2	2	6			$1s^2, 2s^2, 2p^6$

Table 2: Quantum Numbers for Electrons in Neon Atom

n	l	m_l	m_s	Designation of electronic configurations
1	0	0	$+\frac{1}{2}$	$1s$
1	0	0	$-\frac{1}{2}$	$1s$ (two total $1s$ electrons)

2	0	0	$+\frac{1}{2}$	$2s$
2	0	0	$-\frac{1}{2}$	$2s$ (two total $2s$ electrons)
2	1	+1	$+\frac{1}{2}$	$2p$
2	1	+1	$-\frac{1}{2}$	$2p$
2	1	0	$+\frac{1}{2}$	$2p$
2	1	0	$-\frac{1}{2}$	$2p$
2	1	−1	$+\frac{1}{2}$	$2p$
2	1	−1	$-\frac{1}{2}$	$2p$ (six total $2p$ electrons)

The "exclusion" of two electrons having the same set of quantum numbers is called the Pauli exclusion principle. The order in which the quantum numbers occur in the ground, or lowest, energy state is governed by the Hund rule. This rule states that for a set of equal-energy orbitals, each orbital is occupied by one electron before any orbital has two. Therefore, the first electrons to occupy orbitals within a sublevel have parallel spins. The rule is shown in the table above.

Two principal types of bonds, sigma (σ) and pi (π) bonds, can be formed in a molecular orbital. The sigma bond is formed when the molecular orbital (of shared electrons) is symmetrical around the nucleus while the pi bond is formed from parallel p orbitals.

PROBLEM

Write possible sets of quantum numbers for electrons in the second main energy level.

SOLUTION

In wave mechanical theory, four quantum numbers are needed to describe the electrons of an atom. The first or principal quantum number, n, designates the main energy level of the electron and has integral values of 1, 2, 3, ... The second quantum number, l, designates the energy sublevel within the main energy level. The values of l depend upon the value of n and range from zero to $n -$ 1. The third quantum number, m_l, designates the particular orbital within the energy sublevel. The number of orbitals of a given kind per energy sublevel is equal to the number of m_l values $(2l + 1)$. The quantum number m_l can have any integral value from $+l$ to $-l$ including zero. The fourth quantum number, s, describes the two ways in which an electron may be aligned with a magnetic field $(+^1/_2$ or $-^1/_2)$.

The states of the electrons within atoms are described by four quantum numbers, n, l, m_l, s. Another important factor is the Pauli exclusion principle which states that no two electrons within the same atom may have the same four quantum numbers.

To solve this problem one must use the principles of assigning electrons to their orbitals.

If $n = 2$, l can then have the values 0, 1; m_l can have the values of $+1$, 0 or -1; and s is always $+^1/_2$ or $-^1/_2$.

Thus, the answer is:

$$n = 2$$

$$l = 0, 1$$

$$m_l = +1, 0, -1$$

$$s = +\frac{1}{2}, -\frac{1}{2}$$

	n	l	m_l	m_s
2s	2	0	0	$+\frac{1}{2}$
	2	0	0	$-\frac{1}{2}$
2p	2	1	+1	$+\frac{1}{2}$
	2	1	+1	$-\frac{1}{2}$
	2	1	0	$+\frac{1}{2}$
	2	1	0	$-\frac{1}{2}$
	2	1	−1	$+\frac{1}{2}$
	2	1	−1	$-\frac{1}{2}$

PROBLEM

Apply Hund rules to obtain the electron configuration for Si, P, S, Cl, and Ar.

Si	$3p^2$	↑	↑	__
P	$3p^3$	↑	↑	↑
S	$3p^4$	↑↓	↑	↑
Cl	$3p^5$	↑↓	↑↓	↑
Ar	$3p^6$	↑↓	↑↓	↑↓

SOLUTION

The ground state of an atom is that in which the electrons are in the lowest possible energy level. Each level may contain two electrons of opposite spin. When there are several equivalent orbitals of the same energy, Hund rules are used to decide how the electrons are to be distributed between the orbitals:

1) If the number of electrons is equal to or less than the number of equivalent orbitals, then the electrons are assigned to different orbitals.

2) If two electrons occupy two different orbitals, their spins will be parallel in the ground state. Hund rules state that the electrons attain positions as far apart as possible which minimizes the repulsion obtained form interelectronic forces.

To solve this problem one must:

(1) Find the total number of electrons within the atom.

(2) Determine the number of valence electrons.

(3) Find the number of electrons in the highest equivalent energy orbital.

The total number of electrons in an atom is equal to that atom's atomic number.

Thus,   Si   has 14 electrons

P   has 15 electrons

S   has 16 electrons

Cl   has 17 electrons

Ar   has 18 electrons

Next, from the orbital configuration:

$_{14}$Si $1s^2\, 2s^2\, 2p^6\, 3s^2 \,|\, 3p^2$

$_{15}$P $1s^2\, 2s^2\, 2p^6\, 3s^2 \,|\, 3p^3$

$_{16}$S $1s^2\, 2s^2\, 2p^6\, 3s^2 \,|\, 3p^4$

$_{17}$Cl $1s^2\, 2s^2\, 2p^6\, 3s^2 \,|\, 3p^5$

$_{18}$Ar $1s^2\, 2s^2\, 2p^6\, 3s^2 \,|\, 3p^6$

One knows that the highest equivalent energy orbital is the $3p$ orbital. The number of electrons in this orbital increases by 1 starting with 2 for Si, then, 3 for P, 4 for S, 5 for Cl, and 6 for Ar. Thus, Ar closes this orbital. Using Hund rules, the electron configurations can be written as shown in the accompanying figure.

PROBLEM

You are given H, N, O, Ne, Ca, Al, and Zn. Determine which of these atoms (in their ground state) are likely to be paramagnetic. Arrange these elements in the order of increasing paramagnetism.

SOLUTION

Paramagnetic substances possess permanent magnetic moments. An electric current flowing through a wire produces a magnetic field around the wire. Magnetic fields are thus produced by the motion of charged particles. Then, a single spinning electron, in motion around the nucleus, should behave like a current flowing in a closed circuit of zero resistance and therefore should act as if it were a small bar magnet with a characteristic permanent magnetic moment. The magnetism of an isolated atom results from two kinds of motion: the orbital motion of the electron around the nucleus, and the spin of the electron around its axis. Two spin orientations are permitted for electrons, $+ \frac{1}{2}$ and $- \frac{1}{2}$. Two electrons occupy each filled orbital and their opposing spins cancel out the magnetic moments; thus, for an atom to be paramagnetic, it must contain unpaired electrons.

H $(Z = 1)$: $1s^1$ The subshell, s, has only one electron as indicated by the superscript number. In the s subshell, you have only one orbital. Each orbital can hold two electrons. Therefore, this electron is unpaired and H is paramagnetic.

N $(Z = 7)$: $1s^2\, 2s^2\, 2p^3$ The p subshell has three orbitals that contain a total of three electrons. Because electrons have the same charge, they try to avoid each other, if possible. Thus, each electron is in a different orbital. Therefore, they are unpaired and N is paramagnetic.

O $(Z = 8)$: $1s^2\, 2s^2\, 2p^4$ The p subshell has three orbitals with four electrons. Recalling the information given above, this means that one orbital has two electrons. The other two orbitals possess one electron each. Thus, there are two unpaired electrons. O is paramagnetic.

Ne $(Z = 10)$: $1s^2\, 2s^2\, 2p^6$.   Here, all orbitals contain two electrons each. No electron is unpaired. Thus, Ne is not paramagnetic.

Ca $(Z = 20)$: $1s^2\, 2s^2\, 2p^6\, 3s^2\, 3p^6\, 4s^2$.   Again, all orbitals have two electrons. Therefore, calcium is not paramagnetic.

Al $(Z = 13)$: $1s^2\, 2s^2\, 2p^6\, 3s^2\, 3p^1$.   The $3p$ subshell has only one electron for three orbitals. It must be unpaired, as such. It is paramagnetic.

Zn $(Z = 30)$: $1s^2\, 2s^2\, 2p^6\, 3s^2\, 3p^6\, 3d^{10}\, 4s^2$.   Each orbital has two electrons. Thus, no paramagnetism exists.

The order of paramagnetism (increasing) is proportional to the number of unpaired electrons. Thus, H, Al, O, and N is the order of increasing paramagnetism.

Drill 1: Pauli Exclusion Principle, Hund Rules, and Electronic Configuration

1. The quantum numbers for a given orbital are $n = 2$, $l = 1$, $m = 0$. We would usually represent this as

   (A) $1s$.

   (B) $2s$.

   (C) $1p_z$.

   (D) $2p_z$.

   (E) $1d_{xy}$.

2. Which of the following is/are paramagnetic?

   I. Fe^{2+}

   II. Fe^{3+}

   III. CO^{3+}

   IV. Ni^{2+}

   (A) I and IV

   (B) I only

   (C) II only

   (D) I and II

   (E) All the ions listed.

3. Which is the correct electron configuration for a Cr atom?

   (A) [Ar] $4s^2 3d^4$

   (B) [Ar] $4s^1 3d^4$

   (C) [Ar] $4s^2 3d^6$

   (D) [Ar] $4s^1 3d^5$

   (E) [Ar] $4s^2 3d^5$

4. According to Hund rules, how many unpaired electrons does the ground state of iron have?

   (A) 6

   (B) 5

   (C) 4

   (D) 3

   (E) 2

5. The arrangement of electrons shown here in the orbitals of a carbon atom is based on

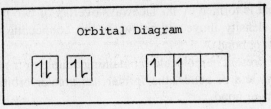

(A) Heisenberg's uncertainty principle.

(B) Lewis' law.

(C) Bohr's model.

(D) Hund rules.

(E) Pauli's exclusion principle.

2. Molecular Orbital Theory

The process of mixing different orbitals of the same atom to form a new set of equivalent orbitals is termed hybridization. The orbitals formed are called hybrid orbitals.

Valence Shell Electron Pair Repulsion (VSEPR) theory permits the geometric arrangement of atoms, or groups of atoms, about some central atom to be determined solely by considering the repulsions between the electron pairs present in the valence shell of the central atom.

Based on VSEPR, the general shape of any molecule can be predicted from the number of bonding and nonbonding electron pairs in the valence shell of the central atom, recalling that nonbonded pairs of electrons (lone pairs) are more repellent than bonded pairs.

Table – Summary of Hybridization

Number of Bonds	Number of Unused e pairs	Type of Hybrid Orbital	Angle between Bonded Atoms	Geometry	Example
2	0	sp	180°	Linear	BeF_2
3	0	sp^2	120°	Trigonal planar	BF_3
4	0	sp^3	109.5°	Tetrahedral	CH_4
3	1	sp^3	90° to 109.5°	Pyramidal	NH_3
2	2	sp^3	90° to 109.5°	Angular	H_2O
6	0	sp^3d^2	90°	Octahedral	SF_6

The bond that is formed by the end-to-end overlap of orbitals to known as a (σ) sigma bond.

The bond that is formed by the sideways overlap of two p orbitals, and that provides electron density above and below the line connecting the bound nuclei, is called a π bond (pi bond).

Pi bonds are present in molecules containing double or triple bonds.

Of the sigma and pi bonds, the former has greater orbital overlap and is generally the stronger bond.

PROBLEM

Compare the bond order of He_2 and He_2^+.

SOLUTION

The bond order, or number of bonds in a molecule, is equal to the difference in the sum of the number of bonding electrons and the number of antibonding electrons divided by two

$$\text{bond order} = \frac{\text{no. of bonding electrons} - \text{no. of antibonding electrons}}{2}$$

This means that the number of bonding and antibonding electrons must be determined. There are two electrons in He, thus in He_2 there are four. These electrons are all in the $1s$ level. For each level, there exists bonding and antibonding orbitals, each of which holds two electrons. Thus, in He_2, two electrons are bonding and two are antibonding. From this, the

$$\text{bond order} = \frac{2-2}{2} = 0.$$

Thus, there are no bonds in He_2; and two He atoms will not bond together to form a molecule of He_2.

In He_2^+, one electron is removed from He_2, which means that there are now three electrons present. They are all in the $1s$ level. This is the lowest energy level that an electron can assume. The three electrons are distributed so that two are in bonding orbitals and one is in an antibonding orbital. Thus,

$$\text{no. of bonds} = \frac{2-1}{2} = 0.5 = \text{bond order.}$$

Because the bond order is not zero, this molecule can form.

PROBLEM

One electron is removed from O_2 and one from N_2. The bonding in O_2 is strengthened, while the bonding in N_2 is weakened. 1) Explain these findings, and 2) predict what happens if an electron is removed from NO.

SOLUTION

There are two types of molecular orbitals: bonding and antibonding. A chemical bond is strengthened by electrons in bonding orbitals and weakened by electrons in antibonding orbitals. For every bonding orbital, there is a corresponding antibonding orbital. Each orbital can hold two electrons. Bond order measures bond strength by giving an indication of the number of electrons in bonding versus antibonding orbitals. Bond order is defined as one-half the number of electrons in the bonding orbital less one-half the number of electrons in the antibonding orbital. Thus, the higher the bond order, the stronger the bond.

To find the original bond order, consider the valence electrons, the outer-most electrons, since they are the only ones that participate in bonding. For N, Z = 7. Its electron configuration is $1s^2 \, 2s^2 \, 2p^3$. The outer-most electrons are in $2p^3$. This means that in N_2 there are a total of six valence electrons. There exist three bonding p orbitals. They can accommodate the six valence electrons. Since no electrons need to be in antibonding orbitals, bond order = $^1/_2(6) - ^1/_2(0) = 3$. When an electron is removed, the bonding orbitals have only five electrons. Thus, the bond order becomes $^1/_2(5) - ^1/_2(0) = 2.5$. Since the bond order went from 3 to 2.5, the bond is weakened by removing the electron from N_2.

For O, $Z = 8$. Since its electron configuration is $1s^2 \, 2s^2 \, 2p^4$, each atom has four valence electrons. O_2 has a total of eight valence electrons. The three bonding p orbitals can hold only six of these electrons. Thus, two electrons are in antibonding orbitals. The bond order is $^1/_2 \, (6) - ^1/_2 \, (2) = 2$. When one removes an electron, it is removed from the antibonding orbitals, if electrons exist in such orbitals. Therefore, the O_2, after removal of the electron, has only one electron in an antibonding electron. This means the bond order becomes $^1/_2 \, (6) - ^1/_2 \, (1) = 2.5$. The bond increased from 2 to 2.5, which means that bond strength increases when one removes an electron from O_2.

To predict what happens to NO bond strength, consider the bond order before and after the electron removal. Recall, an O atom has four valance electrons and N has three valance electrons. The total in NO is seven. The $3p$ bonding orbitals can hold six of these electrons. This means one electron is in an antibonding orbital. Bond order = $^1/_2 \, (6) - ^1/_2 \, (1) = 2.5$. If one removes one electron, the antibonding orbitals contain zero electrons. Thus, the bond order becomes $^1/_2 \, (6) - ^1/_2 \, (0) = 3$. The bond increased from 2.5 to 3, which means the bond strength increases. Therefore, one can predict the chemical bond strength of NO increases when an electron is removed.

PROBLEM

Describe the bonding in linear, covalent $BeCl_2$ and planar, covalent BCl_3. What is the difference in the hybrid orbitals used?

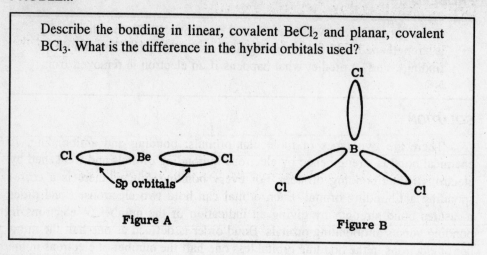

Figure A   Figure B

SOLUTION

The solution to this problem involves the hybridization of orbitals. Once this is clear, the bonding in $BeCl_2$ and BCl_3 will follow.

Quantum theory deals with independent orbitals, such as $2s$ and $2p$. This can be applied to a species, like hydrogen, with only one electron. However, with atoms that contain more than one electron, different methods must be used. For example, the presence of a $2s$ electron perturbs a $2p$ electron, and vice versa, such that a $2s$ electron makes a $2p$ electron take on some s-like characteristics. The result is that the hydrogen-like $2s$ and $2p$ orbitals are replaced by new orbitals, that contain the combined characteristics of the original orbitals. These new orbitals are called hybrid orbitals. The number of hybrid orbitals resulting from hybridization equals the number of orbitals being mixed together. For example, if one mixes an s and p orbital, one obtains two sp hybrid orbitals. One s and two p = three sp^2 orbitals. One s with three p = four sp^3 orbitals. sp orbitals are linear. sp^2 orbitals assume a planar shape, and sp^3 orbitals assume a tetrahedral shape.

Solving: It is given that $BeCl_2$ is linear and covalent. Since sp orbitals are linear, Be undergoes sp hybridization. If something is linear, the bond angle is $180°$. By understanding hybridization, one also knows the geometry of the molecule. A diagram of the bonding resembles Figure 2.1 A.

Given that BCl_3 is planar, and since sp^2 hybridization yields a planar structure, B has sp^2 hybridized bonding with angles of $120°$ (Figure 2.1 B).

Drill 2: Molecular Orbital Theory

1. How many resonance structures contribute to the hybrid structure of the nitrate ion?

    (A) 1

    (B) 2

    (C) 3

    (D) 4

    (E) 5

2. Using molecular orbital theory predict the bond order of the He_2^+.

    (A) $\dfrac{1}{2}$

    (B) 1

    (C) 0

    (D) 3

    (E) $\dfrac{1}{4}$

3. The compound $CH_3CH = CH_2$ has a bond formed by the overlap of which of the following hybrid orbitals?

    (A) $sp^2 - sp^3$

    (B) $sp - sp^2$

    (C) $sp - sp^3$

    (D) $sp^3 - sp^3$

    (E) All of the above

4. The structure of IF_5 is based on which configuration?

    (A) Tetrahedral

    (B) Trigonal bipyramidal

    (C) Square antiprismic

    (D) Octahedral

    (E) Pentagonal bipyramidal

5. Which of the following molecules is linear?

    (A) H_2O

    (B) ClO_2^-

    (C) NO_2^-

    (D) NO_2

    (E) NO_2^+

QUANTUM CHEMISTRY
DRILLS

ANSWER KEY

Drill 1 — Pauli Exclusion Principle, Hund Rules, and Electronic Configuration

1. (D)
2. (E)
3. (D)
4. (C)
5. (D)

Drill 2 — Molecular Orbital Theory

1. (C)
2. (A)
3. (A)
4. (D)
5. (E)

GLOSSARY:
QUANTUM CHEMISTRY

Double Bond

The sharing of two pairs of electrons produces a double bond.

Electronic Configuration

A tabulation of the quantum numbers of each of the electrons in an atom.

Hund Rules

State that for a set of equal-energy orbitals, each orbital is occupied by one electron before any orbital has two.

Pauli Exclusion Principle

The exclusion of two electrons having the same set of quantum numbers.

Pi Bond

Bond formed by parallel *p* orbitals.

Quantum Numbers

A set of four numbers that can be used to describe all electrons in atoms.

Sigma Bond

Formed when the molecular orbital (of shared electrons) is symmetrical around the nucleus.

Triple Bond

The sharing of three pairs of electrons results in a triple bond.

VSEPR (Valence Shell Electron Pair Repulsion) Theory

Permits the geometric arrangement of atoms, or groups of atoms, about some central atom to be determined solely by considering the repulsions between the electron pairs present in the valence shell of the central atom.

CHAPTER 16

Organic Chemistry

➤ Diagnostic Test
➤ Organic Chemistry
Review & Drills
➤ Glossary

ORGANIC CHEMISTRY DIAGNOSTIC TEST

1. Ⓐ Ⓑ Ⓒ Ⓓ Ⓔ
2. Ⓐ Ⓑ Ⓒ Ⓓ Ⓔ
3. Ⓐ Ⓑ Ⓒ Ⓓ Ⓔ
4. Ⓐ Ⓑ Ⓒ Ⓓ Ⓔ
5. Ⓐ Ⓑ Ⓒ Ⓓ Ⓔ
6. Ⓐ Ⓑ Ⓒ Ⓓ Ⓔ
7. Ⓐ Ⓑ Ⓒ Ⓓ Ⓔ
8. Ⓐ Ⓑ Ⓒ Ⓓ Ⓔ
9. Ⓐ Ⓑ Ⓒ Ⓓ Ⓔ
10. Ⓐ Ⓑ Ⓒ Ⓓ Ⓔ
11. Ⓐ Ⓑ Ⓒ Ⓓ Ⓔ
12. Ⓐ Ⓑ Ⓒ Ⓓ Ⓔ
13. Ⓐ Ⓑ Ⓒ Ⓓ Ⓔ
14. Ⓐ Ⓑ Ⓒ Ⓓ Ⓔ
15. Ⓐ Ⓑ Ⓒ Ⓓ Ⓔ
16. Ⓐ Ⓑ Ⓒ Ⓓ Ⓔ
17. Ⓐ Ⓑ Ⓒ Ⓓ Ⓔ
18. Ⓐ Ⓑ Ⓒ Ⓓ Ⓔ
19. Ⓐ Ⓑ Ⓒ Ⓓ Ⓔ
20. Ⓐ Ⓑ Ⓒ Ⓓ Ⓔ

21. Ⓐ Ⓑ Ⓒ Ⓓ Ⓔ
22. Ⓐ Ⓑ Ⓒ Ⓓ Ⓔ
23. Ⓐ Ⓑ Ⓒ Ⓓ Ⓔ
24. Ⓐ Ⓑ Ⓒ Ⓓ Ⓔ
25. Ⓐ Ⓑ Ⓒ Ⓓ Ⓔ
26. Ⓐ Ⓑ Ⓒ Ⓓ Ⓔ
27. Ⓐ Ⓑ Ⓒ Ⓓ Ⓔ
28. Ⓐ Ⓑ Ⓒ Ⓓ Ⓔ
29. Ⓐ Ⓑ Ⓒ Ⓓ Ⓔ
30. Ⓐ Ⓑ Ⓒ Ⓓ Ⓔ
31. Ⓐ Ⓑ Ⓒ Ⓓ Ⓔ
32. Ⓐ Ⓑ Ⓒ Ⓓ Ⓔ
33. Ⓐ Ⓑ Ⓒ Ⓓ Ⓔ
34. Ⓐ Ⓑ Ⓒ Ⓓ Ⓔ
35. Ⓐ Ⓑ Ⓒ Ⓓ Ⓔ
36. Ⓐ Ⓑ Ⓒ Ⓓ Ⓔ
37. Ⓐ Ⓑ Ⓒ Ⓓ Ⓔ
38. Ⓐ Ⓑ Ⓒ Ⓓ Ⓔ
39. Ⓐ Ⓑ Ⓒ Ⓓ Ⓔ
40. Ⓐ Ⓑ Ⓒ Ⓓ Ⓔ

ORGANIC CHEMISTRY DIAGNOSTIC TEST

This diagnostic test is designed to help you determine your strengths and your weaknesses in organic chemistry. Follow the directions and check your answers.

Study this chapter for the following tests:
AP Chemistry, CLEP General Chemistry,
GED, MCAT, MSAT, PRAXIS II Subject Assessment:
Chemistry, SAT II: Chemistry, GRE Chemistry

---

40 Questions

DIRECTIONS: Choose the correct answer for each of the following problems. Fill in each answer on the answer sheet.

1. Isomers differ in

   (A) the number of neutrons in their nuclei.

   (B) their atomic compositions.

   (C) their molecular weights.

   (D) their molecular structures.

   (E) None of the above

2. The functional group $R - \overset{\overset{\displaystyle H}{\displaystyle |}}{C} = O$ indicates a(an)

   (A) alcohol.   (D) ether.

   (B) aldehyde.   (E) ketone.

   (C) ester.

3. Organic compounds of substantial molecular weight are generally insoluble in water because

   (A) they contain large amounts of carbon.

(B) they do not ionize in solution.

(C) they do not dissociate in solution.

(D) they are nonpolar.

(E) they are polar.

$$O$$
$$\|$$
4.   The functional group R – C – H represents

(A) an alcohol.   (D) a ketone.

(B) an ether.   (E) an organic acid derivative.

(C) an aldehyde.

5.   sp bond hybridization for carbon is characteristic in

(A) alkanes.   (D) alkynes.

(B) alkenes.   (E) aromatics.

(C) dienes.

6.   What is the IUPAC name for the structure shown below?

```
     H   H   H   H       O
     |   |   |   |      //
H -  C - C - C - C -  C
     |   |   |   |      \
     H   H   H   H       H
```

(A) Butanal   (D) Pentanol

(B) Pentanal   (E) Butanoic acid

(C) Butanol

7.   Which of the following has the highest boiling point?

(A) $CH_3CH_2CH_2CO_2H$

(B) $CH_3CH_2CH_2CH_2CHO$

(C) $CH_3CH_2CH_2CH_2CH_2OH$

(D) $CH_3(CH_2)_4CH_3$

(E) $CH_3CH_2 - C - CH_2CH_3$
$$\qquad\qquad\ \ \|$$
$$\qquad\qquad\ \ O$$

8. Enzymes, which are organic catalysts, always partly consist of

   (A) carbohydrates.   (D) proteins.

   (B) lipids.   (E) steroids.

   (C) nucleic acids.

9. The reddish color of bromine dissolved in CCl_4 rapidly disappears when a few drops of this reagent is added to the solution

   (A) CH_4.

   (B) $CH_2=CH-CH_2-CH_2-CH_3$.

   (C) $CH_3-CH_2-CH_2-CH_3$.

   (D) $CHCl_3$.

   (E)

$$\begin{array}{ccc} Cl & & Cl \\ | & & | \\ H-C & \!\!\!\!-\!\!\!\! & C-H \\ | & & | \\ H & & H \end{array}$$

10. Which of the following can exhibit optical isomerism?

    (A)
$$\begin{array}{cc} CH_3 & CH_3 \\ \diagdown & \diagup \\ C & = & C \\ \diagup & \diagdown \\ H & H \end{array}$$
              (D) CH_3CBr_2H

    (B)
$$\begin{array}{cc} CH_3 & H \\ \diagdown & \diagup \\ C & = & C \\ \diagup & \diagdown \\ H & CH_3 \end{array}$$
              (E) $HCBrClCN$

    (C) $N_2CH_3CCl_2$

11. Which of the following is acetone?

    (A) $CH_3CH_2CH_2-OH$ (D) CH_3COCH_3

    (B)
$$CH_3CH_2C\diagdown^{\displaystyle H}_{\displaystyle \diagdown\!\!\!O}$$
              (E)
$$\begin{array}{c} CH_3CH_2CH_3 \\ | \\ OH \end{array}$$

    (C) $CH_3-O-CH_2CH_3$

12. Which of the following molecules has two pi bonds?

(A)

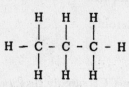

(D) $H - C \equiv C - C \equiv C - H$

(B)

H   H   H
  \ | |
   C = C - C - H
  / |
H   H

(E) $H - C \equiv C - C = C$ with H, H on the terminal carbons

H
|
(C)   H - C ≡ C - C - H
|
H

O
‖
13. What is the functional group $R - C - OH$ representative of?

(A) Ethers   (D) Aldehydes

(B) Alcohols   (E) Esters

(C) Acids

14. Which of the following correctly describes the structure shown below?

H   H   H   H
| | | |
H - C - C - C - C - OH
| | | |
H   H   H   H

(A) Butanol   (D) Both (A) and (B)

(B) Butyl alcohol   (E) (A), (B), and (C)

(C) 4-butanol

15. Which of the following structures indicates a component of an unsaturated fat?

(A)

$$C_{20}H_{41} - C \overset{\displaystyle O}{\underset{\displaystyle OH}{<}}$$

(D)

$$C_{20}H_{39} - C \overset{\displaystyle O}{\underset{\displaystyle OCH_3}{<}}$$

(B)

$$C_{20}H_{41} - \underset{\displaystyle H}{C} \overset{\displaystyle O}{<}$$

(E)

$$C_{21}H_{41} - C \overset{\displaystyle O}{\underset{\displaystyle CH_3}{<}}$$

(C)

$$C_{20}H_{39} - C \overset{\displaystyle O}{\underset{\displaystyle OH}{<}}$$

16. The systematic (IUPAC) name of this structure is

$$
\begin{array}{ccccccc}
& H & H & H & OH & H & H \\
& | & | & | & | & | & | \\
H - & C & - C & - C & - C & - C & - C - H \\
& | & | & | & | & | & | \\
& H & H & H & H & H & H
\end{array}
$$

(A) hexanol.

(D) 4-hexanol.

(B) 3-hydroxyhexane.

(E) isohexanol.

(C) 3-hexanol.

17. Which of the following indicates the functional group of an ether?

(A) $R - OH$

(D)

$$R - \overset{\displaystyle O}{\overset{\displaystyle \|}{C}} - R^1$$

(B) $R - O - R^1$

(E)

$$R - \overset{\displaystyle O}{\overset{\displaystyle \|}{C}} - OH$$

(C) $R - \overset{\displaystyle O}{\overset{\displaystyle \|}{C}} - H$

18. Which of the following structures represents 1,1-dibromoethane?

(A)
```
    Br  H
    |   |
Br-C – C-H
    |   |
    H   H
```

(D) $Br - C \equiv C - Br$

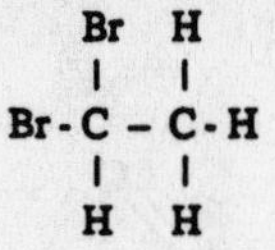

(B)
```
    Br  Br
    |   |
 H-C – C-H
    |   |
    H   H
```

(E)
```
Br        H
  \      /
   C = C
  /      \
Br        H
```

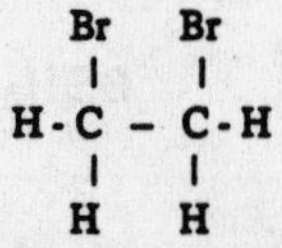

(C) $CH_3 - CH_2 - Br - CH_2 - CH_3$

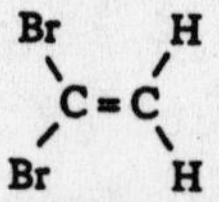

For questions 19–22 choose the answer that best represents the formula given.

(A) Alcohols   (D) Alkynes

(B) Alkanes   (E) Amines

(C) Alkenes

19. C_2H_5OH

20. C_2H_6

21. C_2H_2

22. C_2H_4

23. Which of the following is an outline for a molecular sub-unit for a saturated fat molecular?

(A)
```
      O
      ||
   H-C
      |
   H-C-OH
      |
  HO-C-H
      |
   H-C-OH
      |
   H-C-OH
      |
   H-C -OH
      |
      H
```

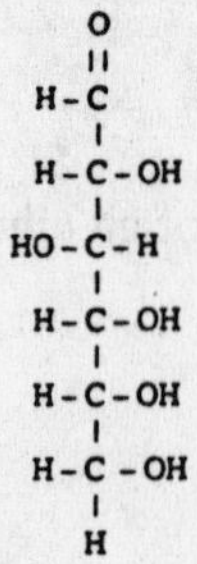

(B)
```
         O
         ||
HO-C-C=C=C=C-C-C-
```

(C)
```
   O  H  H  H  H
   ||  |  |  |  |
HO-C-C-C-C-C
    |  |  |  |
    H  H  H  H
```

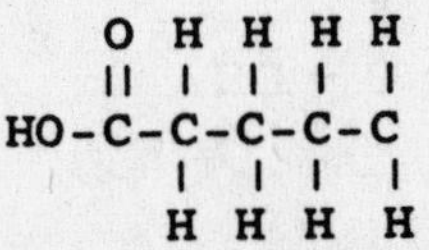

(D)
```
  H       H  O
   \      |  ||
    N – C - C – OH
   /      |
  H       R
```

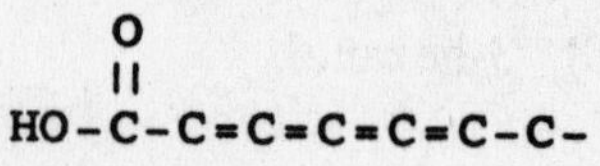

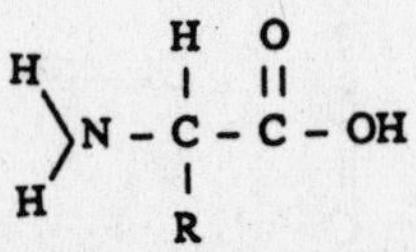

(E)

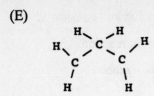

24. What is the functional group of a ketone?

(A)

$$R - \overset{\overset{\displaystyle O}{\|}}{C}O - R^1$$

(D)

$$R - \overset{\overset{\displaystyle O}{\|}}{C} - R^1$$

(B)

$$R - \overset{\overset{\displaystyle O}{\|}}{C} - OH$$

(E) $R - O - R^1$

(C)

$$R - \overset{\overset{\displaystyle O}{\|}}{C} - H$$

25. The structure of the third member of the alkyne series is

(A) $HC \equiv C - H$

(D) $HC \equiv C - C \equiv C - H$

(B) $HC \equiv C - CH_3$

(E) $HC \equiv C - CH = CH_2$

(C) $HC \equiv C - CH_2 - CH_3$

26. Which of the following structures has the IUPAC name propyl butanoate?

(A) $CH_3CH_2CH_2OCH_2CH_2CH_2CH_3$

(B) $CH_3CH_2\overset{\overset{\displaystyle O}{\|}}{C}OCH_2CH_2CH_2CH_3$

(C) $CH_3CH_2CH_2\overset{\overset{\displaystyle O}{\|}}{C}OCH_2CH_2CH_3$

(D) $CH_3CH_2CH_2CH_2\overset{\overset{\displaystyle O}{\|}}{C}OCH_2CH_2CH_3$

(E) None of the above

27. In organic chemistry, the so-called "aromatic behavior" consists of

    (A) addition reactions.

    (B) substitution reactions.

    (C) oxidation reactions.

    (D) reduction reactions.

    (E) polymerization.

28. $C_6H_5NH_2$ is

    (A) benzile ammonia.

    (B) benzyl ammonia.

    (C) hexyl ammonia.

    (D) phenol.

    (E) aniline.

29. Which of the following is most soluble in water?

    (A) Hexanol

    (B) Benzene

    (C) Acetic acid

    (D) Acetylene

    (E) Hexanoic acid

30. A protein can be described as

    (A) an addition polymer.

    (B) an addition copolymer.

    (C) a condensation polymer.

    (D) a polyester.

    (E) a polyvinyl chloride.

31. The reaction

$$R-\overset{\overset{\displaystyle O}{\|}}{C}H + 2Ag^+ + 3OH^- \rightarrow R\overset{\overset{\displaystyle O}{\|}}{C}-O^- + 2Ag + 2H_2O$$

    is the

    (A) Benedict's test.

    (B) Fehling's test.

    (C) Tollen's test.

    (D) protein test.

    (E) Fisher's test.

32. In which of the compounds below may CIS-TRANS isomerism occur?

    (A) CH_4

    (B) C_6H_6

    (C) C_2H_4

    (D) $C_2H_2(CH_3)_2$

    (E) C_2H_6

33. Esterification involves the formation of an ester from an

    (A) acid and a ketone. (D) aldehyde and an alcohol.

    (B) acid and an alcohol. (E) ether and an aldehyde.

    (C) alcohol and an amine.

34. The molecule represented in the structural formula below is

    (A) ammonia.

    (B) methane.

    (C) nitrogen.

    (D) urea.

    (E) uric acid.

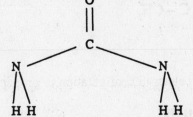

35. Which of the following compounds exhibits tautomerism?

    (A) $CH_3OCH_2CH_3$

    (B) $CH_3-\underset{\underset{H}{|}}{\overset{\overset{OCH_3}{|}}{C}}-OCH_3$

    (C)

    (D) $NH_2-\underset{\underset{NH_2}{|}}{C}=O$

    (E) $CH_2 = CH - CH_2OH$

36. Which of the following compounds contains a chiral carbon?

    (A) $H-\underset{\underset{H}{|}}{\overset{\overset{H}{|}}{C}}-Cl$

    (B) $CH_3-CH_2-C\overset{\overset{O}{\diagup\diagup}}{\diagdown_{OH}}$

(C)

$$CH_3 - \underset{\underset{H}{|}}{\overset{\overset{OH}{|}}{C}} - CH_2 - CH_3$$

(D) $Cl - CH_2 - CH_2 - Cl$

(E)

$$CH_3 - \overset{\overset{O}{\|}}{C} - CH_3$$

37. The expected electron configuration of propanal is

(A)
```
    H H H
H:C:C:C:O:
    H H
```

(B)
```
  H  H H
H:C::C:C:O:
       H
```

(C)
```
   H    H
H:C::C::C:O
```

(D)
```
              H
H:C:::C:C::O
```

(E)
```
    H H H
H:C:C:C::O
    H H
```

38. What type of formula of ethane is depicted below?

(A) Fisher projection formula

(B) Newman projection formula

(C) Lewis projection formula

(D) Kekulé projection formula

(E) Pauling projection formula

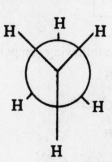

39. The compound shown below is best described as being a(n)

    (A) alcohol.

    (B) alkene.

    (C) ether.

    (D) ester.

    (E) carboxylic acid.

```
      H   H
      |   |
   H-C-O-C-H
      |   |
      H   H
```

40. Which of the following is an acid anhydride?

    (A)
```
      H   H   O       H   H
      |   |   ||      |   |
   H-C - C - C - O - C - C - H
      |   |           |   |
      H   H           H   H
```

    (D)
```
      H   O   H
      |   ||  |
   H-C - C - C - H
      |       |
   H-C-H      H
      |
      H
```

    (B)
```
      H   H   O
      |   |   ||
   H-C - C - C - OH
      |   |
      H   H
```

    (E)
```
      H   H       H
      |   |       |
   H-C - C - O - C - H
      |   |       |
      H   H       H
```

    (C)
```
      H   O       O   H
      |   ||      ||  |
   H-C - C - O - C - C - H
      |           |
      H           H
```

ORGANIC CHEMISTRY: NOMENCLATURE AND STRUCTURE DIAGNOSTIC TEST

ANSWER KEY

1.	(D)	11.	(D)	21.	(D)	31.	(C)
2.	(B)	12.	(C)	22.	(C)	32.	(D)
3.	(D)	13.	(C)	23.	(C)	33.	(B)
4.	(C)	14.	(D)	24.	(D)	34.	(D)
5.	(D)	15.	(C)	25.	(C)	35.	(C)
6.	(B)	16.	(C)	26.	(C)	36.	(C)
7.	(A)	17.	(B)	27.	(B)	37.	(E)
8.	(D)	18.	(A)	28.	(E)	38.	(B)
9.	(B)	19.	(A)	29.	(C)	39.	(C)
10.	(E)	20.	(B)	30.	(C)	40.	(C)

DETAILED EXPLANATIONS
OF ANSWERS

1. **(D)** Isomers differ only in their molecular structures. Isotopes vary in the number of neutrons in the nucleus and thus have different weights.

2. **(B)** The functional group

$$\begin{array}{c} H \\ | \\ R - C = O \end{array}$$

represents an aldehyde. An alcohol is indicated by R – OH, an ester by

$$\begin{array}{c} O \\ \| \\ R - CO - R, \end{array}$$

an ether by R – O – R, and a ketone by

$$\begin{array}{c} O \\ \| \\ R - C - R. \end{array}$$

3. **(D)** Water, being highly polar, cannot dissolve nonpolar compounds (most organic compounds).

4. **(C)** The functional group of an alcohol is indicated by

R – OH;

that of an ether is

$$R - O - R^1;$$

an aldehyde is indicated by

$$\begin{array}{c} O \\ \| \\ R - C - H; \end{array}$$

while that of a ketone is

$$\begin{array}{c} O \\ \| \\ R - C - R^1. \end{array}$$

Derivatives of organic acids; esters, amides, and acid anhydrides, for example, have the following functional groups:

$$R-\overset{\overset{\displaystyle O}{\|}}{C}-OH \qquad \text{carboxylic acid}$$

$$R-\overset{\overset{\displaystyle O}{\|}}{C}-O-R^1 \qquad \text{ester}$$

$$R-\overset{\overset{\displaystyle O}{\|}}{C}-NH_2 \qquad \text{amide}$$

$$R-\overset{\overset{\displaystyle O}{\|}}{C}-O-\overset{\overset{\displaystyle O}{\|}}{C}-R^1 \qquad \text{acid anhydride}$$

5. **(D)** sp bond hybridization of carbon is characteristic of the triple bonds found in alkynes. The double bonds of alkenes, dienes, and aromatics give rise to sp^2 hybridization of carbon while the single bonds of the alkanes characterize sp^3 bond hybridization.

6. **(B)** The carbon skeleton of this molecule contains five carbons so we name it with the prefix pentan-. The molecule contains the functional group of an aldehyde which gives us the suffix -al. Thus, the structure is named pentanal. The correct structures associated with the other choices are

butanal

$$H-\overset{\overset{\displaystyle H}{|}}{\underset{\underset{\displaystyle H}{|}}{C}}-\overset{\overset{\displaystyle H}{|}}{\underset{\underset{\displaystyle H}{|}}{C}}-\overset{\overset{\displaystyle H}{|}}{\underset{\underset{\displaystyle H}{|}}{C}}-\overset{\overset{\displaystyle O}{\diagup\diagup}}{\underset{\underset{\displaystyle H}{\diagdown}}{C}}$$

butanol

$$H-\overset{\overset{\displaystyle H}{|}}{\underset{\underset{\displaystyle H}{|}}{C}}-\overset{\overset{\displaystyle H}{|}}{\underset{\underset{\displaystyle H}{|}}{C}}-\overset{\overset{\displaystyle H}{|}}{\underset{\underset{\displaystyle H}{|}}{C}}-\overset{\overset{\displaystyle H}{|}}{\underset{\underset{\displaystyle H}{|}}{C}}-OH$$

pentanol

$$H-\overset{\overset{\displaystyle H}{|}}{\underset{\underset{\displaystyle H}{|}}{C}}-\overset{\overset{\displaystyle H}{|}}{\underset{\underset{\displaystyle H}{|}}{C}}-\overset{\overset{\displaystyle H}{|}}{\underset{\underset{\displaystyle H}{|}}{C}}-\overset{\overset{\displaystyle H}{|}}{\underset{\underset{\displaystyle H}{|}}{C}}-\overset{\overset{\displaystyle H}{|}}{\underset{\underset{\displaystyle H}{|}}{C}}-OH$$

butanoic acid

$$H-\overset{\overset{\displaystyle H}{|}}{\underset{\underset{\displaystyle H}{|}}{C}}-\overset{\overset{\displaystyle H}{|}}{\underset{\underset{\displaystyle H}{|}}{C}}-\overset{\overset{\displaystyle H}{|}}{\underset{\underset{\displaystyle H}{|}}{C}}-\overset{\overset{\displaystyle O}{\|}}{C}-OH$$

7. **(A)** The compounds listed are approximately the same molecular weight, but the boiling point of the acid is relatively higher as a result of intermolecular hydrogen bonding between two molecules.

$$CH_3CH_2CH_2-C \underset{\ddot{O}-H \cdots :Q}{\overset{\ddot{O}:\cdots H-O}{\Big\langle}} C-CH_2CH_2CH_3$$

8. **(D)** Enzymes are proteins which act as catalysts for biochemical reactions.

9. **(B)** When Br_2 reacts with unsaturated compounds it readily adds across the unsaturated bonds placing a bromine atom on each carbon.

$$R-C=C-R+Br_2 \quad \rightarrow \quad R-\underset{\underset{Br}{|}}{C}-\underset{\underset{Br}{|}}{C}-R$$

$$\text{(red)} \qquad \qquad \text{(colorless)}$$

This reaction is commonly employed as a test for the presence of double bonds in an unknown compound. As the reddish Br_2 solution reacts with the double bonds, the color of the Br_2 dissipates. The only compound listed that is unsaturated is

$$CH_2 = CH - CH_2 - CH_2 - CH_3.$$

10. **(E)** As a rule, an organic compound with an asymmetric carbon atom will display optical activity. An asymmetric carbon atom is a carbon bonded to four different atoms or group of atoms. The only molecule given with four different atoms is (E):

$$\underset{\underset{Br}{|}}{\overset{\overset{H}{|}}{Cl-C-CN}}$$

11. **(D)** Acetone is a ketone. Ketones are distinguished by the presence of $C = O$, the carbonyl group. Acetone has the structural formula

$$CH_3 - \overset{\overset{O}{\|}}{C} - CH_3$$

12. **(C)** A double bond consists of one sigma bond and one pi bond; a triple bond consists of one sigma bond and two pi bonds and a single bond consists of one sigma bond. Therefore, (A) has no pi bonds, (B) has one, (C) has two, (D) has four and (E) has three.

13. **(C)** The functional group is representative of an acid (e.g., acetic acid $CH_3 - COOH$).

Other functional groups such as:

$$R-O-R, \quad R-COH, \quad R-\underset{\underset{O}{\|}}{C}-H, \quad R-\overset{\overset{O}{\|}}{C}-OR$$

represent ethers, alcohols, aldehydes, and esters, respectively.

14. **(D)** This structure may be named butanol, butyl alcohol, or 1-butanol.

15. **(C)** Fats are composed of glycerol and three long chain carboxylic acids. (B) is an aldehyde, (D) is an ester, and (E) is a ketone. (A) and (C) are both long chain acids. An unsaturated carboxylic acid contains one or more double and/or triple bonds. The general formula for an alkane is C_nH_{2n+2}. With C = 20, we find that H = 42. Since we have a carboxy group attached to the carbon skeleton, the saturated fatty acid with C = 20 would have the formula $C_{20}H_{41} - COOH$. Thus (A) is a saturated carboxylic acid. (C), having two fewer hydrogens than (A) must have one double bond and therefore be unsaturated.

16. **(C)** Alcohols are named by replacing the suffix -e of the corresponding hydrocarbon name by the suffix -ol. The position of the hydroxy substituent is numbered from the shorter end of the chain. Thus, the structure is named 3-hexanol. It is a hexanol because the parent hydrocarbon has six carbons and the prefix 3- (not 4-) is used to indicate the location of the hydroxy group on the third carbon.

17. **(B)**

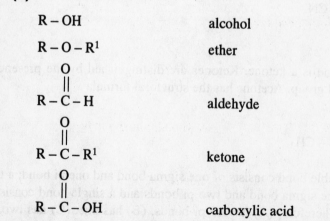

$R - OH$	alcohol
$R - O - R^1$	ether
$R - \overset{\overset{O}{\|}}{C} - H$	aldehyde
$R - \overset{\overset{O}{\|}}{C} - R^1$	ketone
$R - \overset{\overset{O}{\|}}{C} - OH$	carboxylic acid

18. **(A)** Dibromoethane is ethane, CH_3CH_3, with two hydrogen atoms replaced by bromine. The numbers "1,1" indicate that both bromine atoms are on the first carbon.

Thus,

$$
\begin{array}{ccc}
& Br & H \\
& | & | \\
Br - & C - & C - H \\
& | & | \\
& H & H
\end{array}
$$

is the correct structure.

19. **(A)**
20. **(B)**
21. **(D)**
22. **(C)** Alkanes have the general molecular formula C_nH_{2n+2}, such as butane, C_2H_6. Alkenes conform to C_nH_{2n}, for example butene, C_2H_4. Alkynes, C_nH_{2n-2+}, include butyne, C_2H_2. Alcohols, such as butyl alcohol, include an OH in the carbon chain. Amines are nitrogen derivatives and are not represented.

23. **(C)** Illustration (A) shows the molecular formula of glucose. (D) shows an amino acid's formula. (E) is cyclopropane. Both (B) and (C) are outlines of fatty acids. (B), however, is not saturated with hydrogen atoms, covalently bonded to its carbon chain. This is because of the double C-to-C bonds, limiting availability for hydrogen covalent bonding. Carbon's valence is four. The carbon atoms in illustration (C) are single bonded, leaving more bonding sites for hydrogen atoms. It is relatively saturated with these atoms. Fatty acids are the molecular sub-units of fat molecules.

24. **(D)** The functional group of a ketone is indicated by

$$
\begin{array}{c}
O \\
\parallel \\
R - C - R^1
\end{array}
$$

The other choices given are

$$
\begin{array}{c}
O \\
\parallel \\
R - CO - R^1
\end{array}
\qquad \text{ester}
$$

$$
\begin{array}{c}
O \\
\parallel \\
R - COH
\end{array}
\qquad \text{carboxylic acid}
$$

$$
\begin{array}{c}
O \\
\parallel \\
R - CH
\end{array}
\qquad \text{aldehyde}
$$

$$
R - O - R^1 \qquad \text{ether}
$$

25. **(C)** The first member of the alkyne series is acetylene (or ethyne), whose structure is

$$HC \equiv CH$$

The second is propyne:

$$HC \equiv C - CH_3$$

The third is butyne:

$$HC \equiv C - CH_2 - CH_3$$

Note that there are no analogous compounds in the alkene or alkyne series for the first member of the alkane series (methane – CH_4).

26. **(C)** The first word of the ester name is the stem name of the group attached to oxygen (propyl in our case). The second word is the name of the parent carboxylic acid with the suffix -ic replaced by -ate (butanoate in our case). Thus, we are looking for an ester with a propyl group attached to the oxygen of butanoic acid. This structure is

$$\overset{\displaystyle O}{\overset{\displaystyle \|}{CH_3CH_2CH_2COCH_2CH_2CH_3}}$$

$$CH_3CH_2CH_2OCH_2CH_2CH_2CH_3$$

has the assigned name butyl propyl ether.

$$\overset{\displaystyle O}{\overset{\displaystyle \|}{CH_3CH_2COCH_2CH_2CH_2CH_3}}$$

has the assigned name butyl propanoate.

$$\overset{\displaystyle O}{\overset{\displaystyle \|}{CH_3CH_2CH_2CH_2COCH_2CH_2CH_3}}$$

has the assigned name propyl pentanoate.

27. **(B)** The aromatic compounds are very stable, due to resonance, so substitution is the most likely to happen.

28. **(E)** $C_6H_5NH_2$ is aniline NH_2.

29. **(C)** The most polar of the organic substances listed is acetic acid CH_3COOH; therefore, it is the most soluble.

30. **(C)** A protein is the polymer of amino acids. Amino acids have the general formula

$$\underset{\underset{R_1}{|}}{H_2N-\overset{\overset{H}{|}}{C}-\overset{\overset{O}{\|}}{C}-OH}$$

There are two functional groups –NH₂ (amino group) and

$$-\overset{\overset{O}{\|}}{C}-OH - \text{carboxylic group.}$$

In proteins, the amino acids are linked in a long chain,

$$...-\overset{\overset{H}{|}}{\underset{\underset{H}{|}}{N}}-\overset{\overset{O}{\|}}{C}-\overset{\overset{H}{|}}{\underset{\underset{R_n}{|}}{N}}-\overset{\overset{H}{|}}{C}-\overset{\overset{O}{\|}}{\underset{\underset{R_2}{|}}{C}}-...$$

Since water is eliminated in the reaction it is a condensation polymerization.

The bond formed between amino acids

$$...+H_2N-\overset{\overset{H}{|}}{\underset{\underset{R_n}{|}}{C}}-\overset{\overset{O}{\|}}{C}-OH+H-\overset{\overset{H}{|}}{\underset{\underset{R_{n+1}}{|}}{N}}-\overset{\overset{H}{|}}{C}-\overset{\overset{O}{/\!/}}{C}-OH+...\xrightarrow{-H_2O}$$

is called a peptide bond. So the protein can be described as a polypeptide or a condensation polymer.

31. **(C)** Aldehydes are oxidized easily to carboxylic acids by even weak oxidizing agents like the silver ion, Ag⁺. A solution of silver nitrate in ammonium hydroxide when added to an aldehyde results in the oxidation of the aldehyde and the Ag⁺ ion is reduced to metallic silver. This test is the Tollen's test, also referred to as the silver mirror test, because the metallic silver deposits in the form of a mirror on a glass surface.

The Benedict's and Fehling's tests are also tests for aldehydes and also ∝ –hydroxy ketones in sugars. In these cases the reducing agents are solutions containing copper (II) ion complexed with a tartrate (Fehling's solution) or a citrate (Benedict's solution). The copper (II) ion is reduced to copper (I) which precipitates as a brick-red copper (I) oxide.

$$R-\overset{\overset{O}{\|}}{C}-H + Cu^{2+} \rightarrow R-\overset{\overset{O}{\|}}{C}-OH + Cu_2O$$

Fehling's $\qquad$ brick-red ppt.
or Benedict's solution

32. **(D)** CIS-TRANS isomerism occurs only in compounds which divide the molecule into two parts, so that two different configurations are possible. Two

possibilities are shown below:

$$\overset{H}{\underset{H}{\diagdown}}C = C\overset{CH_3}{\underset{CH_3}{\diagup}} \qquad \overset{CH_3}{\underset{H}{\diagdown}}C = C\overset{CH_3}{\underset{H}{\diagup}}$$

33. **(B)** This is one of several classic organic reactions. When an alcohol and an acid combine they form an ester according to the following reaction:

$$\underset{\text{acid}}{R-\overset{O}{\overset{\|}{C}}-OH} + \underset{\text{alcohol}}{R'-OH} \rightleftharpoons \underset{\text{ester}}{R-\overset{O}{\overset{\|}{C}}-OR'} + H_2O$$

34. **(D)** Other choices have the following formulas: methane – CH_4, nitrogen – N, ammonia – NH_3. Urea is $CO(NH_2)_2$. The illustration shows its structural formula.

35. **(C)** Aldehydes and ketones which have ∝-hydrogens exhibit a kind of isomerism called tautomerism in which the ∝-hydrogen migrates to the carbonyl oxygen. The new isomer is both an alkene and alcohol and is called an enol. In the problem there are no aldehydes or ketones given, but if we examine the compounds carefully,, we see that compound (C) is the enol form of a ketone.

enol form   keto form

also

In fact at equilibrium they all co-exist, i.e.,

36. **(C)**   A chiral carbon, or a point of asymmetry, is a carbon which has four different groups attached to it. If a molecule contains a chiral carbon, it can exist in two distinct forms that are non-superimposable mirror images. One form will rotate the plane polarized light to the right whereas the other form will rotate the plane polarized light to the left. The two distinct forms are said to be optically active and constitute what is referred to as a pair of enantiomers. The only compound listed in the answers that contains a chiral carbon is

$$
\begin{array}{c}
OH \\
| \\
CH_3 - C - CH_2CH_3 \\
| \\
H \quad \text{chiral carbon}
\end{array}
$$

37. **(E)**   From the name propanal, we find that we must describe an aldehyde (from the -al suffix) composed of a 3-carbon skeleton (from the prop- prefix) with no multiple bonds (from the -ane root). Remembering that an aldehyde is characterized by the functional group

$$
\begin{array}{c}
O \\
\| \\
- C - H
\end{array}
$$

we obtain

$$
\begin{array}{c}
O \\
\| \\
CH_3 - CH_2 - C - H
\end{array}
$$

as a result. This is equivalent to the electron configuration in (E) since each bond (1) represents two electrons.

38. **(B)**   Three-dimensional shapes of molecules are important to the chemist and biochemist. The Newman projection is a way of showing molecular conformations of alkanes. Conformations are the arrangements of atoms depending on the angle of rotation of one carbon with respect to another adjacent carbon. A Newman projection is obtained by looking at the molecule along the bond on which rotation occurs. The carbon in front is indicated by a point and at the back by a circle, the three remaining valences are placed 120° to each other (but remember that the bond angles about each of the carbon atoms are 109.5° and not 120° as the Newman projection formula might suggest) e.g., for $CH_3^- CH_3$.

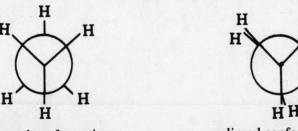

staggered conformation   eclipsed conformation

39. **(C)** The –C–O–C– linkage is characteristic of an ether.

40. **(C)** An acid anhydride is characterized by the functional group

$$
\begin{matrix} O & & O \\ \parallel & & \parallel \\ R-C-O-&C-R' \end{matrix}
$$

(A) is an ester characterized by the functional group

$$
\begin{matrix} O \\ \parallel \\ R-C-O-R' \end{matrix}
$$

(B) is a carboxylic acid characterized by the functional group

$$
\begin{matrix} O \\ \parallel \\ R-C-OH \end{matrix}
$$

(D) is a ketone characterized by the functional group

$$
\begin{matrix} O \\ \parallel \\ R-C-R' \end{matrix}
$$

(E) is an ether characterized by the functional group

$$
R-O-R'
$$

ORGANIC CHEMISTRY REVIEW

1. Alkanes

Structural formula:

$$C_nH_{2n+2}$$

The simplest member of the alkane family is methane (CH_4) which is written as:

$$\text{Acidity} \quad H-F \; < \; H-Cl \; < \; H-Br \; < \; H-I$$
$$H-OH \; < \; H-SH \; < \; H-SeH$$

Figure 1.1

Nomenclature (IUPAC System)

A)   Select the longest continuous carbon chain for the parent name.

EXAMPLE

$$\overset{1}{C}H_3 - \overset{2}{C}H_2 - \overset{3}{C}H - \overset{4}{C}H_2 - \overset{5}{C}H_2 - \overset{6}{C}H_3$$
$$|$$
$$CH_3$$

Figure 1.2

The parent name is hexane.

B)   Number the carbons in the chain, from either end, such that the substituents are given the lowest numbers possible.

EXAMPLE

$$CH_3 - CH_2 - CH - CH_2 - CH_2 - CH_3$$
$$|$$
$$CH_3$$

Figure 1.3

C)   The substituents are assigned the number of the carbon to which they are attached. In the preceding example the substituent is assigned the number 3.

D)   The name of the compound is now composed of the name of the parent chain preceded by the name and the number of the substituents, arranged in alphabetical order. For the same example the name is thus 3-methylhexane.

E)   If a substituent occurs more than once in the molecule, the prefixes"di-," "tri-," "tetra-," etc., are used to indicate how many times it occurs.

F)   If a substituent occurs twice on the same carbon, the number of the substituent is repeated as in 3,3-diethyl-5-isopropyl-4-methyloctane below.

EXAMPLE

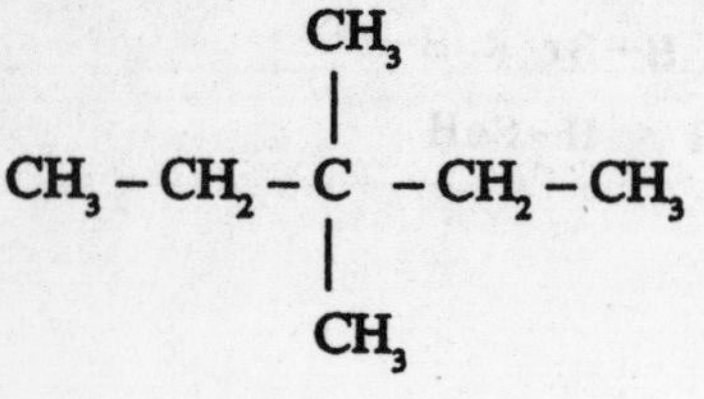

$$CH_3 - CH_2 - \overset{\overset{\displaystyle CH_3}{|}}{\underset{\underset{\displaystyle CH_3}{|}}{C}} - CH_2 - CH_3$$

3, 3–dimethylpentane

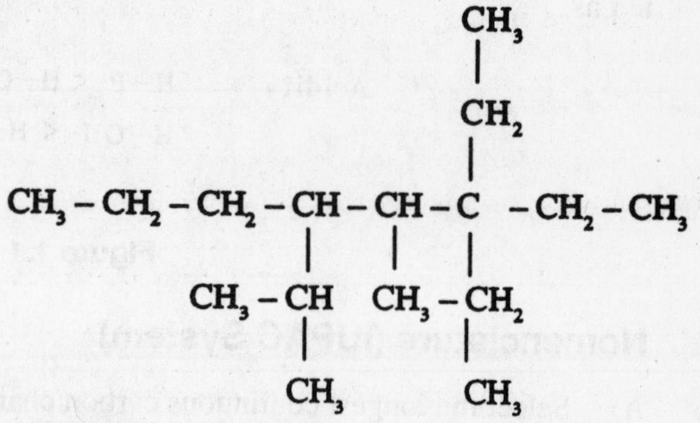

$$CH_3 - CH_2 - CH_2 - \overset{\overset{\displaystyle CH_3 - CH}{|}}{\underset{\underset{\displaystyle CH_3}{|}}{CH}} - \overset{\overset{\displaystyle CH_3 - CH_2}{|}}{\underset{\underset{\displaystyle CH_3}{|}}{CH}} - \overset{\overset{\overset{\displaystyle CH_3}{|}}{\overset{\displaystyle CH_2}{|}}}{C} - CH_2 - CH_3$$

3, 3–diethyl–5–isopropyl–4–methyloctane

Figure 1.4

There are two types of butanes — normal butane and isobutane. They have the same molecular formula, C_4H_{10}, but have different structures. N-Butane is a straight chain hydrocarbon whereas isobutane is a branched-chain hydrocarbon. N-Butane and isobutane are structural isomers and differ in their physical and chemical properties.

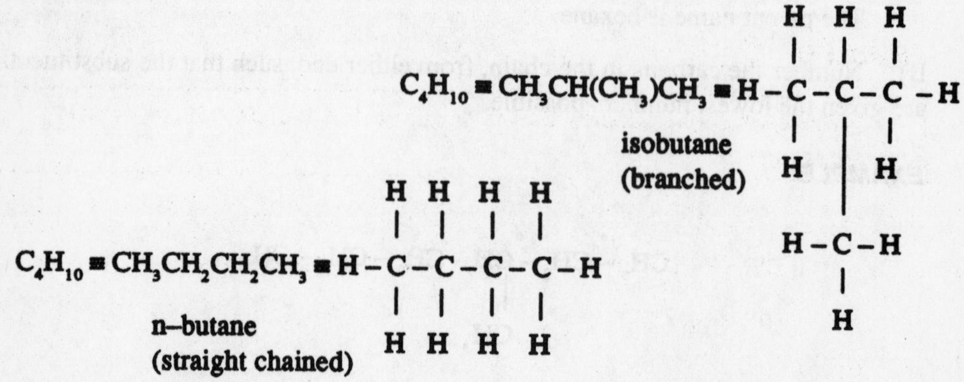

$$C_4H_{10} = CH_3CH(CH_3)CH_3 = H - \overset{\overset{\displaystyle H}{|}}{\underset{\underset{\displaystyle H}{|}}{C}} - \overset{\overset{\displaystyle H}{|}}{C} - \overset{\overset{\displaystyle H}{|}}{\underset{\underset{\displaystyle H}{|}}{C}} - H$$

isobutane (branched)

$$H - \overset{\displaystyle}{\underset{\underset{\displaystyle H}{|}}{C}} - H$$

$$C_4H_{10} = CH_3CH_2CH_2CH_3 = H - \overset{\overset{\displaystyle H}{|}}{\underset{\underset{\displaystyle H}{|}}{C}} - \overset{\overset{\displaystyle H}{|}}{\underset{\underset{\displaystyle H}{|}}{C}} - \overset{\overset{\displaystyle H}{|}}{\underset{\underset{\displaystyle H}{|}}{C}} - \overset{\overset{\displaystyle H}{|}}{\underset{\underset{\displaystyle H}{|}}{C}} - H$$

n–butane (straight chained)

Figure 1.5

In higher homologs of the alkane family, the number of isomers increases exponentially.

PROBLEM

> Explain the terms primary, secondary, and tertiary in regards to covalent bonding in organic compounds.

SOLUTION

If a carbon atom is bound to only one other carbon atom, then the former carbon atom is called primary. If a carbon atom is bonded to two other carbon atoms, then that carbon atom is called secondary. If a carbon atom is bonded to three other carbon atoms, then that carbon atom is called tertiary.

Any group that is attached to a primary, secondary, or tertiary carbon is called a primary, secondary, or tertiary group. For example,

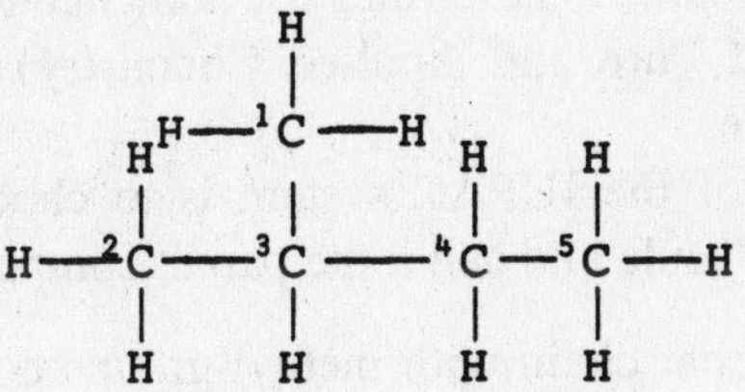

Figure 1.6

Carbons 1, 2, and 5 are primary, carbon 4 is secondary, and carbon 3 is tertiary. The hydrogen atoms attached to carbons 1, 2, and 5 are called primary hydrogens, those attached to carbon 4 are called secondary hydrogens, and those attached to carbon 3 are called tertiary hydrogens. This same principle applies to alcohols, and depending upon where the hydroxyl group ($-OH$) is attached (that is, primary, secondary, or tertiary carbon), the alcohol is called primary, secondary, or tertiary, respectively.

PROBLEM

> Name each of the alkanes shown in Figure 1.7. Indicate which, if any, are isomers.

(A)

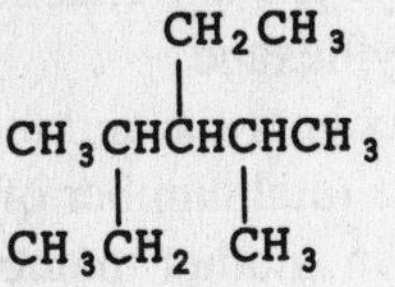

$$CH_3CHCHCHCH_3$$
with CH_2CH_3 on top, and CH_3CH_2 and CH_3 below

(B)

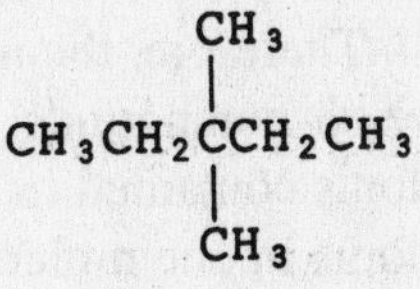

$$CH_3CH_2CCH_2CH_3$$
with CH_3 above and CH_3 below

(C)

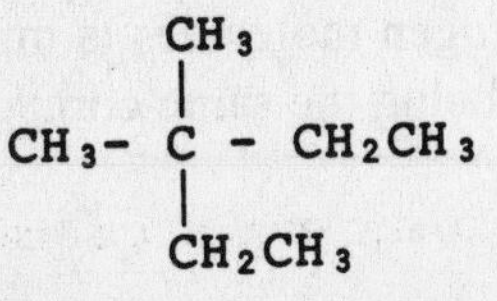

$$CH_3- C - CH_2CH_3$$
with CH_3 above and CH_2CH_3 below

(D)

$$(CH_3)_2CHCHCH_3$$
with CH_2CH_3 above

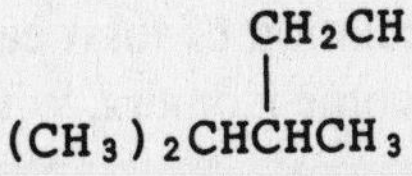

(E) 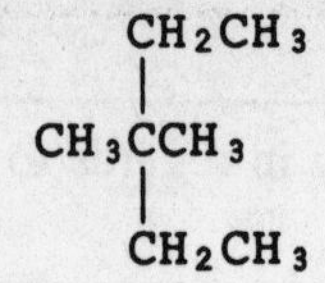 CH_2CH_3 (F) 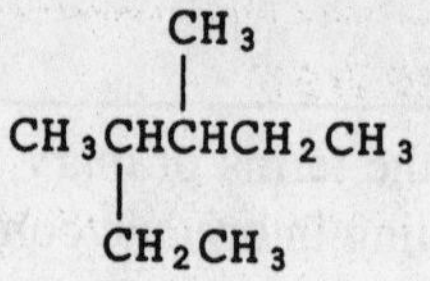 CH_3

           | |

       CH_3CCH_3 $CH_3CHCHCH_2CH_3$

           | |

       CH_2CH_3 CH_2CH_3

Figure 1.7

SOLUTION

Isomers are related compounds that have the same molecular formula but different structural formulas.

Isomerism is not possible among the alkanes until there are enough carbon atoms to permit more than one arrangement of the carbon chain.

To name the above compounds, one uses a set of rules to provide each compound with a clear name. These rules for nomenclature are the IUPAC rules (International Union of Pure and Applied Chemistry), and are referred to as systematic nomenclature.

One of the rules of the IUPAC system is to choose the largest chain of carbon atoms in the molecule and call it the parent compound. Thus,

(A) has a 6 carbon parent chain with methyl groups bonded to the second and fourth carbon atoms of the parent chain. There exists, also, an ethyl group bonded to the third carbon atom.

As such, the name of this organic molecule is 2,4-dimethyl-3-ethylhexane.

(B) has a 5 carbon parent molecule and 2 methyl groups on carbon number 3. Therefore, the name is 3,3-dimethylpentane.

(C) has a 5 carbon parent molecule with 2 methyl groups bonded to the third carbon of the parent chain. Therefore, the name of this structure becomes 3,3-dimethylpentane.

(D) has a 5 carbon parent molecule, and 2 methyl groups attached to carbon numbers 2 and 3. Therefore, the name is 2,3-dimethylpentane.

(E) has a 5 carbon parent molecule, and 2 methyl groups attached to carbon number 3. Therefore, the name is 3,3-dimethylpentane.

(F) has a 6 carbon parent molecule, and 2 methyl groups attached to carbon numbers 3 and 4. Therefore, the name is 3,4-dimethylhexane.

To find which compounds are isomers, one counts the number of carbon and hydrogen atoms contained in the molecule. If the total number of both carbon and hydrogen atoms in one molecule is the same as in another molecule, but they have different structural formulas, they are isomers.

If the calculation of total carbon and hydrogen molecules is made, (B), (C), (D), and (E) become isomers, with (B) and (E) being the same compound.

Drill 1: Alkanes

Questions 1–4. Name the following alkanes.

(A) 3-methylpentane

(B) 2,3-dimethylhexane

(C) 2-methylhexane

(D) 2,2,4-trimethylhexane

(E) None of the above

1.

2.

3.

4.

2.   Alkenes and Alkynes

Alkenes

Alkenes (olefins) are unsaturated hydrocarbons with one or more carbon-carbon double bonds. They have the general formula, CnH_{2n}.

EXAMPLES

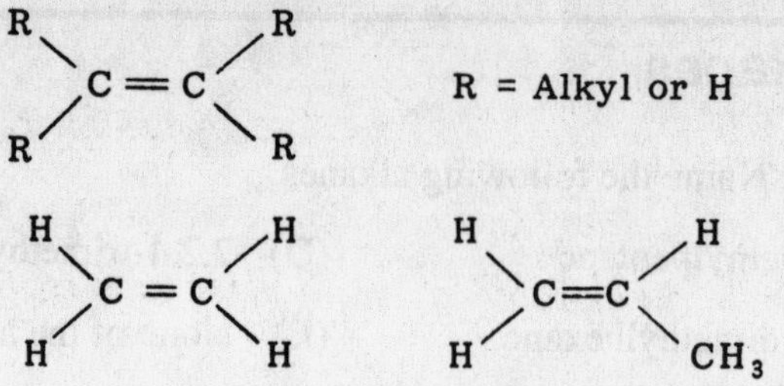

R = Alkyl or H

ethylene   propylene

Figure 2.1

Nomenclature (IUPAC System)

A)   Select the longest continuous chain of carbons containing the double bond. This is the parent structure and is assigned the name of the corresponding alkane with the suffix changed from "-ane" to "-ene."

B)   Number the chain so that the position of the double bond is designated by the lowest possible number assigned to the first doubly bonded carbon.

EXAMPLE

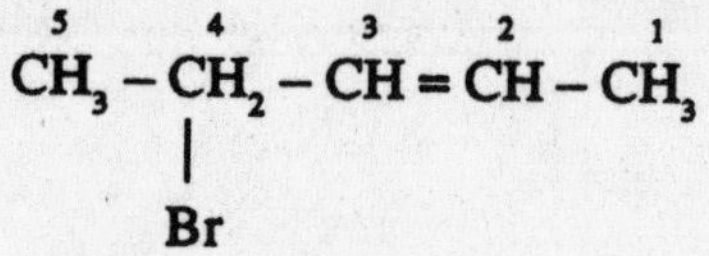

$$\overset{5}{CH_3} - \overset{4}{CH_2} - \overset{3}{CH} = \overset{2}{CH} - \overset{1}{CH_3}$$
$$|$$
$$Br$$

4–bromo–2–pentene

Figure 2.2

Some common names given to families of alkenes are:

$H_2C = CH - R$	vinyl
$H_2C = CH - CH_2 - R$	allyl
$H_3C - CH = CH - R$	propenyl

Dienes

Dienes have the structural formula, C_nH_{2n-2}. In the IUPAC nomenclature system, dienes are named in the same manner as alkenes, except that the suffix "-ene" is replaced by "-diene," and two numbers must be used to indicate the position of the double bonds.

Classification of Dienes

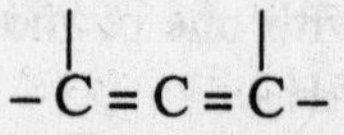

$$-\overset{|}{C} = C = \overset{|}{C}-$$ Cumulated double bonds (allenes)

$$-\overset{|}{C}=\overset{|}{C}-\overset{|}{C}=\overset{|}{C}-$$ Conjugated (alternating) double bonds

$$-\overset{|}{C}=\overset{|}{C}-(CH_2)_n-\overset{|}{C}=\overset{|}{C}-$$ Isolated (non-conjugated) double bonds

Figure 2.3

Alkynes

Alkynes are unsaturated hydrocarbons containing triple bonds. They have the general formula, C_nH_{2n-2}.

$$R - C \equiv C - R \qquad\qquad R = \text{Alkyl or H}$$

EXAMPLE

$$H - C \equiv C - H \qquad\qquad \text{Simplest Alkyne}$$

acetylene

Figure 2.4

Nomenclature (IUPAC System)

Alkynes are named in the same manner as alkenes, except that the suffix "-ene" is replaced with "-yne."

When both a double bond and a triple bond are present, the hydrocarbon is called an alkenyne. In this case, the double bond is given preference over the triple bond when numbering.

EXAMPLE

$$CH_3 - C \equiv CH \qquad\qquad CH_3 - CH = CH - C \equiv CH$$

(A) propyne (B) 1-penten-3-yne

Figure 2.5

PROBLEM

> Classify each of the following as a member of the methane series, the ethylene series, or the acetylene series: $C_{12}H_{26}$, C_9H_{16}, C_7H_{14}, $C_{26}H_{54}$.

SOLUTION

Before beginning this problem, one should first know the general formulas for each of the series. For the alkanes (methane series), the general formula is C_nH_{2n+2}, where n is the number of carbon atoms and $2n + 2$ is the number of hydrogens. Molecules of the ethylene series, also called the alkene series, have two adjacent carbon atoms joined to one another by a double bond. Any member in this series has the general formula C_nH_{2n}. The acetylene series, commonly called the alkyne series, has two adjacent carbon atoms joined to one another by a triple bond. The general formula for this series is C_nH_{2n-2}. With this in mind one can write:

$$C_{12}H_{26} \quad : \quad C_nH_{2n+2} \qquad : \text{alkane series}$$

$$C_9H_{16} \quad : \quad C_nH_{2n-2} \qquad : \text{acetylene series}$$

$$C_7H_{14} \quad : \quad C_nH_{2n} \qquad : \text{ethylene series}$$

$$C_{26}H_{54} \quad : \quad C_nH_{2n+2} \qquad : \text{alkane series}$$

PROBLEM

Draw the structure of 4-ethyl-3,4-dimethyl-2-hexene.

SOLUTION

To draw the structure of more complex compounds, such as this one, certain steps must be followed.

(1) Identify the parent compound that is associated with the longest carbon chain that contains the functional group. In 4-ethyl-3,4-dimethyl-2-hexene, the parent compound is hexene.

(2) Draw the parent carbon skeleton, in this case, a six carbon chain. Do not put any hydrogen atoms in yet.

$$C - C - C - C - C - C$$

Figure 2.6

(3) Number the carbon atoms starting at either end. This is important; otherwise, it may get confusing when one adds the functionality.

$$\begin{array}{cccccc} 1 & 2 & 3 & 4 & 5 & 6 \end{array}$$
$$C - C - C - C - C - C$$

Figure 2.7

(4) Add the suffix functionality, in this case -2-ene. "Ene" tells one that a double bond is present, while "2" indicates that it is at the second carbon.

$$\begin{array}{cccccc} 1 & 2 & 3 & 4 & 5 & 6 \end{array}$$
$$C - C = C - C - C - C$$

Figure 2.8

(5) Add the prefix functionality, starting at the beginning of the name and continuing until the parent name is reached. Here, the prefixes are 4-ethyl-3,4-dimethyl-.

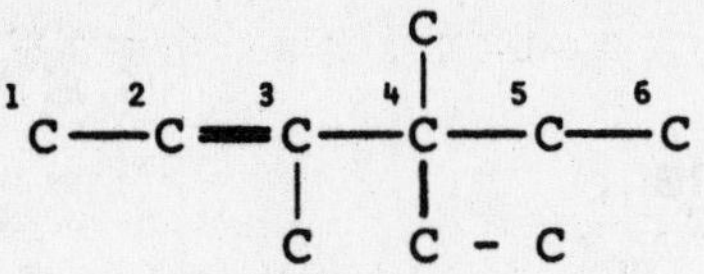

Figure 2.9

Now, the hydrogen atoms can be added to give a complete structure.

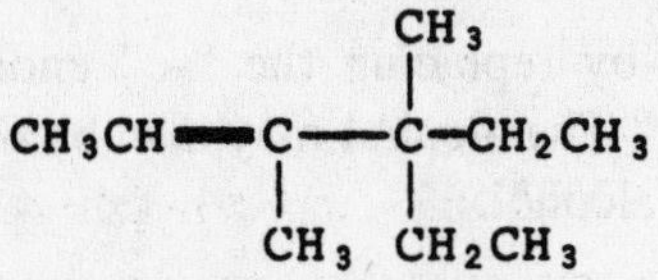

Figure 2.10

Drill 2: Alkenes and Alkynes

1. What is the name of the following structure?

   (A) Benzene

   (B) Propylene

   (C) Ethane

   (D) Ethene

   (E) Methylene

2. What is the structure of 3-penten-1-yne?

   (A) $CH_3 - C \equiv CH$

   (B) $CH_3 - CH = CH_2$

   (C) $CH_3 - CH = CH - C \equiv CH$

   (D) $CH_3 - CH = CH - CH = CH_2$

   (E) $CH_3 - C = CH - CH = CH_2$

3. What is the structure of 2-pentene?

   (A) $CH_3 - CH = CH - CH_3$

   (B) $CH_2 = CH - CH_2 - CH_2 - CH_3$

   (C) $CH_3 - CH = CH - CH_2 - CH_3$

(D) $CH_2 = CH = CH - CH_2 - CH_3$

(E) $CH_3 - CH_2 - CH_2 - CH_2 - CH_3$

3. Alcohols

Alcohols and Glycols

Alcohols and hydrocarbon are derivatives in which one or more hydrogen atoms have been replaced by the –OH (hydroxyl) group. They have the general formula $R - OH$, where R may be either alkyl or aryl.

Nomenclature (IUPAC System)

Alcohols are named by replacing the "-e" ending of the corresponding alkane with the suffix "-ol." The alcohol may also be named by adding the name of the R group to the same alcohol.

EXAMPLE

CH_3CH_2OH Ethanol or methyl alcohol

Depending on what carbon atom the hydroxyl group is attached to, the alcohol is prefixed as follows:

A) Primary (–OH attached to 1° carbon) alcohols are prefixed "n-" or "1-".

B) Secondary (–OH attached to 2° carbon) alcohols are prefixed "sec-" or "2-".

C) Tertiary (–OH attached to 3° carbon) alcohols are prefixed "tert-" or "3-".

EXAMPLE

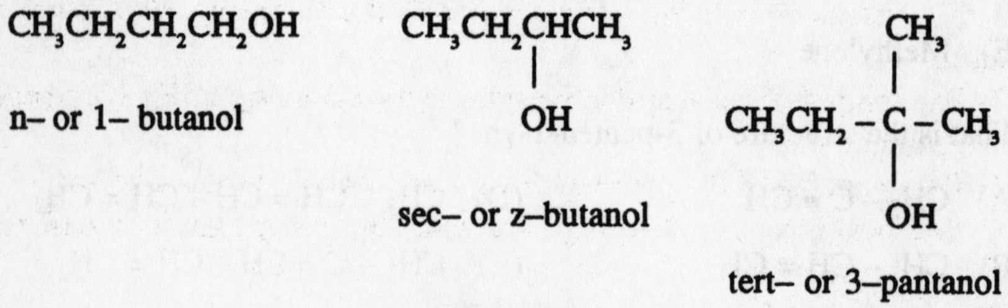

$CH_3CH_2CH_2CH_2OH$

n– or 1– butanol

$CH_3CH_2CHCH_3$
       |
     OH

sec– or z–butanol

CH_3
|
$CH_3CH_2 - C - CH_3$
|
OH

tert– or 3–pantanol

Figure 3.1

Glycols

Alcohols containing more than one hydroxyl group (polyhydroxyalcohols) are represented by the general formula $C_nH_{2n+2} - y\ (OH)y$. Polyhydroxyalcohols containing two hydroxyl groups are called glycols or diols.

EXAMPLE

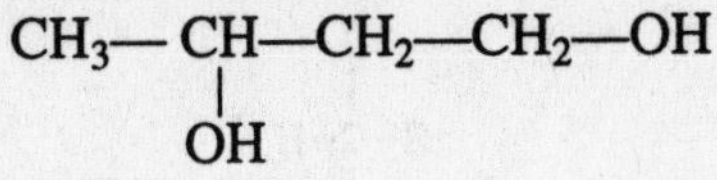

$$CH_3 - CH - CH_2 - CH_2 - OH$$
$$\qquad\quad |$$
$$\qquad\quad OH$$

Figure 3.2

PROBLEM

Name the following systematically:

(a) $CH_3CHOHCH_2CH_2CH_2CH_3$

(b) $CH_3CHCH_2CH_2OH$
$$\qquad\quad |$$
$$\qquad\quad CH_2CH_3$$

(c) $CH_3CH_2CH_2COHCH_3$
$$\qquad\qquad\quad |$$
$$\qquad\qquad\quad CH_3CHCH_3$$

(d) $CH_2OHCHCH_2CHCH_3$
$$\qquad\quad\; | \qquad\; |$$
$$\qquad\quad CH_3 \quad CH_3$$

SOLUTION

All of these compounds are alcohols; they fit into the general formula R–OH, where R is any alkyl group.

In the systematic naming of any alcohol, the following rules should be followed:

(1) The longest chain that contains the hydroxyl group (OH) is considered the parent compound.

(2) The -e ending of the name of this carbon chain is replaced by -ol.

(3) The locations of the hydroxyl and any other groups are indicated by the smallest possible numbers.

Thus, compound (a)

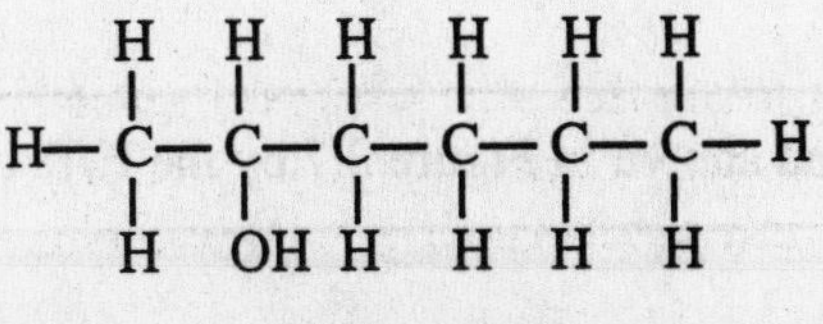

$$H - \overset{\displaystyle H}{\underset{\displaystyle H}{C}} - \overset{\displaystyle H}{\underset{\displaystyle OH}{C}} - \overset{\displaystyle H}{\underset{\displaystyle H}{C}} - \overset{\displaystyle H}{\underset{\displaystyle H}{C}} - \overset{\displaystyle H}{\underset{\displaystyle H}{C}} - \overset{\displaystyle H}{\underset{\displaystyle H}{C}} - H$$

Figure 3.3

has a 6-carbon chain and an hydroxyl group on carbon number 2. The name of this compound is 2-hexanol.

Compound (b)

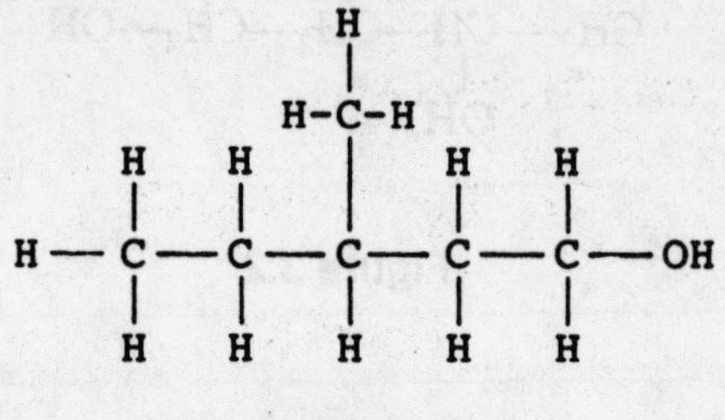

Figure 3.4

has a 5-carbon chain, because this is the longest chain that contains the hydroxyl group. There is a methyl group on carbon number 3, and an hydroxyl group on carbon number 1. The name of this compound is, therefore, 3-methyl-1-pentanol.

Compound (c)

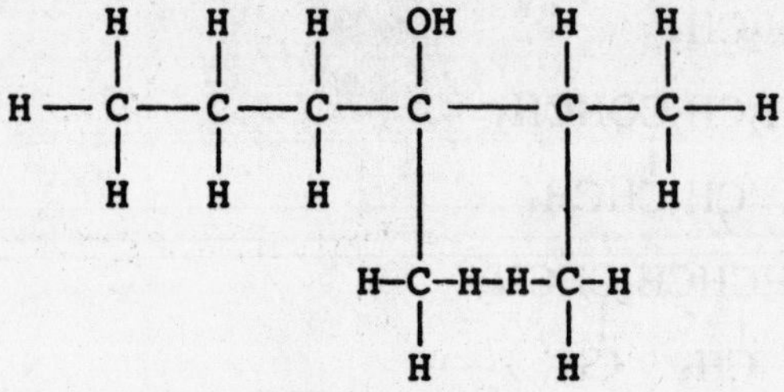

Figure 3.5

has a 6-carbon chain that contains the OH group, two methyl groups on carbons 2 and 3, and an hydroxyl group on carbon number 3. Thus, the name is 2,3-dimethyl-3-hexanol.

Compound (d)

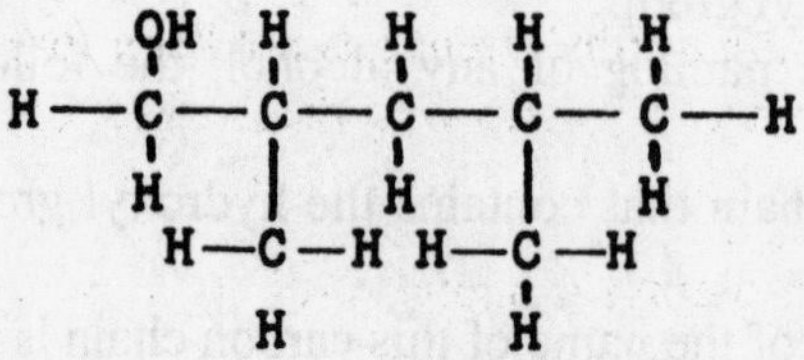

Figure 3.5

has a 5-carbon chain, two methyl groups on carbons 2 and 4, and an hydroxyl group on carbon 1. Thus, the name is 2,4-dimethyl-1-pentanol.

PROBLEM

Name the compound shown in Figure 3.7 by the IUPAC system.

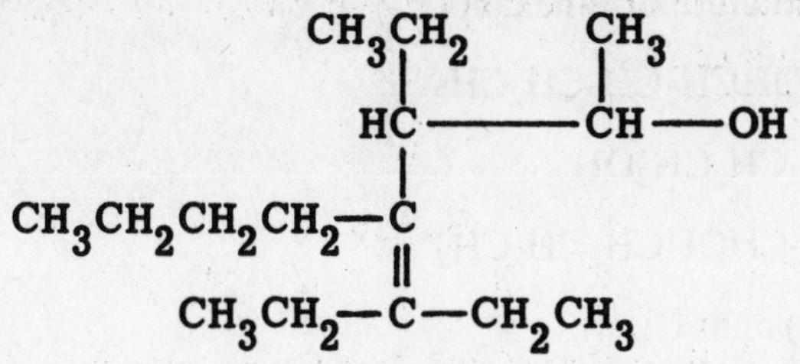

$$CH_3CH_2 \quad\quad CH_3$$
$$HC \text{———} CH \text{——} OH$$
$$CH_3CH_2CH_2CH_2 \text{—} C$$
$$\|$$
$$CH_3CH_2 \text{—} C \text{—} CH_2CH_3$$

Figure 3.7

SOLUTION

In naming complex open chain organic compounds, first find the longest continuous chain containing the functional groups. Write down the parent name. The parent for this particular compound is heptene. Number the chain starting from the end that will give the smallest prefix numbers for the functional groups. This is shown in Figure 3.8.

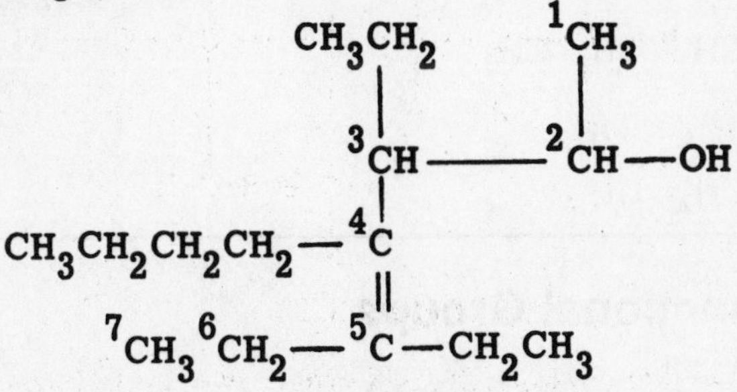

$$CH_3CH_2 \quad\quad {}^1CH_3$$
$${}^3CH \text{———} {}^2CH \text{——} OH$$
$$CH_3CH_2CH_2CH_2 \text{—} {}^4C$$
$$\|$$
$${}^7CH_3 \, {}^6CH_2 \text{—} {}^5C \text{—} CH_2CH_3$$

Figure 3.8

Add the suffix functionality with the appropriate numbering: 4-hepten-2-ol. Add the prefix functionality, remembering to group together like prefixes. Then, double check to make sure a substituent has not been forgotten or one substituted has not been included twice. By following these steps, one arrives at the name for the structure:

4-n-butyl-3,5-diethyl-4-hepten-2-ol

Drill 3: Alcohols

1. Which of the following, if any, are not alcohols derived from the methane series of hydrocarbons: C_6H_5OH, $C_{17}H_{33}OH$, C_4H_8OH, $C_9H_{19}OH$?

    (A) C_6H_5OH

    (B) $C_{17}H_{33}OH$

    (C) C_4H_8OH

    (D) $C_9H_{19}OH$

    (E) None of the above.

2. What is the structure of 2-hexanol?

    (A) $CH_3CHOHCH_2CH_2CH_2CH_2$

    (B) $CH_3CH_2CH_2CH_2OH$

    (C) $CH_3CH_2CHOHCH_2CH_2CH_2$

    (D) Both (A) and (C).

    (E) None of the above

3. What is the structure of isopropanol?

    (A) $CH_3 - CH_2 - CH_2 - CH_2 - OH$

    (B) $CH_3 - CH - CH_3$
                 |
                OH

    (C) $CH_3 - CH_2 - CH_2 - OH$

    (D) $CH_3 - CH_3 - CH - CH_3$
                     |
                   OH

    (E) $CH_3 - CH_2 - OH$

4. Other Functional Groups

Alkyl Halides

Alkyl halides are compounds in which one hydrogen atom is replaced by an atom of the halide family. An important use of alkyl halides is as intermediates in organic synthesis.

Structural formula:

$$C_nH_{2n+1} - X: X = Cl, Br, I, F.$$

Nomenclature (IUPAC System)

Formula	Name
CH_3Cl	chloromethane
CH_3CH_2Br	bromoethane
$CH_3CH_2CH_2I$	1-iodopropane
CH_3CHICH_3	2-iodopropane
$CH_3CH_2CH_2CH_2Cl$	1-chlorobutane
$CH_3CH_2CHBrCH_3$	2-bromobutane
$(CH_3)_3CI$	2-iodo-2-methylpropane
$CH_3CH_2CH_2CH_2CH_2Cl$	1-chloropentane

Cyclic Hydrocarbons

Cyclic hydrocarbons and cyclic alkenes are alicyclic (aliphatic cyclic) hydrocarbons.

Nomenclature

Cyclic aliphatic hydrocarbons are named by prefixing the term "cyclo-" to the name of the corresponding open-chain hydrocarbon, having the same number of carbon atoms as the ring.

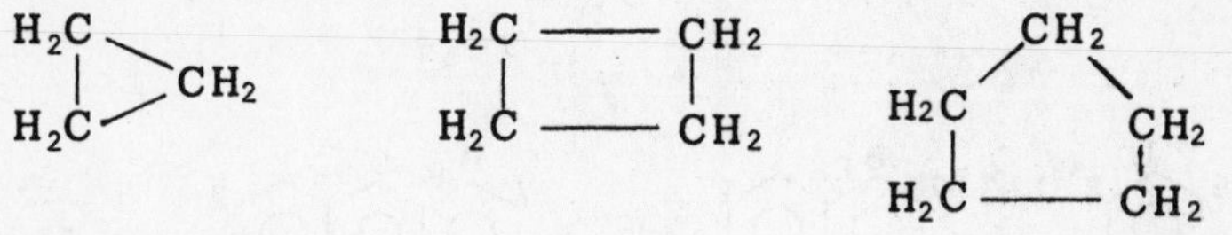

cyclopropane   cyclobutane   cyclopentane

Figure 4.1

Substituents on the ring are named, and their positions are indicated by numbers, the lowest combination of numbers being used.

Aromatic Hydrocarbons

Most aromatic hydrocarbons (arenes) are derivatives of benzene. Examples of benzene derivatives are napthalene, anthracene, and phenanthrene.

Structure

Benzene has a symmetrical structure and the analysis, synthesis, and molecular weight determination indicate a molecular formula of C_6H_6.

Napthalene structure is indicated by the oxidation of 1-nitronapthalene which shows that the substituted ring is a true benzene ring. Reduction and oxidation of the same nucleus indicates that the unsubstituted ring is a true benzene ring.

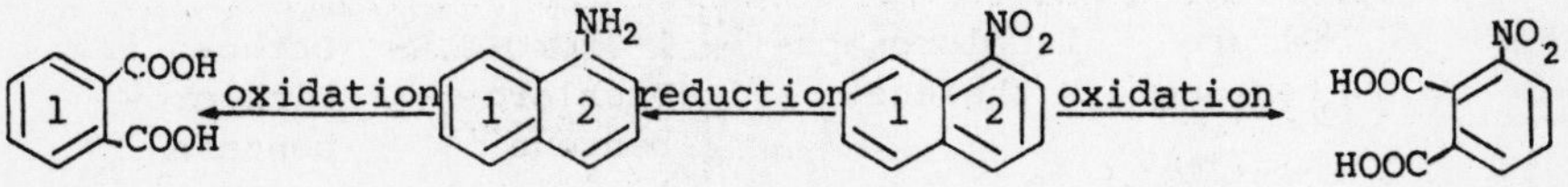

Figure 4.2

Nomenclature (IUPAC System)

Aromatic compounds are named as derivatives of the corresponding hydrocarbon nucleus.

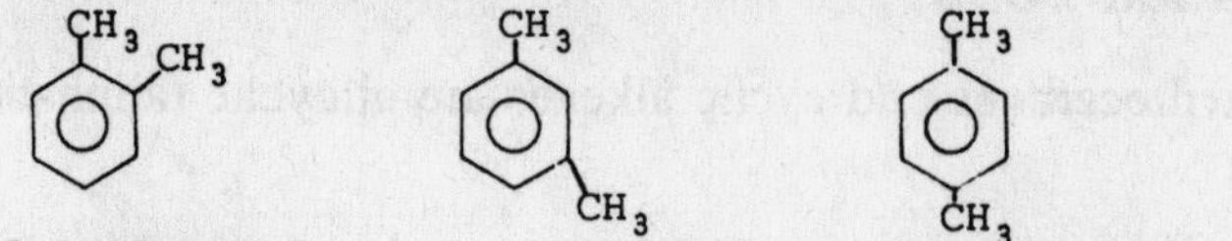

1,2-Dimethyl-benzene (ortho-xylene) | 1,3-Dimethyl-benzene (meta-xylene) | 1,4-Dimethyl-benzene (para-xylene)

Figure 4.3

In the IUPAC system of nomenclature, the position of the substituent group is always indicated by numbers arranged in a certain order.

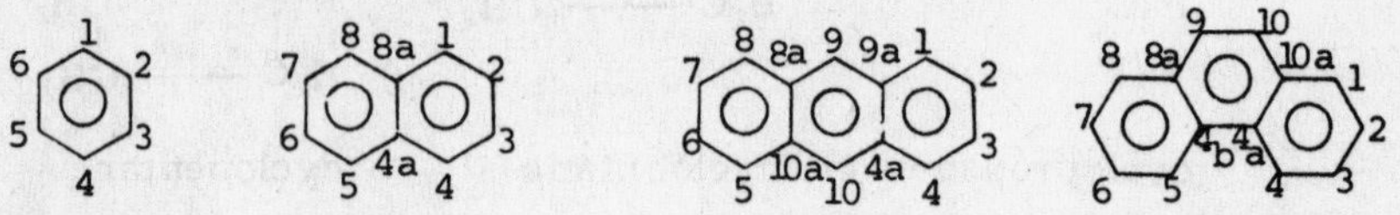

Benzene | Naphthalene | Anthracene | Phenanthrene

Figure 4.4

Aryl Halides

Aryl halides are compounds containing halogens attached directly to a benzene ring. The structural formula is Ar*X*, where the aryl group, Ar, represents phenyl, napthyl, etc., and their derivatives.

Nomenclature

Aryl halides are named by prefixing the name of the halide to the name of the aryl group. The terms meta, ortho, and para are used to indicate the positions of substituents on a dissubstituted benzene ring. Numbers are also used to indicate the positions of the substituents on a benzene ring.

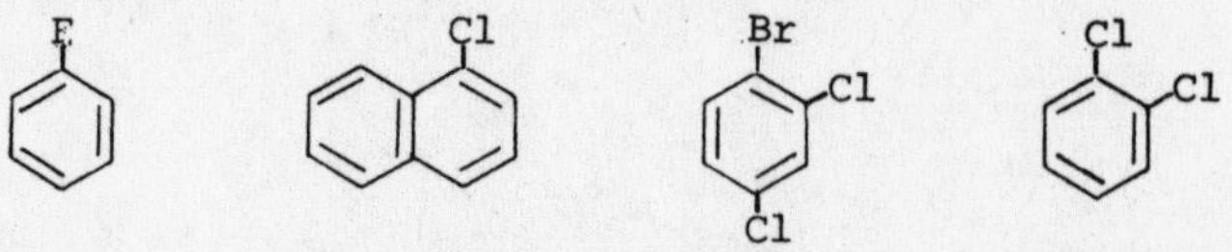

Flouro-benzene | 1-Chloronaph-thalene | 1-Bromo-2,4-Dichloro-benzene | Ortho-Dichloro-benzene

Figure 4.5

Ethers and Epoxides

Ethers are hydrocarbon derivatives in which two alkyl or aryl groups are attached to an oxygen atom. The structural formula of an ether is R – O – R', where R and R' may or may not be the same.

Ethers – Structure

Ethers and alcohols are metameric. They are functional isomers of alcohols with the same elemental composition.

$$CH_3OCH_3 \text{ and } CH_3CH_2OH$$

Nomenclature (IUPAC System) – Common Names

The attached groups are named in alphabetical order, followed by the word ether.

$CH_3CH_2-O-CH_2CH_2CH_3$
Ethyl propyl ether

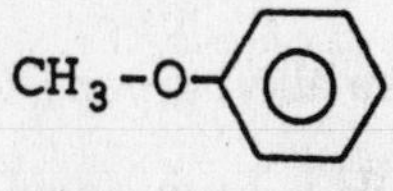

CH_3-O-⬡
Methyl phenyl ether

For symmetrical ethers having the same groups, the compound is named using either the name of the group of the prefix "Di-."

EXAMPLE

$$CH_3 - O - CH_3$$

Methyl ether or Dimethyl ether
Figure 4.6

In the IUPAC system, ethers are named as alkoxyalkanes. The larger alkyl group is chosen as the stem.

EXAMPLE

$$\begin{array}{c} Cl \\ | \\ CH_3 - CH - CH - CH_3 \\ |\quad\quad | \\ Cl \quad OCH_2CH_3 \end{array}$$

2–Ethoxy–3,3–dichlorobutane

Figure 4.7

Epoxides – Structure

Epoxides are cyclic ethers in which the oxygen is included in a three-membered ring.

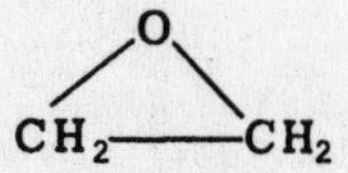

$$\begin{array}{c} O \\ /\,\backslash \\ CH_2\text{———}CH_2 \end{array}$$

An epoxide:
Ethylene oxide

Figure 4.8

Carboxylic Acids

Carboxylic acids contain a carboxyl group

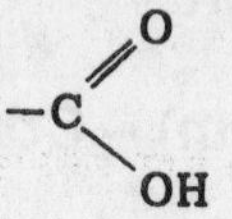

$$-C\overset{\displaystyle O}{\underset{\displaystyle OH}{}}$$

Figure 4.9

bonded to either an alkyl group (RCOOH) or an aryl group (ArCOOH).

HCOOH is formic acid (methanoic acid).

CH_3COOH is acetic acid (ethanoic acid)

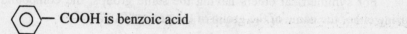 COOH is benzoic acid

Figure 4.10

Nomenclature (IUPAC System)

The longest chain carrying the carboxyl group is considered the parent structure and is named by replacing the "-e" ending of the corresponding alkane with "-oic acid."

$CH_3CH_2CH_2CH_2COOH$	Pentanoic acid

$CH_3CH_2CHCOOH$
    |
    CH_3
	2-Methylbutanoic acid

CH_2CH_2COOH	3-Phenylpropanoic acid
$CH_3CH = CHCOH$	2-Butenoic acid

Figure 4.11

The position of the substituent is indicated by a number, e.g.,

$$\overset{5}{C}-\overset{4}{C}-\overset{3}{C}-\overset{2}{C}-\overset{1}{COOH}$$

Figure 4.12

The name of a salt of a carboxylic acid consists of the name of the cation followed by the name of the acid with the ending "-ic acid" changed to "-ate."

COONa $(CH_3COO)_2Ca$ $HCOONH_4$

Sodium benzoate   Calcium acetate   Ammonium formate

Figure 4.13

Carboxylic Acid Derivatives

Carboxylic acid derivatives are compounds in which the carboxyl group has been replaced by –Cl, –OOCR, –NH$_2$, or –OR'. These derivatives are called acid chlorides, anhydrides, amides, and esters, respectively.

Acid Chlorides – Nomenclature (IUPAC System)

When naming acid chlorides, the ending "-ic acid" in the carboxylic acid is replaced by the ending "-yl chloride."

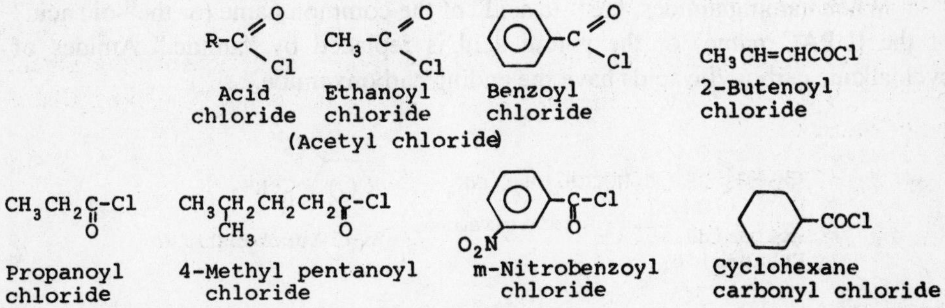

Acid chloride   Ethanoyl chloride (Acetyl chloride)   Benzoyl chloride   2-Butenoyl chloride

Propanoyl chloride   4-Methyl pentanoyl chloride   m-Nitrobenzoyl chloride   Cyclohexane carbonyl chloride

Figure 4.14

Carboxylic Acid Anhydrides – Nomenclature (IUPAC System)

When naming acid anhydrides, the word "acid" in the carboxylic acid is replaced by the word "anhydride."

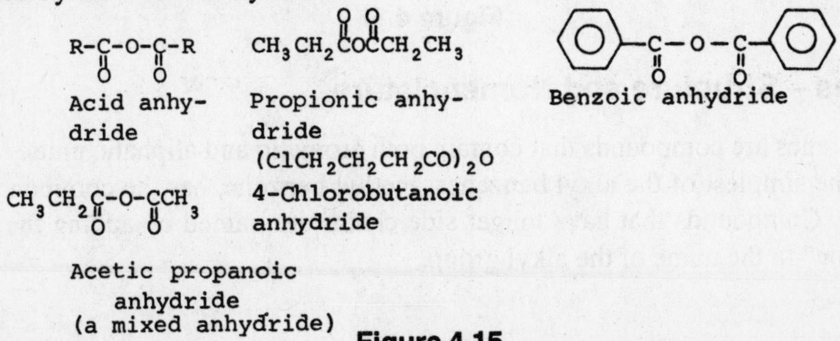

Acid anhydride   Propionic anhydride   Benzoic anhydride

$(ClCH_2CH_2CH_2CO)_2O$
4-Chlorobutanoic anhydride

Acetic propanoic anhydride (a mixed anhydride)

Figure 4.15

Esters – Nomenclature (IUPAC System)

When naming esters, the alcohol or phenol group is named first, followed by the name of the acid with the "-ic" ending replaced by "-ate." Esters of cycloalkane carboxylic acids have the ending "carboxylate."

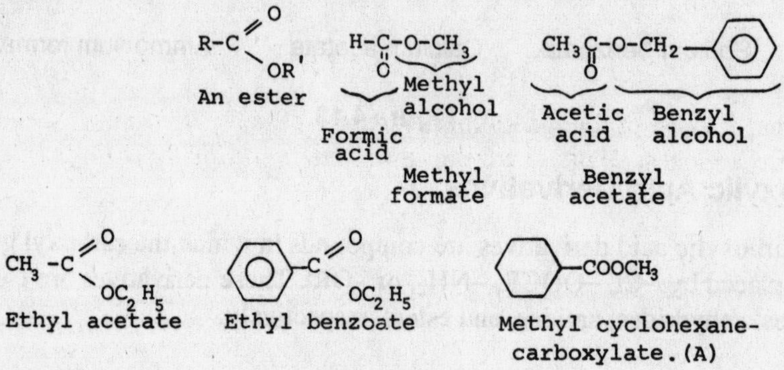

Figure 4.16

Amides – Nomenclature (IUPAC System)

When naming amides, the "-ic acid" of the common name (or the "-oic acid" of the IUPAC name) of the parent acid is replaced by "amide." Amides of cycloalkane carboxylic acids have the ending carbonxamide.

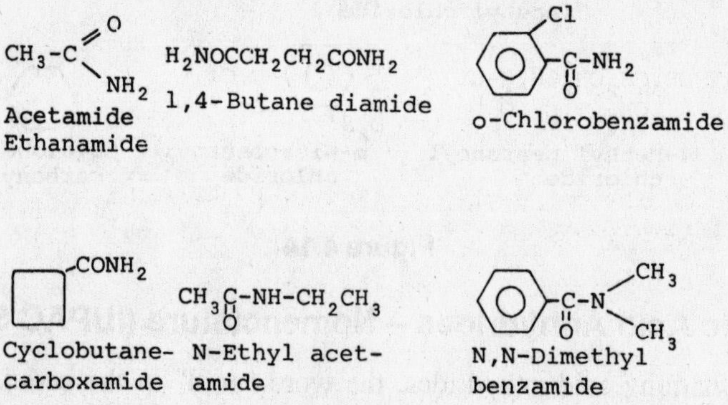

Figure 4.17

Arenes – Structure and Nomenclature

Arenes are compounds that contain both aromatic and aliphatic units.

The simplest of the alkyl benzenes, methyl benzene, has the common name toluene. Compounds that have longer side chains are named by adding the word "benzene" to the name of the alkyl group.

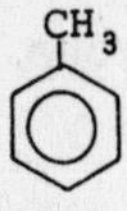
Toluene

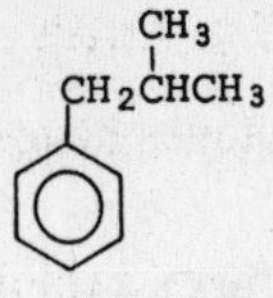
Isobutylbenzene

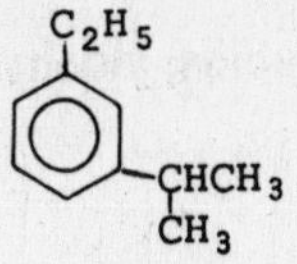
m-Ethylisopropylbenzene

Figure 4.18

The simplest of the dialkyl benzenes, the dimethyl benzenes, have the common name xylenes. Dialkyl benzenes that contain one methyl group are named as derivatives of toluene.

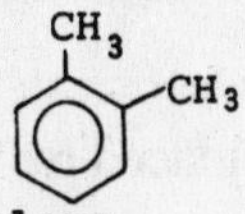
o-Xylene

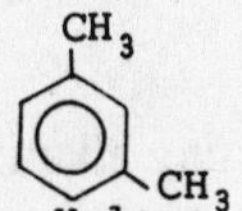
m-Xylene

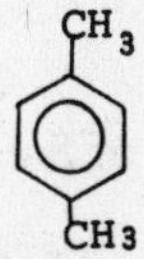
p-Xylene

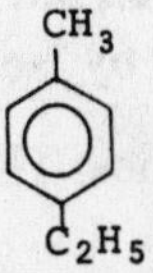
p-Ethyltoluene

Figure 4.19

A compound that contains a complex side chain is named as a phenyl alkane (C_6H_5 = phenyl). Compounds that contain more than one benzene ring are named as derivatives of alkanes.

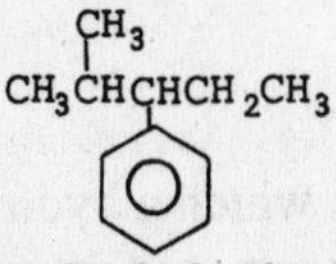
2-Methyl-3-phenylpentane

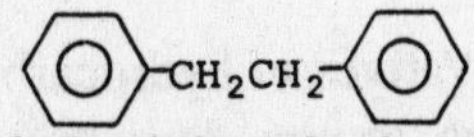
1,2-Diphenylethane

Figure 4.20

Styrene is the name given to the simplest alkenyl benzene. Others are named as substituted alkenes. Alkynyl benzenes are named as substituted alkynes.

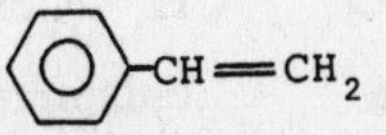
Styrene
(Vinylbenzene)
(Phenylethylene)

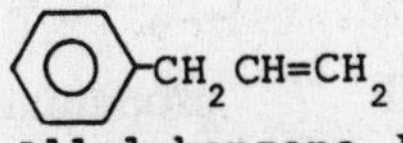
Allyl benzene
(3-Phenylpropene)

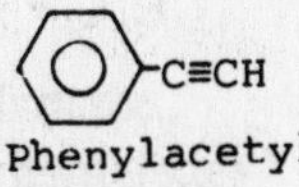
Phenylacetylene

Figure 4.21

Aldehydes and Ketones

Carboxylic acids, aldehydes, and ketones have a carboxylic group,

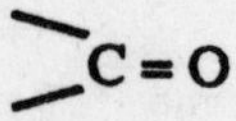

$$\diagdown C = O$$

Figure 4.22

in common. The general formula for aldehydes is RCHO and that for ketones is RCOR.

Nomenclature (IUPAC System)

A) Aldehydes: The longest continuous chain containing the carbonyl group is considered the parent structure and the "-e" ending of the corresponding alkane is replaced by "-al."

B) Ketones: The "-e" ending of the corresponding alkane is replaced by "-one."

Example:

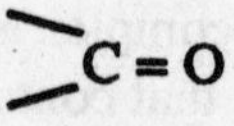

$$\diagdown C = O$$

Figure 4.23

Amines

Amines are derivatives of hydrocarbons in which a hydrogen atom has been replaced by an amino group; derivatives of ammonia in which one or more hydrogen atoms have been replaced by alkyl groups are also known as amines. They are classified, according to structure, as

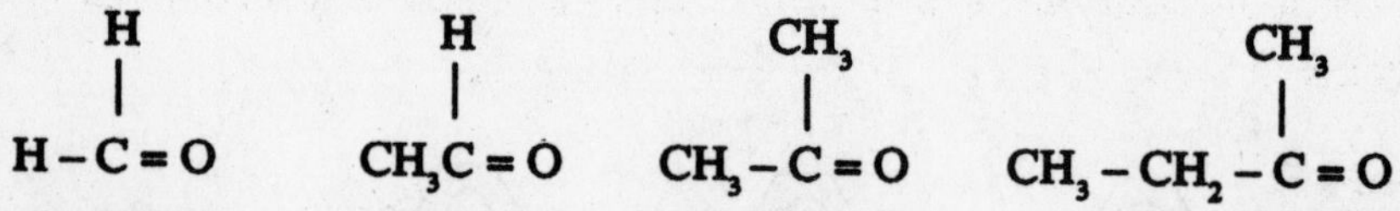

H	H	CH$_3$	CH$_3$
$\|$	$\|$	$\|$	$\|$
H−C=O	CH$_3$C=O	CH$_3$−C=O	CH$_3$−CH$_2$−C=O
methanal	ethanal	propanone	butanone (methyl
(formaldehyde)	(acetaldehyde)	(acetone)	ethyl ketone)

Figure 4.24

Nomenclature (IUPAC System)

The aliphatic amine is named by listing the alkyl groups attached to the nitrogen, and following these by "-amine."

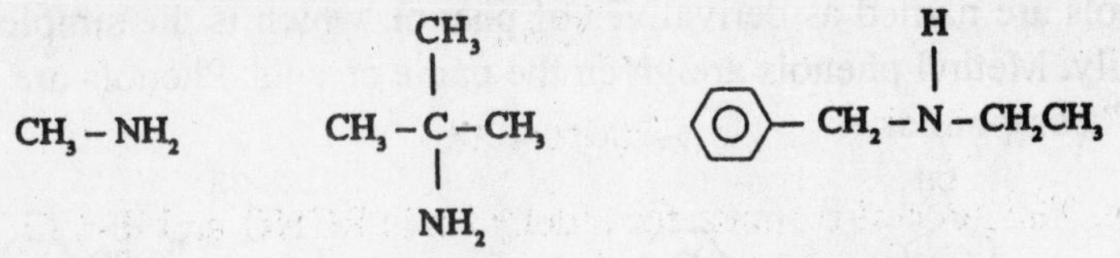

$$CH_3 - NH_2 \qquad CH_3 - \underset{\underset{NH_2}{|}}{\overset{\overset{CH_3}{|}}{C}} - CH_3 \qquad \underset{}{\bigcirc} - CH_2 - \underset{}{\overset{\overset{H}{|}}{N}} - CH_2CH_3$$

Methylamine   tert–Butylamine   Benzyl ethylamine

Figure 4.25

If an alkyl group occurs twice or three times on the nitrogen, the prefixes "di-" and "tri-" are used respectively.

EXAMPLE

$$CH_3 - NH - CH_3 \qquad\qquad CH_3 - \overset{\overset{CH_3}{|}}{N} - CH_3$$

dimethylamine   **trimethylamine**

Figure 4.26

If an amino group is part of a complicated molecule, it may be named by prefixing "amino" to the name of the present chain.

EXAMPLE

$$NH_2CH_2CH_2OH \qquad\qquad CH_3\overset{\overset{NH_2}{|}}{CH} \quad CH_2COOH$$

2–amino ethanol   3–aminobutanoic acid

Figure 4.27

An amino substituent that carries an alkyl group is named as an N-alkyl amino group.

EXAMPLE

$$CH_3NH - CH_2COOH \qquad\qquad CH_3 - NHCH(\overset{\overset{CH_2}{|}}{}CH_2)_4CH_3$$

N–methyl amino acetic acid   2–(N–methylamino) heptane

Figure 4.28

Phenols and Quinones – Nomenclature of Phenols

Phenols have the general formula ArOH. The –OH group in phenols is attached directly to the aromatic ring.

Phenols are named as derivatives of phenol, which is the simplest member of the family. Methyl phenols are given the name cresols. Phenols are also called "hydroxy-" compounds.

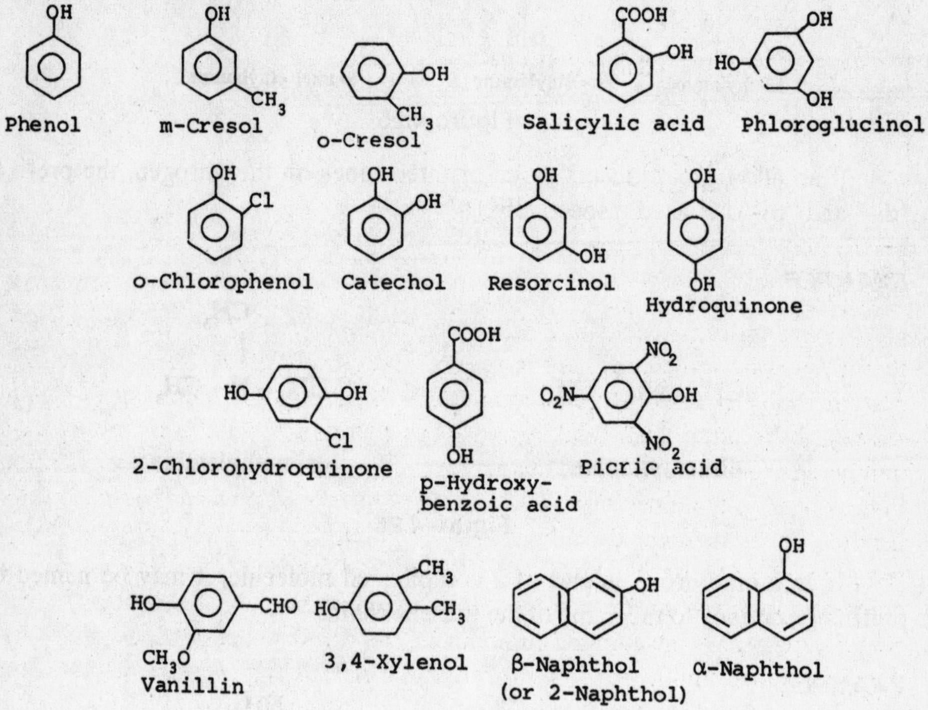

Phenol   m-Cresol   o-Cresol   Salicylic acid   Phloroglucinol

o-Chlorophenol   Catechol   Resorcinol   Hydroquinone

2-Chlorohydroquinone   p-Hydroxy-benzoic acid   Picric acid

Vanillin   3,4-Xylenol   β-Naphthol (or 2-Naphthol)   α-Naphthol

Figure 4.29

Quinones – Nomenclature

Quinones are cyclic, conjugated diketones named after the parent hydrocarbon.

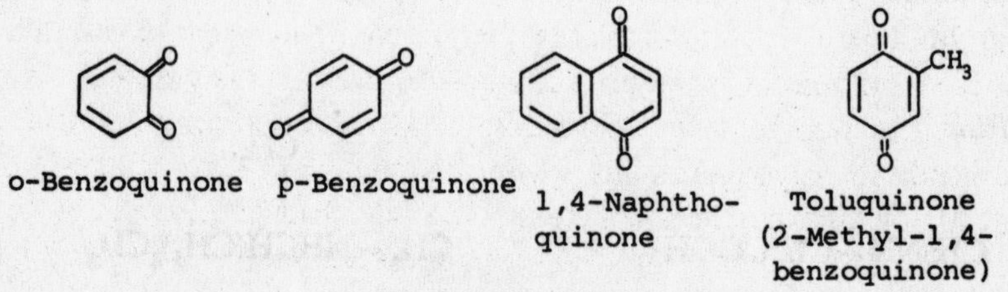

o-Benzoquinone   p-Benzoquinone   1,4-Naphtho-quinone   Toluquinone (2-Methyl-1,4-benzoquinone)

Figure 4.30

PROBLEM

Name each of the organic compounds shown in Figure 4.31 below.

(a) $CH_3 - CH = CH_2$

(b) $CH_3 - \underset{\underset{C_2H_5}{|}}{CH} - CH_2 - CH = CH - CH_3$

(c) $CH_3CH = CHCH_2CH_2CH_3$

(d)   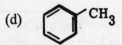 CH_3

Figure 4.31

SOLUTION

A system of nomenclature called the IUPAC system has been formulated so that all organic molecules may have their structures defined adequately. In this system for hydrocarbons (i.e., compounds containing only hydrogen and carbon atoms), the longest chain is taken as the basic structure, and the carbon atoms are numbered from the end of the chain closest to a branch chain or other modification of a simple alkane structure. The position of substituents in the chain are denoted by the number of carbon atom or atoms to which they are attached. In the problem, you are not given alkanes, you are given alkenes, compounds that contain a double bond between a pair of carbon atoms (i.e., unsaturated), and a ring compound. In unsaturated compounds, the rules are the same, except that the position of the double bond is indicated; the numbering starts at the end of the chain nearest the double bond. For rings, name the ring and any substituent present based on which carbon is located. Thus, you proceed as follows:

(a) There is only one chain, and it possesses three carbon atoms. The double bond is located on the first carbon, not the second, since you want to use the lowest possible number. As such, you form the molecule propene. Three carbons suggest propane. It's an alkene, so that you change -ane to -ene. Therefore, propane becomes propene. You have no need to name the position of the double bond here; the double bond can only be in two positions and both yield the same molecule. That is, $CH_2 = CH - CH_3$ is equivalent to $CH_3 - CH = CH_2$.

(b) The longest chain containing the double bond has seven carbon atoms, so that the "hep" prefix is suggested. Because it is a double bond, you add the suffix -ene to obtain heptene. The double bond is located on the second carbon, not the third, since you want the lowest number. You have, therefore, the 2-heptene. Using this numbering system, you see that CH_3, a methyl group, is located on the fifth carbon. Thus, the name of the molecule is 5-metyl-2-heptene.

(c) Has a six carbon chain, which suggests the prefix "hex." There is a double bond so that the molecule has the suffix -ene. Thus, you obtain hexene. The double bond is located on the second carbon, which means the name of the molecule is 2-hexene.

(d) This ring compound has a methyl group (CH_3) positioned on a benzene ring. The compound can be called methyl benzene. It is also given the special name of toluene.

PROBLEM

Write a structural formula for each of the following molecular formulas. Name each structure that you draw. (a) CH_2O; (b) $C_2H_6O_2$; (c) C_2H_6O; (d) $C_2H_4O_2$; (e) C_2H_4O.

SOLUTION

In order to do this problem, one must know how carbon, hydrogen, and oxygen bond. Carbon can share its 4 valence electrons to form 4 covalent bonds, hydrogen can share its 1 valence electron to form 1 covalent bond, and oxygen can share its 6 covalent electrons to form only 2 covalent bonds. Thus, with this knowledge in mind, one can construct several types of molecular structures. A few of these structures have the following functional groups: Alcohols contain the hydroxyl group, – OH, acids contain the carboxyl group,

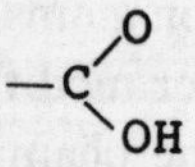

$$-C \overset{\displaystyle O}{\underset{\displaystyle OH}{\diagup}}$$ (represented as - COOH), aldehydes and ketones

(represented as – COOH), aldehydes and ketones contain the carbonyl group,

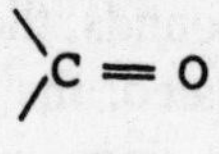

$$\diagup C = O$$

Figure 4.32

Structure (a), CH_2O, is an aldehyde,

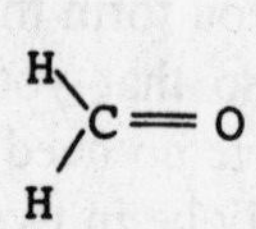

$$\overset{\displaystyle H}{\underset{\displaystyle H}{\diagup}} C = O$$

Figure 4.33

and has the name formaldehyde.

Structure (b), $C_2H_6O_2$, is an alcohol

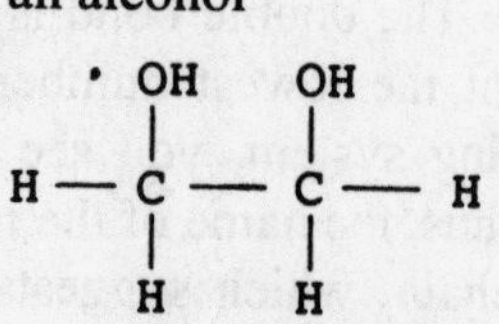

$$\begin{array}{ccc} & \overset{\bullet \ \ OH}{|} & \overset{OH}{|} \\ H - & C - & C - H \\ & | & | \\ & H & H \end{array}$$

Figure 4.34

and has the name 1,2-dihydroxyethane or ethylene glycol.

Structure (c), C_2H_6O, is either an alcohol or an ether

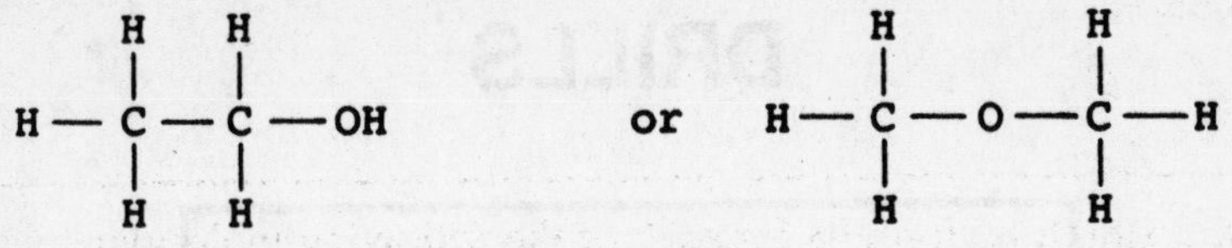

Figure 4.35

and is named ethanol or dimethyl ether, respectively.

Structure (d) $C_2H_4O_2$, is an acid

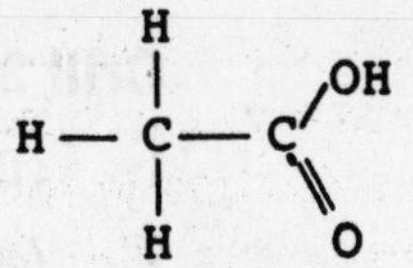

Figure 4.36

and has the name acetic acid.

Structure (e), C_2H_4O, is an aldehyde

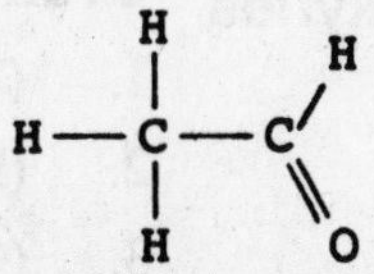

Figure 4.37

and has the name acetaldehyde.

Drill 4: Other Functional Groups

Questions 1–5. Classify (according to functional group) each of the following compounds.

(A) Ether   (D) Alcohol

(B) Ketone   (E) Carboxylic acid

(C) Amine

1. $CH_3COC_4H_9$

2. C_2H_5OH

3. CH_3NH_2

4. $(C_2H_5)_2O$

5. CH_3COOH

ORGANIC CHEMISTRY DRILLS

ANSWER KEY

Drill 1 — Alkanes

1. (A)
2. (B)
3. (C)
4. (D)

Drill 2 — Alkenes and Alkynes

1. (B)
2. (C)
3. (C)

Drill 3 — Alcohols

1. (D)
2. (A)
3. (B)

Drill 4 — Other Functional Groups

1. (B)
2. (D)
3. (C)
4. (A)
5. (E)

GLOSSARY:
ORGANIC CHEMISTRY

Alcohols

Hydrocarbon derivatives in which one or more hydrogen atoms have been replaced by the –OH (hydroxyl) group.

Aldehydes

Compounds that have the general formula RCHO.

Alkanes

Any molecule having the formula C_nH_{2n+2}.

Alkenes

Unsaturated hydrocarbons with one or more carbon-carbon double bonds. They have the general formula C_nH_{2n}.

Alkyl Halides

Hydrocarbon derivatives in which one or more hydrogen atoms have been replaced by an atom of the halide family.

Alkynes

Unsaturated hydrocarbons with one or more carbon-carbon triple bonds. They have the general formula C_nH_{2n-2}.

Amides

Compounds containing the group $CONH_2$.

Amines

Derivatives of hydrocarbons in which a hydrogen atom has been replaced by an amino group.

Aryl Halides

Compounds containing halogens attached directly to a benzene ring.

Carboxylic Acids

Compounds containing the COOH group.

Esters

Compounds containing the COOR group where R is any hydrocarbon.

Ethers

Hydrocarbon derivatives in which two alkyl or aryl groups are attached to an oxygen atom.

Ketones

Compounds that have the general formula RCOR.

Structural Isomers

Compounds having the same molecular formula, but different structures.

CHAPTER 17

Mini Tests

➤ Mini Test 1
➤ Mini Test 2

MINI-TEST 1

1. Ⓐ Ⓑ Ⓒ Ⓓ Ⓔ
2. Ⓐ Ⓑ Ⓒ Ⓓ Ⓔ
3. Ⓐ Ⓑ Ⓒ Ⓓ Ⓔ
4. Ⓐ Ⓑ Ⓒ Ⓓ Ⓔ
5. Ⓐ Ⓑ Ⓒ Ⓓ Ⓔ
6. Ⓐ Ⓑ Ⓒ Ⓓ Ⓔ
7. Ⓐ Ⓑ Ⓒ Ⓓ Ⓔ
8. Ⓐ Ⓑ Ⓒ Ⓓ Ⓔ
9. Ⓐ Ⓑ Ⓒ Ⓓ Ⓔ
10. Ⓐ Ⓑ Ⓒ Ⓓ Ⓔ
11. Ⓐ Ⓑ Ⓒ Ⓓ Ⓔ
12. Ⓐ Ⓑ Ⓒ Ⓓ Ⓔ
13. Ⓐ Ⓑ Ⓒ Ⓓ Ⓔ
14. Ⓐ Ⓑ Ⓒ Ⓓ Ⓔ
15. Ⓐ Ⓑ Ⓒ Ⓓ Ⓔ
16. Ⓐ Ⓑ Ⓒ Ⓓ Ⓔ
17. Ⓐ Ⓑ Ⓒ Ⓓ Ⓔ
18. Ⓐ Ⓑ Ⓒ Ⓓ Ⓔ
19. Ⓐ Ⓑ Ⓒ Ⓓ Ⓔ
20. Ⓐ Ⓑ Ⓒ Ⓓ Ⓔ
21. Ⓐ Ⓑ Ⓒ Ⓓ Ⓔ
22. Ⓐ Ⓑ Ⓒ Ⓓ Ⓔ
23. Ⓐ Ⓑ Ⓒ Ⓓ Ⓔ
24. Ⓐ Ⓑ Ⓒ Ⓓ Ⓔ
25. Ⓐ Ⓑ Ⓒ Ⓓ Ⓔ

MINI TEST 1

25 Questions

DIRECTIONS: Choose the best answer for each of the 25 problems below.

1. An alloy of iron and gallium has 63.8% iron composition. If 351.25 g of the alloy is completely dissolved in H_2SO_4 to produce Fe^{2+} and Ga^{3+} ions, what is the volume of H_2 collected at STP (neglecting the vapor pressure of air)?

    (A) 156.8 l (D) 78.4 l

    (B) 123.2 l (E) 145.6 l

    (C) 246.2 l

2. Inspection of the periodic table allows all of the following *except*

    (A) determining the valence of transition elements.

    (B) finding outer shell electron numbers of alkali and alkali earth elements.

    (C) labeling an element as metal or nonmetal.

    (D) locating atomic number and mass of an element.

    (E) reading outer shell electron numbers of halogen elements.

3. Molecules of sodium chloride

    (A) display ionic bonding.

    (B) display polar covalent bonding.

    (C) are polar.

    (D) dissociate in water solution.

    (E) do not exist.

4. A given atom has an atomic mass of 23 and an atomic number of 11. Select the *incorrect* statement about its atomic structure.

    (A) Eight electrons are in its outermost energy shell.

    (B) Its number of electrons is 11.

    (C) Its number of protons is 11.

(D)   Most of its mass is in the nucleus.

(E)   The number of neutrons is 12.

5.   Select the *incorrect* statement about the element neon.

   (A)   Its chemical symbol is Ne.

   (B)   It is a noble gas.

   (C)   It is a reactive substance.

   (D)   Neon has two energy shells.

   (E)   Neon is found in family VIII of the periodic table.

6.   The attractive force between the protons of one molecule and the electrons of another molecule are strongest

   (A)   in the solid phase.   (D)   during sublimation.

   (B)   in the liquid phase.   (E)   during fusion.

   (C)   in the gas phase.

7. NH_4^+ is the radical named

   (A)   ammonium.   (D)   nitrate.

   (B)   hydrate.   (E)   nitrite.

   (C)   hydroxyl.

8.   Which of the following would produce a highly conductive aqueous solution at equal concentrations?

   (A)   Cyclohexane   (D)   Sucrose

   (B)   Hydrochloric acid   (E)   Acetic acid

   (C)   Benzene

9.   A hydrocarbon was found to contain 92.3% carbon and 7.7% hydrogen. This compound is probably

   (A)   hexane.   (D)   benzene.

   (B)   hexene.   (E)   toluene.

   (C)   octene.

10. Determine the *incorrect* statement.

    (A) In a first order gaseous reaction, if we decrease the volume of the container where the reaction is taking place, the velocity of the reaction will decrease.

    (B) A catalyst creates a new path for the reaction which requires a smaller activation energy.

    (C) Activation energy is constant for a certain reaction.

    (D) By increasing the temperature of a reaction, we increase the amount of molecules with sufficient energy to react.

    (E) All intermolecular collisions result in chemical reactions.

11. Which of the following contains the largest number of atoms or molecules?

    (A) 49 g of Fe   (D) 10 liters of ozone at STP

    (B) 5 liters of H_2 at STP   (E) 80 g of calcium carbide

    (C) 150 g of ethanol

12. A reaction involving oxidation and reduction in which electrons are transferred by means of an external power source is best described as

    (A) exothermic.   (D) electrolytical.

    (B) endothermic.   (E) galvanic.

    (C) photochemical in nature.

13. For a given oxidation-reduction reaction, the equilibrium constant is K, the standard potential is $E°$, and the standard free energy change is $\Delta G°$. If the reaction proceeds spontaneously, which of the following sets of conditions is satisfied?

    (A) $\Delta G° > 0$ $K > 1$ $E° > 0$

    (B) $\Delta G° > 0$ $K < 1$ $E° < 0$

    (C) $\Delta G° < 0$ $K > 1$ $E° < 0$

    (D) $\Delta G° < 0$ $K < 1$ $E° < 0$

    (E) $\Delta G° < 0$ $K > 1$ $E° > 0$

14. Which of the following transition metal complexes or complex ions could exhibit geometric isomerism?

    (A) Octahedral $Co(NH_3)_5I^{2+}$

    (B) Square planar $Pt(NH_3)_2Cl_2$

(C) Tetrahedral $Zn(CN)_4{}^{2-}$

(D) Trigonal bipyramidal $Fe(CO)_6$

(E) Both (A) and (B)

15. The mass of an object is dependent upon its

(A) weight.   (D) acceleration.

(B) volume.   (E) None of the above

(C) density.

16. Which of the following is true of the reaction $2A(g) + B(g)\ C(g)$?

(A) In a reaction starting with equal moles of A and B, reagent B is the limiting reagent.

(B) The rate equation can be expressed as Rate $= k[A]^2 [B]$.

(C) There is an increase in the entropy of the system.

(D) If the reaction is endothermic, a decrease in temperature will favor the formation of a reactant.

(E) The percent yield of the reaction is determined as (moles of C produced/moles of A consumed) $\times$ 100%.

17. The reaction $A(g) + B(g) \leftrightarrow 2C(g) + D(g)$ has K_p of 0.65 at $T = 298K$. If at the beginning $P_A = 2$ atm, $P_B = 5$ atm, and $P_C = 4$ atm, then initially

(A) the free energy of the reaction is zero.

(B) the reaction will proceed from left to right.

(C) the reaction will proceed from right to left.

(D) the system will be at equilibrium.

(E) the free energy of the reaction is negative.

18. A comparison of carbon monoxide and carbon dioxide illustrates

(A) the Law of Definite Composition.

(B) the Law of Multiple Proportions.

(C) conservation of matter.

(D) conservation of energy.

(E) conservation of matter and energy.

19. The Fahrenheit temperature corresponding to 303K is

    (A) −15°. (D) 86°.

    (B) 22°. (E) 1,069°.

    (C) 49°.

20. The sulfate ion, SO_4^{2-} is different from a water molecule in that

    (A) the sulfate ion is a cation while the water molecule is neutral.

    (B) the sulfate ion has several resonance structures.

    (C) the water molecule is a linear triatomic molecule.

    (D) All of the above.

    (E) None of the above.

21. The Law of Conservation of Matter states that matter is conserved in terms of

    (A) pressure. (D) mass.

    (B) volume. (E) density.

    (C) temperature.

22. For the molecules listed below, the resultant dipole moments are oriented as (from left to right)

    (A) O, →, ←.

    (B) ↑, O, ↓.

    (C) ↑, O, ↑.

    (D) ↓, O, ↑.

    (E) ↑, ←, ↓.

23. A compound that when dissolved in water barely conducts electrical current can probably be

    (A) a strong electrolyte. (D) a strong base.

    (B) an ionic salt. (E) none of the above.

    (C) a strong acid.

24. Choose the correct statement pertaining to the following energy diagram.

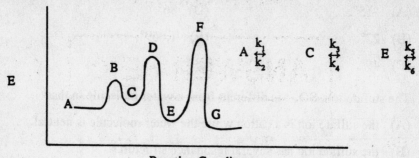

Reaction Coordinate

(A) The reaction $A \rightarrow G$ is endothermic.

(B) k_3 is larger than k_1.

(C) Species B is called an intermediate product.

(D) Species E is kinetically and thermodynamically more stable than species C.

(E) The rate determining step involves reaction $C \rightarrow E$.

25. Nonmetals when compared to metals

(A) are less electronegative and have smaller atomic radii.

(B) have greater ionization energies and larger atomic radii.

(C) are more electronegative and have smaller atomic radii.

(D) have larger atomic radii and are more electronegative.

(E) have smaller atomic radii and lower ionization energies.

MINI TEST 1

ANSWER KEY

1.	(B)	8.	(B)	15.	(E)	22.	(B)
2.	(A)	9.	(D)	16.	(D)	23.	(E)
3.	(E)	10.	(E)	17.	(C)	24.	(D)
4.	(A)	11.	(C)	18.	(B)	25.	(C)
5.	(C)	12.	(D)	19.	(D)		
6.	(A)	13.	(E)	20.	(B)		
7.	(A)	14.	(B)	21.	(D)		

DETAILED EXPLANATIONS
OF ANSWERS

1.   **(B)**   351.25 g of alloy contain:

Fe: $0.638 \times 351.25 = 224$ g of Fe

or $\dfrac{224}{56}$

    $56 \approx$ atomic weight of Fe

Ga: $351.25 - 224 = 127.25$ g of Ga = 4 moles of Fe

or   1 mole of Ga since atomic wt. of Ga = 127.25

    H_2SO_4 supplies $2H^+$ ions in solution per H_2SO_4

$\therefore$ $Fe + 2H^+ \rightarrow Fe^{2+} + H_2 \uparrow$

    $2Ga + 6H^+ \rightarrow 2Ga^{3+} + 3H_2$

One mole of Fe gives 1 mole of H_2; hence, 4 moles of Fe will yield 4 moles of H_2.

Two moles of Ga gives 3 moles of H_2; hence, 1 mole of Ga will yield 1.5 moles of H_2.

$\therefore$ 351.25g of alloy gives 5.5 moles of H_2

At STP one mole of a gas occupies 22.4 *l*.

$\therefore$ The volume occupied by 5.5 moles of $H_2 = 22.4 \times 5.5$ gives $V = 123.2$ *l*.

2.   **(A)**   The wide span of intermittent transition elements in the table are not under a vertical group (i.e., 1, 2, 6, or 7) to name the outer shell electron number or valence. Alkali, alkali earth, or halogen families are groups 1, 2, or 7 to convey this fact. Metals are demarcated to the left of the step line with nonmetals to the right. Atomic number and mass are statistics within the cell for each element.

3.   **(E)**   Molecules of sodium chloride do not exist individually. Rather sodium and chloride ions occupy points in a crystal lattice structure.

4.   **(A)**   By definition, atomic number is the number of protons or electrons. With 11 electrons, its electron arrangement is 2-8-1 over three energy levels. If the atomic number is 23, however, *12* neutrons must add to 11 protons for this total mass by simple subtraction. Note its outer level has *one* electron.

5. **(C)** Neon, Ne, is in family VIII of the periodic table. The outer of its two electron shells is filled with 8, obeying the octet rule.

6. **(A)** The attractive force between the protons of one atom and the electrons of another is inversely proportional to the distance between the atoms, i.e., (F $\alpha 1/d$) where F = force and d = distance between two atoms. This shows that the attraction is strongest at small distances (as in solids) and weakest at large distances (as in gases).

7. **(A)** Hydroxyl is OH^-, nitrate is NO_3^- and nitrite is NO_2^-.

8. **(B)** The conductivity of a solution is directly related to the number of ions in solution. Hydrochloric acid, being a strong acid, dissociates completely while acetic acid, a weak acid, is only slightly dissociated. Cyclohexane, benzene, and sucrose may be said to be undissociated in solution.

9. **(D)** A 100 g sample of this compound contains 92.3 g of carbon and 7.7 g of hydrogen as determined by the given percentages. Dividing each weight by the respective atomic weight gives the number of moles of each atom present:

$$\frac{92.3}{12} = 7.7 \text{ moles of carbon}$$

$$\frac{7.7}{1} = 7.7 \text{ moles of hydrogen}$$

This gives $C_{7.7} H_{7.7}$ as a result. Dividing each term by 7.7 gives CH. This is the empirical formula of the compound. Benzene, C_6H_6, is the only compound given with this empirical formula. The other choices are hexane (C_6H_{14}), hexene (C_6H_{12}), octene (C_8H_{16}), and toluene (C_7H_8).

10. **(E)** Not all collisions result in chemical reactions: only the collisions between molecules with an amount of energy greater than or equal to the activation energy result in such chemical reactions.

11. **(C)**

(A) Atomic weight of Fe = 55.8 g mole^{-1}.

$\therefore$ There are $\dfrac{49}{55.8} \approx 0.88$ moles of Fe.

One mole element will contain Avogadro's number of atoms which is 6.026 $\times 10^{23}$

$\therefore$ No. of atoms is 55.8 g of Fe= $0.88 \times 6.026 \times 10^{23}$

$$= 5.3 \times 10^{23} \text{ atoms}$$

(B)   At STP 1 mole of gas occupies 22.4 *l*

$$\text{No. of moles of 5 l of gas is } \frac{5}{22.4} \times 1 = 0.22 \text{ moles.}$$

No. of molecules present is

No. of moles × Avogadro's number $= 0.22 \times 6.026 \times 10^{23}$

$$= 1.3 \times 10^{23} \text{ molecules}$$

(C)   Ethanol has the molecular formula C_2H_5OH.

Molecular weight of ethanol $= (2 \times 12) + 6 + 16 = 24 + 22 = 46$

$\therefore$ No. of moles in 150 g of $C_2H_5OH = \dfrac{150}{46} = 3.26$

No. of molecules $= 3.26 \times 6.026 \times 10^{23} = 19.65 \times 10^{23}$

(D)   10 l of O_3 at STP contains 10/22.4 moles of O_3.

$\therefore$ No. of moles of $O_3 = \dfrac{10}{22.4} \times 6.026 \times 10^{23} = 2.69 \times 10^{23}$

(E)   The no. of molecules in 80 g of calcium carbide is

$$\frac{80}{\text{Mol. Wt. of Ca}_2\text{C}} \times 6.026 \times 10^{23} = \frac{80}{92.16} \times 6.026 \times 10^{23}$$

$$= 5.23 \times 10^{23}$$

Mol. Wt. of $Ca_2C = (2 \times 40.08) + 12 = 92.16$

$\therefore$ The answer is (C) 150 g of ethanol contains the largest number of molecules (19.65×10^{23} molecules).

12.   **(D)**   Electrolysis involves chemically induced changes by the passage of an external current through a substance. This method can be used to convert H_2O to H_2 and O_2, or to produce Na metal from molten salt.

In contrast to an electrolytic cell, a galvanic or voltaic cell spontaneously generates a current by virtue of the electron transfer reactions within the cell. In essence, an electrolytical cell requires electricity, while a voltaic cell produces it.

13.   **(E)**   K is related to $E°$ by:

$$E° = \frac{RT}{F} \ln K$$

K is related to $\Delta G°$ by:

$$\Delta G° = -RT \ln K$$

Only (B) and (E) are consistent with these relations. In order for the reaction to proceed spontaneously, $\Delta G°$ must be negative; therefore, (E) is the correct answer.

14. **(B)** When a compound exists in different geometric forms, each of which has the same molecular formula, the relationship between the forms are referred to as geometric isomers. The only possible isomers of the compounds and geometries given are the cis and trans geometric isomers of $Pt(NH_3)_2Cl_2$. Writing the possible structural arrangements of the groups coordinated to the platinum, we find there are only two distinct possibilities:

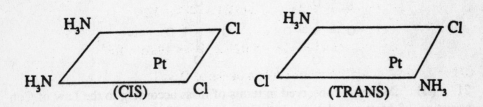

The one on the left is referred to as the cis isomer in which the two similar groups are adjacent to each other. Whereas in the structure on the right the similar groups are opposite each other, which is referred to as the trans isomer. For the other compounds listed, no geometric isomers exist since there exists only one possible structural arrangement of the groups around the metal.

15. **(E)** The mass of an object is independent of all of these qualities.

16. **(D)** The application of Le Chatelier's principle provides a solution: heat + 2A + B → C. The removal of heat (decrease in temperature) shifts the reaction equation toward the formation of reactants. Note that (B) would only be a true statement if this reaction was identified as the rate-limiting (or slowest) step.

17. **(C)** The direction of a reaction is determined by ΔG. If $\Delta G < 0$, then the reaction proceeds spontaneously as written. If $\Delta G > 0$, then it will proceed in the reverse direction of the written equation.

18. **(B)** This illustrates the Law of Multiple Proportions: there are some compounds formed by the same two elements where the weight of one element is constant and the weight of the other varies. The weights of the other element are always in the ratio of small whole numbers.

19. **(D)** Converting to the Celsius scale ($t = T - 273$) we have

$$t = 303 - 273 = 30°.$$

Converting to the Fahrenheit scale

$$°F = \frac{9}{5}(°C) + 32$$

$$°F = \frac{9}{5}(30) + 32$$

$$°F = 86$$

20. **(B)** The water molecule has a bent geometry, and the sulfate ion is an anion with several resonance structures:

$$
\overset{\displaystyle O^-}{\underset{\displaystyle O^-}{O = S = O}} \quad \leftrightarrow \quad \overset{\displaystyle O}{\underset{\displaystyle O}{^-O - S - O^-}} \quad \text{etc.}
$$

21. **(D)** Matter is conserved in terms of mass according to the Law of Conservation of Matter. All the other choices may change due to a chemical or physical change.

22. **(B)** The dipole moment of a bond is directed from the partial positive charge to the partial negative charge or from the less electronegative to the more electronegative atom in the bond.

Example: H – Cl

$\propto + \quad \propto -$ direction of dipole moment.

If the molecule contains more than one bond moment, then the resultant dipole moment is the vector sum of all the bond moments.

Thus, for H_2O, we have:

bond moments

i.e.: resultant dipole moment is oriented according to the resultant.

For CO$_2$, we have:

$$\overset{\alpha-}{\underset{\cdot\cdot}{\overset{\cdot\cdot}{O}}} = C = \overset{\alpha-}{\underset{\cdot\cdot}{\overset{\cdot\cdot}{O}}}$$

bond moments $\longleftarrow + \longrightarrow$

The vector sum is zero; therefore, the resultant dipole moment is 0.

For SO$_2$,

bond moments

The resultant vector sum

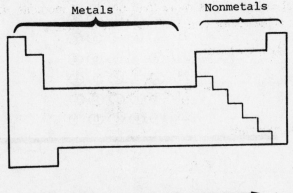

is oriented as ↓ so the combination of resultant dipole moments is ↑, O, ↓.

23. **(E)** Most common ionic salts, strong acids (HCl, nitric acid, sulfuric acid, etc), and strong bases (most hydroxides of IA and IIA metals, except Be) are strong electrolytes because they can conduct electrical current when dissolved in water.

24. **(D)** Species E is lower in energy than species C and is thermodynamically more stable. It is also kinetically more stable because of the larger activation energies (proportional to the height of the "hills") for the forward and reverse reactions.

25. **(C)** Knowledge of the periodic trends make the answer to this question clear.

Metals   Nonmetals

Increasing ionization energy
Decreasing atomic radius
Increasing electronegativity

MINI-TEST 2

1. Ⓐ Ⓑ Ⓒ Ⓓ Ⓔ
2. Ⓐ Ⓑ Ⓒ Ⓓ Ⓔ
3. Ⓐ Ⓑ Ⓒ Ⓓ Ⓔ
4. Ⓐ Ⓑ Ⓒ Ⓓ Ⓔ
5. Ⓐ Ⓑ Ⓒ Ⓓ Ⓔ
6. Ⓐ Ⓑ Ⓒ Ⓓ Ⓔ
7. Ⓐ Ⓑ Ⓒ Ⓓ Ⓔ
8. Ⓐ Ⓑ Ⓒ Ⓓ Ⓔ
9. Ⓐ Ⓑ Ⓒ Ⓓ Ⓔ
10. Ⓐ Ⓑ Ⓒ Ⓓ Ⓔ
11. Ⓐ Ⓑ Ⓒ Ⓓ Ⓔ
12. Ⓐ Ⓑ Ⓒ Ⓓ Ⓔ
13. Ⓐ Ⓑ Ⓒ Ⓓ Ⓔ
14. Ⓐ Ⓑ Ⓒ Ⓓ Ⓔ
15. Ⓐ Ⓑ Ⓒ Ⓓ Ⓔ
16. Ⓐ Ⓑ Ⓒ Ⓓ Ⓔ
17. Ⓐ Ⓑ Ⓒ Ⓓ Ⓔ
18. Ⓐ Ⓑ Ⓒ Ⓓ Ⓔ
19. Ⓐ Ⓑ Ⓒ Ⓓ Ⓔ
20. Ⓐ Ⓑ Ⓒ Ⓓ Ⓔ
21. Ⓐ Ⓑ Ⓒ Ⓓ Ⓔ
22. Ⓐ Ⓑ Ⓒ Ⓓ Ⓔ
23. Ⓐ Ⓑ Ⓒ Ⓓ Ⓔ
24. Ⓐ Ⓑ Ⓒ Ⓓ Ⓔ
25. Ⓐ Ⓑ Ⓒ Ⓓ Ⓔ

MINI TEST 2

25 Questions

DIRECTIONS: Choose the best answer for each of the 25 problems below.

1. In the Van der Waal's equation:

$$\left(P + \frac{n^2 a}{V^2}\right)(V - nb) = nRT$$

The constant a is best described as a correction factor due to

(A) temperature.

(B) intermolecular attractions of real gases.

(C) the molecular weights of gases.

(D) the volume of the actual gas molecules.

(E) the specific heat of the gas molecules.

2. What volume (in liters) does 34 grams of ammonia gas occupy at standard temperature and pressure? (assume ideal behavior)

(A) 22.4
(B) 44.8
(C) 11.2

(D) 0.1
(E) None of the above

For questions 3 and 4 choose the answer choice which best describes the gas law that follows the graphs shown.

(A) Graham's Law
(B) Boyle's Law
(C) Charles' Law

(D) Gay–Lussac's Law
(E) Avogadro's Law

3.

PRESSURE ➡

4.

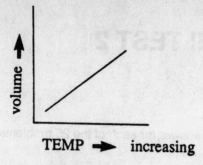

5. What do the following ions have in common?

Mg^{2+} O^{2-} F^- Na^+

(A) They are isoelectronic.

(B) The same number of protons.

(C) They are metal ions.

(D) They have the same atomic radius.

(E) None of the above

6. Select the organic compound

(A) C_3H_8 (D) HCl

(B) CO_2 (E) NaCl

(C) CO

7. Which of the following is false?

(A) A process with $\Delta H < 0$ is more likely to be spontaneous than one with $\Delta H > 0$.

(B) The rate law for a reaction is an algebraic expression relating the forward reaction rate to product concentration.

(C) Ammonia is an amphoteric substance.

(D) The products of a Brönsted-Lowry acid-base reaction are always a new acid and a new base.

(E) One faraday is the total charge of one mole of electrons.

8. All of the following are nonelectrolytes except

(A) sodium chloride crystals.

(B) a benzene solution of ethanol.

(C) a water solution of sucrose.

(D) a nickel-zinc alloy.

(E) liquid nitrogen.

9. Which electronic subshell do the lanthanides have incompletely filled?

(A) 4f

(B) 5f

(C) 6f

(D) 5g

(E) 6g

10. What is the electron configuration of Ir^{+3} ion in its ground state?

(A) $[Xe]4f^{14}5d^{4}6s^{2}$

(B) $[Xe]4f^{14}5d^{6}$

(C) $[Xe]4f^{14}5d^{5}6s^{1}$

(D) $[Xe]4f^{11}5d^{7}6s^{2}$

(E) $[Xe]4f^{13}5d^{6}6s^{1}$

11. The VSEPR method is used to

(A) predict the geometries of an atom.

(B) estimate the energy levels of orbitals in an atom.

(C) estimate electronegativities of elements.

(D) predict the geometries of molecules and ions.

(E) There is no such thing as a VSEPR method.

12. Which of the following elements is the most electronegative?

(A) Li

(B) Na

(C) K

(D) Be

(E) Ca

13. The species O_2^-

(A) has 2 unpaired electrons.

(B) has 1 unpaired electron.

(C) has 0 unpaired electrons.

(D) has a bond order of $2^1/_2$.

(E) has a bond order of 2.

14. Select the metal with the highest ionization energy.

(A) Cesium

(B) Lithium

(C) Potassium

(D) Rubidium

(E) Sodium

15. A reaction is usually spontaneous if

    (A) $\Delta H < 0.$

    (B) $\Delta S > 0.$

    (C) $\Delta G < 0.$

    (D) $\Delta G > 0.$

    (E) $\Delta G = 0.$

16. In the reaction

    $$^{223}_{87}Fr \rightarrow {}^{223}_{88}Ra + X$$

    particle X is

    (A) a neutron.

    (B) a proton.

    (C) an alpha particle.

    (D) an electron.

    (E) a positron.

17. The equation for the electrolysis of water is

    $$H_2O(l) + 68.3 \text{ kcal} \rightarrow H_2(g) + \frac{1}{2}O_2(g).$$

    How many liters of gaseous product are produced by the addition of 273.2 kcal of electrical energy to excess water?

    (A) 22.4 liters

    (B) 44.8 liters

    (C) 134.4 liters

    (D) 96.8 liters

    (E) 119.2 liters

18. What is the oxidation number of chlorine in ClO_4^-?

    (A) +1

    (B) +3

    (C) +5

    (D) +7

    (E) +8

19. Which of the following would you expect to precipitate first if an $AgNO_3$ solution is added to an aqueous solution containing NaCl, NaBr, NaI?

    (A) Ag

    (B) AgCl

    (C) AgBr

    (D) AgI

    (E) NaI

20. The heat of a reaction may be found by

    (A) adding the heats of formation of the reactants.

(B) adding the heats of formation of the products.

(C) determining (A) + (B).

(D) determining (A) – (B).

(E) determining (B) – (A).

21. Fifty ml of a solution of H_2SO_4 with unknown concentration is titrated to the phenolphthalein endpoint with 10.0 ml of 1.00 M of NaOH. The concentration of the H_2SO_4 is:

(A) 5.0 M.

(B) 1.0 M.

(C) 0.5 M.

(D) 0.2 M.

(E) 0.1 M.

22. The k_{sp} for BaF_2 is 1.7×10^{-6}. What is the solubility of fluoride in moles per liter?

(A) 3.4×10^{-6}

(B) 1.3×10^{-3}

(C) 1.5×10^{-2}

(D) 7.5×10^{-3}

(E) 1.7×10^{-6}

23. Which of the following usually forms diatomic molecules?

(A) He

(B) B

(C) C

(D) Cl

(E) Ba

24. What is the volume of H_2O which can be condensed after the complete combustion of 2,240 l of ethane (volume of gas measured at STP?

(A) 10.8 ml

(B) 10.8 l

(C) 5.4 l

(D) 5.4 ml

(E) 3.0 l

25. How many grams of chlorine gas is equivalent to 12×10^{23} molecules?

(A) 9 g

(B) 18 g

(C) 35.5 g

(D) 71 g

(E) 142 g

MINI TEST 2

ANSWER KEY

1. (B)	8. (D)	15. (C)	22. (C)
2. (B)	9. (A)	16. (D)	23. (D)
3. (C)	10. (B)	17. (C)	24. (C)
4. (B)	11. (D)	18. (D)	25. (E)
5. (A)	12. (D)	19. (D)	
6. (A)	13. (B)	20. (E)	
7. (B)	14. (B)	21. (E)	

DETAILED EXPLANATIONS
OF ANSWERS

1. **(B)**   The Ideal Gas Law equation,

$$PV + nRt$$

derived from kinetic and molecular theory, neglects two important factors in regard to real gases. First, it neglects the actual volume of gas molecules, and second, it does not take into account the intermolecular forces that real gases exhibit.

The factor b in the Van der Waal equation is a correction for the actual volume of the gas molecules and the factor a is a correction of the pressure due to the intermolecular attractions that occur in real gases.

2. **(B)**   Since one mole of ideal gas occupies a volume of 22.4 liters at STP, we have:

$$\frac{34 \text{ grams of } NH_3}{17 \text{ grams (M. W. of } NH_3)} = 2 \text{ moles of } NH_3 \times \frac{22.4 \text{ liters}}{1 \text{ mole of } NH_3}$$

$$= 44.8 \text{ liters of } NH_3$$

M.W. = molecular weight.

3. **(C)**   The graph in question 3 is a representation of Charles' Law which states that at constant pressure the volume of a given quantity of a gas varies directly with temperature:

$$V \propto T.$$

4. **(B)**   The graph in question 4 is a representation of Boyle's Law which states that at constant temperature the volume of a fixed quantity of a gas is inversely proportional to the pressure:

$$V \propto \frac{1}{P}.$$

It can be seen from the graph that as the pressure increases the volume decreases as stated in Boyle's Law.

5. **(A)**   The only commonality among the ions listed is that they all contain 10 electrons and have the same electron configuration:

$$1s^2 2s^2 2p^6.$$

Ions or atoms that have the same electron configurations are said to be isoelectronic.

6. **(A)** This is the formula for propane. Its structural formula is:

$$H-\overset{\displaystyle H}{\underset{\displaystyle H}{\overset{|}{\underset{|}{C}}}}-\overset{\displaystyle H}{\underset{\displaystyle H}{\overset{|}{\underset{|}{C}}}}-\overset{\displaystyle H}{\underset{\displaystyle H}{\overset{|}{\underset{|}{C}}}}-H$$

Note that its carbon skeleton, indicative of organic compounds, is lacking in the other molecules.

7. **(B)** The statement in (B) is false, because the rate law is an equation which relates the rate of the reaction to the concentrations of the reactants.

Statement (A) is true, because a process is spontaneous if $\Delta G < 0$. Since $\Delta G = \Delta H - T\Delta S$, if $\Delta H < 0$, ΔG is more likely to be negative than if $\Delta H < 0$, because ΔS then has to be large and positive.

Statement (C) is true, because NH_3 can act as a base $NH_3 + H^+ \to NH_4^+$ and also as an acid $2NH_3 \to NH_4^+ + NH_2^-$ amide ion.

Statement (D) is also true. A Brönsted acid is a substance that gives protons (hence it must contain hydrogen), and the Brönsted base is a substance that accepts a proton (i.e., must have an unshared electron pair). The products of a Brönsted acid-base reaction are always a new acid and a new base, e.g.,

$$CH_3COOH + OH^- \quad \to \quad CH_3COO^- \quad + \quad H_2O$$
$$\text{acid} \qquad\qquad \text{base} \qquad\quad \text{conjugate base} \qquad \text{conjugate}$$
$$\text{of } CH_3COOH \qquad \text{acid of } OH^-$$

Statement (E) is true, because the total charge of a mole of electrons (i.e., Avogadro's number of electrons) is defined as a faraday.

8. **(D)** An electrolyte may be defined as a substance that conducts an electric current. Metal alloys conduct electricity since they are characterized by positive nuclei surrounded by free moving electrons. Sodium chloride crystals are not electrolytes but molten sodium chloride or a water solution of sodium chloride are electrolytes since the sodium and chlorine ions are free to move. Neither benzene nor sucrose ionizes in solution, so they are nonelectrolytes. The same holds true for liquid nitrogen.

9. **(A)** The inner transition series, the lanthanides and actinides, correspond to incompletely filled f subshells. The first of these, the lanthanides, correspond to the first f subshell, which is $4f$, although the lanthanides are in the sixth period.

10. **(B)** One may assume that when removing electrons to form stable cations, the electrons would leave the orbitals in the reverse order in which they were filled. This is true to some extent; but there are many exceptions especially in the transition elements. A good rule to follow is that the electrons that are removed from an atom or ion are those with the maximum value of the principal quantum number n; and of this set of electrons the easiest to remove are those with the maximum L.

Ir^{+3} is formed by removal of three electrons from Ir which has an electron configuration $[Xe]4f^{14}5d^{7}6s^{2}$. So by the rule given, the first two electrons would leave the $6s^{2}$ and the third from the $5d^{7}$ orbitals to give a ground state configuration of $[Xe]4f^{14}5d^{6}$.

11. **(D)** VSEPR is an acronym for valence-shell electron-pair repulsion. It states that bonding electron pairs and lone electron pairs of an atom will arrange themselves in space so as to minimize electron-pair repulsion around that atom. It is used to predict the geometries of various molecules and ions.

12. **(D)** Electronegativity increases as we move up and to the right on the periodic table. Be is in the upper right-hand corner of the periodic table; therefore, it is the most electronegative element of the choices given.

13. **(B)** The MO (molecular orbital) bonding schemes for simple diatomic species must be used to answer this question. For O_2^-, the MO scheme is

$$(\sigma 1s)^2(\sigma*1s)^2(\sigma 2s)^2(\sigma*2s)^2(\sigma 2p)(\pi 2p)^4(\pi*2p)^3.$$

O_2^- has three net bonding electrons and a BO (bond order) of $1\frac{1}{2}$. It also has one unpaired electron (two electrons in one of the $\pi*$ orbitals and one unpaired electron in the other $\pi*$ orbital).

14. **(B)** Ionization energy is the energy needed to move one or more electrons from the neutral atoms. All choices are alkali metals from group #1 of the periodic table. Although all these metals tend to lose their single valence electron and ionize, lithium has only two energy shells, with its electron close to the nucleus attracting sphere. Therefore, more energy is needed to attract its valence electron from the atom.

15. **(C)** A reaction is spontaneous if ΔG is negative ($\Delta G < 0$), and non-spontaneous if $\Delta G > 0$. At equilibrium, $\Delta G = 0$.

16. **(D)** X must have negligible mass, since the mass of the reacting nucleus remains unchanged. X must have a charge of -1, since the nuclear charge increases by one. An electron is the only choice which fits this description.

17. **(C)** 1.5 moles of gaseous product (one mole of H_2 and 0.5 mole of O_2) are produced for every 68.3 kcal of energy added to water. Since 4×68.3 kcal = 273.2 kcal, 4×1.5 or 6 moles of gaseous product are formed. Since molar volume is given as 22.4 liters at STP, 6×22.4 liters = 134.4 liters of gaseous product are evolved.

18. **(D)** Since oxygen has an oxidation number of –2, chlorine must have an oxidation number of +7 for the chlorate ion to have the sum of $1Cl(+7) + 4O(-2)$ = –1. Thus, the oxidation number of Cl is +7.

19. **(D)** The most insoluble of the salts would precipitate out of the solution. AgI is the least soluble salt present. In general, solubility of the AgI is less than the solubility of AgBr and AgCl. AgI < AgBr < AgCl.

20. **(E)** The heat of a reaction is determined by subtracting the sum of the heats of formation of the reactants from the sum of the heats of formation of the products.

$$\Delta H_{rxn} = \Sigma \Delta H_{f \text{ products}} - \Sigma \Delta H_{f \text{ reactants}}$$

21. **(E)** 10.0 ml of 1.00 M NaOH contains:

$$(10.0 \text{ ml}) (1.00 \text{ mole/liter}) = 10.0 \text{ mmoles OH}^-$$

This should neutralize 5.0 mmoles of H_2SO_4 since H_2SO_4 is a diprotic acid (two H^+'s). The concentration is thus:

$$\frac{5.0 \text{ mmoles}}{50 \text{ ml}} = 0.1 \text{ mole/liter}$$

22. **(C)**

$$BaF_2 \rightarrow Ba^{2+} + 2F^-; \; k_{sp} = 1.7 \times 10^{-6} = [Ba^{2+}] [F^-]^2$$

Let $\quad x^2 = [Ba^{2+}]$ and $2x = [F^-]$;

$\quad\quad (x) (2x)^2 = 4x^3 = 1.7 \times 10^{-6}$;

$\quad\quad x = 7.5 \times 10^{-3}$

and $\quad 2x = [F^-] = 1.5 \times 10^{-2}$.

23. **(D)** Although fluorine is known to occur as a diatomic gas, it is unstable and prefers to undergo the reaction $F_2 + 2e^- \rightarrow 2F^-$.

24. **(C)** At STP 2,240 l of ethane contains

$$\frac{2,240}{22.4} = 100 \text{ moles of ethane}$$

The equation for the complete combustion of ethane is

$$2C_2H_2 + 14O_2 \rightarrow 4CO_2 + 6H_2O$$

1 mole of C_2H_6 yields 3 moles of H_2O

∴   100 moles of C_2H_6 yields 300 moles of H_2O

1 mole of H_2O is 18 g

∴   300 moles weighs $300 \times 18 = 5,400$ g $= 5.4$ Kg

1 Kg of H_2O occupies a volume of 1 l

∴   Volume of water condensed $= 5.4$ l

25. **(E)**   Using the unit conversions

$$12 \times 10^{23} \text{ molecules} \times \frac{1 \text{ mole}}{6 \times 10^{23} \text{ molecules}} \times \frac{71 \text{ grams of } Cl_2}{1 \text{ mole}}$$

we find that 142 g of Cl_2 is equivalent to 12×10^{23} molecules.

THE PERIODIC TABLE

METALS — NONMETALS

KEY

Atomic Number → 22
Symbol → Ti
Atomic Weight → 47.88

4 IVA / IVB ← Group Classification

( ) indicates most stable or best known isotope

TRANSITIONAL METALS

Group	1 IA	2 IIA	3 IIIB	4 IVB	5 VB	6 VIB	7 VIIB	8 VIII	9 VIII	10 VIII	11 IB	12 IIB	13 IIIA	14 IVA	15 VA	16 VIA	17 VIIA	18 VIII / 0
	1 H 1.008																	2 He 4.003
	3 Li 6.941	4 Be 9.012											5 B 10.811	6 C 12.011	7 N 14.007	8 O 15.999	9 F 18.998	10 Ne 20.180
	11 Na 22.990	12 Mg 24.305											13 Al 26.982	14 Si 28.086	15 P 30.974	16 S 32.066	17 Cl 35.453	18 Ar 39.948
	19 K 39.098	20 Ca 40.078	21 Sc 44.956	22 Ti 47.88	23 V 50.942	24 Cr 51.996	25 Mn 54.938	26 Fe 55.847	27 Co 58.933	28 Ni 58.693	29 Cu 63.546	30 Zn 65.39	31 Ga 69.723	32 Ge 72.61	33 As 74.922	34 Se 78.96	35 Br 79.904	36 Kr 83.8
	37 Rb 85.468	38 Sr 87.62	39 Y 88.906	40 Zr 91.224	41 Nb 92.906	42 Mo 95.94	43 Tc (97.907)	44 Ru 101.07	45 Rh 102.906	46 Pd 106.4	47 Ag 107.868	48 Cd 112.411	49 In 114.818	50 Sn 118.710	51 Sb 121.757	52 Te 127.60	53 I 126.905	54 Xe 131.29
	55 Cs 132.905	56 Ba 137.327	57 La 138.906	72 Hf 178.49	73 Ta 180.948	74 W 183.84	75 Re 186.207	76 Os 190.23	77 Ir 192.22	78 Pt 195.08	79 Au 196.967	80 Hg 200.59	81 Tl 204.383	82 Pb 207.2	83 Bi 208.980	84 Po (208.982)	85 At (209.982)	86 Rn (222.018)
	87 Fr (223.020)	88 Ra (226.025)	89 Ac (227.028)	104 Unq (261.11)	105 Unp (262.114)	106 Unh (263.118)	107 Uns (262.12)	108 Uno (265)	109 Une (266)	110 Uun (272)								

Alkali Metals — Alkaline Earth Metals — Halogens — Noble Gases

LANTHANIDE SERIES

58 Ce 140.115	59 Pr 140.908	60 Nd 144.24	61 Pm (144.913)	62 Sm 150.36	63 Eu 151.965	64 Gd 157.25	65 Tb 158.925	66 Dy 162.50	67 Ho 164.930	68 Er 167.26	69 Tm 168.934	70 Yb 173.04	71 Lu 174.967

ACTINIDE SERIES

90 Th 232.038	91 Pa 231.036	92 U 238.029	93 Np (237.048)	94 Pu (244.064)	95 Am (243.061)	96 Cm (247.070)	97 Bk (247.070)	98 Cf (251.080)	99 Es (252.083)	100 FM (257.095)	101 Md (258.1)	102 No (259.101)	103 Lr (262.11)

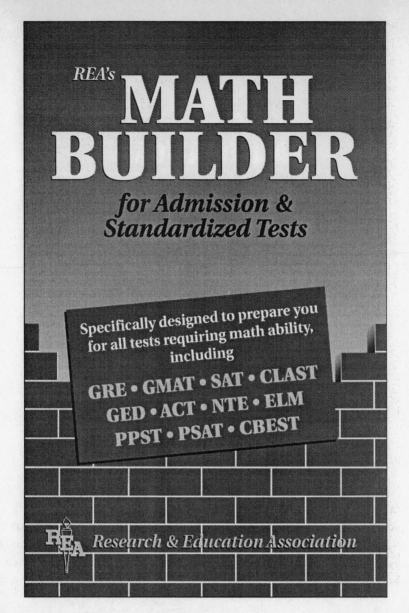

HANDBOOK of
**MATHEMATICAL,
SCIENTIFIC, &
ENGINEERING**
FORMULAS, FUNCTIONS,
GRAPHS, TABLES,
& TRANSFORMS

RESEARCH & EDUCATION ASSOCIATION

A particularly useful reference for those in math, science, engineering and other technical fields. Includes the most-often used formulas, tables, transforms, functions, and graphs which are needed as tools in solving problems. The entire field of special functions is also covered. A large amount of scientific data which is often of interest to scientists and engineers has been included.

Available at your local bookstore or order directly from us by sending in coupon below.

REA's Test Preps
The Best in Test Preparation

- REA "Test Preps" are **far more** comprehensive than any other test preparation series
- Each book contains up to **eight** full-length practice exams based on the most recent exams
- **Every** type of question likely to be given on the exams is included
- Answers are accompanied by **full** and **detailed** explanations

REA has published over 60 Test Preparation volumes in several series. They include:

Advanced Placement Exams (APs)
Biology
Calculus AB & Calculus BC
Chemistry
Computer Science
English Language & Composition
English Literature & Composition
European History
Government & Politics
Physics
Psychology
Statistics
Spanish Language
United States History

College-Level Examination Program (CLEP)
Analyzing and Interpreting Literature
College Algebra
Freshman College Composition
General Examinations
General Examinations Review
History of the United States I
Human Growth and Development
Introductory Sociology
Principles of Marketing
Spanish

SAT II: Subject Tests
American History
Biology
Chemistry
English Language Proficiency Test
French
German

SAT II: Subject Tests (continued)
Literature
Mathematics Level IC, IIC
Physics
Spanish
Writing

Graduate Record Exams (GREs)
Biology
Chemistry
Computer Science
Economics
Engineering
General
History
Literature in English
Mathematics
Physics
Political Science
Psychology
Sociology

ACT - American College Testing Assessment

ASVAB - Armed Services Vocational Aptitude Battery

CBEST - California Basic Educational Skills Test

CDL - Commercial Driver's License Exam

CLAST - College Level Academic Skills Test

ELM - Entry Level Mathematics

ExCET - Exam for Certification of Educators in Texas

FE (EIT) - Fundamentals of Engineering Exam

FE Review - Fundamentals of Engineering Review

GED - High School Equivalency Diploma Exam (US & Canadian editions)

GMAT - Graduate Management Admission Test

LSAT - Law School Admission Test

MAT - Miller Analogies Test

MCAT - Medical College Admission Test

MSAT - Multiple Subjects Assessment for Teachers

NJ HSPT- New Jersey High School Proficiency Test

PPST - Pre-Professional Skills Tests

PRAXIS II/NTE - Core Battery

PSAT - Preliminary Scholastic Assessment Test

SAT I - Reasoning Test

SAT I - Quick Study & Review

TASP - Texas Academic Skills Program

TOEFL - Test of English as a Foreign Language

TOEIC - Test of English for International Communication

RESEARCH & EDUCATION ASSOCIATION
61 Ethel Road W. • Piscataway, New Jersey 08854
Phone: (732) 819-8880

Please send me more information about your Test Prep books

Name _____

Address _____

City _____ State _____ Zip _____

REA's Problem Solvers

The "PROBLEM SOLVERS" are comprehensive supplemental text-books designed to save time in finding solutions to problems. Each "PROBLEM SOLVER" is the first of its kind ever produced in its field. It is the product of a massive effort to illustrate almost any imaginable problem in exceptional depth, detail, and clarity. Each problem is worked out in detail with a step-by-step solution, and the problems are arranged in order of complexity from elementary to advanced. Each book is fully indexed for locating problems rapidly.

ACCOUNTING
ADVANCED CALCULUS
ALGEBRA & TRIGONOMETRY
AUTOMATIC CONTROL
  SYSTEMS/ROBOTICS
BIOLOGY
BUSINESS, ACCOUNTING, & FINANCE
CALCULUS
CHEMISTRY
COMPLEX VARIABLES
COMPUTER SCIENCE
DIFFERENTIAL EQUATIONS
ECONOMICS
ELECTRICAL MACHINES
ELECTRIC CIRCUITS
ELECTROMAGNETICS
ELECTRONIC COMMUNICATIONS
ELECTRONICS
FINITE & DISCRETE MATH
FLUID MECHANICS/DYNAMICS
GENETICS
GEOMETRY

HEAT TRANSFER
LINEAR ALGEBRA
MACHINE DESIGN
MATHEMATICS for ENGINEERS
MECHANICS
NUMERICAL ANALYSIS
OPERATIONS RESEARCH
OPTICS
ORGANIC CHEMISTRY
PHYSICAL CHEMISTRY
PHYSICS
PRE-CALCULUS
PROBABILITY
PSYCHOLOGY
STATISTICS
STRENGTH OF MATERIALS &
  MECHANICS OF SOLIDS
TECHNICAL DESIGN GRAPHICS
THERMODYNAMICS
TOPOLOGY
TRANSPORT PHENOMENA
VECTOR ANALYSIS

If you would like more information about any of these books,
complete the coupon below and return it to us or visit your local bookstore.

RESEARCH & EDUCATION ASSOCIATION
61 Ethel Road W. • Piscataway, New Jersey 08854
Phone: (732) 819-8880

Please send me more information about your Problem Solver books

Name _____

Address _____

City _____ State _____ Zip _____